HOW TO USE THIS BOOK

Heinemann Advanced Science: Biology has been written to accompany your advanced level biology course, and contains all the core syllabus material you will need during your period of study. The book is divided into six sections of associated material which have been made easy to find by the use of colour coding. At the beginning of each section a Section Focus explores the real-life context of an area of the science to come. You may use this to whet your appetite for the chapters that follow, or read it when you have made some progress with the science to emphasise the relevance of what you are doing. However you use it, I hope it will make you want to read on!

In addition to the main text, the chapters of the book contain two types of boxes. The grey *information boxes* contain basic facts or techniques which you need to know. This often includes key facts you will have already met at GCSE – the boxes then carry these ideas forward to A level. *Information boxes* are sometimes referred to in the main text, and can be read either as you meet them or when you have finished reading the chapter. The pink headed *extension boxes* contain more advanced information which is not referred to in the main text. You do not need to address the contents of these *extension boxes* until you have got to grips with the rest of the material in the chapter.

When you have completed the work in a chapter of the book, there are questions to help you find out how much of the material you have understood and to help you with your revision. Summaries at the end of each chapter provide further help with revision. At the end of each section of the book there is a selection of A level questions.

Throughout the book the Institute of Biology's recommendations on biological nomenclature have been followed.

This book has been written to be an accessible, clear and exciting guide to A level biology. I hope that it will help to maintain your interest in the subject you have chosen to study, and that it will play a valuable role in developing your knowledge of biology – and with it, an increased understanding of all life on Earth.

Acknowledgements

Many people have been extremely generous with their time and expertise while I have been writing this book. In particular, I should like to thank Dr Jennifer Clay, Jonathan Wiles, Christine Newsome, Alan Pritchard and Don Nicholson. Dr Francis Gilbert of Nottingham University and Dr Eric Turner were of immense help in checking the veracity and relevance of the text. Of course, as author I accept full responsibility for the final content of the book.

I should like to thank Eluned Harries for her enthusiasm and commitment in starting the *Heinemann Advanced Science* series, and Clare Farley and Lindsey Charles, for ensuring the books made it into bound copy form. My copy editor Ruth Holmes deserves special mention for her creative editing and moral support. Without her the book would undoubtedly be poorer.

Finally, I should like to thank my friends and family who have supported and encouraged me throughout the writing of this book, especially my husband Patrick who has provided invaluable help and advice. I am also extremely grateful to my father for his support, and to my mother, without whom the whole project would have been impossible. They have my wholehearted thanks.

Ann Fullick, 1994

Dedication

For William, Thomas, James and Edward

CONTENTS

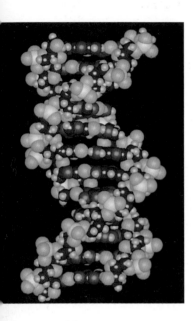

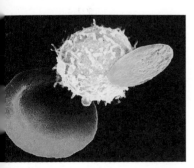

3 ENERGY SYSTEMS

4 CONTROL SYSTEMS

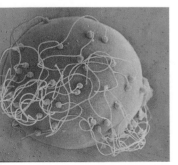

5 REPRODUCTION AND GENETICS

6 POPULATION BIOLOGY

BUILDING BLOCKS
OF LIFE

Kill or cure – disease and the human race

*The headlines of the world media are frequently focused on violent death and destruction. Whilst the horrors of wars, bombs and habitat destruction cannot and should not be denied, on a biological or evolutionary timescale **disease** has caused the premature deaths of far more living organisms than any disaster, natural or artificial. Disease causes immense devastation and loss both to individuals and to communities, both by its effect on people and by the damage it causes to agricultural plants and animals.*

Causes of disease

Disease is the result of an abnormality in or attack on the normal body cells of an organism, be it plant or animal. Abnormalities in cells may have resulted from changes or mutations in the genetic material of the parents of an organism. Such mutations can give rise to cell misfunctions which can have dramatic effects on the entire organism, such as the genetic disorder cystic fibrosis. Cell abnormalities may also result from external factors such as carcinogenic (cancer-forming) chemicals coming into contact with particular groups of cells. The affected cells may then begin to change, giving rise to tumours. A classic example is the response of the lung cells in some people when exposed to the carcinogens present in cigarette smoke.

Figure 1 The destruction by diseases of crops and domestic animals, as well as the human suffering caused, has fuelled the drive to find cures for the whole spectrum of illnesses.

More frequently, diseases arise as a result of the invasion of the body cells by other organisms. These other organisms may be worms or protozoans, but often they are microorganisms such as viruses and bacteria. The cells of these microorganisms are fundamentally different from the cells they are invading. When the body of an animal or plant reacts to invasion by these foreign cells, the symptoms of disease arise.

Many diseases have plagued us for a very long time – the remains of humans who died several thousands of years ago show signs of diseases still recognisable today. Figure 2 and table 1 give information about the types of organisms that cause disease, and also about how diseases are spread. Knowledge about the transmission of disease is vital for the development of preventative measures as well as possible cures.

Figure 2 Diseases range from the trivial to the life threatening. They are caused by a range of **pathogenic** (disease-causing) organisms and are transmitted from person to person in a wide variety of ways.

Malaria kills over 2 million people each year. It is caused by a protozoan which invades and destroys the blood, spread by the bite of the female anopheles mosquito. This scanning electron micrograph shows the mouthparts that spread the disease.

Bacteria such as this *Salmonella* are well known disease-causing agents. They give rise to a range of diseases from common complaints such as tonsillitis to much rarer and frequently lethal conditions such as tetanus.

Many diseases caused by both bacteria and viruses are spread from one person to another by droplet infection. A sneeze like this expels countless droplets containing disease-causing organisms into the air at high speed.

Name of disease	Causative agent	Effect on body	Method of spread
Influenza ('flu)	Myxovirus A, B or C	Affects respiratory passages and inflames epithelial lining cells of mouth and trachea	Droplet infection from coughs and sneezes
Poliomyelitis (polio)	Polio virus	Affects the throat (pharynx) and intestines as well as the motor neurones of the spinal cord, causing paralysis	Droplet infection or by the contamination of water by human faeces – swimming pools used to be a common source of infection in this country
Yellow fever	Arthropod-borne RNA virus	Affects the lining of blood vessels and the liver	Carried by vectors, e.g. ticks, mosquitoes
Tuberculosis (TB)	*Mycobacterium tuberculosis*, a rod-shaped bacterium	Affects mainly the lungs, may affect lymph glands and bone	Droplet infection and by drinking milk from infected cows
Syphilis	*Treponema pallidum*, a spirochaete	Initial effect on reproductive organs, then spreads to bones, joints, eyes, central nervous system, heart and skin	Sexual contact
Salmonellosis (bacterial food poisoning)	*Salmonella* spp., rod-shaped bacteria	Affects the alimentary canal	Eating undercooked meat from infected animals – mainly poultry and pigs – or infected eggs, also by faecal contamination
Malaria	*Plasmodium* spp.	'Flu-like symptoms followed by invasion of liver and red blood cells	Female anopheles mosquito is the vector transmitting *Plasmodium* between people
Schistosomiasis	Trematode worms	Body tissues react to eggs laid within them	Water snails act as vectors
Onchocerciasis	Filarial nematode worms	Skin and eyes (if eyes affected, causes river blindness)	*Simulium* black flies act as vectors

Table 1

The lessons of history

The occurrence and spread of human diseases, and to a lesser extent those of their animals and crops, have been documented for almost as long as written records are available. Changes in the structure of society have frequently led to 'plagues' – massive outbreaks of diseases such as the Black Death, cholera and dysentery. Squalid, cramped and unhygienic living conditions such as those found in city slums have always been and continue to be a fertile ground for infections.

In the developed world many infectious diseases are largely a thing of the past, because of a combination of higher public health standards, improved knowledge of how diseases are transmitted and effective medicines. Infectious diseases have been largely replaced by the chronic diseases of middle age such as heart disease and cancer, which are increasingly seen as being linked to both diet and lifestyle. In the developing world, teeming cities with shanty towns around the edges are still the sites of great epidemics of 'old-fashioned' infectious diseases. Dysentery and cholera as well as tropical diseases such as yellow fever, malaria and dengue cause immense suffering and debilitation of the population, and often result in death. And in both developed and developing countries a new disease is rapidly spreading which is leaving millions infected and millions more dead. It is predicted that AIDS will overtake malaria as the most prevalent disease of the human population by the turn of the century – around 267 million people are currently affected by malaria.

The arrival of AIDS has highlighted many areas of concern and interest in our current state of understanding of disease. We know the infective agent and the method of transmission from one individual to another, yet have failed so far to halt the spread of the disease. Major research laboratories here and in many other countries have been working feverishly to develop a cure or a vaccine – as yet without real success. By considering in some detail the science of AIDS we can learn much of both disease and how people attempt to overcome it.

AIDS – the new plague

In 1981 doctors in America reported a new disease. It first became apparent in the homosexual community of Los Angeles, where unusually high numbers of young men developed a type of cancer, Kaposi's sarcoma, which is usually only seen in very elderly patients. Eventually a whole range of common symptoms (fevers, persistent diarrhoea and weight loss, along with many opportunistic infections, TB, a rare type of pneumonia and Kaposi's sarcoma) were recognised as resulting from infection by one particular virus. The end result for those showing symptoms was usually death. The disease first recognised in Los Angeles was **AIDS** (**Acquired immunodeficiency syndrome**), and the pathogen responsible is known as **HIV** (**Human immunodeficiency virus**). Infection by HIV does not necessarily lead to full-blown AIDS. People infected by the virus who do not display the symptoms of AIDS are referred to as HIV positive. This is because when their blood is tested it shows the presence of HIV antibodies.

How is HIV transmitted?

The human immunodeficiency virus is very fragile and cannot survive in the air; it must be contained in human body fluids. The source of HIV infection may be either an individual who is 'HIV positive' (who will be without symptoms and quite possibly unaware of their infective status) or someone with AIDS.

The virus can be transmitted from person to person in one of three ways, each of which involves contact between the body fluids of the two individuals. The first and most common means of transmission is by sexual contact, between both homo- and heterosexuals. So far in the developed world homosexual sex has been the most striking risk behaviour, with relatively few cases of infection via heterosexual intercourse, but on a global scale heterosexual intercourse is the major infective route.

The second way in which HIV is spread is through infected blood. Intravenous drug users have appeared as a high risk group, and this has been assumed to be due to infection from shared needles. Some evidence now suggests that the prostitution often used to pay for the drugs might be more pertinent. However, without doubt AIDS *has* been spread through the use of infected blood products. Haemophiliacs have received factor 8 (needed to enable their blood to clot normally) carrying HIV and numbers of them, along with their families, became

Figure 3 AIDS is a modern plague for which we have not, so far, found a cure. The means of transmission are known, and ordinary social contact between AIDS sufferers and healthy individuals is perfectly safe. The virus is only passed from one person to another when body fluids come into contact.

infected with the virus before the risk was recognised. In America and much of Europe blood is now treated to destroy HIV before it is used for transfusions. However, in much of the developing world blood for transfusions is bought from individuals who may well be infected, it is frequently not treated in any way and the numbers of transfusions are increasing all the time. This is the result of increasing numbers of people moving to live in cities where they use the hospital resources available.

Thirdly, HIV can cross from a mother to her fetus in the early stages of pregnancy. Infection may also occur during birth, and the virus has been shown to be present in breast milk. Infants may be born free from HIV but become infected through breast feeding.

How does HIV cause AIDS?

Most of the clinical symptoms of AIDS are the result of the profound effect HIV has on the functioning of the immune system of the body. It leaves individuals vulnerable to a whole range of infections which might be trivial in a healthy person but which, in an AIDS sufferer, may mean death. The immune system has evolved to destroy and eliminate invading pathogens such as HIV, yet doesn't produce sufficient antibodies to overwhelm the virus and escape long-term damage. Why not? The answer to this appears to be in the action of the virus itself. An important part of the immune system is a group of cells known as **T4** or **helper T-cells**. They have been termed 'the leaders of the immunological orchestra' because of their central role in the immune response. Helper T-cells support and amplify the responses of other cells within the immune system, the killer T-cells and some B-cells (see section 1.6) in particular.

However, on the surface of the lipoprotein cell membranes of the helper T-cells is a specific glycopro-

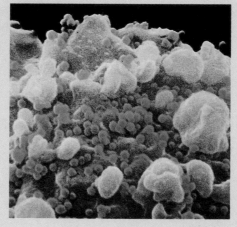
A scanning electron micrograph showing the AIDS virus (green) attacking a helper T-cell

tein, CD4. This acts as a cellular receptor recognised by the HIV. Once the viruses are attached to the receptor they can infect the cells, as shown in figure 4. As a result of the HIV infection, the normal functioning of the helper T-cells is lost or impaired. This undermines the ability of the entire immune system to react to the invasion of the body by other pathogens. The infections that sweep the body as a result of the ineffective immune system are the most common cause of death in AIDS patients.

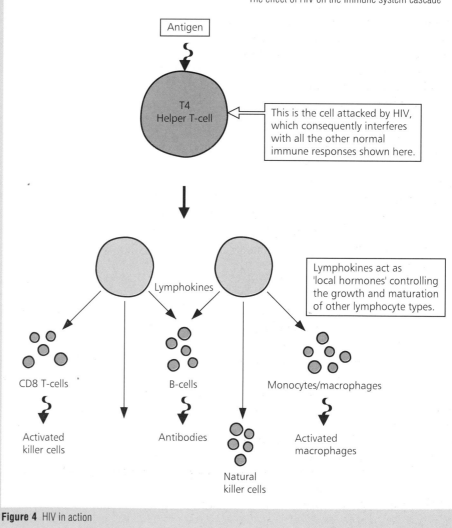

The effect of HIV on the immune system cascade

Antigen

T4 Helper T-cell

This is the cell attacked by HIV, which consequently interferes with all the other normal immune responses shown here.

Lymphokines

Lymphokines act as 'local hormones' controlling the growth and maturation of other lymphocyte types.

CD8 T-cells

B-cells

Monocytes/macrophages

Activated killer cells

Antibodies

Natural killer cells

Activated macrophages

Figure 4 HIV in action

Can AIDS be cured?

At the moment AIDS is an incurable disease. Although a variety of methods of attack are currently under investigation no cure has yet been found. Perhaps the most important of these methods is the attempt to limit the spread of the disease. As can be seen in figure 5, AIDS and HIV positive individuals now number millions. Though the problem was first identified in America, the numbers affected are highest in the African and Asian continents. Education programmes help people to understand the ways in which HIV is spread, and how to prevent it. Less promiscuous sex, the use of condoms to prevent internal contact between sexual partners and the inadvisability of sharing needles in drug abuse are three of the main messages being given in the battle to contain this modern-day plague.

Many specialists feel that the development of an effective vaccine is crucial to the containment of the AIDS epidemic, but producing a vaccine against an organism whose target is part of the immune system is very difficult. Vaccines would need to give protection against all strains of the AIDS virus, and HIV is a very variable virus indeed. So far insufficient detail is known about the infective mechanism of HIV to decide which part of the virus should be used as the basis for a vaccine. Added to this, AIDS is a peculiarly human disease. Chimpanzees are the only other animals which seem to be susceptible to the virus and so work on animals to develop vaccines is very limited. An effective vaccine remains a distant hope on which much research is currently focused.

Finally, drug therapy has a role to play. Although a variety of drugs have been tried so far, none has been found to be truly effective either in preventing the shift from HIV positive status to full-blown AIDS, or in slowing the progress of AIDS once it begins. Antiviral drugs are largely ineffective in the treatment of AIDS, and although immunostimulatory drugs have produced a transitory improvement in killer T-cell functions, this has not as yet been maintained.

Thus we can see that to combat a disease such as AIDS an understanding of the structure of both human cells and viruses is important. The biochemistry of the cells and their microscopic structure contain vital clues to the progress of the disease. And a full understanding of the intricacies of the immune system is needed in the planning of a defence strategy against this or any other disease. This first section of the book covers just these areas – the structure of cells, their biochemistry and the systems of cell recognition that play an important role in both the maintenance of health and the progress of a disease.

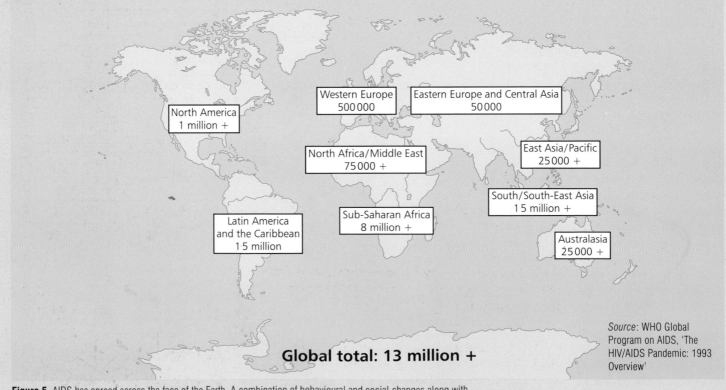

Western Europe
500 000

Eastern Europe and Central Asia
50 000

North America
1 million +

North Africa/Middle East
75 000 +

East Asia/Pacific
25 000 +

Latin America
and the Caribbean
1·5 million

Sub-Saharan Africa
8 million +

South/South-East Asia
15 million +

Australasia
25 000 +

Global total: 13 million +

Source: WHO Global Program on AIDS, 'The HIV/AIDS Pandemic: 1993 Overview'

Figure 5 AIDS has spread across the face of the Earth. A combination of behavioural and social changes along with possible new vaccines and drugs may yet be able to prevent HIV from becoming even more destructive than at present.

1.1 The units of life: Cells

Biology is the study of life. The planet which we inhabit teems with a great variety of living animals and plants, all interdependent on each other and on the environment in which they live. Biologists are concerned with studying the internal and external mechanisms of life, including the balances within individual cells, in whole organisms and between the members of specific ecosystems. To begin to understand an animal or plant, alone or as part of an interconnected system of organisms, we need to know what it is made up of, and this is where we shall begin – with cells.

WHAT IS A CELL?

The idea of cells is familiar to most of us. Radio, newspapers and television often discuss cells in relation to cancer, test-tube babies, drug testing and other topics. But in spite of this, most of us have only a vague idea of what a cell is.

Cells were first seen over 300 years ago. In 1665 Robert Hooke, an English scientist, designed and put together one of the first working optical microscopes. Amongst the many objects he examined were thin sections of cork. Hooke saw that these sections were made up of many tiny, regular compartments which he called **cells**.

It took many years of further work for the full significance of Hooke's work to emerge. In 1676 Anton van Leeuwenhoek, a Dutch draper who ground lenses in his spare time, used his lenses to observe a wide variety of living unicellular organisms in drops of water. He called these organisms 'animalcules'. By the 1840s it was recognised that cells are the basic units of life, an idea that was first expressed by Matthias Schleiden and Theodor Schwann in their 'cell theory' of 1839.

In the years since 1839 knowledge about cells has progressed a very long way, helped by developments in technology which have made ever-increasing detail available.

HOW WE SEE CELLS

There are some cells which can be seen very easily with the naked eye. Unfertilised birds' eggs are single cells. The largest single cell in the world is an ostrich egg. But of the vast majority of cell types little or nothing can be seen without some kind of magnification.

The light microscope

Ever since it was first developed, the **light** or **optical microscope** has been the main method of observing cells, and in spite of the development of newer instruments like the electron microscope, it is still of immense value.

Making the image clearer

Living, untreated cells can be observed under a light microscope, but the image is not easy to see. Instead, slides are made of tissues or individual organisms. These are very thin slices of biological material which have been specially treated and stained so that particular features are easier to see. It is important to remember when looking at stained cell samples that the cells are

How does a light microscope work?

A specimen or thin slice of biological material is placed on the stage and illuminated from underneath. Figure 1.1.1 shows how the objective lens produces a magnified and inverted image, which the eyepiece lens focuses at the eye.

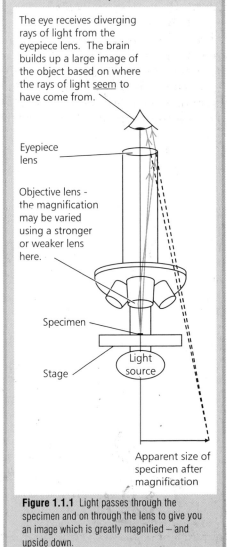

The eye receives diverging rays of light from the eyepiece lens. The brain builds up a large image of the object based on where the rays of light <u>seem</u> to have come from.

Eyepiece lens

Objective lens - the magnification may be varied using a stronger or weaker lens here.

Specimen

Stage

Light source

Apparent size of specimen after magnification

Figure 1.1.1 Light passes through the specimen and on through the lens to give you an image which is greatly magnified — and upside down.

usually dead, and that the processes of fixing the material so that it does not decay and then slicing and staining it are potentially destructive. It is impossible to tell whether the features seen on such slides are present in the living cell, or are the result of these processes.

The information we get from the light microscope can be improved by using the light in different ways. **Dark field illumination**, where the background is dark and the specimen illuminated, can be useful for showing tiny structures inside cells. **Phase contrast microscopy** uses the fact that different parts of the cell refract light differently. A special *phase plate* in the microscope allows light from different regions to form interference patterns which result in very sharp contrasts. This is particularly useful for observing transparent objects such as living cells.

Advantages of the light microscope

One of the biggest advantages of using a light microscope is that living plants and animals or parts of them can be seen directly. This is of considerable value in itself, and it also allows us to check whether what we see on prepared slides is at all like the living thing. However, it is not usually possible to magnify living cells as much as dead tissue.

The other big advantage is that light microscopes do not necessarily cost a lot of money. Any biologist working in a hospital, industrial or research laboratory will have a light microscope readily available.

Disadvantages of the light microscope

The biggest problem with the light microscope is the limited detail it can show. A minimum distance is necessary between two objects for them to be seen clearly as separate. If they are closer together than this minimum distance they merge. This distance is known as a microscope's **limit of resolution** – the smaller the limit of resolution, the greater is the **resolving power** of the microscope and the more detail it can show. For the optical microscope the limit of resolution is approximately 0.2 μm. In comparison, the unaided eye can resolve down to about 0.1 mm, as figure 1.1.2 shows.

The limit of resolution for any system depends ultimately on the wavelength of the radiation passing through it, so the resolution of a light microscope with the very best quality lenses is limited by the wavelength of light. A magnification of 1500 times is about the greatest that a light microscope can give with a clear image. At this magnification the average person would become over 2.5 kilometres tall so it might seem quite adequate – but to see the details of the inside of a cell a further technological leap is needed.

The electron microscope

In the 1950s the **electron microscope** was developed. This microscope has made great advances in biological knowledge possible. Instead of relying on light with its limit of resolving power, an electron beam is used. The image is formed as electrons are scattered by the biological material, in much the same way as light is scattered in the light microscope. The electrons behave like light waves, but have a much smaller wavelength. Resolving power is increased as the wavelength gets smaller, and as a result the electron microscope can resolve detail many thousand times better than the light microscope.

Preparation of specimens

Samples of material are prepared for examination under an electron microscope in a way which is quite similar to that for the light microscope, although the details are different, as table 1.1.1 shows.

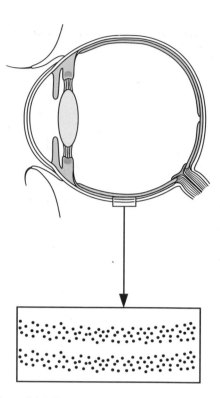

Figure 1.1.2 The resolving power of your eyes means that a mass of dots on the page looks like a clear line – because you can't resolve the dots individually. In the same way, what you can see through the light microscope is limited by the resolving power of the microscope itself.

Treatment of specimen	For light microscopes	For electron microscopes
Fixation – the material has to be preserved in as life-like a state as possible and hardened for sectioning.	Living specimens are not fixed or subjected to most of the other stages. For non-living specimens a mixture of ethanol and glacial ethanoic acid is often used (proportions 99:1).	Specimens always fixed. Often glutaraldehyde or a mixture of glutaraldehyde and osmic acid are used – osmic acid also stains lipids black.
Dehydration – water is removed from the specimen to be replaced with the embedding medium.	Immersion in increasing concentrations of ethanol or propanone.	Immersion in increasing concentrations of ethanol or propanone.
Embedding – supports the tissue for sectioning.	Tissue embedded in wax.	Tissue embedded in resin such as araldite.
Sectioning – produces very thin slices for mounting.	Sections cut on a microtome with a metal knife to produce sections a few micrometres thick.	Sections cut on an ultramicrotome with a diamond or glass knife to produce sections 20–100 nm thick.
Staining – gives contrast between different structures and makes structures easier to see.	Coloured dyes are used to reflect visible light.	Heavy metals such as lead and uranium are used to reflect electrons.

Table 1.1.1 The preparation of specimens for light microscopy and electron microscopy

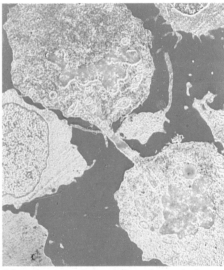

Figure 1.1.3 The electron microscope has revealed, amongst other things, the secrets of how plants provide us with energy, the start of a human life and how our cells divide.

False-colour transmission electron micrograph of a lymphocyte dividing

Seeing the image

Complex electronics produce an image on a television screen which can then be recorded as an **electron micrograph** or **EM**. This can have a magnification of up to 500 000 times. Our average person becomes over 830 km tall!

The most common type of electron micrograph you will see is produced by a **transmission electron microscope**, but the **scanning electron microscope** produces spectacular images of the surfaces of cells and organisms. Figure 1.1.3 shows both kinds of electron micrograph.

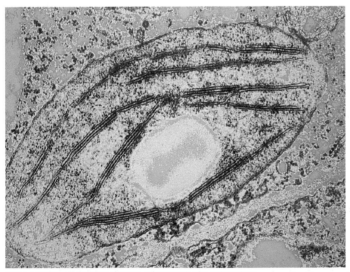

False-colour transmission electron micrograph of a chloroplast

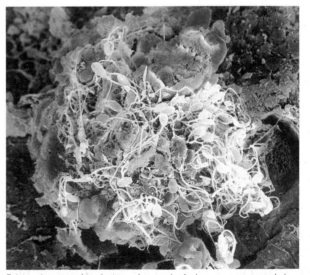

False-colour scanning electron micrograph of a human egg surrounded by sperms (yellow)

Advantages of the electron microscope

The amount of detail that can be seen using an electron microscope is immense. This is its biggest advantage. Many structures have been seen for the first time since electron microscopes were developed. Others that were known to exist from light microscope studies have been shown to have complicated substructures inside them which have helped us to understand how they work.

Disadvantages of the electron microscope

There are several disadvantages to the electron microscope. One is that air in the microscope would scatter the electron beam and so make the image of the tissue fuzzy. This means that all the specimens are examined in a vacuum – and so it is impossible to look at living material.

This leads to the second problem. How realistic a picture do we get from tissue which has been processed in many ways, including being sliced extremely thinly and put in a vacuum? There has been considerable argument among scientists about this over the years.

Finally, electron microscopes are very expensive, fill a room, have to be kept at a constant temperature and pressure and need to maintain an internal vacuum. As a consequence, relatively few scientists outside research laboratories have easy access to such equipment.

Observing cells

You have probably looked at cells using a light microscope. With increasing practice and developing skills you may find that more becomes visible. As mentioned before, cells can be observed either unstained or stained.

There is one particular problem to bear in mind when you are working with microscopes. Unless you are looking at living material, or have the use of a scanning electron microscope, all the cells that you see appear flat and two dimensional. But cells are actually three dimensional – spheres, cylinders, strange asymmetrical shapes – so try to exercise some imagination in your consideration of cells and view them as complete living things.

In both the animal and plant kingdoms there is a very wide range of different types of cells, each carrying out different functions. But for both animals and plants, there are certain features of their cells which turn up again and again – so much so that we can put them together in a **typical cell**. Remember that this typical cell does not really exist, but it acts as a useful guide to what to look for in an animal or a plant cell.

THE TYPICAL ANIMAL CELL

A typical animal cell is shown in figure 1.1.5. It is surrounded by a membrane known as the **cell surface** or **plasma membrane**. Inside this membrane is a jelly-like liquid called **cytoplasm**, containing a **nucleus**. The cytoplasm and nucleus together are known as **protoplasm**. The cytoplasm contains much of what is needed to carry out the day-to-day tasks of living, whilst the nucleus is vital to the long-term survival of the cell.

When the light microscope was the only tool biologists had to observe cells, they thought that the cytoplasm was a relatively structureless, clear jelly. But the electron microscope changed all that by revealing that the cytoplasm was full of all manner of complex and detailed structures, now known as **organelles**. This detailed organisation is known as the **ultrastructure** of the cell. The cell's ultrastructure is explain below.

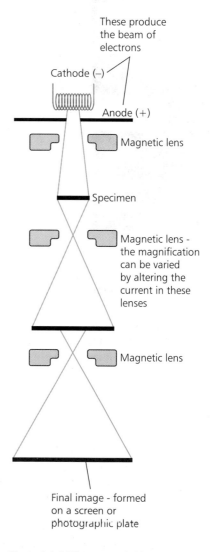

How does an electron microscope work?

Electron beams are focused by magnetic lenses, as figure 1.1.4 shows. A series of magnifications results in a final image which cannot be seen with the eye, but is formed on a screen or photographic plate.

These produce the beam of electrons

Cathode (–)

Anode (+)

Magnetic lens

Specimen

Magnetic lens - the magnification can be varied by altering the current in these lenses

Magnetic lens

Final image - formed on a screen or photographic plate

Figure 1.1.4 What goes on inside an electron microscope

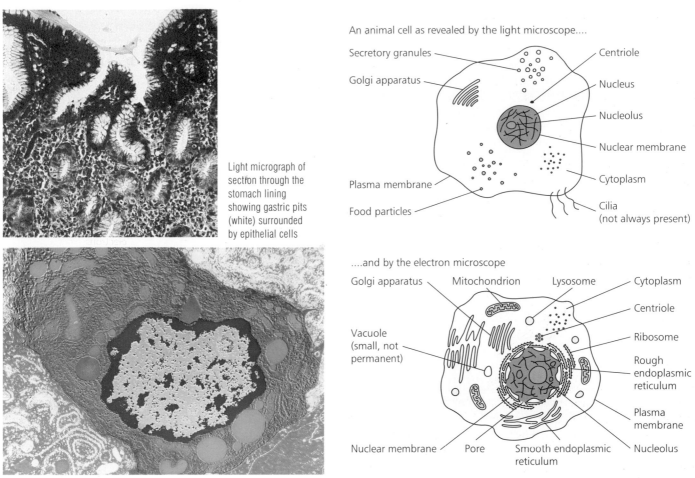

Light micrograph of section through the stomach lining showing gastric pits (white) surrounded by epithelial cells

False-colour transmission electron micrograph of a single epithelial cell

An animal cell as revealed by the light microscope....

Secretory granules
Golgi apparatus
Centriole
Nucleus
Nucleolus
Nuclear membrane
Cytoplasm
Plasma membrane
Food particles
Cilia (not always present)

....and by the electron microscope

Golgi apparatus
Mitochondrion
Lysosome
Cytoplasm
Centriole
Ribosome
Vacuole (small, not permanent)
Rough endoplasmic reticulum
Plasma membrane
Nuclear membrane
Pore
Smooth endoplasmic reticulum
Nucleolus

Figure 1.1.5 Many regions of an animal cell which under a light microscope appear to have no particular features, or to be blurred areas, are revealed as complex structures by electron microscopy.

The structure of each organelle is closely related to its function within the cell. The basic ultrastructure of the typical animal cell gives rise to an enormous variety of animal cells suited for the different functions that arise within the animal kingdom.

The ultrastructure of the animal cell

Membranes

One of the most important structures in any cell is its membranes. The cell surface membrane acts as a boundary to the cell itself, and there is also a multitude of internal (intracellular) membranes. Much work has gone into producing a model of the structure of membranes. Over the years improvements in technology have made possible different, improved models until we have reached the level of understanding that we have today. We shall be considering the functions of membranes in almost every part of the cell in turn, but will leave the details of its structure until we know more about the chemicals which make it up.

The nucleus – the information centre

The **nucleus** is usually the largest organelle in the cell and can be seen with the light microscope. In experiments nuclei have been removed from some large amoebae. They can survive for a short time, but then die. No cell replication occurs. So although most of the reactions important to life go on in the cytoplasm, the nucleus is vital to organise and direct these reactions and to bring about reproduction.

Electron micrographs show us that the nucleus, which is usually spherical in shape, is surrounded by a **nuclear membrane** or envelope. This is a double membrane containing many holes or pores. In order to control events in the cytoplasm, the nucleus needs to communicate with it. That is the reason for the pores – they enable chemicals to travel easily in and out of the nucleus.

Inside the nuclear membrane is the **nucleoplasm**. This consists of two main substances, **nucleic acids** and **proteins.** The nucleic acids are **deoxyribonucleic acid** (**DNA**), which is the basic genetic material, the blueprint for the cell to be passed on when it divides, and **ribonucleic acid** (**RNA**), which translates the genetic code into instructions for making proteins in the cytoplasm of the cell. When the cell is not actually dividing the DNA is bonded to the protein to form **chromatin**, which appears as tiny granules.

Within the nucleus will be at least one **nucleolus**, an extra dense area of almost pure DNA which appears to be involved in producing RNA. When the cell is about to divide, the genetic material associates into long chains. These rapidly become more coiled, shorter and denser, readily taking up stain and becoming visible under the light microscope. These chains are called **chromosomes**, which means 'coloured bodies'. It can be seen that chromosomes have distinctive shapes and occur in pairs. Each species of living organism has a characteristic number of chromosomes in the nucleus.

The endoplasmic reticulum – manufacture and transport

This spreads throughout the whole cytoplasm. It is a three-dimensional network of cavities, some sac-like and some tubular, bounded by membranes. The network links with the membrane around the nucleus.

The outside of much of the endoplasmic reticulum (ER) membrane is encrusted with granules called **ribosomes**. Because of this it is known as **rough endoplasmic reticulum** or **RER**. The function of the ribosomes is to make proteins. The function of the rough endoplasmic reticulum is to isolate and transport these proteins once they have been made. Some of the proteins, such as digestive enzymes and hormones, are required outside the cell which has made them, so they have to be **secreted** or moved out of the cell. Many other proteins are needed within the cell itself.

The rough endoplasmic reticulum has a large surface area, giving space for the synthesis of all the proteins needed, and making possible their storage and transport both within the cell and from the inside to the outside. Any cell which secretes large amounts of material (such as those producing the digestive enzymes in the lining of the gut) will have a large amount of rough endoplasmic reticulum.

Not all endoplasmic reticulum is covered in ribosomes. **Smooth endoplasmic reticulum** (**SER**) is also involved in synthesis and transport. However, in this case it is the synthesis and transport of fatty molecules known as steroids and lipids. A large quantity of smooth endoplasmic reticulum is found in the testes and the liver, for example. By looking at the amount and type of endoplasmic reticulum in a cell we can get an idea of its function.

The Golgi apparatus – packaging the products

The **Golgi apparatus** can be seen using a light microscope, although it simply looks like a rather dense area of cytoplasm. It was first spotted almost 100 years ago by the Italian scientist Camillo Golgi, but it is only since the arrival of the electron microscope that we have been able to see it clearly. The Golgi apparatus is made up of stacks of parallel, flattened membrane pockets. They

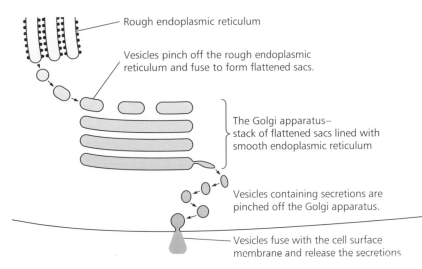

Rough endoplasmic reticulum

Vesicles pinch off the rough endoplasmic reticulum and fuse to form flattened sacs.

The Golgi apparatus– stack of flattened sacs lined with smooth endoplasmic reticulum

Vesicles containing secretions are pinched off the Golgi apparatus.

Vesicles fuse with the cell surface membrane and release the secretions

Figure 1.1.6 A production line for cell secretions – the Golgi apparatus takes the raw materials, assembles them, packages them and transports them to the surface of the cell.

link with, but are not joined to, the rough endoplasmic reticulum. It has taken a long time to discover exactly what the Golgi apparatus does. Radioactively labelled materials have been followed through the system to try and find out. Figure 1.1.6 shows the result.

It seems that protein is transported from the rough endoplasmic reticulum in small vesicles. These pinch off from the rough endoplasmic reticulum and fuse to form the pockets of the Golgi apparatus. Then carbohydrate is combined with the proteins to form glycoproteins, the main substances secreted from the cell. Mucus is one example of a glycoprotein. The Golgi apparatus also appears to produce materials for plant cell walls and insect cuticles.

Lysosomes – digestion and destruction

Some of the vesicles pinched off the rough endoplasmic reticulum come to enclose digestive enzymes and form another organelle known as a **lysosome**. When food is taken into the cell of a single-celled animal such as *Amoeba*, it must be broken down into simple chemicals that can then be used. When organelles get worn out they need to be destroyed. Lysosomes carry out these functions.

Lysosomes appear as dark, spherical bodies in the cytoplasm of most cells and contain a powerful mix of digestive enzymes. They frequently fuse with each other and with membrane-bound vacuoles containing either food or an obsolete organelle. The enzymes then break down the contents into smaller molecules that can often be reused. The word 'lysis' from which they get their name means 'breaking down'.

Lysosomes can also self-destruct. If an entire cell is damaged or running down the lysosomes may rupture, releasing their enzymes to destroy the entire contents of the cell. Problems can arise if this starts to happen before the proper time, as in diseases like rheumatoid arthritis, when the cartilage self-destructs.

Vacuoles – dealing with food and fluids

Vacuoles are not a permanent feature in animal cells. These membrane-lined enclosures of cell sap are formed and lost as needed. Many simple animals form food vacuoles around the prey they engulf. White blood cells in higher animals form similar vacuoles around pathogens which are engulfed. Contractile vacuoles are an important feature in simple animals living in fresh water as they allow control of the water content of the cytoplasm. But apart from these examples, vacuoles are not a major feature of animal cells.

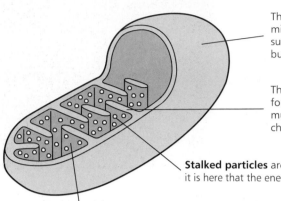

The outer membrane of the mitochondrion allows small molecules such as glucose to pass through freely, but larger molecules are excluded.

The inner membrane has many folds called **cristae**. These give a much increased surface area for chemical reactions to occur on.

Stalked particles are found on the cristae and it is here that the energy store is formed.

The matrix of the mitochondrion contains enzymes to carry out the reactions of respiration, and its own genetic material so that a mitochondrion can reproduce itself.

Figure 1.1.7 The mitochondrion – a vital organelle whose structure is closely related to its function

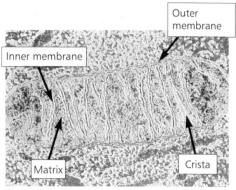

A transmission electron micrograph of the interior of a mitochondrion

Mitochondria – the cellular power station

The name **mitochondrion** means 'thread granule'. This describes what mitochondria look like under the light microscope – tiny rod-like structures in the cytoplasm of almost all cells. In recent years biologists have been able to sort out not only their complex structure but also their functions.

Mitochondria are the 'powerhouse' of the cell. Here, in a series of complicated biochemical reactions, energy is released from the respiration of food. This energy is in a form which can be used for all the other functions of the cell and indeed the organism. Just as the amount and type of endoplasmic reticulum can give us valuable clues to the function of a particular cell, so can the number of mitochondria present. Cells which require little energy to carry out their functions, for example fat storage cells, have very few mitochondria. Any cell which does an energy-demanding job, such as muscle cells, will contain large numbers of mitochondria.

Mitochondria have a double membrane. They contain their own genetic material, so that when a cell divides, the mitochondria replicate. Mitochondria also replicate at times other than cell division, for example when the long-term energy demands of a cell increase. Figure 1.1.7 shows their internal arrangement which is perfectly adapted for their function.

The cytoskeleton

Work in recent years has shown that a **cytoskeleton** is a feature of all eukaryotic cells (see pages 16–17). This skeleton is a dynamic, three-dimensional web-like structure that fills the cytoplasm. It is made up of **microfilaments**, which are protein fibres, and **microtubules**, which are tiny tubes about 20 nm in diameter. They too are made of protein and are found, both singly and in bundles, throughout the cytoplasm. The cytoskeleton performs several vital functions. It gives the cytoplasm structure and keeps the organelles in place. Many of the proteins involved are closely related to actin and myosin, the contractile proteins in muscle (described in more detail in section 2.6) and the cytoskeleton is closely linked with cell movements and with transport within cells. Figure 1.1.8 shows the cytoskeleton.

Centrioles, cilia and flagella

A pair of **centrioles** is found near to the nucleus, made up of bundles of nine tubules. These centrioles pull apart during cell division to produce a spindle made of microtubules which are involved in the movement of chromosomes.

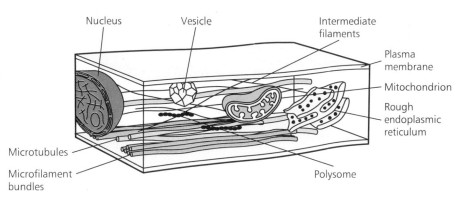

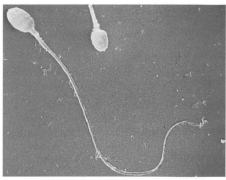

Single flagellum on spermatozoon

Figure 1.1.8 The cytoskeleton forms a tangled web of structural and contractile fibres that hold the organelles in place and enable cell movement to occur.

Cilia and **flagella** are quite common features of cells. They are very similar to each other in structure, as figure 1.1.9 shows, but cilia, with a length of 5–10 μm, are shorter than flagella, with an average length of about 100 μm. Cilia are also found in much greater numbers.

The importance of both structures is that they lash backwards and forwards. In flagella this action is used to produce movement of a cell or organism, but in cilia it may be put to other uses. For example, great borders of cilia waft mucus and other substances along the tubes of our bodies, whereas a single flagellum helps move a spermatozoon to an egg.

THE TYPICAL PLANT CELL

In many ways the typical plant cell, shown in figure 1.1.10, bears a strong resemblance to the typical animal cell. In common with animal cells, plant cells have many membranes and contain cytoplasm and a nucleus. You will find rough and smooth endoplasmic reticulum spreading throughout the cytoplasm, and Golgi apparatus as well. Mitochondria release energy from the respiration of food which is as vital to the working of the plant as it is to the animal cell. However, of the organelles found in animal cells, plant cells do not have centrioles. More importantly, there are several more organelles peculiar to plant cells, which are described overleaf.

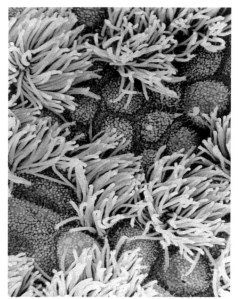

Mass of cilia in respiratory tract

Nine pairs or doublets of microtubules around the outside

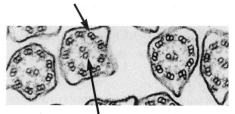

Central pair of microtubules

Internal structure of cilia

Figure 1.1.9 It is not known exactly how cilia and flagella move. They all have this 9 + 2 arrangement of microtubules in the centre.

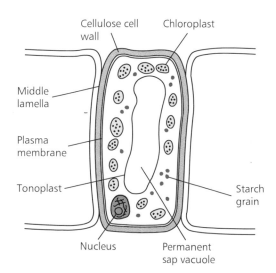

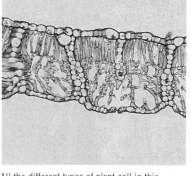

All the different types of plant cell in this section of a leaf share the same basic structure.

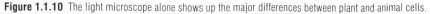

Figure 1.1.10 The light microscope alone shows up the major differences between plant and animal cells.

The ultrastructure of the plant cell

The plant cell wall

Animal cells can be almost any shape. Plant cells tend to be more regular and uniform in their appearance, largely because each cell is bounded by a rigid **cell wall**. We can visualise a plant cell as a jelly-filled balloon inside a shoe box. The cell wall gives plants their strength and support. It is made up of insoluble cellulose fibres which are meshed in a matrix of carbohydrates called pectates and hemicelluloses. The cell wall can become impregnated with suberin in cork tissues, or with lignin to produce wood. Unless it is affected by these substances the cell wall is freely permeable – it does not act as a barrier to anything which might enter the cell in solution.

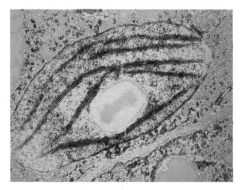

The vacuole

A **vacuole** is a fluid-filled space in the cytoplasm surrounded by a membrane. Vacuoles occur quite frequently in animal cells, but they appear only when needed, and then disappear. In plants the vacuole is a permanent feature of the structure of the cell, with an important role to play. The vacuole of a plant cell is surrounded by a membrane called the **tonoplast**. It is filled with **cell sap**, a solution of various substances in water. This solution causes the movement of water into the cell by a process called **osmosis** and the result is that the cytoplasm is kept pressed against the cell wall. This keeps the cells firm and contributes to keeping the whole plant upright. Osmosis and the role it plays in plants will be considered in more detail in section 1.4.

Chloroplasts

Of all the differences between plant and animal cells, it is perhaps the presence of **chloroplasts** in plant cells which is the most important, because they enable plants to make their own food. Chloroplasts have a structure similar in some ways with that of mitochondria. Chloroplasts are large organelles, the shape of biconvex lenses with a diameter of 4–10 μm and 2–3 μm thick. They contain their own DNA and have a double membrane. Also like mitochondria, chloroplasts have an enormously folded inner membrane which gives a greatly increased surface area on which biochemical reactions take place.

Figure 1.1.11 shows the structure of a chloroplast. The membranes are arranged in stacks called **grana**. This is where **chlorophyll** is found. Chlorophyll is a green pigment which is largely responsible for trapping the energy of light and so making it available for the plant to use. In a matrix surrounding the membrane stacks, called the **stroma**, are all the enzymes needed to complete the process of photosynthesis and provide the plant with its food. This food is stored as **starch grains** in the chloroplast.

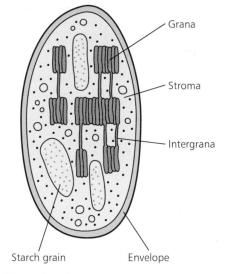

Figure 1.1.11 The structure of a chloroplast is shown in this electron micrograph.

Eukaryotic and prokaryotic cells

Animals, plants, fungi and protoctists (single-celled organisms such as *Amoeba*) all have cells containing the structures we have just looked at. These organisms are known as **eukaryotes** and are made up of **eukaryotic cells**.

There is another group of organisms, probably the most ancient on Earth, which do not fall into this category. These are the bacteria, the photosynthetic blue-green bacteria, and the viruses, collectively known as the monerans – and they have a very different structure. Organisms such as bacteria are called **prokaryotes** and are made up of **prokaryotic cells** which lack much of the structure and organisation of the eukaryotic cells. They do not have a membrane-bound nucleus – the genetic material is a single strand coiled up in the centre to form the **nucleoid**. Sometimes there are small additional pieces of genetic material within the cell called **plasmids**. The cytoplasm contains enzymes, ribosomes and food storage granules but there is no endoplasmic reticulum, no Golgi apparatus, no mitochondria and no chloroplasts. Respiration takes place on a special piece of the plasma membrane called a **mesosome**. Those prokaryotes which can photosynthesise have a form of chlorophyll but no complex structures to hold it. Figure 1.1.12 shows a typical prokaryotic cell.

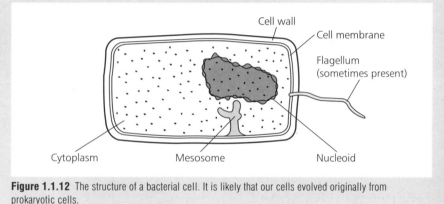

Figure 1.1.12 The structure of a bacterial cell. It is likely that our cells evolved originally from prokaryotic cells.

By looking at the features of a typical plant and animal cell we have familiarised ourselves with the main working parts of the most fundamental unit of life. Now we can move on to look even more closely at these basic units of existence – to the chemicals which work together to make up a living thing.

SUMMARY

- Living organisms are made up of individual units called **cells**.
- Cells can be observed using a light microscope (which magnifies up to ×1500) or an electron microscope (up to ×500 000). Cells may be altered by the various preparations which make them visible under a microscope.
- **Eukaryotic** cells comprise a **nucleus** and **cytoplasm** enclosed by a **cell surface membrane**. The **ultrastructure** of a cell consists of the organelles visible with the electron microscope.
- A typical animal cell has the following organelles:
  a **nucleus** – which controls cell function
  rough endoplasmic reticulum – a network of membrane-bound cavities linking with the nucleus
  ribosomes – structures on the rough endoplasmic reticulum involved with protein synthesis
  smooth endoplasmic reticulum – involved with the synthesis of steroids and lipids

Golgi apparatus – stacks of flattened membrane pockets, involved in the synthesis of glycoproteins

lysosomes – vesicles which digest food and worn-out organelles

vacuoles – temporary membrane-bound pockets of cell sap, such as food vacuoles formed around prey

mitochondria – rod-shaped organelles with a double membrane, the inner one greatly folded, involved in the release of energy from food by the process of respiration

cytoskeleton – a system of **microtubules** and proteins which maintain the internal structure of the cell

centrioles, **cilia** and **flagella** – concerned with movement in the cell. Centrioles pull apart and form a spindle of microtubules during cell division. Cilia and flagella have a similar 9 + 2 arrangement of microtubules. Banks of cilia waft substances past them, while flagella aid the swimming motion of a cell.

- A typical plant cell is similar to an animal cell *except*:
  it has no centrioles
  it has a rigid cellulose **cell wall** which is freely permeable; this supports the cell
  it has a permanent **vacuole** filled with cell sap, involved with water balance in the cell
  it has **chloroplasts** – structures similar to mitochondria, which contain chlorophyll and are responsible for photosynthesis.

- **Prokaryotic** cells do not have this high level of organisation. Their genetic material is grouped in a **nucleoid**, and sometimes on **plasmids**. Respiration takes place on a piece of membrane called a **mesosome**.

QUESTIONS

1 Draw up a table to summarise the appearance of a typical plant cell under
  a the light microscope
  b the electron microscope.

2

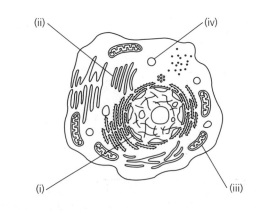

(ii)   (iv)

(i)   (iii)

Figure 1.1.13

a Name the structures marked **i**, **ii**, **iii** and **iv**. For each, give a brief explanation of its function.

b What is the main function of this type of cell? Which features of the cell suggest its function?

3 Write a brief account of the impact of the development of the electron microscope on the study of cell biology.

1.2 The molecules of life: Carbohydrates, lipids and proteins

Cells are the fundamental units of life. But cells are themselves made up of a vast array of molecules, some very simple, some extremely complex. To understand how cells function, and so in turn to understand the way whole living organisms function, we need to have a grasp of the building blocks which make up the units of life.

ORGANIC COMPOUNDS IN LIVING THINGS

We have looked at the ultrastructure of cells and seen the range of organelles that work together to continue the processes of life. But what are these organelles made of? What is the composition of cytoplasm? Around 65% of cells is water – we shall consider later why this is. There are also other important groups of molecules, many of which are **organic compounds**. This term is explained in the box below. There are three main types of organic compounds found in living cells – **carbohydrates**, **lipids** and **proteins**. We shall look at each in turn.

Organic molecules

Almost all material in a cell which is not water is made up of organic molecules. Organic compounds all contain carbon and hydrogen atoms. Most organic molecules also contain oxygen atoms, and some include nitrogen, sulphur and phosphorus.

Each carbon atom can make four bonds, so has the potential to join up with four other atoms – we say it has a **valency** of four. This is shown in figure 1.2.1. Carbon atoms bond very strongly to other carbon atoms. This means that many carbon atoms can join together to build very large molecules. Long chain molecules can be formed, with hydrogen and oxygen atoms bonded to the carbon atoms in the chain. Branched chains, or rings, or any number of three-dimensional shapes are also possible. Often a relatively small molecule (a **monomer**) will be joined to many other similar monomers to make a very large molecule called a **polymer**.

This ability to make large molecules is what makes carbon such an important element. The very many types of organic molecules provide the great biochemical variety and complexity needed in living things.

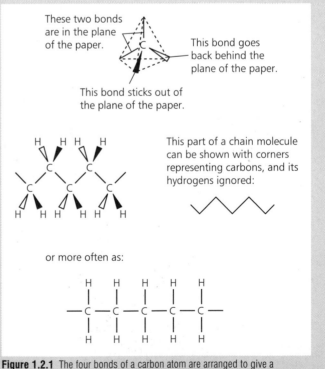

Figure 1.2.1 The four bonds of a carbon atom are arranged to give a tetrahedral shape. This means that organic molecules have a complicated three-dimensional structure. As this becomes extremely difficult to draw and confusing to look at, organic molecules are often drawn 'flat'.

CARBOHYDRATES

The main job of carbohydrates within an organism is to provide energy. Carbohydrates are also involved in storing energy, and in plants they form an important part of the cell wall.

The most commonly known carbohydrates are the sugars and starches. We have all come across **sucrose** as 'sugar'. **Glucose** is the energy supplier in sports and health drinks, and you will be familiar with **starch** in flour and potatoes. But the group of chemicals known to the biochemist as carbohydrates is much wider than this.

The basic structure of all carbohydrates is the same. They are made up of carbon, hydrogen and oxygen. They fall into three main types, depending on the complexity of the molecules. These are the **monosaccharides**, **disaccharides** and **polysaccharides**.

Monosaccharides – the simple sugars

In these simple sugars there is one oxygen atom and two hydrogen atoms for each carbon atom present in the molecule. A general formula for this can be written:

$$(\textbf{CH}_2\textbf{O})_n$$

n can be any number, but is usually low. The **triose** ($n = 3$) sugars such as glyceraldehyde are important in the biochemistry of respiration. These sugars have the formula $C_3H_6O_3$. The **pentose** ($n = 5$) sugars are most important in the structure of the nucleic acids which make up genetic material. Ribose is a pentose with which you will become familiar. It has the formula $C_5H_{10}O_5$. The best known monosaccharides, including glucose, are the **hexose** sugars with $n = 6$. They have the general formula $C_6H_{12}O_6$.

These general formulae tell us how many and what types of atoms there are in the molecule, but do not tell us what the molecule looks like and why it behaves as it does. To do this we look at displayed formulae, for example those shown in figure 1.2.2. Although these do not follow every kink in the carbon chain, they do give an idea of how the molecules are arranged in space. This three-dimensional structure is very important in understanding how biological systems work.

α–glucose

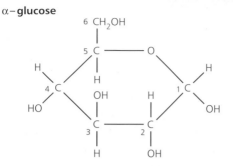

β–glucose

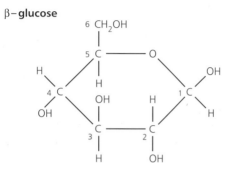

Fructose

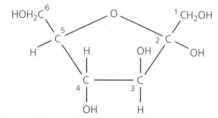

Figure 1.2.2 Hexose sugars often have a ring structure. The arrangement of the atoms on the side chains can make a great difference to the way in which the molecule is used by the body. This is why the carbon atoms are numbered. Look carefully at the differences between α-glucose and β-glucose. These two molecules are known as **structural isomers**.

Stereoisomerism – the same but different

A triose sugar such as glyceraldehyde, with only three carbon atoms, sounds a fairly simple molecule. But the three-dimensional arrangement of the bonds means that there are two different forms, which are mirror images of each other. These are shown in figure 1.2.3. However hard you try you cannot superimpose the image of D-glyceraldehyde onto the image of L-glyceraldehyde. This phenomenon is called **stereoisomerism**, and the two forms of glyceraldehyde are called **stereoisomers**.

Glyceraldehyde is the standard molecule with which all others are compared in order to determine whether they are the D or L form. It is important to know which is which, because biological systems are sensitive to the difference between stereoisomers. For example, it is largely only the D forms of sugars and the L forms of amino acids that are found in living things.

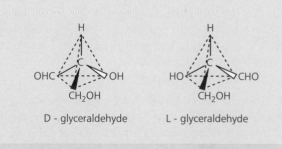

D - glyceraldehyde   L - glyceraldehyde

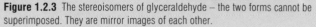

Figure 1.2.3 The stereoisomers of glyceraldehyde – the two forms cannot be superimposed. They are mirror images of each other.

Stereoisomerism and optical activity

To the eye, solutions of L- and D-glyceraldehyde (see figure 1.2.3) look identical. The only simple way to tell them apart is by shining a beam of polarised light through a solution of glyceraldehyde in water (an aqueous solution). The D form of glyceraldehyde turns polarised light to the right (it is **dextro-rotatory**, indicated by (+)). The L form of glyceraldehyde turns polarised light to the left (it is **laevo-rotatory**, indicated by (–)).

Other molecules can be separated into L and D forms by comparing the position of their hydroxyl group with that of the stereoisomers of glyceraldehyde. Their optical activity can also be checked. D-isomers of any stereoisomeric compound other than glyceraldehyde may rotate polarised light to either the right *or* the left (may be + or –). The same is true for L-isomers. The D- or L- prefix indicates *the stereoisomerism of the molecule compared with a molecule of glyceraldehyde*. This is *not* the same as the optical activity of the molecule, shown as (+) or (–), although the optical activity results from the stereoisomerism. For example, D-fructose is laevorotatory (–), unlike D-glyceraldehyde. The only way of telling which way a particular stereoisomer will rotate polarised light is to do the experiment – you cannot tell from the structure of the molecule. But you *can* tell whether it is the D- or L-isomer, by comparing it with glyceraldehyde.

Biological systems can be sensitive to stereoisomerism – for example, some enzymes will act on only the D-isomer of a substrate – but not to optical activity.

Disaccharides – the double sugars

The molecules of disaccharides are made up of two monosaccharide molecules joined together. Sucrose is the result of a molecule of α-glucose joining with a molecule of β-fructose. The two monosaccharides join in a **condensation reaction** to form a disaccharide, and a molecule of water (H_2O) is removed. The bond between the two monosaccharides which results is known as a **glycosidic link**, shown in figure 1.2.4.

This joining of monosaccharides gives a different general formula for disaccharides, and indeed for chains of monosaccharides of any length:

$$(C_6H_{10}O_5)_n$$

When different monosaccharides join together, different disaccharides result. Table 1.2.1 shows some of the more common ones.

Disaccharide	Source	Monosaccharide units
Sucrose	Stored in plants such as sugar beet and sugar cane	Glucose + fructose
Lactose	Milk sugar – the main carbohydrate found in milk	Glucose + galactose
Maltose	Malt sugar – found in germinating seed such as barley	Glucose + glucose

Table 1.2.1 Some common disaccharides

Figure 1.2.4 The forming of a glycosidic link. The reaction between two monosaccharides results in a disaccharide and a molecule of water. Although these examples look relatively simple, remember that isomers of all different sorts are possible. Also, a link between C1 and C4 of two monosaccharides will result in a different disaccharide from a C1–C6 link between the same two monosaccharides.

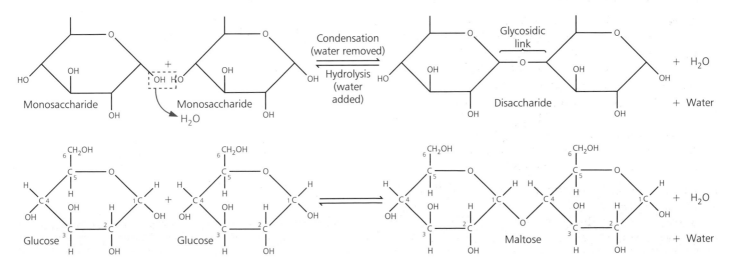

Polysaccharides

The most complex carbohydrates are the polysaccharides. The sweet taste which is characteristic of both mono- and disaccharides is lost when many single sugar units are joined to form a polysaccharide. Polysaccharides generally result from the linking of glucose molecules in different ways, and they form very compact molecules which are ideal for storing energy. The glucose units can then be released when they are needed to supply energy. Polysaccharides are physically and chemically very inactive, so their presence in the cell does not interfere with other cell functions. Figure 1.2.5 shows polysaccharide stores in both plant and animal cells.

Starch is one of the best known polysaccharides. It is particularly important as an energy store in plants. The sugars produced by photosynthesis are rapidly converted into starch. Storage organs such as potatoes are particularly rich in starch. Starch is made up of long coiled chains of α-glucose, joined with 1–4 glycosidic links. The chains are branched – the more branches there are, the less easily the starch 'dissolves' in water. (In fact, starch forms a colloid in water. This is explained in section 1.4, page 47.)

Glycogen is sometimes referred to as 'animal starch'. It is the only carbohydrate energy store found in animals. Chemically it is very similar to starch, also being made up of many α-glucose units. The difference is that in glycogen there are some 1–6 glycosidic links as well as 1–4 links. Glycogen is found mainly in muscle tissue and particularly in liver tissue, which is very active and needs a readily available energy supply at all times.

Cellulose is an important structural material in plants. As we have seen, it is the main constituent in plant cell walls. Like starch and glycogen, it consists of long chains of glucose, but in this case β-glucose held together by 1–4 glycosidic links. This is shown in figure 1.2.6. Mammals, and indeed most animals, do not possess the enzymes needed to break β-1–4 linkages and so they cannot digest cellulose. (Herbivores can digest cellulose because of gut bacteria. This is discussed in section 3.3.)

The way that the monomers are joined in cellulose means that –OH groups protrude from the molecule, giving rise to hydrogen bonds which hold neighbouring chains together (see the box 'Holding together' on page 30). Cellulose is therefore a material with considerable strength, which is exploited in many ways, as figure 1.2.7 shows. In the cell wall, groups of about 2000 chains of cellulose form interweaving microfibrils which can be seen by the electron microscope.

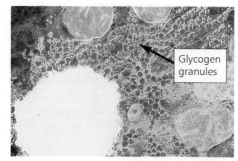

Liver cell

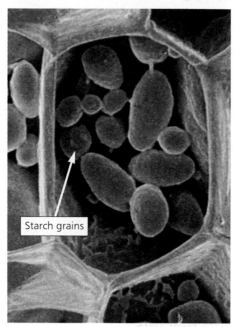

Potato cell

Figure 1.2.5 Storage carbohydrates play important roles in both plant and animal cells.

Figure 1.2.6 These seemingly small differences in the molecules of starch, glycogen and cellulose make all the difference to where they are found and what they do.

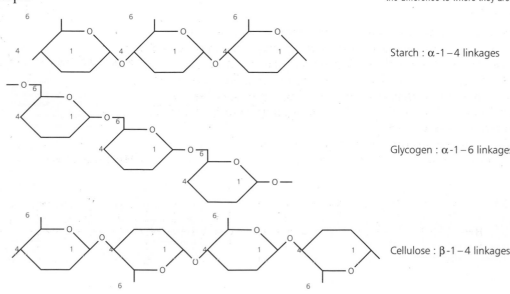

Starch : α-1–4 linkages

Glycogen : α-1–6 linkages

Cellulose : β-1–4 linkages

Figure 1.2.7 Cellulose forms the basis of some major human industries. We use cellulose to make paper and cellophane, as well as using derivatives to produce explosives and films. Cellulose in the form of pure cotton is worn throughout the world.

LIPIDS

Another group of organic chemicals which make up cells are the **lipids**. Lipids include some of the highest profile chemicals in public health issues – **cholesterol** and **fat**. The media constantly remind us of the importance of a low-fat diet and the dangers of high cholesterol levels. But are fats really harmful, and what *is* cholesterol?

Lipids are an extremely important group of chemicals which play major roles in living systems. They are an important source of energy in the diet of many animals and the most effective form for living things to store energy – they contain more energy per gram than carbohydrates or proteins. Many plants and animals convert spare food into fats or oils for use at a later date. Combined with other molecules lipids also play vital roles in cell membranes and in the nervous system.

All lipids dissolve in organic solvents but are insoluble in water. This is important because it means they do not interfere with the many reactions which go on in aqueous solution in the cytoplasm of a cell.

As in carbohydrates, the chemical elements that make up lipids are carbon, hydrogen and oxygen. In lipids, however, there is a considerably lower proportion of oxygen than in carbohydrates.

Fats and oils

One of the main groups of lipids are the **fats** and **oils**. They are chemically extremely similar, but fats (for example, butter) are solids at room temperature and oils (for example, olive oil) are liquids. Fats and oils are made up of combinations of two types of organic chemicals: **fatty acids** and **glycerol** (or propane-1,2,3-triol).

Glycerol has the chemical formula $C_3H_8O_3$, which can be shown more clearly as in figure 1.2.8.

There is a wide range of fatty acids. Over 70 different ones have been extracted from living tissues. All fatty acids have a long hydrocarbon chain – a pleated backbone of carbon atoms with hydrogen atoms attached – and a carboxyl group (–COOH) at one end. Fatty acids vary in two main ways. The length of the carbon chain can differ, although in living organisms it is frequently between 15 and 17 carbon atoms long. More importantly, the fatty acid may be **saturated** or **unsaturated**.

In a saturated fatty acid each carbon atom is joined to the next by a *single* bond. The example shown in figure 1.2.9 is stearic acid.

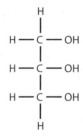

Figure 1.2.8 Displayed formula of glycerol

Figure 1.2.9 Displayed formula of stearic acid

$CH_3(CH_2)_{16}COOH$

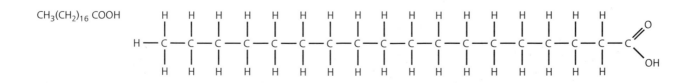

In an unsaturated fatty acid the carbon chains contain one or more *double* carbon–carbon bonds. An example shown in figure 1.2.10 is linoleic acid, which is an essential fatty acid in the diet of mammals, including ourselves, as it cannot be synthesised within mammalian metabolic pathways.

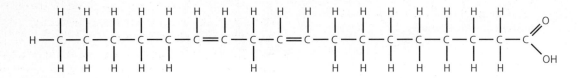

Figure 1.2.10 Displayed formula of linoleic acid

A fat or oil results when glycerol combines with one, two or three fatty acids to form a **mono-**, **di-** or **triglyceride**. A bond is formed between the carboxyl (–COOH) group of a fatty acid and one of the hydroxyl (–OH) groups of the glycerol. A condensation reaction takes place, involving the removal of a molecule of water, and the resulting bond is known as an **ester link**. This type of reaction is called **esterification** and is illustrated in figure 1.2.11.

The nature of a fat or oil depends largely on the fatty acids present in it, as figure 1.2.12 shows.

Figure 1.2.11 Esterification – and the molecules that result

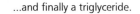
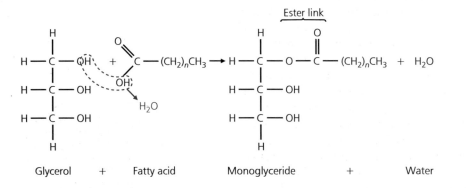

Glycerol   +   Fatty acid   Monoglyceride   +   Water

The process is repeated to give a diglyceride...   ...and finally a triglyceride.

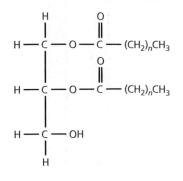
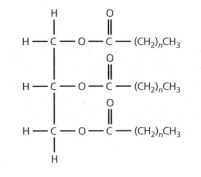

Figure 1.2.12 It is the combination of fatty acids in a triglyceride that determines what it is like. Saturated fatty acids give solid fats like butter and blubber, whereas unsaturates give liquids such as olive oil.

Biochemistry and affairs of the heart

Recent medical research seems to indicate that high levels of fat, particularly saturated fat, in our diet are not good for our long-term health. Fatty foods are very high in energy, and so a diet high in fats is likely to result in obesity. Worse than this, however, is the implication that saturated fats – found particularly in animal products such as dairy produce and meat – can cause problems in the metabolism leading to fatty deposits in the arteries. In the long term this can lead to all sorts of problems, including heart disease and death. Unsaturated fats, found mainly in plants, do not seem to have this effect and so people are being encouraged to replace saturates in their diets with unsaturates whenever possible.

But there is a further twist to this tale. Unsaturates themselves can be further divided into **polyunsaturates** and **monounsaturates**. Most of the fatty acids in polyunsaturates have two or more double bonds in their carbon chain. It seems that these do not have the damaging effects of saturated fats. The majority of the fatty acids in monounsaturates have only one double bond in the carbon chain – and these seem to have a positively beneficial effect, helping the body to cope better with saturated fats. The story is a long and complicated one, which will doubtless take many years to unravel.

Phospholipids

Inorganic phosphate, PO_4^{3-}, is present in the cytoplasm of every cell. Sometimes one of the hydroxyl groups of glycerol undergoes an esterification reaction with phosphate instead of with a fatty acid, and a simple **phospholipid** is formed. The phosphate element of the phospholipid may go on to react further and combine with other chemicals to form substances such as choline phosphoglyceride, needed for the formation of neurotransmitter substances and thus the successful functioning of the nervous system.

As we have seen, fats and oils are insoluble in water – they are not polar molecules. This makes them useful as inert storage materials, but limits their use elsewhere. Phospholipids are important because the lipid and phosphate parts of the molecule give it very different properties. The lipid part is neutral and is insoluble in water – it is known as **hydrophobic** or water-hating. In contrast, the phosphate part is highly polar and dissolves readily in water – it is **hydrophilic** or water-loving. This structure is shown in figure 1.2.13. It means that part of the molecule can be dissolved in fatty material such as the membrane, whilst the other part interacts with substances dissolved in water. Phospholipids are a vital component of cell membranes, and as such are present in almost all living things.

Waxes

Waxes are lipids made up of very long chain fatty acids joined to alcohols by ester links. The difference between fats and waxes is that there is only one fatty acid joined to each alcohol in a wax, as the alcohols only have one hydroxyl group, whereas glycerol has three. Waxes are very insoluble and are largely used by both plants and animals for waterproofing. Some waxes produced by insects to protect their cuticles from water loss can withstand extremely high temperatures without melting.

Steroids

Steroids are not typical lipids – apart from the fact that they are insoluble in water, they have little in common with the others. But steroids are of great biological importance, particularly as hormones in both animals and plants.

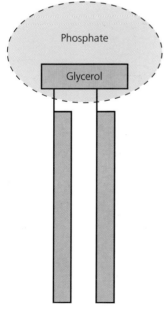

Fatty acid chains

Figure 1.2.13 The fatty acid chains of a phospholipid are neutral and insoluble in water. The phosphate head carries a negative charge and is soluble in water.

Steroids are made up of very large numbers of carbon atoms arranged in complex ring structures. Figure 1.2.14 gives an example.

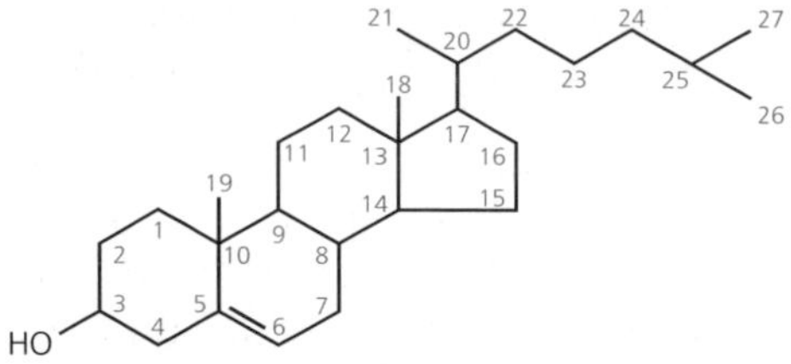

PROTEINS

About 18% of the human body is made up of protein – a high proportion second only to water. Our hair, skin, nails, the enzymes which control all the reactions in our cells, the enzymes which digest our food and many of the hormones that control our organs and their functions are made of protein. Protein molecules are also responsible for the contraction of muscle fibres, protection from disease in the form of antibodies, the clotting of blood by fibrinogen and prothrombin, the transport of oxygen in our blood by haemoglobin and much, much more. Proteins are extremely important molecules with a variety of functions throughout the living world. An understanding of how protein molecules are made up and the factors that affect their shapes and functions gives an insight into the biology of all living things.

Like carbohydrates and fats, proteins are made up of the elements carbon, hydrogen and oxygen, but in addition they all contain nitrogen. Many proteins also contain sulphur, and some also have phosphorus and various other elements. Proteins are very large molecules known as **macromolecules**. They are polymers, made up of many small units joined together. These small units are called **amino acids**. Amino acids combine in long chains to produce proteins in the same way as monosaccharide units join together to form polysaccharides. However, whilst each polysaccharide is made up of one or two different types of monosaccharide, there are about 20 different naturally occurring amino acids that combine to form proteins – so the potential variety of proteins is vast.

Amino acids – the building blocks of proteins

All amino acids have the basic structure shown in figure 1.2.15 of an **amino** (–NH$_2$) group and a **carboxyl** (–COOH) group attached to the same carbon atom.

Figure 1.2.14 One of the best known steroids – cholesterol. A high blood cholesterol level can mean heart trouble ahead.

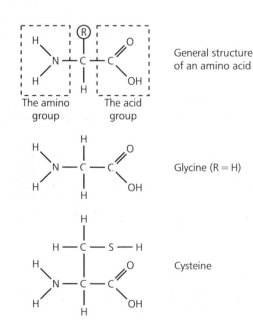

General structure of an amino acid

The amino group   The acid group

Glycine (R = H)

Cysteine

Figure 1.2.15 The R group varies from one amino acid to another. In the simplest amino acid, glycine, R is a single hydrogen atom. In a larger amino acid such as cysteine, R is a much more complex group.

L- and D-amino acids

The carbon atom to which the amino, carboxyl and R groups are joined is called the α-carbon. Draw out a general amino acid structure as in figure 1.2.16 and you will see that it looks similar to glyceraldehyde – it is an asymmetrical system. This means that there are stereoisomers of amino acids, and just as with carbohydrates, living systems are sensitive to the different stereoisomers. In contrast to the carbohydrates it is the L-isomers of amino acids which are mainly found in living organisms.

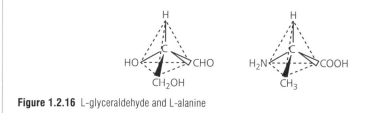

Figure 1.2.16 L-glyceraldehyde and L-alanine

Proteins from amino acids

When looking at how amino acids link to form long chains, we can ignore the R group and concentrate entirely on the amino and carboxyl groups. To make the diagrams easier to follow, the central carbon atom with the hydrogen atom and R group attached is shown simply as a striped circle, as in figure 1.2.17.

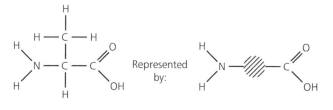

Alanine

Amino acids join together by a reaction between the amino group of one amino acid and the carboxyl group of another amino acid. They join in a condensation reaction and a molecule of water is lost. The bond formed is known as a **peptide link** and when two amino acids join, a **dipeptide** is the result, as in figure 1.2.18. More and more amino acids can join together to

Are amino acids really acids?

The carboxyl end (–COOH) of an amino acid is acidic in nature. It will ionise in water to give hydrogen ions. However, the amino end (–NH$_2$) is basic in nature. It attracts hydrogen ions in solution. In acidic solutions an amino acid acts like a base, and in alkaline solutions it acts like an acid. In the mainly neutral conditions found in the cytoplasm of most living organisms it can act as both. This ability means an amino acid is said to be **amphoteric**. The R group also affects how the amino acid behaves – some R groups are more acidic in nature than others. The combination of all these things means that different amino acids can be separated by a sophisticated form of electrolysis which takes place in silica gel or on paper and is called **electrophoresis**. This is explained further in the box 'How to unravel a protein' on page 31.

Figure 1.2.17 Simple representation of alanine

Figure 1.2.18 Amino acids can be joined in a seemingly endless variety of orders to produce an almost infinite variety of polypeptides.

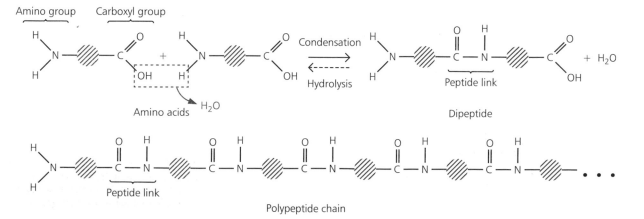

form **polypeptide** chains, which may be from around ten to many thousands of amino acids long. A polypeptide can fold or coil or become associated with other polypeptide chains to form a **protein**.

Protein structure

Proteins are described by a **primary**, a **secondary**, a **tertiary** and a **quaternary structure**, illustrated in figure 1.2.19.

The primary structure of a protein describes the sequence of amino acids which make up the polypeptide chain. But this is only the beginning of the story. The secondary structure describes the three-dimensional arrangement of the polypeptide chain. In many cases it forms a right-handed (α-) helix or spiral coil with the peptide links forming the backbone and the R groups sticking out from the coil. Hydrogen bonds hold the structure together. Most fibrous proteins have this sort of structure. In other proteins, the polypeptide chains fold up into pleated sheets, again with the pleats held together by hydrogen bonds. Sometimes there is no regular secondary structure and the polypeptide forms a sort of random coil.

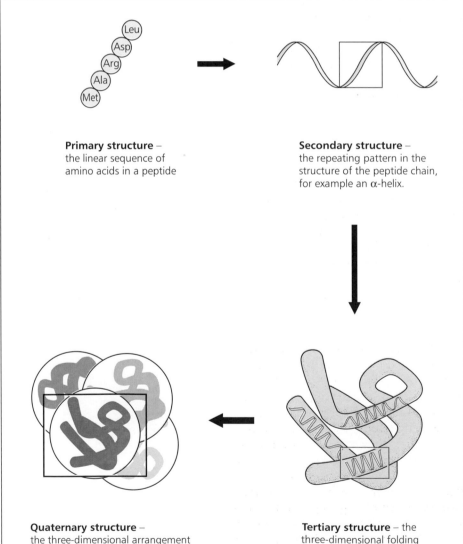

Primary structure –
the linear sequence of
amino acids in a peptide

Secondary structure –
the repeating pattern in the
structure of the peptide chain,
for example an α-helix.

Quaternary structure –
the three-dimensional arrangement
of more than one tertiary polypeptide

Tertiary structure – the
three-dimensional folding
of the secondary structure

Figure 1.2.19 It is not only the sequence of amino acids but also the arrangement of the polypeptide chains which determines the characteristics of a protein.

Some proteins are very large molecules indeed, consisting of thousands of amino acids joined together. The globular proteins (described below) such as the oxygen-carrying blood pigment haemoglobin and enzyme proteins are so large that they need a further level of organisation. The α-helices and pleated sheets are folded further into complicated shapes. These three-dimensional shapes are held in place by hydrogen bonds, sulphur bridges and ionic bonds. This organisation is the **tertiary structure** of the protein.

And finally, some enzymes and haemoglobin are made up of not one but several polypeptide chains. The **quaternary structure** describes the way these polypeptide chains fit together.

Fibrous and globular proteins

Proteins fall into two main groups – fibrous and globular proteins. **Fibrous proteins** have little or no tertiary structure. They are long parallel polypeptide chains with occasional cross-linkages making up fibres. They are insoluble in water and are very tough, which makes them ideally suited to their mainly structural functions within living things. They are found in connective tissue, tendons and the matrix of bones (collagen), in the structure of muscles, in the silk of spiders' webs and silkworm cocoons and as the keratin making up hair, nails, horns and feathers.

Globular proteins have complex tertiary and sometimes quaternary structures. They are folded into spherical (globular) shapes. Globular proteins make up the immunoglobulins (antibodies) in the blood. They form enzymes and some hormones and are important for maintaining the structure of the cytoplasm.

Conjugated proteins

Sometimes a protein molecule is joined with or **conjugated** to another molecule, called a **prosthetic group**. For example, the **glycoproteins** are proteins with a carbohydrate prosthetic group. Many lubricants used by the human body, such as mucus and the synovial fluid in the joints, are glycoproteins, as are some proteins in the cell membrane. **Lipoproteins** are proteins conjugated with lipids and they too are most important in the structure of cell membranes. Haemoglobin, the complex oxygen-carrying molecule in the blood, is a conjugated protein with an inorganic iron prosthetic group.

The properties of proteins

The secondary, tertiary and quaternary structures of proteins can be relatively easily damaged or **denatured**. Although the strong covalent bonds between the amino acids in the polypeptide chains are not readily destroyed, the relatively weak forces holding the different parts of the chains together can be disrupted very easily. As the functions of most proteins rely very heavily on their three-dimensional structure, this means that the entire biochemistry of cells and whole organisms is very sensitive to changes that might disrupt their proteins, such as a rise in temperature or a change in pH which will distort the internal balance of charges.

The large size of protein molecules affects their behaviour in water. Because they have ionic properties due to their carboxyl and amino groups, and also to many of the R groups, we might expect them to dissolve in water and form a solution. In fact, the molecules are so big that they form a colloidal suspension (see section 1.4, page 47) and play an important role in holding molecules in position in the cytoplasm.

Holding together

Peptide chains are held in the secondary, tertiary and quaternary structures of proteins by different sorts of bonds. Proteins often change their shape and then return to normal. What sorts of bond will allow this to happen?

Hydrogen bonds

Tiny negative charges are present on the oxygen atoms of carboxyl groups, and tiny positive charges are found on the hydrogen atoms of amino groups, OH– groups and indeed any polar group. When these charged groups are close enough to each other, the opposite charges attract, forming a **hydrogen bond**. Hydrogen bonds are weak compared with covalent bonds, but a large number can hold a structure very firmly. They are easily broken and reformed, for example by varying pH conditions, and are of great importance in biological molecules.

Sulphur bridges

Sulphur bridges are formed when two cysteine molecules are close together in the structure of a polypeptide. An oxidation reaction takes place between the two sulphur-containing groups, resulting in a strong covalent bond known as a sulphur bridge or disulphide link. Sulphur bridges are much stronger than hydrogen bonds.

Blow-drying and perming

A simple demonstration of the difference in strength between these two types of bond is shown by blow-drying and perming hair. When you blow-dry your hair you break the hydrogen bonds in the protein and reform them with your hair curled in a different way. Next time you wash your hair it returns to its natural style as the original hydrogen bonds reform. If you have a perm, the chemicals of the perm solution break the sulphur bridges between the polypeptide coils and reform them in a different place. It is an involved procedure and the effect on that piece of hair is permanent.

Figure 1.2.20 Some very complex molecule shapes result in our hair, the myoglobin in our muscles and the enzymes vital for all forms of life. The shapes of the molecules are vital to their functions, and those shapes are maintained by hydrogen bonds and the sulphur bridges shown here.

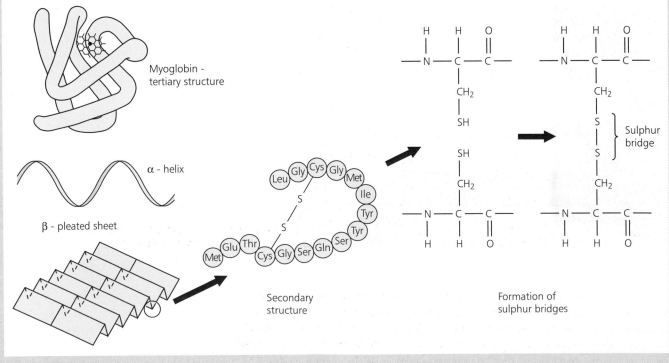

How to unravel a protein

There are many different tools used by scientists to discover the molecular structure of proteins, but two techniques in particular have played a large part in helping us to understand them.

Primary structure – electrophoresis

The first is called **electrophoresis**. Chromatography can be used to separate amino acids quite successfully, but electrophoresis gives even better results. Known amino acids are placed on a special support medium in a buffering solution (to keep the pH constant) and an electric current is passed through it. The amino acids move on the medium at different rates according to the charge on their R group, as figure 1.2.21 shows. Once the medium has dried the amino acids can be revealed using a ninhydrin spray. This reacts with amino acids to form a coloured product. The distance each amino acid has travelled under these known conditions can then be measured.

This technique can be used to find out exactly which amino acids make up a particular protein. First the protein is broken down into its component amino acids. This is done using enzymes which break the peptide links in the protein. After electrophoresis the distance the amino acids have travelled can be measured and compared with how far known 'marker' amino acids have travelled under the same conditions.

The precise order of the amino acids in a particular protein can also be worked out. This involves special enzymes which break peptide links one at a time, starting from either the amino or the acid end of the molecule. It is a very time-consuming process – when Frederick Sanger did it for the first time in 1958, working out the sequence of amino acids in the relatively small protein insulin, he was awarded the Nobel Prize!

Secondary and tertiary structure – X-ray crystallography

X-ray crystallography is a relatively simple technique. X-rays are fired into a crystal of the pure protein. The X-rays are reflected by the atoms of the protein molecules and this scattered pattern is recorded on a photographic plate. The crystal is then turned round and the process repeated many times around the structure to build up a complete record. Interpreting the resulting images is very skilled work. X-ray crystallography has revealed regular and predictable protein molecule shapes – most frequently helices and globular shapes.

Figure 1.2.21 The amino acids revealed. Electrophoresis enables us to work out the amino acids present in a protein.

Of the three most commonly found biological molecules – carbohydrates, lipids and proteins – proteins are present in the largest amounts. They are also the largest of the three types of molecule, and have the most complex structures. But where do they come from? How do cells know what proteins to make? The answer to that lies with another group of the molecules of life – the **nucleic acids**.

SUMMARY

- **Organic compounds** contain carbon and hydrogen atoms, usually with oxygen and sometimes with other atoms. Carbon atoms bond strongly to each other forming large complex chain and ring structures. Important organic compounds include **carbohydrates**, **lipids** and **proteins**.

- **Carbohydrates** provide energy and are divided into **monosaccharides**, **disaccharides** and **polysaccharides**.

- **Monosaccharides** have the general formula $(CH_2O)_n$. They include the trioses ($n = 3$) such as glyceraldehyde, pentoses ($n = 5$) such as ribose and hexoses ($n = 6$) such as glucose. Glyceraldehyde is a molecule which has **stereoisomers**, and other stereoisomers are named L or D by comparison with the structure of glyceraldehyde.

- **Disaccharides** are made up of two monosaccharide units joined by a condensation reaction and have the general formula $(C_6H_{10}O_5)_n$. Glucose is a disaccharide.

- **Polysaccharides** are polymers of glucose and are used for food storage. They include starch, glycogen and cellulose.

- **Lipids** store energy and occur in cell membranes. They include **fats** and **oils**, **phospholipids**, **waxes** and **steroids**.

- **Fats** are solids at room temperature and **oils** are liquids. Fats and oils are made up of fatty acids and glycerol. Fatty acids may be saturated or unsaturated. Glycerol forms ester links with one, two or three fatty acids to form mono-, di- or triglycerides respectively.

- **Phospholipids** are formed when one hydroxyl group of glycerol esterifies with a phosphate group. Phospholipids have a hydrophobic fatty acid chain and a hydrophilic phosphate part, and are important in cell membranes.

- **Waxes** are made up of fatty acids joined to an alcohol with one hydroxyl group, and are used as waterproofing. **Steroids** such as cholesterol have complex ring structures and are important as hormones.

- **Proteins** contain nitrogen in addition to carbon, hydrogen and oxygen. They are large macromolecules which are polymers of **amino acids**.

- There are about 20 naturally occurring amino acids that combine to form proteins, each with a carboxyl group and an amino group but with different R groups.

- Amino acids join by peptide links to form **dipeptides** (two amino acids) and **polypeptides** (up to thousands of amino acids). Polypeptides fold up to form proteins.

- The **primary** protein structure is the order of amino acids in the polypeptide chain.

- The **secondary** structure is the arrangement of the polypeptide chain into a helix or pleated sheet.

- The **tertiary** structure is the three-dimensional shape formed by folding of the secondary structure.

- **Quaternary** structures occur when two or more polypeptide chains associate to form a protein.

- **Fibrous** proteins, such as collagen, have no tertiary structure but form tough fibres.

- **Globular** proteins have tertiary and sometimes quaternary structures and include enzymes.
- **Conjugated** proteins are joined with another molecule called a prosthetic group.

QUESTIONS

1 a i How do fats and oils differ?
  ii Which chemical elements are found in fats?
  iii What is meant by an *unsaturated* fatty acid?
  iv Name two cell structures in which fatty acids occur.
  v Give an example of another type of lipid, and one example of its role in living organisms.
 b Compare the structure of lipids with the structure of polysaccharides.

2 Discuss how the properties of amino acids and proteins are suited to their variety of roles in living systems.

3 Produce a table suitable for inclusion in a revision guide summarising the three groups of macromolecules discussed in this chapter. It will need to contain as much information as possible, yet be compact and easy to read, understand and remember.

1.3 The molecules of life: Nucleic acids

All living things have a biological drive to reproduce. From the tiniest microscopic organism to the largest mammal, replacement individuals must be made before the original wears out. But how is this reproduction of a living thing brought about? Somewhere within each cell must be a pattern, a set of instructions for the assembling of new cells – to form offspring, and also to form identical cells for growth. Over the last 50 or so years scientists have made enormous strides towards understanding the form of these instructions. In the unravelling of the secrets of the genetic code, we have come closer than ever before to understanding the mystery of life.

HOW CELLS REPRODUCE

The processes that form new cells from an original are controlled by enzymes, which as we know are proteins. It is enzymes that control the biochemistry of the sort of cell that will form – whether it will be a muscle cell or a skin cell, for example. But what determines which enzymes will be produced in a cell, and how are the enzymes made?

This information is carried by molecules called **nucleic acids** in the cell. One kind of nucleic acid is **DNA**, or **deoxyribonucleic acid**, which makes up the chromosomes in the nucleus of the cell. DNA carries a code in its molecules which is read by another nucleic acid – **RNA** or **ribonucleic acid**. RNA is involved in protein synthesis, a production line which turns out enzymes to carry out the instructions and build a new cell.

Years of research and vastly expensive technology was needed to bring to light the sequence of events summarised in these few words. The first step in understanding this fascinating piece of molecular biology is to understand the structure of the main molecules involved.

THE STRUCTURE OF DNA AND RNA

DNA and RNA are both polymers. The chemical structure of the simple monomer units making up both DNA and RNA is very similar, though the polymer molecules themselves are quite different. The monomers are called **nucleotides**. Each nucleotide has three parts – a five-carbon or **pentose sugar**, a **nitrogen-containing base** and a **phosphate** group.

The pentose sugars

The sugar will be one of two very similar pentose rings. Ribonucleic acids contain the sugar **ribose**. Deoxyribonucleic acids contain the sugar **deoxyribose**. The only difference between these two sugars is that deoxyribose, as its name suggests, contains one oxygen atom less than ribose. Figure 1.3.1 shows their structures.

The nitrogen-containing bases

There are two types of bases found in nucleic acids. The **purine** bases have two nitrogen-containing rings, while the **pyrimidines** have only one. Each nucleic acid molecule contains combinations of four different bases with equal numbers of purines and pyrimidines. In DNA the purines are **adenine (A)** and

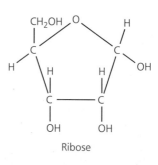

Ribose

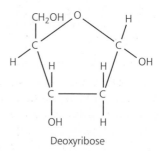

Deoxyribose

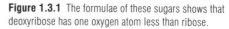

Figure 1.3.1 The formulae of these sugars shows that deoxyribose has one oxygen atom less than ribose.

guanine (**G**) and the pyrimidines are **cytosine** (**C**) and **thymine** (**T**). In RNA the purine bases are the same as in DNA but the pyrimidines are cytosine and **uracil** (**U**). These rings, shown in figure 1.3.2, have the chemical property of being bases because of the nitrogen atoms they contain.

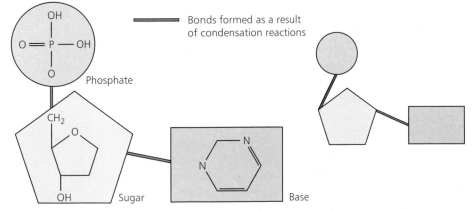

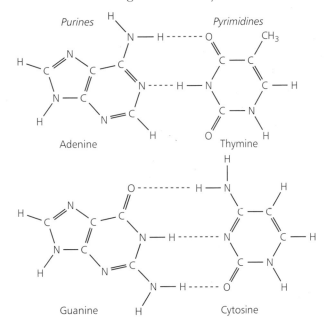 Figure 1.3.2 section

Figure 1.3.2 Look carefully at the shapes of these molecules – the importance of the way they pair will soon become clear.

Phosphate

We met phosphate groups (usually derived from phosphoric acid) when we looked at phospholipids. A phosphate group makes up the third component of a nucleotide. It gives nucleotides, and the nucleic acids that they make up, their acidic character.

Putting it together – the nucleotide and nucleic acids

Figure 1.3.3 shows how the sugar, the base and the phosphate are joined to form the nucleotide monomer.

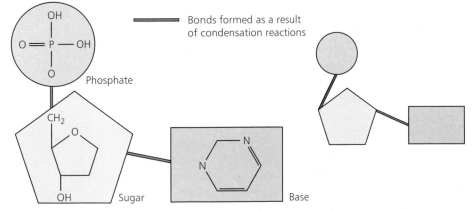

Figure 1.3.3 The three parts of a nucleotide are joined by condensation reactions. Two molecules of water are removed in the process.

    To form DNA, nucleotides containing the bases A, T, C and G join together to make chains which can be millions of units long. RNA is made up of long chains of nucleotides containing A, C, G and U. Knowledge of how these units join together, and the three-dimensional structures that are produced in DNA in particular, is the basis of our understanding of molecular genetics. The structure of DNA is taken as common knowledge now, yet the story of how it was worked out is very recent history.

The story of the double helix

Gregor Mendel's ideas of genetics were years ahead of their time. Once they had been accepted and absorbed by the scientific community, the next step was to discover what 'genes' actually were. It was almost 100 years after Mendel's time before the answers were found. For many years there were arguments between biologists about which of the many complex chemicals in cells might be the one that carried the genetic information. By the late 1940s most people agreed that it must be DNA. It was known to occur in almost every living cell. It was usually contained in the nucleus. It was a large and complex molecule. But how did it work? To understand the genetic code, the structure of DNA itself had to be understood, and so far no one had managed to work that out.

All sorts of threads of information were available. In 1951 Erwin Chargaff analysed DNA from a wide range of species. He found that in every case the proportions of cytosine and guanine were the same. In the same way he found that the proportion of adenine was always the same as that of thymine. But there was no relationship between the two groups.

At King's College in London, Maurice Wilkins and Rosalind Franklin were working on the X-ray crystallography of DNA. This proved to be rather difficult. It was very hard to get pure crystals of DNA to work with, as DNA does not crystallise easily, and the pictures were so complicated that interpreting them proved a further major hurdle. One such picture is shown in figure 1.3.4.

Meanwhile at Cambridge James Watson and Francis Crick were trying a different approach. They gathered all the available information about DNA and kept trying to build a model that fitted with all the facts. They worked with space-filling models and also with simpler representations of the known components of DNA. Any model they produced had to explain all the available data about the structure of the molecule and how it behaved. By a process of assimilating information from other researchers, long discussions and hours of manipulating the models, an idea emerged which seemed to work.

What finally took shape was the now famous **double helix**. The patterns from the X-ray crystallography suggested a helix measuring 3.4 nm for every complete turn. The idea of a double or parallel helix emerged – but how was the structure maintained? Watson noticed that if in every case cytosine was paired with guanine, and thymine with adenine, hydrogen bonds would hold them together. The two sets of base pairs (cytosine/guanine and thymine/adenine) are roughly the same size and fit within the measured dimensions of the molecule. Two purines would be too large to fit and two pyrimidines too small as hydrogen bonds would not hold them tightly enough together. The realisation that the bases are always paired in this way was a major breakthrough in understanding the structure of the DNA molecule. This model explained the relationship between the bases shown by Chargaff's results.

The base pairs occupied a length along the helix of 0.34 nm – meaning that ten of them would neatly make up one complete twist of the helix (3.4 nm) as measured by the team at King's College London. And best of all, the two complementary chains of DNA could 'open up' along the line of hydrogen bonds between the base pairs. They could then replicate to produce two identical double helices, making possible the vital role of DNA in reproduction. A Nobel prize was the reward for this masterly piece of molecular detection – and the DNA double helix is the basis of all our modern molecular genetics.

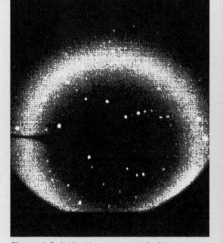

Figure 1.3.4 Working out a model of the large three-dimensional molecule that produced this picture proved to be no easy task.

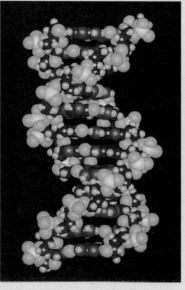

Figure 1.3.5 To the untrained eye this model may not seem much easier to understand than the X-ray crystallograph. Within this complex arrangement of atoms is the whole basis of genetics.

HOW DNA REPLICATES

The complex model of DNA in figure 1.3.5 does not help most of us to understand how it functions. However, when the structure is looked at more closely it becomes easier to see how the DNA molecule works.

To be sure that the genetic code is passed on to each new cell, DNA needs to **replicate** – it needs to produce another molecule exactly the same as itself. Figure 1.3.6 shows how it does this. The hydrogen bonds holding the two strands of the helix together break, so the strands are no longer joined. Nucleotides that are present free in the cell come in and pair with the nucleotides in the 'unzipped' strands – again, A pairs with T and G with C. These nucleotides are then joined by condensation reactions, controlled by the enzyme DNA polymerase, to form a new complementary strand which is hydrogen bonded to the original strand. The result is two double helices, identical with the original.

Figure 1.3.6 We can understand how DNA functions by looking at a small piece of it.

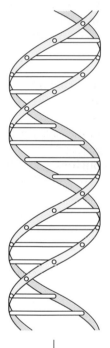

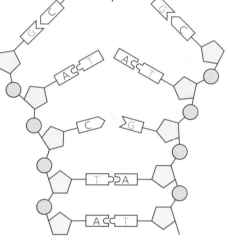

(1) The double helix is held together by the hydrogen bonds between the matching base pairs - 10 pairs for each complete twist of the helix. The strands are known as the 5' (5 prime) and 3' (3 prime) strands. They are named according to the number of the carbon atom in the pentose sugar to which the phosphate group is attached in the first nucleotide of the chain.

(3) To replicate, the two strands of the DNA molecule 'unzip' along the line of hydrogen bonds and unravel. Free nucleotides move in and new hydrogen bonds are formed between matching base pairs. The nucleotides join to form new strands.

(2) The chains of nucleotides fit together perfectly – as long as cytosine and guanine, adenine and thymine are always matched together. As you can see, any other arrangement would be bound to fail.

(4) The result is two new strands of DNA which are identical with the original one.

Conservative and semiconservative replication

Several years after Watson and Crick had produced their double helix model for the structure of the DNA molecule, there were two main ideas about how it could replicate. One was known as **conservative replication**. It said that the original double helix remained intact and in some way instructed the formation of a new, identical double helix made up of new material. The other was known as **semiconservative replication**. This assumed that the DNA 'unzipped' and new nucleotides aligned along each strand. This would mean that each new double helix contained one strand of the original DNA and one strand made up of new material. As the result of a very elegant set of experiments by M. S. Meselson and F. W. Stahl, semiconservative replication became the accepted model.

The experiments of Meselson and Stahl

In the late 1950s M.S. Meselson and F.W. Stahl performed a classic series of experiments which showed very clearly that semiconservative replication was the way DNA worked. Their experiment depended on the use of an **isotope** of nitrogen. Nitrogen 15 is a 'heavy isotope' – it has one more proton in its atoms than the more common nitrogen 14. Both isotopes have the same number of electrons and protons and therefore react in the same way. Isotopes, particularly radioactive ones, can be used to 'label' biological molecules and follow their progress through an organism. In this experiment, the nitrogen 15 isotope used was not radioactive, but its density helped Meselson and Stahl to find out what happens when DNA replicates.

(1) Meselson and Stahl grew several generations of the gut bacteria *Escherichia coli* in a medium whose only source of nitrogen was the isotope nitrogen 15, in labelled ammonium chloride $^{15}NH_4Cl$. The bacteria were grown on this medium until all the bacterial DNA was labelled with nitrogen 15.

(2) The bacteria were then moved to a medium containing normal $^{14}NH_4Cl$ as their only nitrogen source, and the density of their DNA was tested as they reproduced.

(3) If DNA reproduces by conservative replication, some of the DNA would have the density expected if it contained nothing but nitrogen 15 (the original strands), and some of it would have the density expected if it contained nothing but nitrogen 14 (the new strands). If DNA reproduces by semiconservative replication then all the DNA would have the same density, half-way between that of nitrogen 15- and nitrogen 14-containing DNA.

The DNA was all found to have the same density – and so Meselson and Stahl concluded that DNA must replicate semiconservatively.

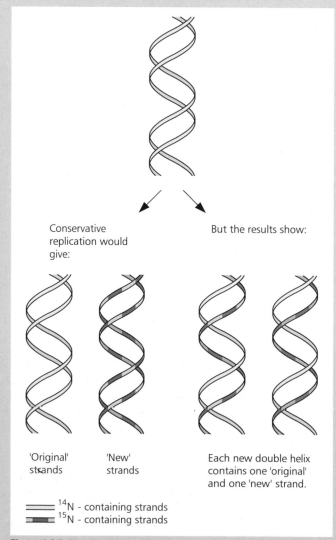

Conservative replication would give:

But the results show:

'Original' strands   'New' strands

Each new double helix contains one 'original' and one 'new' strand.

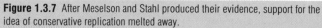

^{14}N - containing strands
^{15}N - containing strands

Figure 1.3.7 After Meselson and Stahl produced their evidence, support for the idea of conservative replication melted away.

How DNA controls cell division and cell function

Now we have seen the structure of DNA and the way in which it replicates, the next question must be *how* does it act as the genetic code? Each different type of living organism must have a distinctive genetic message which produces, for example, sea urchin cells rather than daffodil cells. Added to that, within each type of living organism are millions of individual cells, each of which is unique and which must therefore have its own unique genetic message. There will often be a variety of types of cells within an organism, containing different structures and enzymes and performing a range of jobs. The information needed to give this enormous variety is found within the DNA. The sequence of the base pairs in the molecule is used as a code. The code determines which amino acids are joined together to form proteins.

Proteins are the key to how the genetic code works. Almost all enzymes are proteins, and enzymes control the synthesis and biochemistry of everything in the cell and in the larger organism. So DNA carries its instructions for the make-up of a cell in the form of instructions for particular proteins. Once the proteins (usually enzyme proteins) are made, they in turn construct the rest of the cell.

The genetic code

Proteins are made up of amino acids. There are about 20 amino acids that occur in proteins, but joined together in different combinations they make up an almost infinite variety of proteins. The amino acids which are put together to make a protein are arranged as a result of the genetic code.

In a double helix of DNA, the components which vary along the structure are the bases. Thus it was deduced that it is the arrangement of the bases which carries the genetic information. There are only four bases, but there are at least 20 possible amino acids. This means that one base obviously cannot code for one amino acid. Even two bases do not give a large enough selection of amino acids – the possible arrangements of four bases into groups of two are $4 \times 4 = 16$. Thus at least three bases are needed to code for one amino acid.

A **triplet code** gives $4 \times 4 \times 4 = 64$ possible combinations – more than enough for all the possible amino acids. By the early 1960s it had been shown that it was indeed a triplet code of bases which was the basis of the genetic code. Each sequence of three bases along a strand of DNA codes for something very specific. Most code for a particular amino acid, but some 'nonsense triplets' do not code for an amino acid at all – they signal the beginning or the end of one particular amino acid sequence. A sequence of three bases on a molecule of DNA or RNA is known as a **codon**.

Cracking the code

Once the structure of DNA had been worked out many scientists, including Francis Crick, started work on the genetic code. The codons of DNA are difficult to work out because the molecule is so vast, so most of the work has been done on the codons of a type of RNA called messenger RNA. This RNA is formed as complementary strands to DNA, but its molecules are much smaller, usually carrying the instructions for a single polypeptide. Once the RNA sequence is known, the DNA sequence is simple to deduce.

A dictionary of the genetic code

The result of all this work on sequencing DNA was a sort of dictionary of the genetic code as shown in table 1.3.1. Much of the original work, done in the 1960s, used the gut bacteria *E. coli*, but all subsequent studies suggest that the code is very similar throughout the living world.

First base	Second base								Third base
	U		**C**		**A**		**G**		
U	UUU	Phe	UCU	Ser	UAU	Tyr	UGU	Cys	U
	UUC	Phe	UCC	Ser	UAC	Tyr	UGC	Cys	C
	UUA	Leu	UCA	Ser	UAA	c.t.	UGA	c.t.	A
	UUG	Leu	UCG	Ser	UAG	c.t.	UGG	Try	G
C	CUU	Leu	CCU	Pro	CAU	His	CGU	Arg	U
	CUC	Leu	CCC	Pro	CAC	His	CGC	Arg	C
	CUA	Leu	CCA	Pro	CAA	Gln	CGA	Arg	A
	CUG	Leu	CCG	Pro	CAG	Gln	CGG	Arg	G
A	AUU	Ile	ACU	Thr	AAU	Asn	AGU	Ser	U
	AUC	Ile	ACC	Thr	AAC	Asn	AGC	Ser	C
	AUA	Ile	ACA	Thr	AAA	Lys	AGA	Arg	A
	AUG	Met*	ACG	Thr	AAG	Lys	AGG	Arg	G
G	GUU	Val	GCU	Ala	GAU	Asp	GGU	Gly	U
	GUC	Val	GCC	Ala	GAC	Asp	GGC	Gly	C
	GUA	Val	GCA	Ala	GAA	Glu	GGA	Gly	A
	GUG	Val	GCG	Ala	GAG	Glu	GGG	Gly	G

Table 1.3.1 The triplet code which underpins all work on genetics. The code shown is that for messenger RNA, the most commonly used because it is simplest to work out.
U uracil C cytosine A adenine G guanine
c.t. These codons code for the termination of polypeptide chain synthesis (stop codes).
* Under some conditions this codon codes for the initiation of polypeptide chain synthesis (start code).

Even this analysis of the genetic code did not answer all the questions. It appears that the code is **degenerate** – that is it contains more information than it needs to. If you look carefully at the 'dictionary' you will find that often only two of the three nucleotides seem to matter in determining which amino acid results. A possible reason for this is as follows: if each amino acid had only one codon, then any accidental change in the genetic code (**mutation**) would cause havoc. With the code as it stands, an error is more likely to produce another amino acid which may make little or no difference to the functioning of the final protein. There is even a chance that a change could still result in the same amino acid. Mutations are relatively common – the degenerate code at least partly protects living organisms from their effects.

HOW PROTEINS ARE SYNTHESISED IN THE CELL

DNA is contained within chromosomes in the nucleus of the cell. Proteins are synthesised on ribosomes, which are found on the rough endoplasmic reticulum in the cytoplasm. Nuclear DNA has never been detected in the cytoplasm, so the message about the order of amino acids in a protein cannot be carried direct. Messages are relayed from the nuclear DNA to the active synthetic enzymes on the ribosomes by ribonucleic acids (RNA).

Economy or freedom of choice

Another question was raised about the genetic code: do codons overlap or not? Take any sequence of RNA bases, for example UUUAGC. This could code for two amino acids:

phenylalanine (UUU) and serine (AGC)

On the other hand, if the codons overlap, it could code for four amino acids:

phenylalanine (UUU), leucine (UUA), a nonsense or stop codon (UAG) and serine (AGC)

An overlapping code would be very economical as relatively short lengths of DNA could carry the instructions for many different proteins. However, the amino acids which could be coded for side by side would be limited. In the example given, only leucine out of the 20 or so available amino acids could ever follow phenylalanine, because only leucine has a codon starting with UU-.

Some mutations which cause recognisable diseases (for example sickle cell disease) have been shown to change only *one* amino acid in a protein chain, rather than three. This is evidence against an overlapping code as a change in one codon would affect three amino acids in that model. As a result of all the evidence available, and the simple common sense of the argument, the model of a *non-overlapping* code is now fully accepted.

What does RNA do?

As we have already seen, RNA is closely related in structure to DNA. However, it contains a different sugar, ribose, and a different base, uracil instead of thymine. It consists of a single helix and does not form enormous and complex molecules like DNA. The sequence of bases along a strand of RNA is related to the sequence of bases on a small part of the DNA in the nucleus.

RNA carries out three main jobs in the cell:

1 It carries the instructions about which amino acids will be in a polypeptide from the DNA in the nucleus to the ribosomes where proteins are made.

2 It picks up specific amino acids from the protoplasm and carries them to the surface of the ribosomes.

3 It makes up the bulk of the ribosomes themselves.

To perform these three different functions, three distinct forms of RNA exist. These are **messenger RNA**, **transfer RNA** and **ribosomal RNA**.

Messenger RNA

Messenger RNA (**mRNA**) is formed in the nucleus in a similar way to the replication of DNA. Figure 1.3.8 shows the process. A part of the DNA double helix 'unzips' and the RNA nucleotides pair up with those on one of the DNA strands. They are joined to form a strand of mRNA. Whereas a double helix of DNA carries information about a vast array of proteins, a piece of mRNA usually carries instructions for just one polypeptide. The mRNA is formed on the 5' strand of the DNA. The bases that line up are complementary to those on this DNA strand.

Parts of the DNA are said to be **transcribed** on to strands of mRNA. This means that the mRNA carries the same code as the DNA. The process is brought about by an enzyme called **DNA-directed RNA polymerase**. The name of this enzyme is usually shortened to **RNA polymerase**, though the full

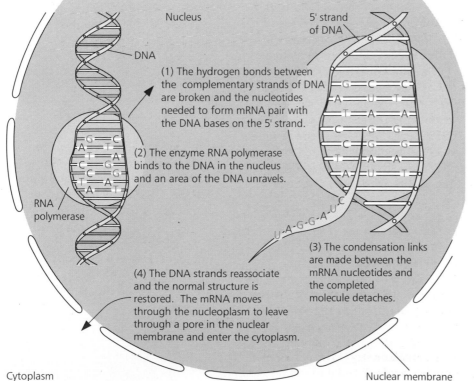

Nucleus

DNA

5' strand
of DNA

(1) The hydrogen bonds between the complementary strands of DNA are broken and the nucleotides needed to form mRNA pair with the DNA bases on the 5' strand.

(2) The enzyme RNA polymerase binds to the DNA in the nucleus and an area of the DNA unravels.

RNA polymerase

(3) The condensation links are made between the mRNA nucleotides and the completed molecule detaches.

(4) The DNA strands reassociate and the normal structure is restored. The mRNA moves through the nucleoplasm to leave through a pore in the nuclear membrane and enter the cytoplasm.

U-A-G-G-A-U-C

Cytoplasm

Nuclear membrane

Figure 1.3.8 The transcription of the DNA message. Any mistakes in this process result in the wrong proteins being made, which can have fatal consequences for the cell or even the whole organism.

name does tell us precisely what it does. It polymerises the nucleotide units to form RNA in a way which is determined by the DNA. Just as in the DNA, the bases of the mRNA form a triplet code and each triplet of bases is known as a codon.

The relatively small mRNA molecules pass easily through the pores in the nuclear membrane, so carrying the genetic message from the nucleus to the cytoplasm. They then move to the surface of the ribosomes, transporting the instructions to the site of protein synthesis.

Transfer RNA

Transfer RNA (tRNA) is found in the cytoplasm. It picks up particular amino acids from the vast numbers that are always free there and carries them to the ribosome where they are joined to form a protein. Each amino acid has its own specific tRNA molecule. tRNA molecules have a structure closely related to their function, as figure 1.3.9 shows.

Ribosomal RNA

Ribosomal RNA (rRNA) makes up about 50% of the structure of a ribosome and it is the most common form of RNA found in cells. It is made in the nucleus, under the control of the nucleoli, and then moves out into the cytoplasm where it binds with proteins to form ribosomes. The ribosomes are made up of a large and a small subunit. They surround those parts of the mRNA which are being used to make a protein, and then move along to the next part of the mRNA. Their job is to hold together the mRNA, the tRNA and the enzymes controlling the process of protein synthesis.

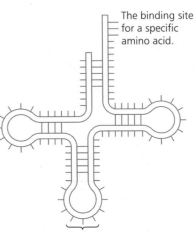

The binding site for a specific amino acid.

The anticodon - these three bases determine which piece of mRNA the tRNA will join to. This in turn decides the exact order of the amino acids in the resulting polypeptide chain.

Figure 1.3.9 There are almost 60 different tRNA molecules found in the cytoplasm of cells – more than enough to carry all the necessary amino acids to the surface of the ribosomes ready for synthesis into protein molecules.

Protein synthesis – what happens?

The genetic code of the DNA of the nucleus is transcribed onto messenger RNA. This mRNA moves out of the nucleus into the cytoplasm and becomes attached to a ribosome. Molecules of tRNA carry individual amino acids to the surface of the ribosome. tRNA molecules complementary to the codons in the mRNA strand line up, and enzymes link the amino acids together. This process is called **translation**. Its job done, the individual tRNA molecules return to the cytoplasm to pick up another amino acid. The ribosome moves along the molecule of mRNA until the end is reached, leaving a completed polypeptide chain. The process is illustrated in figure 1.3.10. The message may be read repeatedly to make many strands of the same polypeptide.

Protein synthesis, like many other events in living things, is a continual process. However, it makes it simpler to understand if we look at the two main aspects of it separately. The events in the nucleus involve the *transcription* of the DNA message onto the mRNA molecule. In the cytoplasm that message is *translated* into polypeptide molecules and hence into proteins.

Mass production

A common sight within the cytoplasm of cells are **polysomes**. These are groups of ribosomes joined by a thread of mRNA, and they appear to be a form of mass production of particular proteins. Instead of one ribosome moving steadily along a strand of mRNA to produce its polypeptide and then repeating the process, ribosomes attach in a steady stream to the mRNA and move along one after the other producing lots of identical polypeptides.

Mutation

Thus the genetic code carried on the DNA is translated into living cellular material by the synthesis of proteins. The nucleic acids are vital to the process,

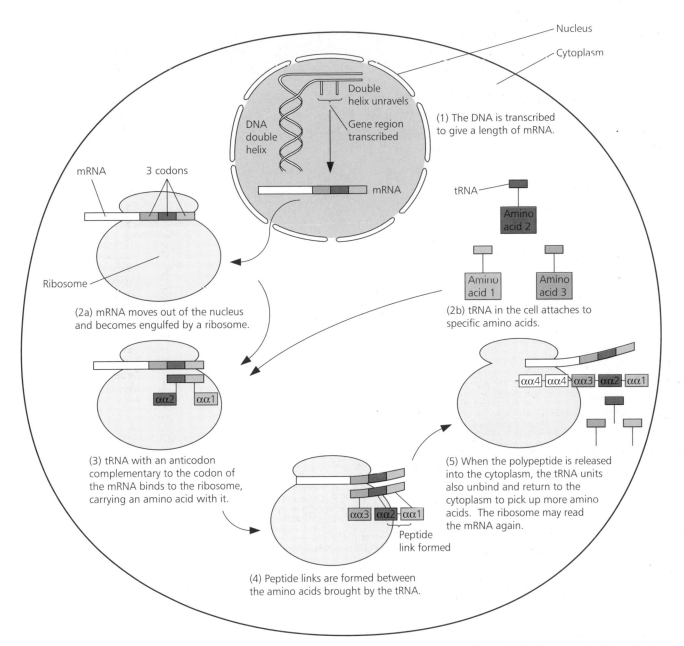

Figure 1.3.10 The stages of protein synthesis

as both the carriers and the translators of the genetic code. If a single codon is changed by a mutation, then the likelihood is that it will code for a different amino acid, so the whole polypeptide chain and indeed the final protein will be altered. Such a tiny alteration at this molecular level may have no noticeable effect at all, but equally it can have devastating effects on the whole organism. Examples include some human genetic diseases where the blood proteins are not manufactured correctly or where certain enzymes do not function properly. Yet when the complexity of the process is considered, it is perhaps surprising that it does not go wrong more often. The precision of the various stages and of the enzymes such as DNA polymerase associated with them, the unvarying association of the base pairs with each other and the degenerate nature of the triplet code are just some of the factors which ensure that for most organisms, most of the time, the genetic messages in DNA are faithfully converted into the systems of life through the mechanism of protein synthesis.

SUMMARY

- All cell processes including cell replication and the reproduction of individuals are controlled by enzymes. The synthesis of these protein enzymes is determined and controlled by the **nucleic acids**: **deoxyribonucleic acid** (**DNA**) and **ribonucleic acid** (**RNA**).

- DNA and RNA are polymers with a **nucleotide** monomer made up of a pentose sugar, a base and a phosphate group.

- In DNA the pentose sugar is deoxyribonucleic acid and the bases are adenine and guanine (purines) and cytosine and thymine (pyrimidines).

- In RNA the sugar is ribose and the bases are adenine and guanine (purines) and cytosine and uracil (pyrimidines).

- The DNA molecule is a double helix with two strands being held together by hydrogen bonds between complementary purine and pyrimidine base pairs.

- DNA replicates by the unzipping of the two strands, and nucleotides with complementary bases pair with those in the existing strands. These nucleotides are joined by condensation reactions catalysed by the enzyme DNA polymerase and two new identical double helices result.

- The order of bases along the DNA molecule gives a code for the synthesis of proteins. Three bases (a **codon**) code for one amino acid. The code is read and converted into proteins in a process known as **protein synthesis** which involves RNA.

- A strand of **messenger RNA** (**mRNA**) is synthesised in the nucleus, complementary to a strand of DNA. The code is **transcribed** onto the mRNA molecule. The mRNA passes out of the nucleus into the cytoplasm and onto a ribosome.

- **Transfer RNA** (**tRNA**) molecules have triplets of bases that pair with the codons on the mRNA. They each carry a specific amino acid to the mRNA molecule. The amino acids are joined to form a polypeptide, in an order determined by the mRNA strand. The code is **translated** into the polypeptide.

- **Ribosomal RNA** (**rRNA**) in the ribosomes holds together the mRNA, tRNA and enzymes during protein synthesis.

- A change in a base along the DNA or rRNA molecule is a **mutation**. This may lead to a change in an amino acid in the polypeptide chain.

QUESTIONS

1 DNA and RNA are the information molecules of the cell. Explain clearly the differences in the basic structure of these two molecules.

2 **a** How does DNA replicate?
  b What is the evidence for this?

3 The genetic code is said to be a *triplet code*. What does this mean and why is it important?

4 The DNA is contained within the nucleus of a eukaryotic cell. The proteins for which it codes are needed within the cytoplasm. Explain the roles of the following in the translation of the genetic code into an active enzyme in the cytoplasm.
  a DNA
  b messenger RNA
  c transfer RNA
  d ribosomal RNA
  e polysomes

1.4 Cellular processes

You now have a picture of the basic make-up of the cells and major groups of chemicals found in living things. The next stage is to consider how these different elements interact in the processes of cellular life.

But before moving on, there is one other chemical which needs careful consideration. Water makes up the largest proportion of all cells – in fact your body mass is at least 60% water. It is a molecule of enormous biological importance which has many vital roles.

WATER, WATER EVERYWHERE ...

Around two-thirds of the surface of the Earth is covered by water. It is believed that the conditions for the beginnings of life developed in the oceans many millions of years ago, and that much of the subsequent evolutionary process took place in water. In fact, each cell can be regarded as a membrane-bound drop of the primaeval broth of those early seas.

Even those species which dwell completely on the land need a water-based environment both for their reproductive cells and for the development of embryos. This environment is provided by the eggs of reptiles and birds, and in the more complex reproductive arrangements of the mammals.

Water is the medium in which the chemicals of life are dissolved, and in which all the reactions in living cells take place. It is the basis of the transport systems found in most complex organisms. It is one of the reactants in the process of photosynthesis, on which all life depends. And water is a major habitat – it supports more life than any other area of the planet.

The chemistry of water

The ability of water to play its wide variety of roles, and the reason for its importance in biological systems, is due to the basic chemistry of the molecule. The simple chemical formula of water is H_2O, which tells us that two atoms of hydrogen are joined to one atom of oxygen to make up each water molecule, as figure 1.4.1 shows.

The water molecule is slightly polarised. This means it has a very slightly *negative* end – the oxygen atom – and a very slightly *positive* end – the hydrogen atoms. This separation of electrical charge is called a

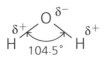

Figure 1.4.1 The water molecule

dipole, and the tiny charges (represented as δ+ and δ–) give the water molecule its very important properties. One of the most important results of this charge separation is the tendency of water molecules to form **hydrogen bonds**, shown in figure 1.4.2.

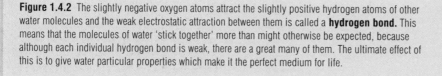

Figure 1.4.2 The slightly negative oxygen atoms attract the slightly positive hydrogen atoms of other water molecules and the weak electrostatic attraction between them is called a **hydrogen bond**. This means that the molecules of water 'stick together' more than might otherwise be expected, because although each individual hydrogen bond is weak, there are a great many of them. The ultimate effect of this is to give water particular properties which make it the perfect medium for life.

Why is water so important?

A variety of the properties of water are important in biological systems. Some of the most important ones are given here.

(1) An unusual and excellent solvent

Many other substances will dissolve in water. The fact that the water molecule has a dipole, with slightly positive and negative parts, means that **polar** substances with positive and negative regions, and particularly **ionic** substances such as sodium chloride (salt), made up of positive and negative ions, will dissociate (separate) and dissolve in water. Polar substances will not dissolve in organic solvents. Once the ions or polar molecules have dissolved in water they become surrounded by water molecules which keep them in solution, as figure 1.4.3 shows.

Water can also act as a solvent to many **non-polar** substances. As all the chemical reactions that go on within cells take place in aqueous solution, the ability of water to act as a solvent is vitally important for the processes of life.

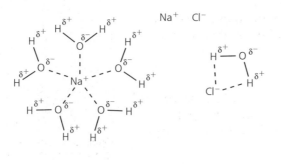

Figure 1.4.3 Once salt has dissolved, the ions become surrounded by water molecules which hold them in solution.

(2) The change of density with temperature

As water cools to below 4 °C the molecules take on an arrangement which occupies more space than the arrangement at room temperature. When freezing takes place at 0 °C, this new arrangement becomes rigid, so *ice is less dense than liquid water*. This makes water unique. The fact that ice floats on water means that living things can survive in ponds and rivers when the temperatures fall below freezing. The ice acts as an insulating layer, helping to prevent the rest of the water mass below from freezing. If ice formed from the bottom up, freshwater life would only be found in those areas where the water never freezes. Ice also thaws quickly because it is at the top, nearest to the warming effect of the sun.

(3) Slow to absorb and release heat

The **specific heat capacity** of a substance is a measure of the amount of energy needed to raise the temperature of a fixed amount of that substance by 1 °C. The specific heat capacity of water is high – it takes a lot of energy to warm it up. This makes water, particularly large bodies of water like lakes, seas and oceans, a thermally stable environment. Therefore the very wide variety of aquatic organisms do not have to contend with their surroundings getting quickly hotter or colder depending on the weather. This thermal stability is also seen within the water-based protoplasm of individual cells, and allows the biochemistry of life to be carried out at fairly constant rates.

(4) Taking large amounts of energy to turn from a liquid to a gas and from a solid to a liquid

The **latent heat of vaporisation** is a measure of the energy needed to overcome the attractive forces between the molecules of a liquid and turn it into a gas. For water the amount of energy needed for this is very high, because the hydrogen bonds holding the molecules together have to be broken before it can become a gas. Although each hydrogen bond is very weak, there are a great number of them. There are two major values of this to biological systems. One is that there is a large amount of water left on the surface of the Earth – it does not all vaporise and disappear into the atmosphere. The other is that the evaporation of water uses up a large amount of heat energy, and so can be put to good use as a cooling mechanism. The sweating mechanism of the mammals has capitalised on this particularly.

The **latent heat of fusion** of water is also high – it takes a great deal of energy to melt ice. Conversely, water needs to lose a lot of energy to form ice. This means that, in most circumstances, the contents of cells are unlikely to freeze. This is very important because the formation of ice in cells, with the accompanying increase in volume, is almost always damaging if not fatal.

(5) A high surface tension

The hydrogen bonds between water molecules tend to pull them down and together where water and air meet, giving water one of the highest known surface tensions, illustrated in figure 1.4.4.

(6) An amphoteric molecule

The water molecule is **amphoteric** – it acts as both an acid and a base. An acid gives up (donates) electrons to form H^+ ions. An alkali gains (receives) electrons to form OH^- ions. Water molecules can do both, as the following reaction shows:

$$2H_2O \rightarrow H_3O^+ + OH^-$$

Does it dissolve in water?

You know that if you stir a spoonful of sugar crystals into water they disappear – they dissolve. The molecules of the sugar are **polar** – they have positive and negative regions. They become surrounded by the water molecules with their tiny dipoles and form a **true solution**. But not all substances dissolve completely in water like this. Some form **colloids**, whilst others exist as **suspensions**.

Colloidal solutions are formed when the solute particles are much bigger than those of the solvent, but they do not separate out under the influence of gravity – they are distributed evenly through the solvent. Many of the plasma proteins of the blood are in colloidal solution, and cannot pass through the capillary walls.

In **suspensions** the solute particles are so large that they separate out under the influence of gravity unless the suspension is constantly moved and stirred. The blood itself also falls into this category – if it is left to stand the cells and platelets sink to the bottom.

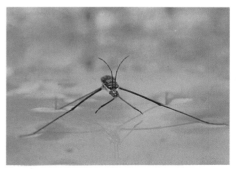

Figure 1.4.4 Surface tension is of great importance in plant transport systems, and also affects life at the surface of ponds, lakes and other water masses.

This ability of water molecules to both donate and receive electrons means that it is the perfect medium for the biochemical reactions occurring in cells. It acts as a **buffer**. A buffer helps to prevent changes in the pH of a solution when an acid or an alkali is added, by neutralising any excess H$^+$ or OH$^-$ ions that are introduced. Thus the water in cells minimises any changes in pH which might result from products of reactions in progress, and so prevents any interference with the smooth running of the metabolism.

Armed with an understanding of the unique role of water in biological systems, we can now begin to consider the most basic processes of life. A vital step in the beginning of life was the evolution of the **cell membrane**. The processes of life are closely bound up with the structure of the membranes of the cell, so this is where we shall begin.

Figure 1.4.5 The fox needs water to drink, and the stable cellular environment of all these living things depends on the unique chemistry of water. Its 'skin' supports the pondskater and water makes possible the transport systems and photosynthesis of plants.

CELL MEMBRANES

The functions of membranes

Membranes are ubiquitous in cells. The cell surface membrane acts as the boundary of the cell – anything that enters or leaves the cell must pass through it. If there was no barrier between the contents of a cell and its surrounding medium, there would obviously be no difference between the cytoplasm and the medium. Equally, if the cell surface membrane was completely permeable to all the molecules and ions in the external medium, the chemical make-up of the cytoplasm would be the same as that of its surroundings. But in fact the make-up of the cytoplasm of a cell is very different from the constituents of the surrounding fluid. This is a clear indication that the cell surface membrane controls and regulates what gets into and out of the cell. This is the most important function of membranes. Both polar and non-polar substances must have access to the cell, and the entry and exit of substances from specialised membrane-bound areas of the cell such as the nucleus, mitochondria and chloroplasts must be controlled.

Membranes perform many other functions too. Cells, even plant cells, are not rigid structures. A cell may need to change shape very slightly as the water content changes, or it may need to change shape quite dramatically as when a single-celled organism finds and engulfs food. The membrane must be flexible enough to allow this to happen. We have seen that chemical secretions made by the cell are packaged into vesicles, so the membrane must be capable of breaking and fusing readily.

How does the structure of the membrane allow it to perform these various functions?

The structure of the membrane

It has taken a number of years of research to arrive at our current model of the structure of cell membranes. In time there may well be further refinements, but the overall picture seems unlikely to change dramatically. The membrane is made up of two main types of molecules – **lipids** and **proteins** – arranged in a very specific way.

The lipid part

The lipids are of a particular type called **polar lipids**. These are lipid molecules with one end joined to a polar group. The phospholipids, already met on page 25, are a good example with a phosphate group as the polar part.

When polar lipids come into contact with water, the two parts of the molecule behave differently. As we have seen, the polar area is **hydrophilic** (water-loving) and is attracted to the water. The lipid tails are **hydrophobic** (water-hating) and move away from the water. If the molecules are tightly packed this gives a **monolayer** as shown in figure 1.4.6.

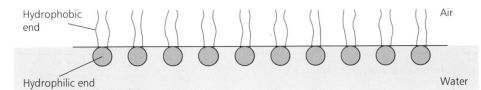

Hydrophobic end

Air

Hydrophilic end

Water

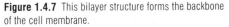
Figure 1.4.6 Polar lipids form a monolayer at an air–water interface.

However, a surface between air and water is a fairly rare situation in living cells. Much more usually the environment is entirely water based, and in that situation a **bilayer** is formed by the lipid molecules with the hydrophilic heads pointing out into the water while the hydrophobic tails are together in the middle. This structure, shown in figure 1.4.7, is the basis of the cell surface membrane or **unit membrane**.

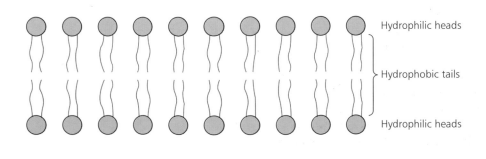

Hydrophilic heads

Hydrophobic tails

Hydrophilic heads

Figure 1.4.7 This bilayer structure forms the backbone of the cell membrane.

A lipid bilayer alone would allow the passage of organic molecules which are fat soluble, but there are many ionic chemicals needed in cells. Ionic compounds dissolve in water but cannot dissolve in or pass through lipid. This, and the microscopic appearance of the cell membrane, is explained by the presence of proteins in the bilayer.

The protein part

The unit membrane is regarded as a fluid lipid system, with protein parts floating within it like icebergs. This is called the **fluid mosaic model** of the membrane. Generally the proteins have a hydrophobic part which is buried in the lipid bilayer and a hydrophilic part which can be involved in a variety of activities. Some of these proteins travel about freely, others are fixed in place. Some of them penetrate all the way through the lipid, others only part of the way. Thousands of different proteins have been found associated with the membrane. What do they do?

One of the main functions of the membrane proteins is to let substances into and out of the cell. They form **pores** – some permanent, some temporary – which allow different molecules and ions to pass in or out. Some protein pores are **active carrier systems** – they use up energy to move materials into or out of cells. Others are simply gaps in the lipid layer which give access to ionic substances in particular. Figure 1.4.8 shows the routes for substances through the membrane.

Some proteins are not pores but can act as specific receptor molecules, for example, making cells sensitive to a particular hormone. Others act as enzymes, particularly on the internal cell membranes. Still other proteins are

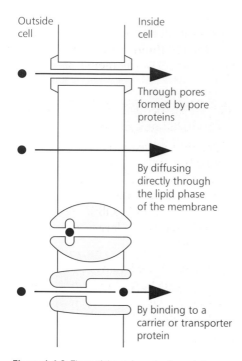

Figure 1.4.8 Three of the main routes through the membrane

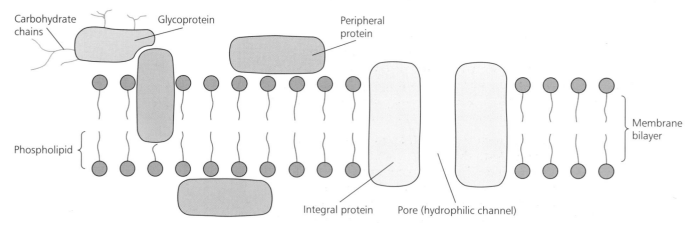

Transmission electron micrograph showing the two layers of the membrane

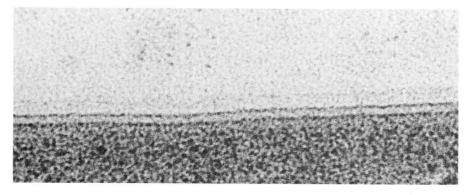

Figure 1.4.9 Whether acting as the boundary of a cell or as a major element of its internal make-up, the complex structure of the membrane is closely linked to its wide variety of functions. This model of the floating proteins in a lipid sea is known as the fluid mosaic model and was first proposed by Jonathon Singer and Garth Nicholson.

glycoproteins, with a carbohydrate part added to the molecule. These are very important on the surface of cells as part of the cell recognition set-up and we shall look at them in more detail in section 1.6. Figure 1.4.9 shows the fluid mosaic model of the membrane.

Where does our picture of the membrane come from?

The first indications that lipids are important components of cell surface membranes came at the end of the nineteenth century when E. Overton made a series of observations on how easily substances passed through various membranes. As a result of these observations, along with the fact that lipid-soluble substances enter cells more easily than any others, he concluded that a large part of the cell surface membrane structure must be lipid.

Studies on the behaviour of membranes when cells join together, and the way in which most membranes seal themselves if they are punctured with a fine needle, led to the idea that cell membranes are not rigid structures but are much more like a fluid.

In the early twentieth century I. Langmuir demonstrated the lipid monolayer we have mentioned, and developed equipment for collecting lipid monolayers, known as the Langmuir trough.

In 1925 two Dutch scientists, Gorter and Grendel, set out to measure the total monolayer film size of lipids extracted from human red blood cells (erythrocytes). They estimated the total surface area of an erythrocyte and found that the area of monolayer they measured was about twice the estimated area. As a result they concluded that the cell membrane was in fact a lipid bilayer. We now know that their results were wrong on two counts – they did not extract all the membrane, and they miscalculated the surface area of the erythrocytes, thinking they were discs rather than biconcave. In spite of this, their conclusions were correct – by an amazing coincidence the two errors cancelled each other out!

By 1935 H. Davson and J. F. Danielli had produced a further model of the membrane which is broadly the basis of our current ideas. They suggested a membrane with a lipid centre coated on each side by protein.

This hypothesis was backed up in the 1950s by electron microscope work by J. D. Robertson. He found ways of staining the membrane which showed it up as a three-layered structure – two distinct lines with a gap in the middle. When the membrane was treated with propanone to extract the lipid, the two lines remained intact, suggesting that these were the protein layers and that they played an important role.

More recently still, techniques such as X-ray diffraction and freeze-etching, which allows us to see both the surface and the inside of the membrane structure, have been used to add to our knowledge of the structure of cell membranes. The combination of all this work and thought resulted in the fluid mosaic model devised by J. Singer and G. Nicholson of a lipid sea with many and various proteins floating in and on it, a structure which is compatible with the enormous variety of functions associated with this ubiquitous organelle.

Diffusion

In physical terms, diffusion is the movement of the molecules of a liquid or gas from an area where they are more highly concentrated to an area where they are at a lower concentration. They move along a **concentration gradient**. This occurs because of the random motion of molecules – the more tightly packed they are, the more likely they are to move apart. The end result of diffusion is a uniform concentration of the molecules. Although the molecules do not stop moving once they reach a uniform concentration, the movement no longer causes a change in concentration. Figure 1.4.10 illustrates the process of diffusion.

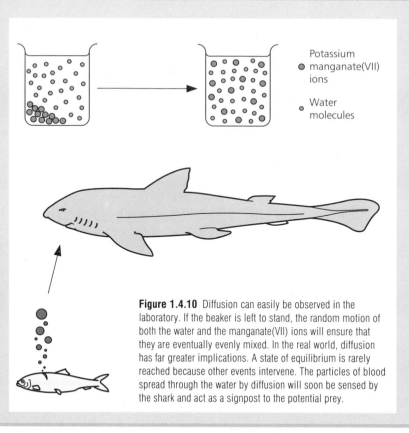

Potassium
● manganate(VII) ions

· Water molecules

Figure 1.4.10 Diffusion can easily be observed in the laboratory. If the beaker is left to stand, the random motion of both the water and the manganate(VII) ions will ensure that they are eventually evenly mixed. In the real world, diffusion has far greater implications. A state of equilibrium is rarely reached because other events intervene. The particles of blood spread through the water by diffusion will soon be sensed by the shark and act as a signpost to the potential prey.

DIFFUSION AND OSMOSIS

How does the membrane exercise its control over the passage of substances into and out of the cell?

Some substances, particularly those which dissolve very easily in lipids, simply pass through the membrane as though it was not there. Other very small molecules, water in particular, seem to pass freely in and out of cells through minute hydrophilic pores in the membrane. In both these cases the molecules move by **diffusion**.

Osmosis – a special case of diffusion

Diffusion takes place when there is no barrier to the free movement of molecules or ions. Water molecules have complete access to the cytoplasm of the cell through the membrane. This means that as a result of random motion, the water molecules will tend to move into or out of the cell along a concentration gradient.

The situation becomes more complicated when other substances are also considered, because although water can move freely across the membrane of a cell, other substances cannot. The membrane is **partially permeable**. Some molecules cannot cross the membrane, others can but only slowly, some move quite rapidly and some (like water) enter freely.

We can make a model of this situation using an artificial membrane which is permeable to some molecules – in particular water – and impermeable to others such as glucose. There are many experiments which show the movement of water in these circumstances, and one of the simplest is illustrated in figure 1.4.11. The presence or absence of glucose in the various solutions can be shown by carrying out Benedict's tests on the solutions.

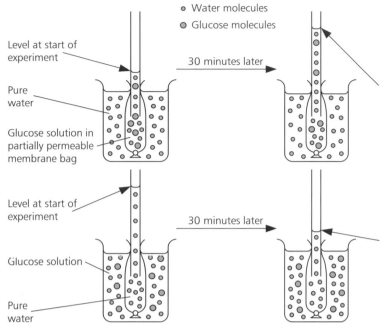

• Water molecules
• Glucose molecules

Level at start of experiment

30 minutes later

Pure water

Glucose solution in partially permeable membrane bag

Random motion of the water molecules leads to a net movement of water *into* the membrane bag along the concentration gradient of water molecules. Random motion of the glucose molecules would lead to a movement of glucose molecules *out of* the bag along the concentration gradient of glucose molecules if the bag was freely permeable. This cannot happen because the pores in the membrane are too small to allow glucose molecules through, so the glucose stays inside the bag and water continues to move in, causing the level in the capillary tubing to rise.

Level at start of experiment

30 minutes later

Glucose solution

Pure water

Random motion of the water molecules leads to a net movement of water *out of* the bag along the concentration gradient of the water molecules. Glucose molecules cannot cross the membrane and so stay in the external solution. As a result the water level in the tubing falls.

This simple model of osmosis gives us a picture of what is happening in a cell. The cell membrane is partially permeable. If the solution bathing the cell has a *lower* concentration of dissolved substances (**solutes**) than the solution inside the cell, there will be a concentration gradient of water molecules *into* the cell. If the opposite is true and the solution bathing the cell has a *higher* concentration of solutes than the cell contents, water will move *out* of the cell. In the context of living cells, the movement of water by osmosis and its control is very important. In animal cells in particular, it is vital that water does not simply move continually into the cells along a concentration gradient, because the cells would eventually swell up and burst.

Figure 1.4.11 When water is moving freely across a membrane through which the other molecules involved cannot move, we say that the water is moving by **osmosis**.

Terms of osmosis

Osmosis may be defined as the net movement of solvent molecules from a region where they are at a higher concentration to a region where they are at a lower concentration *through a partially permeable membrane*.

When describing osmosis, certain terms are commonly used. These include:

• **Solvent** – the substance in which molecules or ions of other chemicals are dissolved. The only solvent in living systems is water.
• **Solute** – the chemical dissolved in a solvent. (For example, in a salt solution, water is the solvent and salt ions are the solute.)
• **Partially permeable membrane** – a membrane which allows solvent molecules to pass through it freely, and some other molecules to pass through as well, although in a selective way.
• **Osmotic concentration** – the concentration of a solution, taking into account only those dissolved substances which have an osmotic effect. Many large molecules found in the cytoplasm do not affect the movement of water and so are ignored when calculating osmotic concentration.

- **Isosmotic** – solutions with the same osmotic concentration. There would be no net movement of water between two isosmotic solutions separated by a partially permeable membrane.
- **Osmotic pressure** – if a solution is separated from pure water by a partially permeable membrane, the osmotic pressure is the hydrostatic pressure which would have to be applied to prevent any movement of water across the membrane. It can only be measured in a special instrument called an osmometer and as such is not a very useful measure in itself.
- **Solute potential** – the potential of a solution to cause water movement into it across a partially permeable membrane as a result of dissolved solutes. As a solute is dissolved in water, in osmotic terms it effectively reduces the concentration of water molecules and so lowers the **water potential** (see box 'Terms of osmosis in living cells'). It is always given a negative sign, and is measured in kilopascals (kPa). The symbol for solute potential is Ψ_s.

Osmosis in living cells

It is most important to animal cells that the net movement of water in or out is kept to a minimum. The problems of too much water moving out of a cell, although not as dramatic as those associated with too much going in, are equally damaging to the organism. The cells shrivel and the concentrated cytoplasm loses its internal structure and ceases to function. This is called **plasmolysis**.

In plant cells the situation is rather different because of the presence of the cellulose cell wall. Although it is freely permeable, it exerts an inward pressure on the cell and so has an effect on osmosis. Think of a balloon. You can tell by looking at it how far it is blown up. Keep blowing it up and eventually it bursts. Now imagine fitting a balloon inside a shoe box, sealing down the lid and then blowing up the balloon. The balloon will inflate so far, but the point will come when you cannot force any more air into it because of the inward pressure of the walls of the box. The outward appearance of the box will be little altered whether the balloon is completely empty or fully inflated. This gives you some idea of the difference between animal and plant cells. The outer box represents the cellulose cell wall.

Terms of osmosis in living cells

Most work on osmosis in cells is done on plant cells. They are generally larger and easier to see with the light microscope than animal cells, and also changes are easier to see and measure than changes in animal cells. In order to consider osmosis in living cells, we need to introduce some more terms.

- **Water potential** – the potential for water to move *out* of a solution by osmosis. Water potential has the symbol Ψ. Pure water has the highest possible water potential. Water molecules will always move from pure water into any solution on the other side of a partially permeable membrane. This maximum water potential is given as zero. *All* solutions have a lower water potential than pure water, because their concentration of water molecules must obviously be lower than that in pure water, and so Ψ always has a negative value. Water always moves from a region of higher water potential to an area of lower water potential. This means we can redefine osmosis as:

> The movement of water molecules from an area of higher water potential to an area of lower water potential through a partially permeable membrane.

In a plant cell the water potential is the result of two elements:
- **Solute potential** – the potential of a solution to cause water movement into it across a partially permeable membrane as a result of dissolved solutes. It reflects how much the solute molecules have lowered the water potential of the cell sap. But this alone does *not* give the water potential of the cell because another factor is involved.
- **Pressure potential** – the influx of water by osmosis into a plant cell immersed in pure water is affected by the inward pressure of the cell wall. As water moves into the cell, the vacuole expands and pushes out the protoplasm. This means an inward pressure is exerted by the cell wall as the cell contents expand and press outwards. (Think of the balloon in the box.) This pressure is known as the pressure potential. It is usually a positive figure and has the symbol Ψ_p.

During osmotic experiments cells are often immersed in solutions of varying concentrations. When working with animal cells, these have the following definitions:
- **Hypotonic solution** – a solution in which the osmotic concentration of solutes is lower than that in the cell.
- **Isotonic solution** – a solution in which the osmotic concentration of solutes is the same as that in the cell.
- **Hypertonic solution** – a solution in which the osmotic concentration of solutes is higher than that in the cell.

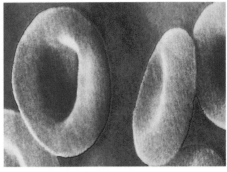

Normal red blood cells in isotonic solution

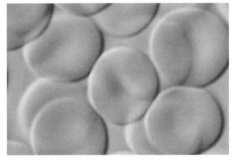

Red blood cells in hypotonic solution

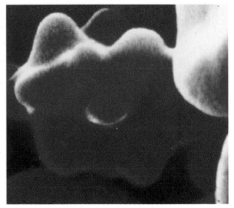

Red blood cells in hypertonic solution

Figure 1.4.12 The effects of osmosis on plant and animal cells. Remember that these situations are almost entirely confined to the laboratory – they do not usually happen in living animals and plants. Even when a plant wilts, for example, it is not because of plasmolysis but because the entire cells, wall and all, have shrunk due to lack of water. But seeing these extreme effects of osmosis helps us to understand why living things have so many complex systems designed to control their water balance.

If the cytoplasm of a plant cell contains more solutes than the surrounding fluids, water will enter the cell by osmosis – but not indefinitely. The inward pressure of the cell wall builds up as the cytoplasm swells until the pressure cancels out the tendency for water molecules to move in by osmosis. At this point the cells are full and rigid, in a state known as **turgor**. This is the normal and desirable state for plant cells. Figure 1.4.12 shows osmotic effects in plant and animal cells.

Plant cells in a solution with a higher water potential than the cell sap – turgor

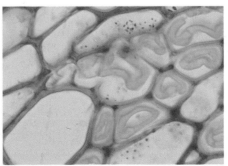

Plant cells in a solution with a lower water potential than the cell sap – plasmolysis

The water potential in a plant cell

The water relationships in a plant cell can be summed up using the ideas met so far. The water potential of a cell is the sum of the solute potential of the cell sap and the pressure potential of the cell wall. Thus:

Water potential = solute potential + pressure
of cell .... **of cell sap** .... **potential**
(Ψ, usually –ve) (Ψ_s, always –ve) (Ψ_p, usually +ve)

Combining the changes we can see in actual cells with the changes we know are occurring in Ψ, Ψ_s and Ψ_p, we can develop the picture shown in figure 1.4.13 of what happens when plant cells are immersed in different solutions.

Figure 1.4.13 The changes which take place in plant cells as they pass from full turgor to plasmolysis as a result of osmosis.

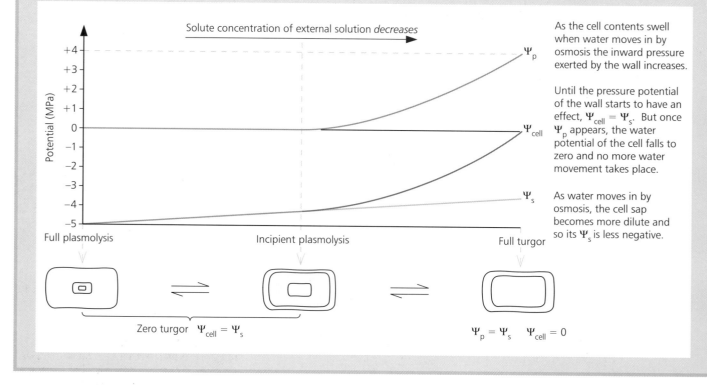

As we have seen, many of the substances which pass into and out of the cell are not lipid soluble and are too big to fit through the minute pores available to water. Also, water movement by osmosis depends largely on the relative concentrations of various solutes in the cytoplasm of the cells and in the surrounding fluids. Thus it is of prime importance that solutes can cross the cell surface membrane in the right proportions. This may at times even mean moving them *against* a concentration gradient – and through a system which is naturally impenetrable to ionic substances. How is all this made possible?

MEMBRANE 'FERRIES'

In general, membranes are very specific about the substances which can cross them. Usually only water molecules have free access to cells along a concentration gradient. There seem to be two main ways in which other substances are transported across the cell membrane. Both involve membrane proteins. One does not use up energy and is called **facilitated diffusion**. The other needs energy supplied by the cell and is called **active transport**.

Facilitated diffusion

This depends on carrier molecules floating on the surface of the membrane – for example, red blood cells seem to have a carrier to help glucose move into the cells rapidly. The carriers are found on the *outside* surface of the membrane structure when a substance is to be moved *into* the cell or organelle, and on the *inside* for transport *out* of the cell. The protein carriers are specific for particular molecules or groups of molecules. Once they have picked up a substance they rotate through the membrane, carrying the new molecule with them. Once on the other side of the membrane, the carrier releases the substance. The movement through the membrane takes place because of the change in the shape of the carrier once it is carrying something. The process can only take place along a concentration gradient – from a higher to a lower concentration of a molecule. Facilitated diffusion simply helps diffusion to occur in a situation where it would otherwise be impossible because the molecule involved could not pass through the cell membrane – it provides a route through the membrane for that molecule, as figure 1.4.14 illustrates.

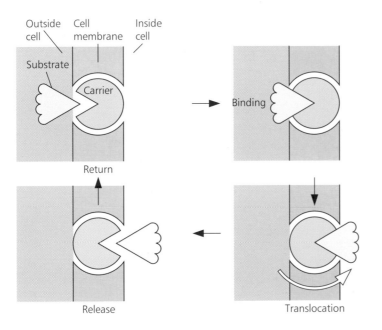

Figure 1.4.14 Facilitated diffusion acts as a ferry across the lipid membrane sea. But this is a boat with no oars, sail or engine – it can only work when the tide (the concentration gradient) is in the right direction.

Active transport

Diffusion and facilitated diffusion both rely on a concentration gradient in the right direction. But what happens if a chemical needs to be moved *against* a concentration gradient? This often happens, and cells have several ways of dealing with the problem.

One is simply to 'mop up' the chemical as soon as it arrives inside the cell, either by immediately starting to metabolise it or by using another carrier molecule to remove it from the pool of free ions and molecules. Another is to chemically change the molecule immediately it enters the cell. These two methods effectively change the concentration gradient so that the substance continues to move into the cell by diffusion. The third alternative is to use a transport system which can move substances *against* a concentration gradient. To do this needs energy, and so the process is known as **active transport**.

The transport of the molecule needed by the cell is linked with that of another particle – often a sodium ion. One of the best known examples of active transport is the **sodium pump** found in the membranes of nerve cells, amongst others. This pump actively moves potassium ions into the cell and sodium ions out. As we shall see in section 4, this is vital for the working of the nervous system.

The carrier often spans the whole membrane, as figure 1.4.15 shows. It may be very specific, only picking up one type of ion or molecule, or it may work for several relatively similar substances. In this case, the particles have to compete with each other for a place on the carrier, so the one which is present in the highest concentration usually wins.

The energy needs of the cell are met by the molecule **ATP** or **adenosine triphosphate**. This is explained more fully in section 3.1. The carrier system in the membrane involves the enzyme **ATPase** which breaks down a molecule of ATP into **ADP** or **adenosine diphosphate**. This breakdown releases energy which is used to move a substance against a concentration gradient. The energy may be used to move the carrier system in the membrane or to release the transported substances and return the system to normal. Active transport is a one-way system, so that substances are only moved in the direction required by the cell – they cannot move back down the concentration gradient which has just been overcome.

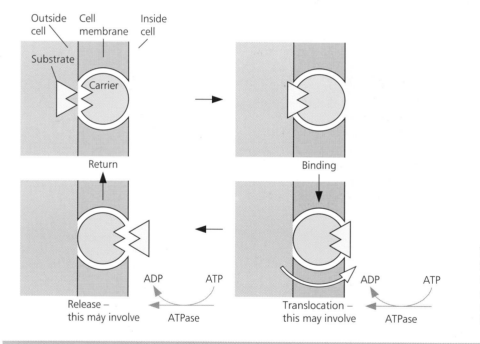

Figure 1.4.15 Active transport – a far superior 'ferry' in many ways. Because it has its own power, molecules can be transported in any direction at any time. The only disadvantage is that the energy must be provided – cells which carry out a lot of active transport generally also have large numbers of mitochondria.

Evidence for active transport

Active transport requires energy in the form of energy-rich ATP. This is produced in the process of cellular respiration, as explained in section 3.4. Much of the evidence for *active* transport is concerned with linking these two processes together, showing that without ATP active transport cannot take place:

(1) Active transport only takes place in living, respiring cells.
(2) The rate of active transport is affected by temperature and oxygen concentration. These

also have a big impact on the rate of respiration and so on the production of ATP.
(3) Many cells known to carry out a lot of active transport contain very large numbers of mitochondria – the sites of cellular respiration and ATP production.
(4) Poisons which stop respiration or prevent ATPase from working also stop active transport. For example, cyanide inhibits ATP production. It also stops active transport. However, if ATP is added artificially, transport starts again.

The combination of diffusion, facilitated diffusion and active transport means that the cell surface membrane acts as an excellent 'barrier' between the contents of the cell and its surroundings. The concentrations of ions and molecules within the cell can be maintained at levels very different from those of the external fluids. Yet there is still communication between the two environments – materials can be exchanged between them in a variety of ways. The same role is performed by the internal cell membranes, providing a range of microenvironments within the cell, each suited to different functions and yet all in communication. The partially permeable nature of the cell membrane is a major contributor to the processes of life.

CELL 'EATING' AND 'DRINKING'

We have considered how the membrane allows ions and molecules to pass into and out of the cell. But there are times when larger particles need to enter or leave the cell. Millions of simple animals, along with cells in the bodies of most large multicellular creatures, need on occasion to take in large particles. White blood cells ingest bacteria, and amoebae engulf their prey. Membrane transport systems cannot do this type of job – but the membrane itself can.

Endocytosis and exocytosis

Endocytosis is the term used when materials are surrounded by and taken up into membrane-lined vesicles. This can occur at a relatively large scale, for example when bacteria are ingested. In this case it is called **phagocytosis** – 'cell eating'. Endocytosis also appears to happen at a microscopic level, when minute vacuoles taking in the external medium are formed. This is called **pinocytosis**. Recent work with the electron microscope shows that pinocytosis is very common, even in some mammalian cells.

Exocytosis is the term for the emptying of a membrane-lined vesicle at the surface of the cell or elsewhere. For example, in cells producing hormones, vesicles containing the hormone fuse with the cell surface membrane to release their contents.

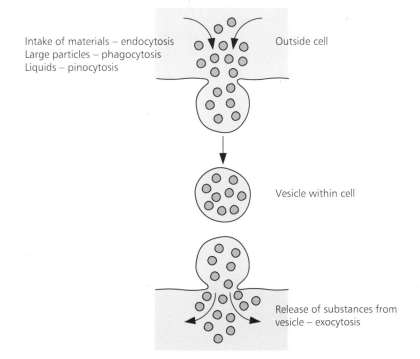

Intake of materials – endocytosis
Large particles – phagocytosis
Liquids – pinocytosis

Outside cell

Vesicle within cell

Release of substances from vesicle – exocytosis

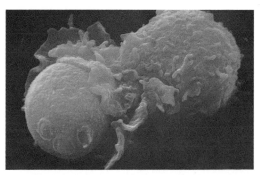

Figure 1.4.16 The properties of the membrane allow cells to take in large particles or release secretions. The scanning electron micrograph shows a lymphocyte engulfing a yeast cell.

These processes, about which you will find out more in later sections of the book, are made possible by the fluid mosaic nature of the membrane. It is capable of flowing around a particle, sealing up on the surface and fusing with internal membranes in ways in which a rigid structure could not do – another example of the way the structure of the membrane is intimately related to its functions.

SUMMARY

- Water is the medium in which all cellular reactions take place. Water has unique chemical properties, generally due to its ability to form hydrogen bonds between the slightly positive hydrogens and slightly negative oxygens of adjacent molecules:
  Water is a good solvent.
  Ice is less dense than liquid water.
  Water has a high specific heat capacity.
  Water has a high latent heat of vaporisation and a high latent heat of fusion.
  Water has a high surface tension.
  Water molecules are amphoteric.

- Cell surface membranes provide a **partially permeable** barrier between the cell contents and the external medium, allowing some materials to pass freely but not others.

- Cell surface membranes consist of a **phospholipid bilayer** with the hydrophilic ends of the molecules pointing outwards and the hydrophobic ends together in the centre. **Proteins** float in the lipid bilayer. This is the **fluid mosaic model**.

- **Diffusion** is the movement of a substance from an area of higher concentration to an area of lower concentration by random movement of the molecules or ions of the substance.

- **Osmosis** is the movement of water molecules from a region of higher concentration of water to a region of lower concentration through a partially permeable membrane.

- The **water potential** of a system is the difference between the potential of water in that system and the potential of pure water at the same temperature and pressure. Pure water has a water potential of zero, and all other water potentials have negative values. Water moves from a solution of higher water potential to an area of lower water potential.

- Osmosis in living systems is the movement of water from a region of higher water potential to a region of lower water potential.

- Water potential Ψ combines the effects of the **solute potential** Ψ_s (the osmotic effect of the solutes in the cell) and the **pressure potential** Ψ_p (the hydrostatic pressure, for example that exerted by the plant cell wall):

$$\Psi = \Psi_s + \Psi_p$$

- The proteins in cell surface membranes allow molecules other than water to cross the membrane through pores, by **facilitated diffusion** or by **active transport**.

- In **facilitated diffusion**, specific protein carriers pick up a molecule or ion. This changes the conformation of the protein and causes it to rotate through the membrane. It releases the molecule or ion on the inside of the cell. Facilitated diffusion takes place *along* a concentration gradient.

- **Active transport** moves molecules or ions through the cell surface membrane *against* a concentration gradient. The specific carrier protein picks up a molecule or ion and rotates, again releasing it on the other side. The process involves energy, which is provided by the conversion of **ATP** to **ADP**, catalysed by **ATPase**.

- **Endocytosis** is the taking up of materials in membrane-bound vesicles. **Phagocytosis** involves large vesicles, for example around a bacterium, and **pinocytosis** is on a smaller scale, taking in minute vacuoles of medium.

- **Exocytosis** is the emptying of a membrane-bound vesicle at the surface of the cell, as in secretions.

QUESTIONS

1 Water is a molecule of great biological importance. Explain why this is the case, using three of the main properties of water to illustrate your answer.

2 a Describe the fluid mosaic model of the structure of the cell membrane.
  b How has the development of the electron microscope helped in the production of this model?

3 Osmosis is a physical process which has a great effect on the biology of both plants and animals. It occurs in cells as a result of the properties of the cell membrane.
  a What is osmosis?
  b How does osmosis affect animal cells?
  c How does osmosis affect plant cells?

4 List, with a simple explanation, the main ways in which substances may be transported across the cell membrane.

1.5 Enzymes

Inside every cell, hundreds of chemical reactions are occurring simultaneously on the complex infrastructure of the cytoplasm. In a science laboratory it is a matter of skill and coordination to produce the desired products from a single reaction. The process is usually relatively inefficient, with material wasted along the way and many attempts needed to produce the desired end products.

How is the efficiency and control of living cells achieved?

ENZYMES – THE ENABLERS

Under the conditions of temperature and pH found in living cells, the majority of the reactions needed to provide cells with energy and produce new biological material would naturally take place very slowly indeed – too slowly for life to exist. But within cells is a group of molecules which speed up chemical reactions without changing the conditions in the cytoplasm. These are the **enzymes**.

What is an enzyme?

An enzyme is a protein which speeds up one or more biological reactions. Within any cell a great range of chemical reactions is going on at any one time. Those reactions which build up new chemicals are known as **anabolic** reactions. Those which break substances down are **catabolic** reactions. The combination of these two processes results in the complex array of biochemistry which we call **metabolism**. Most of the reactions of metabolism occur not as single events but as part of a sequence of reactions known as a **metabolic chain** or **pathway**.

Anabolism + catabolism = metabolism

A **catalyst** is a substance which speeds up a reaction without changing the substances produced, and without itself being changed. Enzymes are powerful biological catalysts. Each enzyme will only catalyse a particular reaction or group of reactions – they show varying levels of **specificity**. For metabolism to proceed at an appropriate rate, every reaction needs catalysing. To achieve this each cell contains several hundred different enzymes controlling a multitude of reactions.

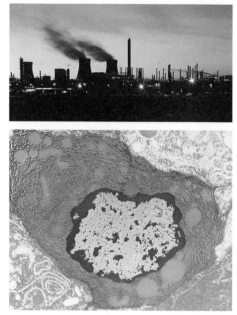

Figure 1.5.1 Living cells are like miniature, complex biochemical factories. They cannot afford the luxury of mistakes or wastage.

How do we know about enzymes?

In modern biology and medicine, knowledge of enzymes is very much taken for granted. But our present knowledge has developed over a long period of time, with much information being relatively new.

In 1835 it was noticed that the hydrolysis of starch was brought about more effectively by malt than by sulphuric acid. This suggested that there was a catalyst present in the living malt which was more effective than the inorganic acid. Also, for a long time it was suspected that there was a biological catalyst in yeast which brought about the fermentation of sugar to alcohol, though nobody could prove it. Initially called 'ferments', it was in 1877 that the name **enzyme** (literally 'in yeast') was first used for these theoretical chemicals.

1897 brought a major landmark in enzyme research. Eduard Buchner extracted from yeast cells the enzyme responsible for fermenting sugar, and showed it could work independently of the living cell structure.

The first pure, crystalline enzyme, extracted from jack beans, was produced in 1926 by J. Sumner. It was the enzyme urease, which catalyses the breakdown of urea. Sumner showed that the crystals were protein and concluded that enzymes must be proteins. Unfortunately no one believed him!

In 1930–36 the protein nature of enzymes was firmly established as the protein-digesting enzymes pepsin, trypsin and chymotrypsin were extracted from the gut and crystallised by J. Northrop.

Work has continued since then, with more and more enzymes being extracted and a variety of techniques being developed to look at their structure.

Name that enzyme

Most enzymes have several names:
- a relatively short **recommended** name, which is often the name of the **substrate** (the molecule that the enzyme works on) with '-ase' or 'kinase' added, for example sucrase, creatine kinase
- a longer **systematic** name describing the type of reaction which is being catalysed, for example ATP:creatine phosphotransferase
- a classification number, for example EC 2.7.3.2.

It is easy to tell what some enzymes do, such as urease, ribonuclease and lipase, from their recommended names. But some enzymes are still known by old and distinctly uninformative names, such as trypsin and pepsin.

How do enzymes speed up reactions?

To answer this question, we first need to understand what is happening in a chemical reaction. Reactions involve breaking and remaking chemical bonds. In order to react, molecules first need sufficient energy – they have to get over an 'energy hill', known as the **activation energy** for the reaction. Imagine pushing a large boulder up a steep hill. Once the boulder gets to the top (achieves activation energy), it will roll down the other side quite easily.

Increasing the temperature is one way of increasing the rate of chemical reaction. It increases the numbers of reactant molecules which have sufficient energy to react. However, living cells could not survive the high temperatures needed to speed up many cellular reactions to the required rate, and the energy demands needed to do this would be enormous. Instead, enzymes solve the problem by lowering the activation energy needed for a reaction to take place (they make the hill smaller). Figure 1.5.3 uses an energy diagram to show this.

Figure 1.5.2 The chemicals in the cardboard tube and the oxygen in the air form a potentially explosive mixture. But it is not until heat from the lighted fuse supplies the activation energy that the sparks begin to fly.

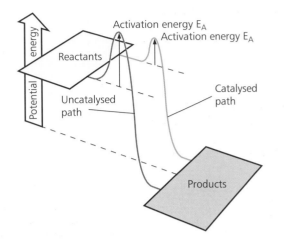

Figure 1.5.3 This energy diagram shows the difference in activation energy between a catalysed and an uncatalysed reaction.

Apart from speeding up reactions, enzymes have another important function. They control and regulate the sequences of reactions occurring in metabolic chains. We shall be returning to this later in this section.

THE CHARACTERISTICS OF ENZYMES

All enzymes share certain characteristics. By considering these in turn we can learn a great deal about how enzymes work.

1 All enzymes are *globular proteins*. Within these large molecules is found an **active site** which is the part responsible for the functioning of the enzyme. Anything which affects the three-dimensional shape of a protein molecule affects its ability to function.

2 Enzymes only change the *rate* of a reaction. They do not alter the end products that are formed, or affect the equilibrium of the reaction, that is, the ratio of reactants to products. This shows that they act purely as catalysts and not as modifying influences in any other way.

3 Enzymes are present in very small amounts. This reflects their great *efficiency as catalysts*. Minute amounts of enzymes can speed up reactions enormously. In order to explain this more fully, we need to introduce the idea of **molecular activity** or **turnover number**. This is a measure of the number of substrate molecules transformed per minute by a single enzyme molecule. The number of molecules of hydrogen peroxide broken down by the enzyme catalase in one minute is 6×10^6! A more usual turnover number would be thousands of molecules per minute rather than millions – catalase, extracted from liver cells, is the fastest known enzyme. It has been calculated that the enzyme urease can break down in one second an amount of urea that would hydrolyse spontaneously in 3 million years! Enzymes generally increase reaction rates by factors from 10^8 to 10^{26}.

4 Enzymes are *specific* to the reactions that they catalyse. They are unlike inorganic catalysts which can frequently be used to catalyse a wide range of reactions, although often at extremes of temperature and pressure. Some enzymes will only catalyse one particular reaction. Others are specific to a particular group of molecules of similar shape, or to a type of reaction which always involves the same groups. Many enzymes show great **stereospecificity** – they will only catalyse reactions involving either L- or D-isomers. This suggests that a site in the enzyme has a particular shape into which a specific substrate will fit.

5 Enzyme-catalysed reactions are affected by the *amount of substrate* which is present. Take a simple reaction in which the substrate A is converted to the product Z. If the concentration of A is gradually increased, the rate of the enzyme-catalysed reaction:

$$A \rightarrow Z$$

will increase – but only for so long. Then the enzyme becomes **saturated** and a further increase in substrate concentration will not further increase the rate of the reaction. The graph in figure 1.5.4 illustrates this. This observation reinforces the idea of an active site in an enzyme. Once all the available active sites are involved in the reaction, further increases in the concentration of substrate molecules will not increase the rate any further. (In rate experiments like this, every other factor – temperature, pressure, pH and the enzyme concentration – must be kept the same, so that any changes can only be the result of the change in substrate concentration.)

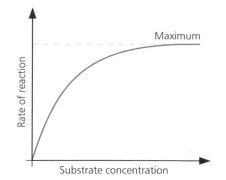

Figure 1.5.4 The effect of substrate concentration on an enzyme-catalysed reaction. The enzyme becomes saturated with substrate and the reaction rate levels off.

6 The rate of an enzyme-catalysed reaction is affected by *temperature* in a characteristic way. The effect of temperature on the rate of any reaction can be expressed as the temperature coefficient, Q_{10}. This is given as:

$$Q_{10} = \frac{\text{rate of reaction at } (x + 10)\,°C}{\text{rate of reaction at } x\,°C}$$

Between about 0 and 40 °C, Q_{10} for any reaction is 2. In other words, in that temperature range every 10 °C rise in temperature produces a doubling of the rate of reaction. Outside this range, however, Q_{10} for enzyme-catalysed reactions decreases markedly, whilst Q_{10} for other reactions remains relatively unchanged. The rate of enzyme-catalysed reactions falls and by about 60 °C the reaction has stopped completely in most cases, as figure 1.5.5 demonstrates.

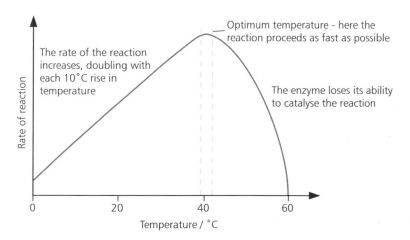

Figure 1.5.5 The effect of temperature on the rate of a typical enzyme-catalysed reaction. All factors other than temperature must be kept constant.

We know that at temperatures over 40 °C, most mammalian proteins lose their tertiary and quaternary structures – they unravel or **denature**. As enzymes denature, they lose their ability to catalyse reactions. This confirms our idea of an active site in the three-dimensional structure of the globular protein.

There are a few enzymes that do not denature at such low temperatures. The enzymes of thermophilic bacteria, found in hot springs at temperatures of up to 85 °C, can obviously function at higher temperatures. They are made of unusual temperature-resistant proteins.

7 *pH* has a major effect on enzyme activity. pH is known to affect the shapes of protein molecules as hydrogen bonds and sulphur bridges are broken or formed. This again confirms the importance of the three-dimensional structure. Different enzymes work in different ranges of pH, as figure 1.5.6 shows. pH may also affect the substrate molecule.

ENZYME MECHANISMS

We have seen a great deal of evidence suggesting that the function of an enzyme is closely tied in with its three-dimensional structure. Now we shall take a closer look at models for enzyme action, and also at more evidence to support the picture.

Enzymes catalyse reactions by lowering the activation energy for the reaction. To bring this about the enzyme forms a **complex** with the substrate or substrates of the reaction. Thus we have a simple picture of enzyme action:

Substrate + enzyme → enzyme/substrate complex → enzyme + products

Once the products of the reaction are formed they are released and the enzyme is free to form a new complex with another substrate molecule.

The lock and key mechanism

How does this relate to the structure of the enzyme? The basic picture is summed up in the **lock and key mechanism** shown in figure 1.5.7.

Induced fit

The lock and key mechanism fits most of our evidence about enzyme characteristics. It illustrates how enzymes can become saturated when the concentration of substrate molecules rises above a certain level. It also explains how any change in the protein structure, such as those brought about by changes in pH or temperature, could affect enzyme action by altering the shape of the active site.

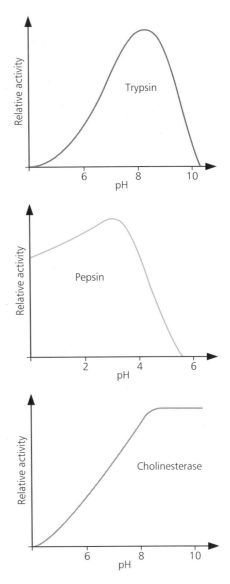

Figure 1.5.6 Different enzymes work best at different pHs. Again, all other factors must be kept constant. Interestingly, the optimum pH for an enzyme is not always the same as the pH of its normal surroundings. It is thought that cells may control the activity levels of their intracellular enzymes by minute changes in pH.

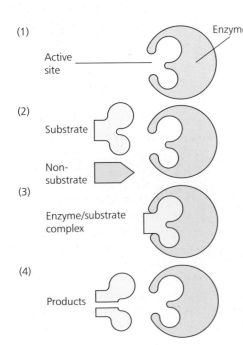

(1) Within the structure of each enzyme is an area known as the **active site**. It may involve only a small number of amino acids. It has a very specific shape, which gives each enzyme its specificity, as only one substrate or type of substrate will fit into the gap.

(2) Here we can see the difference between the shape of the enzyme substrate and another biological molecule. Only a molecule of the right shape can be a substrate for the enzyme.

(3) The enzyme and substrate slot together to form a complex, as a key fits into a lock. In this complex the substrate is enabled to react at a lower activation energy. This may be due to bonds within it being deformed and stressed in the complex, so making them more likely to react.

(4) Once the reaction has been catalysed, the products are no longer the right shape to stay in the active site and the complex breaks up, releasing the products and freeing the enzyme for further catalytic action.

Figure 1.5.7 The lock and key mechanism – the basis of our understanding of enzyme action

However, it is now thought that the lock and key mechanism is a slight simplification. Evidence from X-ray crystallography, chemical analysis of active sites and other techniques suggests that the active sites of enzymes are not the rigid shapes once supposed. In the **induced fit theory**, which is generally accepted as the best current model, the active site is thought of as having a distinctive but flexible shape. Thus, once the substrate enters the active site, the shape of that site is modified around it to form the complex. Once the products have left the complex the enzyme reverts to its inactive, relaxed form until another substrate molecule binds, as illustrated in figure 1.5.8.

Inhibitors

Our understanding of enzymes and how they work can be increased by evidence gained from **enzyme inhibitors** which stop enzymes from working or reduce their catalytic power. There are two main types of inhibition, **reversible** and **irreversible**.

Reversible inhibition

Reversible inhibition of enzymes occurs when an inhibitor affects an enzyme in a way which does not permanently damage it, so that when the inhibitor is removed, the enzyme can function normally again. Reversible inhibition is quite a common feature of metabolic pathways, as we shall see. There are two major forms of reversible inhibition – **competitive** and **non-competitive**.

In competitive inhibition, the inhibitor is similar in shape to the substrate molecule. It competes with it to bind at the active site of the enzyme, forming an **enzyme/inhibitor complex**. If the amount of inhibitor is fixed, the percentage of inhibition can be reduced by increasing the substrate concentration. As the two molecule types are competing, the more substrate molecules there are, the less likely is it that inhibitor molecules will bind to the active site. An example of competitive inhibition is found in the enzymes involved in the Krebs cycle, the pathway for cellular respiration, described in section 3.4.

In non-competitive inhibition, the inhibitor may form a complex with the enzyme itself, with the enzyme/substrate complex or with a prosthetic group (see page 69). The inhibitor is not competing for the active site but joins to the enzyme molecule elsewhere. This is confirmed by the fact that the concentration of the substrate makes no difference to the level of inhibition – only the concentration of inhibitor affects that. What appears to happen in most cases is that the presence of the inhibitor on the enzyme or enzyme/substrate complex deforms the active site so that it can no longer catalyse the reaction.

Figure 1.5.9 shows the differences between competitive and non-competitive inhibition.

Irreversible inhibition

In irreversible inhibition the inhibitor combines with the enzyme by permanent covalent bonding to one of the catalytic groups. This changes the shape and structure of the molecule in such a way that it cannot be reversed – the enzyme is inactivated permanently. Irreversible inhibition tends to occur more slowly than other forms of inhibition, but its effects are much more devastating and it is never used within the cells to control metabolism. Arsenic, cyanide and mercury are poisonous because they exert irreversible inhibition on enzyme systems. Some of the nerve gases used in chemical warfare also work in this way. They combine with and completely inactivate acetylcholinesterase. The normal function of this enzyme is to destroy the neurotransmitter acetylcholine as soon

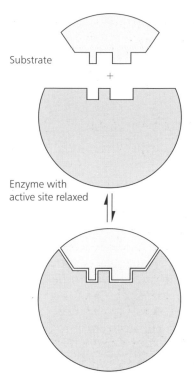

Substrate

+

Enzyme with active site relaxed

Enzyme/substrate complex showing the induced form of the active site, fitting snugly round the substrate

Figure 1.5.8 The induced fit theory of enzyme action proposes that the catalytic groups of the active site are not brought into their most effective positions until a substrate molecule is bound onto the site, *inducing* a change in shape (conformation).

An example of competitive inhibition

Succinate dehydrogenase removes two H^+ ions from succinate, an intermediate compound of the Krebs cycle (explained in section 3.4), to form malonate. Malonate and other similar molecules inhibit succinate dehydrogenase. However, an increase in the concentration of succinate reduces the level of inhibition and allows the reaction to continue. This type of inhibition is used to give very precise controls on the rates at which metabolic pathways progress.

Competitive
inhibition

Non-competitive
inhibition

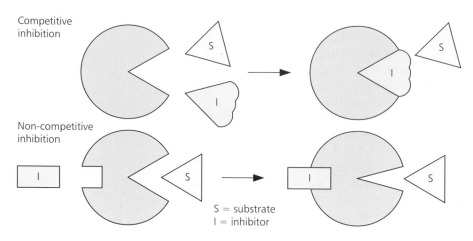

S = substrate
I = inhibitor

Figure 1.5.9 Competitive inhibitors bind at the active site, unlike non-competitive inhibitors.

as a message has been passed from a nerve to a muscle. When the enzyme is inhibited the muscles go into prolonged spasms and death results as breathing and swallowing become impossible.

Feedback control

How do cells control the hundreds of reactions going on inside them? There are many factors involved. Membrane compartments keep reactants apart. Variations in pH can be used to change the rate of enzyme-catalysed reactions, and the amount of substrate available is another mechanism at work. One of the most important methods of control is that exerted by the **regulatory enzymes**.

Regulatory enzymes are often **allosteric enzymes**. 'Allosteric' literally means 'another place'. These enzymes have another site, separate from the active site, to which another molecule can bind and act either as an inhibitor or as an activator. This ability to be activated as well as inhibited makes allosteric control different from non-competitive inhibition. Regulatory enzymes are usually found within metabolic pathways. The regulatory enzyme is found near the beginning of the pathway and is acted on by one of the end products of the chain. This is known as **end product control** or **feedback control**, shown in figure 1.5.10. There are some important examples of this in the respiratory pathways (see section 3.4).

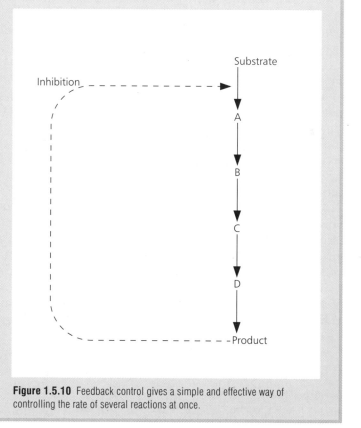

Figure 1.5.10 Feedback control gives a simple and effective way of controlling the rate of several reactions at once.

Enzyme cofactors

Many enzymes need a non-protein part to function properly. These non-protein components can be one of several types of ion or molecule and are given the general term **cofactors**. Sometimes cofactors are unchanged at the end of a reaction, but sometimes they are changed and then regenerated by another process. When an enzyme is linked with its cofactor to form a complex it may be referred to as a **holoenzyme**. The protein enzyme alone is known as the **apoenzyme**.

Simple inorganic ions are one type of cofactor. They may bind to the enzyme/substrate complex in a way that makes it function more effectively, for example, salivary amylase works better in the presence of chloride ions. Equally, they may form an integral part of the enzyme molecule.

Prosthetic groups are another type of cofactor. The word 'prosthetic' is derived from the Greek meaning 'added on, giving additional power'. These are organic molecules which are very tightly bound to the enzyme itself. A good example is haem, which is a flat ring molecule containing iron. Haem is the prosthetic group for a number of enzymes, including cytochrome oxidase which is involved in cellular respiration (see section 3.4), catalase and peroxidase. Haem is a permanent part of the structure of these enzymes.

Coenzymes are cofactors which are complex non-protein organic molecules loosely associated with an enzyme. They transfer chemical groups, atoms or electrons from one enzyme to another. Many of them are vitamins or derivatives of vitamins. Nicotinamide adenine dinucleotide (NAD) is a coenzyme which is very important in cellular respiration and other metabolic pathways.

Enzyme technology

Enzymes have much to offer industry. Unlike most inorganic catalysts, they work at low temperatures, normal pressures and in a very easily achieved range of pH levels. This means that an industrial process which can use an enzyme catalyst may well be relatively cheap to run. As our ability to use genetic engineering to produce specific enzymes increases, so will the use of enzymes. Much enzyme-based technology already exists and enzymes are used for a wide variety of processes.

The food industry is traditionally a main user of enzyme technology. Rennin is used to clot milk in cheese-making and enzymes from yeast are used widely in both the brewing and the baking industries, as they have been for many, many years past. Cellulases and pectinases are used to clear hazes in fruit juice production. Other newer uses include trypsin to predigest baby foods, proteases in biscuit manufacture to lower the protein content of the flour, and a variety of enzymes to make sweet syrups from starch.

Some other industrial uses of enzymes are more surprising. They are used in detergents to digest particular types of dirt, particularly the protein elements of food and sweat (see figure 1.5.11). Enzymes are also being developed which will nip off the 'pilling' – little bobbles that form on cotton and woollen clothing when they are washed – and leave the fabric smooth. Enzymes are used in the rubber industry to produce oxygen to convert latex to foam rubber, in the paper industry and in the photographic industry. They also have many uses in medical fields – for example, glucose oxidase is used for detecting glycosuria (excess sugar in urine, as in diabetes). With improvements in technology and advances in genetic engineering making 'designer enzymes' increasingly possible, the use of enzymes in industry can only increase in the future.

Figure 1.5.11 The enzymes which are effective at attacking protein dirt also attack people's skin. Until these tiny capsules were developed to contain the enzymes, many of the workers in detergent factories suffered from allergic reactions.

Ribozymes

Throughout this section the assumption has been made that all enzymes are proteins. However, in very recent years Thomas Cech and his colleagues at the University of Colorado have shown that RNA can act as a biological catalyst. Small sections of a variety of tRNAs, mitochondrial RNAs and nuclear RNAs have been shown to have enzymatic activity and are known as **ribozymes**. The study of biological catalysts has been opened up by this discovery – in years to come our present view of enzymes may be shown to have been far too narrow.

SUMMARY

- The biochemistry of a cell is known as **metabolism**, made up of **anabolism** (building up complex molecules) and **catabolism** (breaking down complex molecules). Metabolism is controlled by enzymes.

- Enzymes lower the **activation energy** of a reaction, making it possible for more molecules to react under normal conditions of temperature, pH, etc.

- Enzymes have certain properties in common:
  Enzymes are globular proteins with an active site within their three-dimensional structure.
  They are **catalysts** – they speed up reactions without affecting the products or position of equilibrium, and are not affected themselves by the reaction.
  Enzymes are very efficient – small quantities may speed a reaction greatly. The **turnover number** reflects the number of substrate molecules transformed per molecule of enzyme.
  They show varying degrees of **specificity**, from stereospecificity (acting on only the D- or L-form of a substrate) to those which catalyse a type of reaction involving a particular group.
  The action of enzymes is affected by substrate concentration to a maximum rate where they become **saturated**.
  The rate of an enzyme-catalysed reaction increases with increasing temperature until the enzyme starts to **denature** – it loses its tertiary and quaternary structure and then no longer catalyses the reaction.
  Different enzymes work in different ranges of pH. pH affects the shape of protein molecules and therefore their active sites.

- The **lock and key mechanism** of enzyme action says that the substrate molecule fits into the three-dimensional shape of the active site to form an enzyme/substrate complex. This holds the substrate in such a way that the activation energy is lower, and the products are then released.

- The **induced fit theory** is a refinement of the lock and key mechanism which says that the active site has a flexible shape which is modified around the substrate molecule once bound. On release of the products, the inactive shape is resumed.

- Enzymes are inhibited in two ways – by **reversible inhibitors** and by **irreversible inhibitors**.

- **Reversible inhibitors** may be **competitive** or **non-competitive**. **Competitive inhibitors** are similar in shape to the substrate and bind to the active site to form an enzyme/inhibitor complex, blocking the site for substrate molecules. Increasing the concentration of substrate increases the rate of the reaction. **Non-competitive inhibitors** form a complex with another part of the enzyme but still affect the active site. The rate of reaction is not affected by substrate concentration – only by inhibitor concentration.

- **Irreversible inhibitors** such as some poisons change the shape of the enzyme permanently by covalent bonding.

- **Regulatory enzymes** in a metabolic pathway regulate that pathway. **Allosteric enzymes** have a site separate from the active site to which an inhibitor or an activator can bind. These are involved in feedback control.

- **Cofactors** are non-protein parts that enable certain enzymes (**apoenzymes**) to function. The enzyme and cofactor together form a **holoenzyme**. Cofactors include inorganic ions, **prosthetic groups** such as haem which are organic molecules tightly bound to the enzyme molecule, and **coenzymes** which are organic molecules more loosely associated with the enzyme.

QUESTIONS

1 What is an enzyme and why are enzymes important?

2 a What are the main characteristics of an enzyme?
  b How are enzyme-catalysed reactions affected by
   i temperature
   ii substrate concentration?
  c How do enzymes work?

3 a Explain the main types of enzyme inhibition
  b Apart from their role as an investigative tool, what is the importance of enzyme inhibitors in biochemical pathways?

4 Give four examples of enzyme technology in industry. What do you think would be the main difficulties of setting up an industrial process which relied on enzyme catalysis?

1.6 Cell recognition

RECOGNITION OF CELLS BY OTHER CELLS

Many living organisms, both plant and animal, consist of single cells. For them, simple diffusion is adequate for supplying the oxygen needed for cellular respiration and for removing the waste products of metabolism. Osmosis can supply all the water required. The surface area of the cell compared to its volume (the **surface area : volume ratio**) is relatively large for small organisms, as figure 1.6.1 shows, and complex methods of gaining oxygen and water or getting rid of waste are not required.

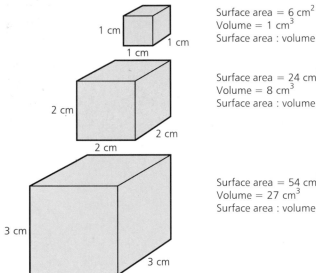

Surface area = 6 cm^2
Volume = 1 cm^3
Surface area : volume ratio = 6 : 1

Surface area = 24 cm^2
Volume = 8 cm^3
Surface area : volume ratio = 24 : 8 = 3 : 1

Surface area = 54 cm^2
Volume = 27 cm^3
Surface area : volume ratio = 54 : 27 = 2 : 1

Figure 1.6.1 As the linear dimensions of an organism increase, the ratio of surface area:volume decreases. This sets limits on the lifestyle of an organism, or forces evolutionary change to overcome the restrictions which result.

It might seem that for unicellular organisms, cell recognition would not be important. However, they need to sense the presence of potential food, and those which undergo a primitive form of sexual reproduction need to recognise the cells of others of the same species.

For larger organisms, the surface area:volume ratio is such that simple diffusion is not adequate to supply all the necessities of life. Cells within a multicellular animal or plant become increasingly specialised, differentiating further and further away from the generalised cell to perform very individual functions, be it causing movement, obtaining food or oxygen or carrying nerve impulses to allow for coordinated action. In such an organism it is obviously important that individual cells recognise each other as members of the same living entity, and also that they recognise other cells performing the same function so that they can operate as a group. It is also important to be able to recognise foreign cells, not simply potential prey but also invaders such as bacteria and viruses within the organism. This aspect of cell recognition is particularly important for animals. Their motile way of life and relatively vulnerable cells make them particularly open to attack by invading organisms. Plants, without the need to move about, can produce much stouter external barriers to prevent invasion.

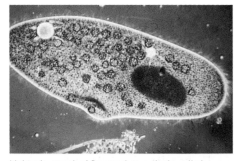

Light micrograph of *Paramecium*, a single-celled organism

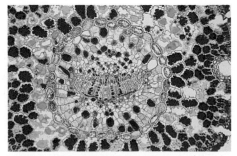

Light micrograph of a section through a cedar leaf, showing the many different types of cells that make up the leaf

Figure 1.6.2 The more cells there are, the more important cell recognition becomes.

Evidence for cell recognition

Sponges are the most primitive of the multicellular animals. Until 1765, they were thought to be plants. They are made up of many simple cells grouped together, and do not move about. Their cells show some differentiation into different types concerned with food extraction and water flow, but they do not have high levels of organisation.

In the early years of this century E. Wilson split sponges into their individual cells by passing them through sieves (sponges have no nervous system and so cannot feel pain!). The individual cells began to show amoeboid movement, and continued to move until they encountered another cell. Aggregates of sponge cells formed, clearly demonstrating cell recognition, and these aggregates went on to develop into new animals. So even in simple organisms we see evidence of cell recognition, and this recognition plays a vital role at all levels of organisation.

Figure 1.6.3 For a long time thought of as plants, sponges are in fact simple animals. These primitive organisms clearly display cell recognition.

How do cells recognise each other?

As we have seen, protruding from the outer surface of the cell surface membrane are many proteins, in particular **glycoproteins**, protein molecules with a carbohydrate component. These chains of sugar molecules can be very varied, and they seem to be important in cell recognition in several ways. Similar sugar recognition sites may bind to each other, holding similar cells together. When tissues and organs are forming in embryonic development this kind of recognition and binding is of great significance.

THE IMMUNE SYSTEM

Apart from the association and differentiation of cells in the developing embryo, perhaps the most striking example of cell recognition in living organisms is the **immune system**. The immune system enables the body to recognise foreign material – that is, anything which is not part of itself – and to produce the appropriate response to remove it from the system as quickly and efficiently as possible. Various aspects of the immune system are accepted as normal parts of life. We speak of being immune to a particular illness. We are vaccinated against a wide variety of diseases. But the immune system is extremely complex. We shall look here most closely at the human immune system, since this is of greatest relevance to us in terms of the prevention of disease.

How does the body recognise foreign material?

The protein and glycoprotein markers on the surface of cell surface membranes are known as **antigens**. Each organism carries its own unique set of antigens – some of them common to every member of a particular species, others specific to a particular individual within a species. The more closely related two individuals are, the more antigens they are likely to have in common. But the only individuals who will have matching antigens are genetically identical twins.

The term antigens, which refers specifically to the markers on the membranes, is often used loosely to refer to the whole cells that carry the antigens.

Lymphocytes

The body recognises the presence of a foreign antigen by means of **lymphocytes**. Lymphocytes are white blood cells, made in the white bone marrow of the long bones. Every individual possesses a wide range of lymphocytes. Each lymphocyte is capable of recognising *one* specific antigen. There are two types of lymphocytes involved in the immune response. **T-cells** (activated by the thymus gland) are involved in what is called the **cell-mediated response**. **B-cells** (activated, it is thought, by the cells of the bone marrow, liver and spleen) are involved in the **humoral response**.

The cell-mediated response

T-cells congregate at the site of an infection and mount a direct attack on the foreign organism or tissue. The T-cells present in the blood have receptor proteins attached to their cell surface membranes. When a T-cell comes into contact with a **complementary antigen**, an antigen it recognises on another cell, the antigen is bound to the receptor proteins and the cell destroyed. As a result of this reaction, the T-cell undergoes a rapid series of cell divisions to produce a **clone** of identical T-cells all capable of recognising and destroying the same type of antigen. These cells are sometimes known as **killer lymphocytes** or **killer T-cells** because they destroy the invading material.

The humoral response

B-cells in the lymph glands and around the body make substances to be carried around which attack the invader. Just like the T-cells, B-cells have receptor proteins on the cell surface membrane. However, when B-cells detect a complementary antigen and bind it to their receptor protein, the effect is very different. Like the T-cell, the B-cell undergoes a rapid series of cell divisions, but *two* different types of new cells result – plasma cell clones and memory cells.

The majority of these new cells are **plasma cell clones**. These have the ability to produce large amounts of **antibodies**, special proteins which are released into the circulation and which bind to the antigen, causing its destruction in one of several ways. The B-cells and the plasma cell clones are not themselves directly involved in the destruction of the antigen. Plasma cell clones only live for a few days but can produce up to 2000 antibody molecules per second.

Memory cells are also produced by the divisions of the B-cell. These are important in allowing the body to respond very rapidly to a second invasion by the same antigen. Once you have had a disease, you do not usually catch it again. This is because when you encounter the disease-causing antigen again, your body can produce antibodies against it so rapidly that it is destroyed before the symptoms of the disease develop. It is the memory cells that make this possible. As yet no one is entirely sure how memory cells provide this **immunological memory**.

Figure 1.6.5 summarises the immune response and compares the cell-mediated response and the humoral response.

Figure 1.6.4 In a newborn infant, lymphocytes cannot immediately recognise foreign tissue or perform any other useful function – they have to be activated. Once we understand how cell recognition is activated, we should be able to suppress it and solve many of the problems of rejection associated with the technique of transplantation.

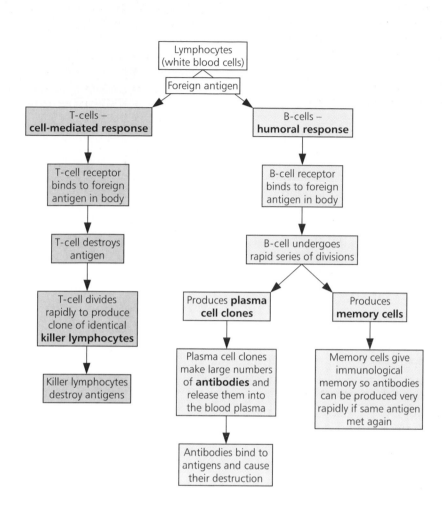

How antibodies destroy antigens

Antibodies are large protein molecules produced by plasma cell clones in the humoral response. They are collectively known as the **immunoglobulins**. They contribute to the defence of the body against the invasion of foreign material in a variety of ways.

1 Antibodies reduce the ability of most **pathogens** (disease-causing organisms) to invade the host cells.

2 Antibodies bind to antigens and **agglutinate** or clump together which helps to prevent their spread through the body.

3 The antigen–antibody complex is readily engulfed and digested by phagocytes – another type of white blood cell – which travel through the circulatory system ingesting foreign material. Figure 1.6.6 shows how this happens.

4 The antigen–antibody complex may stimulate other reactions within the body, such as the destruction of the membrane of the antigen if it has one, or the release of the chemical histamine by the invaded cells, causing inflammation.

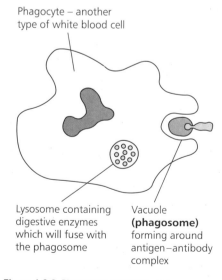

Phagocyte – another type of white blood cell

Lysosome containing digestive enzymes which will fuse with the phagosome

Vacuole **(phagosome)** forming around antigen–antibody complex

Figure 1.6.6 Phagocytes engulf the whole antigen–antibody complex.

Types of immunoglobulins

Although immunoglobulins are all different, each being specific to one antigen, they all have the same basic structure. They are globular proteins made up of four polypeptide chains, two *light* or short chains and two *heavy* or long chains. Five different types of immunoglobulin have been isolated, each of which appears to have a slightly different function.

IgM is the first type of immunoglobulin to appear in the blood in response to an antigen. It is produced for a week or two and seems to be involved in the **primary response** of the body to an antigen – that is, the way it responds the first time it meets an antigen.

IgG is the main immunoglobulin type found circulating in the blood – it makes up about 80% of immunoglobulin. IgG is the workhorse of the immune system, conferring most of the benefits of immunity and being very heavily involved in the **secondary response** of the body to an antigen – the way it responds on repeated exposures to a particular antigen. IgG cannot be synthesised during the first couple of months of human life, and so newborn babies are dependent on immunoglobulins which they have received across the placenta and continue to receive through their mother's milk to protect them from a wide variety of diseases.

IgA is found in relatively low levels in the blood, but is the main immunoglobulin to be found in body secretions such as tears, saliva, nasal drippings, colostrum (the special antibody-rich form of milk produced during pregnancy and the first few days of lactation) and milk. The presence of IgA in colostrum and milk appears to be an important way of providing the infant with immunity to disease. IgA is one of the few proteins that can survive the pH conditions and enzymes of the stomach and intestines and be absorbed into the body with its biological activity intact.

IgD and **IgE** are found in relatively small amounts. The functions of these two immunoglobulins are not yet clearly understood. IgE seems to have a role in allergic reactions.

Different types of immunity

Natural immunity

Normally, where the body comes into contact with a foreign antigen, the immune system is activated and antibodies are formed which result in the destruction of the antigen. This is known as **natural active immunity**. The other natural type of immunity is that passed between mother and offspring, where preformed antibodies are provided to the infant either before or after birth, providing it with temporary immunity until its own system becomes active. This is called **natural passive immunity**, and tends to be quite short-lived because the antibodies are used up and are not replaced.

Figure 1.6.7 Natural passive immunity ensures that young mammals survive until their own immune system becomes active.

Induced immunity

The value of immunity to disease has long been recognised and we have managed to induce it artificially in a number of ways. Beginning with Edward Jenner's observation that those who had suffered from cowpox were less likely to suffer smallpox, the development and introduction of vaccinations has been a major preoccupation of the medical profession. Smallpox, diphtheria, polio and tuberculosis have all been major killers in their time, but have now been either eradicated (smallpox) or greatly reduced in incidence, in the developed world at least.

Acquired passive immunity is given when antibodies are formed in one individual, extracted and injected into another individual. This does not confer long-term immunity, but can be valuable if someone is suspected of having been exposed to a dangerous antigen. A good example of this is in the treatment of tetanus, also known as lockjaw. Tetanus results from a toxin, produced by a microorganism, which causes all the muscles to go into spasm (or tetanus), making swallowing and breathing impossible and so causing death. People who may have been exposed to the tetanus microorganism, for example from a deep cut whilst gardening or whilst working with horses, are injected with antibodies against tetanus extracted from the blood of horses. This prevents the development of the disease in the short term, but does not give prolonged immunity.

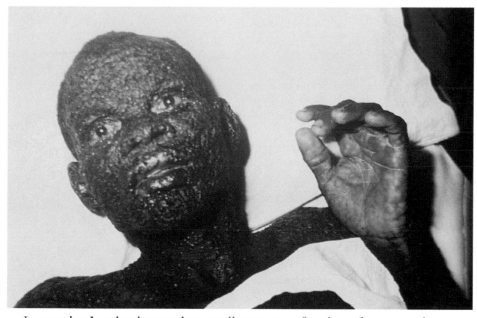

Figure 1.6.8 Smallpox has been officially eradicated from the globe by an intensive vaccination programme.

In **acquired active immunity**, small amounts of antigen (known as the **vaccine**) are used to produce immunity in an individual. The antigen is not usually the normal live microbe, as this might have fatal results. It is made safe without reducing its ability to act as an antigen. This can be done in a number of ways. If a toxin produced by the microorganism causes the symptoms, a **detoxified** form that cannot produce the symptoms will be injected. Sometimes dead viruses or bacteria are used as vaccines, and in other cases **attenuated** organisms (living but modified so they cannot produce disease) are used. The body will then produce antibodies against the antigen, and appropriate memory cells will be formed. Should the individual subsequently come into contact with the active antigen, the body will destroy the antigen without experiencing the symptoms of the disease it causes. Figure 1.6.9 illustrates the process.

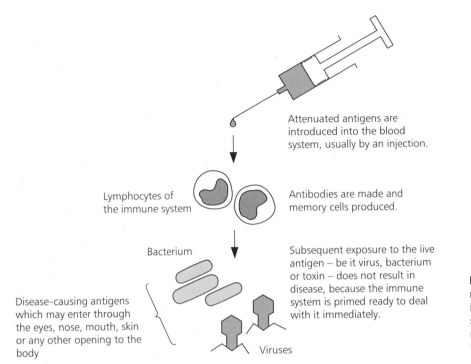

Attenuated antigens are introduced into the blood system, usually by an injection.

Lymphocytes of the immune system

Antibodies are made and memory cells produced.

Bacterium

Subsequent exposure to the live antigen – be it virus, bacterium or toxin – does not result in disease, because the immune system is primed ready to deal with it immediately.

Disease-causing antigens which may enter through the eyes, nose, mouth, skin or any other opening to the body

Viruses

Figure 1.6.9 The implications of this artificial manipulation of the immune system are enormous. The introduction of vaccines worldwide gives hope that some of the epidemics seen in the past, leaving millions dead and crippled, may one day be consigned to history.

BLOOD GROUPS

The ABO system

A specialised and unique aspect of the immune response is seen in the human blood. There is a variety of different **blood groups**, but the one most familiar to us and in most common usage is the **ABO system**. There are four possible blood groups within this system.

On the surface of the red blood cells of any individual are molecules which act as antigens. In this case they are mucopolysaccharides known as **agglutinogens** and there are two different ones, **A** and **B**. There are two antibodies (**agglutinins**) in the blood plasma called **a** and **b**. These antibodies are not made in direct response to an antigen as normal antibodies are – they are present regardless of any exposure to the antigen. If red blood cells carrying a particular agglutinogen come into contact with plasma containing the complementary agglutinin, the reaction between them causes the red blood cells to **agglutinate** (stick together). This means that they can no longer do their job and are likely to block capillaries or even larger vessels. Table 1.6.1 shows the agglutinogens and agglutinins present in the different blood groups.

Blood group	Agglutinogen (antigen on red blood cell)	Agglutinin (antibody in the plasma)
A	A	b
B	B	a
AB	A and B	–
O	–	a and b

Table 1.6.1 The ABO system of blood groups

In the UK group O is the most common blood group (46% of the population), with group A a close second (42%). Groups B (9%) and AB (3%) are much less common, although within specific ethnic groups the proportions may vary quite markedly.

Why does blood group matter?

Imagine an accident victim with blood group A who has haemorrhaged severely and needs several pints of blood. Figure 1.6.10 shows which blood groups are compatible. Blood from donors of groups B or AB cannot be given to the victim. Group A blood has the plasma antibody b. If this comes into contact with antigen B on the surface of red blood cells, the plasma would cause cell agglutination.

Recipient / Donor	O (Antibodies a and b)	A (Antigen A, antibody b)	B (Antigen B, antibody a)	AB (Antigens A and B)
O (Antibodies a and b)	✓	✓	✓	✓
A (Antigen A, antibody b)	✗	✓	✗	✓
B (Antigen B, antibody a)	✗	✗	✓	✓
AB (Antigens A and B)	✗	✗	✗	✓

Figure 1.6.10 Correct matching of the blood makes the difference between life and death when a blood transfusion is needed.

For most of the population most of the time, blood groups have little importance. As long as your blood remains within your own body system and you avoid major blood loss or the need for surgery, it is quite possible that no one will know your blood group. But blood group information becomes vital when blood needs to be given from one person to another. Blood donors' blood groups are checked and the donated blood is clearly labelled. Before undergoing surgery or giving birth, or in any situation where the possibility of a blood transfusion arises, the blood group needs to be ascertained.

When a blood transfusion is given, it is the cells of the donated blood that will be affected by any adverse reaction. The blood of the recipient is usually present in much greater amounts and so any adverse reaction is relatively minor. But if the donated blood agglutinates as it flows into the recipient's system, then difficulties can arise.

Blood group O can be given to anybody in a transfusion. Because it carries no antigens on its red blood cells it cannot stimulate an agglutination reaction. It is often referred to as the **universal donor**. On the other hand, group AB blood cannot be given to anyone other than an AB recipient, because all the other blood groups contain antibodies in their plasma which would cause some level of agglutination. However, people with blood group AB can receive any type of blood as their plasma contains no antibodies – so group AB is known as the **universal recipient**.

The rhesus factor

The **rhesus factor** is an agglutinogen (antigen) which is found on the surface of some red blood cells whatever their ABO grouping. In fact, 85% of us possess this particular feature of our red blood cells and are known as **rhesus**

positive. The remainder, who do not have the agglutinogen, are **rhesus negative.** Normally neither rhesus positive nor rhesus negative blood possesses rhesus agglutinins in the plasma. Rhesus positive blood never forms rhesus agglutinins, otherwise it would coagulate itself. But in certain circumstances, rhesus negative blood will form plasma agglutinins, usually during pregnancy if a rhesus negative mother carries a fetus that is rhesus positive.

In theory this should not matter at all. Every fetus is genetically different from its mother. By some suspension of the immune system which we do not fully understand, this foreign genetic material is allowed to grow and thrive in the uterus for some 40 weeks without being attacked and destroyed. The placenta forms a barrier between the cells of the mother and the cells of the fetus so that an inappropriate immune response is not triggered. However, the placenta leaks. This leakage is usually only slight in the first pregnancy, but a few fetal red blood cells will get into the bloodstream of the mother. If they are of a different ABO blood group they will probably be destroyed. However, if mother and fetus have the same ABO group or a compatible one, the fetal blood cells will survive long enough to stimulate the production of rhesus antibodies in the mother's plasma.

In subsequent pregnancies the placenta gets progressively more leaky. The build-up of maternal agglutinins to any further rhesus positive fetal blood cells is much larger and more rapid. These agglutinins can cross the placenta. The red blood cells of the fetus are then attacked and agglutinated. In the past this led to many babies dying, either before they were born or shortly afterwards. In more recent years, once the rhesus incompatibility was recognised, affected babies were given transfusions as soon as they were born. Now blood transfusions are carried out whilst the fetus is still within the uterus.

It is worth noticing that rhesus incompatibility only matters in one direction – that is, if the mother makes antibodies against the fetus. If the developing baby is rhesus negative and the mother rhesus positive, the same thing will happen in reverse. But the amounts of antibodies the fetus makes against its mother's blood are so tiny that their effect is not noticed.

In order to prevent the necessity for intrauterine blood transfusions, women who are rhesus positive and have carried one rhesus negative child are usually given an injection shortly after the birth. This contains anti-rhesus agglutinins and is known as anti-D. It prevents the antibody-forming process, which means that a second fetus is at no higher risk than the first.

TRANSPLANTS AND THE IMMUNE SYSTEM

In recent years medical advances have highlighted the immune system in a new way. The development of the transplantation of organs and tissues from one person to another has led to much work being done on the immune system and its suppression. However closely the tissues of a donor organ – for example, a kidney – are matched to the tissues of the patient who is to receive the transplant, a perfect match is not possible unless the donor is an identical twin. This means that, to a greater or lesser extent, the immune system of the recipient will set out to destroy or **reject** the donated organ.

The problem is how to prevent the recipient from rejecting the transplanted organ without reducing the ability of the immune system to the extent that the patient dies from a succession of infections which the body cannot fight. Rejection is prevented by a cocktail of **immunosuppressant drugs** which endeavour to get the balance right. Transplant patients have to take these drugs for the rest of their lives.

The immune system is capable of 'holding off'. As we have already mentioned, during the development of the fetus the mother's immune system does not destroy this foreign genetic material. At an even earlier stage of reproduction, sperm are allowed to travel and live within the female reproductive tract for days at a time without triggering an immune reaction. Once scientists can understand how these suspensions of the normal immune response are brought about, they may be in a position to develop a more specific and effective way of preventing rejection.

SUMMARY

- Unicellular organisms have a large surface area:volume ratio and can gain the materials they need by diffusion. However, they need to recognise potential food and recognition may also be necessary for sexual reproduction.

- Multicellular organisms have a variety of specialised cells so cell recognition is needed for the organism's organisation, and also to deal with invasion by foreign cells.

- Cells are recognised by glycoproteins on their cell surface membranes. In a foreign organism, these glycoproteins are called **antigens**. The **immune response** enables the body to recognise and remove foreign antigens.

- **Lymphocytes** are white blood cells that each recognise one particular antigen. There are two types of lymphocytes – **T-cells** and **B-cells**.

- T-cells bring about the **cell-mediated response** to invasion by a foreign antigen. A T-cell binds to and destroys its complementary antigen and then divides to produce a clone of **killer T-cells** which all destroy the same antigen.

- B-cells bring about the **humoral response** by producing **plasma cell clones** and **memory cells**. Plasma cell clones produce **antibodies** which bind to the antigens and destroy them. Memory cells convey **immunological memory** so that the next encounter with the antigen results in rapid production of plasma clone cells.

- Antibodies are proteins known as **immunoglobulins** which prevent pathogens entering host cells, cause antigens to **agglutinate** (clump together) so facilitating their digestion by phagocytes, and stimulate histamine production.

- **Natural active immunity** results from activation of the immune system. **Natural passive immunity** results from absorbing antibodies from another individual. It is passed from mother to child before and shortly after birth, and is short lived.

- Immunity may be induced artificially. **Acquired passive immunity** results from the injection of ready-made antibodies, for example to tetanus. **Acquired active immunity** is gained in response to injection with a vaccine containing an attenuated organism or detoxified toxin which stimulates the immune response without producing symptoms of the disease.

- The **ABO blood system** groups people according to antigens or **agglutinogens** (A and B) found on the red blood cells, and antibodies or **agglutinins** (a and b) found in the blood plasma. The blood groups are A, B, AB and O.

- In a blood transfusion, donated blood with particular agglutinogens cannot be given to a recipient who has complementary agglutinins, as the donated blood would agglutinate. Group O is the **universal donor** since it lacks any agglutinogens, and group AB is the **universal recipient** since it lacks any agglutinins.

- The **rhesus factor** is an agglutinogen found on the red blood cells of **rhesus positive** people. If a **rhesus negative** mother carries a rhesus positive child, she may develop agglutinins in her plasma to any fetal red blood cells that leak through the placenta. Subsequent pregnancies with further rhesus positive fetuses may result in the mother's agglutinins causing the fetal blood to agglutinate.

- Patients who receive transplanted organs have to take **immunosuppressant drugs** to prevent the donated organ being rejected.

QUESTIONS

1 Why is cell recognition important in
   a simple organisms
   b complex multicellular organisms?

2 When the body is invaded by a foreign organism, it responds by identifying and destroying the invader. This is brought about by the lymphocytes in a two-pronged attack involving a *cell-mediated response* and a *humoral response*.
   What is
   a the cell-mediated response
   b the humoral response?

3 Explain the differences between
   a natural active immunity
   b natural passive immunity
   c acquired active immunity
   d acquired passive immunity.

4 What is meant by the human ABO blood system? What is the importance of the blood groups in the context of a blood transfusion between one individual and another?

Questions

1 'The electron microscope has transformed our understanding of cell structure, and has made major contributions to biological research.' Discuss this statement. **(20 marks)**

(ULEAC January 1992)

2 Mammalian red blood cells were added to two different solutions of sodium chloride. Photograph A below shows the cells in 0.9% sodium chloride solution and photograph B shows the cells in a sodium chloride solution of a different concentration.

A

B

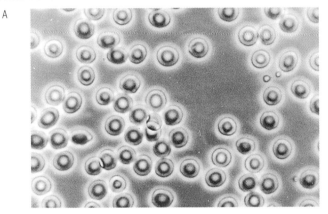

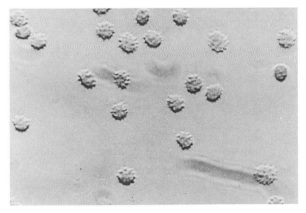

a i Describe the appearance of the cells in the photographs A and B. **(2 marks)**
ii Explain what has caused the differences you have observed. **(3 marks)**
b A similar sample of red blood cells was added to distilled water and examined under a microscope. No cells were seen. Explain this result. **(3 marks)**

(ULEAC June 1990)

3 The figure shows a three-dimensional model of the enzyme lysozyme, a small globular protein of 129 amino acids.
a Explain what determines the three-dimensional shape of such a protein. **(4 marks)**

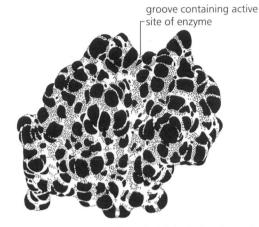

groove containing active site of enzyme

b Briefly explain, with reasons, how the biological activity of such an enzyme may be affected by:
i changes in pH,
ii increasing temperature. **(10 marks)**
c Suggest *two* consequences of the shape of the enzyme lysozyme for its function. **(2 marks)**
d State precisely where protein synthesis takes place in the cell. **(1 mark)**
e Describe briefly the role of the nucleus in directing protein synthesis. **(5 marks)**

(Cambridge June 1991)

4 a State what is meant by each of the following terms.
i Peptide linkage **(2 marks)**
ii Conjugated protein **(2 marks)**
b Give the role in protein synthesis of each of the following.
i Transfer RNA **(3 marks)**
ii Ribosomes **(3 marks)**

(ULEAC January 1991)

5 An enzyme catalysing the breakdown of certain types of cell walls was found in two different forms, called X and Y. The activity of the two forms of the enzyme was investigated. The graph shows the results of an experiment in which the two forms of the enzyme were added separately to suspensions of the cell wall material. All results were obtained at pH 7.5 and 22 °C, with standard volumes and concentrations of substrate and enzyme.

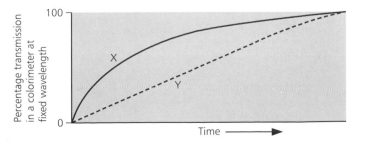

		86	87	88	89	90	91	92	93	
protein X	N terminal . . . end	Lys	Ser	Pro	Ser	Leu	Asn	Ala	Ala	. . . C terminal end
DNA sequence for protein X		TTT	TCA	GGT	AGT	GAA	TTA	CGA	CGA	
mRNA sequence for protein X										
protein Y	N terminal . . . end	Lys 86	Val 87	His 88	His 89	Leu 90	Met 91	Ala 92	Ala 93	. . . C terminal end

a i Explain how change in light transmission through the enzyme–substrate mixture allows the course of the reaction to be followed.

ii Suggest why the cell wall suspensions were shaken thoroughly before being placed in the colorimeter. **(3 marks)**

b Which of the following, **A** to **D**, would provide the best control for the investigation? Explain your answer.

A enzyme and water **C** boiled enzymes and water
B cell walls and water **D** cell walls and boiled enzymes **(2 marks)**

The amino acid sequences of the two proteins (enzymes X and Y) were determined. Each contained 136 amino acids joined together in a single chain. The only differences in sequence were noticed in a central part of the proteins. In the diagram above, the amino acids concerned are given, together with their numbered individual positions in the proteins. The sequence of 'bases' (nucleotides) corresponding to this central sequence is given for protein X.

c Copy the grid above and enter the appropriate mRNA sequence for this central part of protein X. **(2 marks)**

Further investigations showed that enzyme Y was the result of two mutations to the DNA coding for enzyme X. In the first mutation a single nucleotide was deleted (lost) from the DNA. In the second mutation a single nucleotide was inserted (added) to the DNA.

GENETIC DICTIONARY
(giving only the mRNA codons of the amino acids shown in X and Y)

Amino acid	Codons which can be used
Lys	AAA
Ser	AGU or UCA
Pro	CCA
Leu	CUU or UUA
Asn	AAU
Ala	GCU
Val	GUC
His	CAU or CAC
Met	AUG

d Using the genetic dictionary provided here, work out as accurately as you can

i the sequence of bases in the mRNA coding for protein Y. (copy the grid)

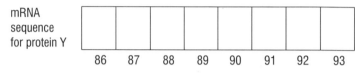

mRNA sequence for protein Y

86 87 88 89 90 91 92 93

ii the identity of the **DNA** base (nucleotide) deleted in the first mutation.

iii the identity of the **DNA** base (nucleotide) inserted in the second mutation. **(4 marks)**

e Using the information provided in the question, suggest a reason for the difference in the enzymic activities of X and Y. **(2 marks)**

(NEAB June 1991)

6 a Describe the structure of

i the cell wall, and

ii the cell membrane in plant cells.

Include in your answer the arrangement of the chemical constituents and explain how they affect the functions of the two structures. **(12 marks)**

b Discuss the role and distribution of membranes in

i mitochondria,

ii endoplasmic reticulum,

iii nucleus. **(8 marks)**

(NEAB June 1991)

7 a Relate the structure of DNA to its functions. **(9 marks)**

b How do the structures of mRNA and tRNA differ from that of DNA? Relate these differences in structure to the functions of mRNA and tRNA. **(6 marks)**

c Explain how the constituent amino acids contribute to the final three-dimensional shape of a protein molecule. **(5 marks)**

(NEAB June 1989)

8 A student was asked to find the osmotic potential of onion cell sap by the method of incipient plasmolysis. Strips of epidermal cells were removed from an onion bulb and put on to separate microscope slides. A series of sucrose solutions of different molarity was then made and sterilised. As soon as a drop of one solution was added to a strip of epidermal cells, it was viewed immediately under the microscope in order to count the number of plasmolysed cells. This procedure was repeated for each concentration of sucrose.

a i Give **two** ways in which the practical technique used was incorrect or unnecessary, explaining why in each case.

ii Given distilled water and a molar solution of sucrose, describe how you would make up 10 cm^3 of each of a 0.3 M and a 0.2 M solution.

iii Describe how a plasmolysed cell could be distinguished from a turgid cell. **(7 marks)**

The experiment was carried out again in an improved form and gave the results shown in the graph opposite.

b From the graph, what molarity of sucrose has the same osmotic potential as the onion cell sap? **(1 mark)**

c From the results of such experiments it was calculated that the osmotic potential of the onion cell sap is −1050 kPa. Explain what is meant by the term osmotic potential, and why it is a negative value. **(2 marks)**

(NEAB June 1989)

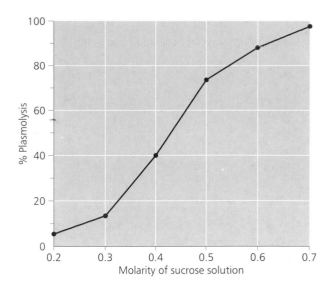

SYSTEMS
OF LIFE
2

All systems go – the physiology of exercise

When Roger Bannister first ran a mile in under 4 minutes, sport was very much an amateur business, with systematic training regarded as somehow 'ungentlemanly'. Thirty years ago anyone jogging, weight training or taking regular exercise classes would have been regarded as a health freak. Today, although many people are still relatively inactive, the idea that exercise is good for us has taken a hold in the population – it has been estimated that around 20 million Americans go jogging or running every day! And for the serious athlete, rigorous training is now an absolute necessity to compete successfully in top level competitions

Why exercise?

Our bodies are made up of a number of interacting systems supplying us with all our basic physical needs. In many areas of the world it takes all the resources and ingenuity of the population to eke out sufficient food to keep themselves alive and to produce families. In the developed world an adequate food supply is available to most people, in fact for many of us too much food is the problem rather than too little. We have the time and resources to consider additional elements of existence and one of these is the development of **physical fitness**.

What do we mean by physical fitness? It can be considered as 'the capacity to meet successfully the pre-

Figure 1 Sporting activities are undertaken by people of every age, and at every level. But what effects do the demands of exercise have on the human body?

sent and potential challenges of life'. In other words, our body systems are ideally in a condition to cope not only with the physical demands of our current way of life but also with any increased demands which might be placed upon us. Most animals maintain physical fitness simply through their natural way of life. However, the average human being in Europe and America lives a far more sedentary existence than our evolutionary ancestors and so needs to make an effort to attain a reasonable level of fitness. The value of exercise is that it allows us to build up physical fitness, and at the same time reduce our likelihood of developing some diseases.

Responding to exercise

At rest the systems of the body provide the cells with sufficient food and oxygen to maintain the basal metabolism – to survive and grow. The skeleton and muscles support the body and allow for movement, and air is moved in and out of the lungs, providing oxygen and removing carbon dioxide. The digestive system breaks down the food eaten into smaller chemical molecules which can be absorbed into the blood and carried to the cells. The heart beats at a rate sufficient to carry the blood around the body, supplying the nutrients and oxygen needed and removing the waste products of metabolism before they build up to toxic levels. However, once the body begins a period of exercise the demands on the body change dramatically. The body systems need to adapt to the new demands or the body will not function.

Undertaking a single, isolated piece of physical exercise – for example, running for a bus – will be rather stressful in the short term, but in the long term it has no effect on the body. The immediate needs of the body are dealt with as well as possible, and

then the systems return to their original functioning. But a regular, sustained programme of exercise brings about modifications in the body systems, having long-term benefits to the overall health of the individual. Athletes take this process to its extremes, and frequently develop a very specific fitness suited to their own particular sport, but physical fitness is a goal which can be achieved by anyone who is prepared to make a regular effort.

The internal environment of the body is maintained in a very stable state. The levels of glucose and other substances in the blood are kept as constant as possible, the core temperature is held within very narrow limits and action is taken as soon as events threaten to disturb the status quo. This is known as homeostasis, and will be considered in some detail in section 4. For now, figure 2 summarises how the body reacts to the changes occurring when, for example, a period of exercise begins. We shall look at the effects of exercise on

three main systems – the heart, the lungs and the skeleton and muscles.

Exercise and the heart

The human heart is endowed with many attributes in popular culture. We love with all our hearts, our hearts are broken, we may be big-hearted, we make a heart-felt plea. From childhood we assume that the seat of life is in the heart, and that when our heart ceases to beat, we shall cease to be. In spite of this, most people are very careless about their hearts.

In reality the heart is not even heart shaped. It is a bag of powerful, reddish-brown muscle divided into four chambers. It begins to beat long before we are born, and whilst the cessation of the heartbeat does not necessarily mean death, the two are closely linked. The heart contracts or beats to force blood around the body, and the blood carries oxygen and food for the cells and removes the waste products of metabolism.

In the cells of the body food molecules, in particular glucose and the

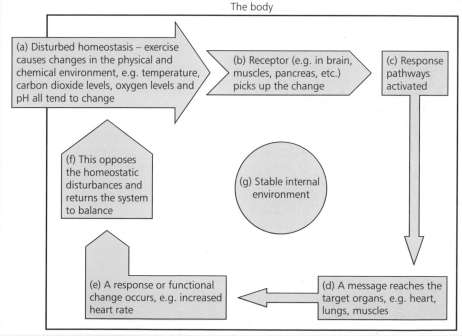

The body

(a) Disturbed homeostasis – exercise causes changes in the physical and chemical environment, e.g. temperature, carbon dioxide levels, oxygen levels and pH all tend to change

(b) Receptor (e.g. in brain, muscles, pancreas, etc.) picks up the change

(c) Response pathways activated

(f) This opposes the homeostatic disturbances and returns the system to balance

(g) Stable internal environment

(e) A response or functional change occurs, e.g. increased heart rate

(d) A message reaches the target organs, e.g. heart, lungs, muscles

Figure 2 A simplified scheme showing the physiological response of the body to a period of exercise

carbohydrate store glycogen, are broken down in the process of cellular respiration to produce energy for cellular activity and muscular contraction (see section 3). This energy is in the form of ATP. Efficient cellular respiration requires oxygen, and the waste products of the process are carbon dioxide and water. ATP is vital for the process of muscle contraction, and so during periods of exercise the demand for ATP rises steeply. As a result the demand of the cells for oxygen for respiration rises steeply too. The heart pumps blood at a higher rate to supply this oxygen, and to remove the extra carbon dioxide.

At rest in a normal individual the heart beats about 70 times and pumps between 4 and 6 litres of blood each minute. In a trained athlete this blood flow can be increased to around 30 litres a minute during exercise. The increased blood flow is brought about both by an increase in the rate of the heartbeat and by an increase in the amount of blood pumped each time the heart contracts. The response of the heart to exercise is very rapid, and when exercise is anticipated in a fit individual the heart rate will begin to increase before the exercise begins, as

shown in figure 3. The heart rate is controlled by a series of complex interactions which are considered in more detail in section 2.5. At the end of exercise in a fit individual, the heartbeat drops back to normal very rapidly. In those of us who are less fit, the pounding continues for several minutes!

A healthier heart

Heart disease affects many people, particularly men, from their late thirties onwards. Heart problems are frequently associated with the 'furring up' of the coronary arteries (arteries supplying the heart with oxygenated blood) by a fatty substance called cholesterol. This leads to an increasingly inefficient heart and can result in heart attacks and death.

As a result of increased exercise, the heart muscle becomes stronger and more efficient and the resting heart rate drops. The coronary arteries become larger and more extensive as a result of the increased blood flows they carry to the heart during exercise. Regular exercise has been shown to increase blood levels of **high-dens-**

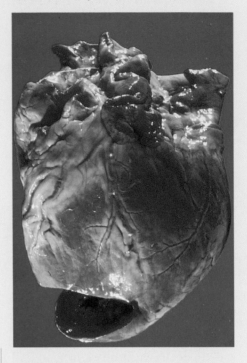

Figure 4 Whilst exercise cannot prevent coronary heart disease, it is likely to reduce it, and a fit individual is more able to overcome any heart problems which do arise.

ity lipoproteins which help to remove cholesterol from the blood, and also reduce the tendency of fatty deposits to build up on the artery walls. Most studies show that individuals who exercise regularly reduce their likelihood of suffering from heart disease, even if they have a family history of heart problems, smoke or are overweight.

Exercise and the lungs

We have already seen that the oxygen needs of the body increase with exercise. The oxygen is taken into the body through the lungs, and during exercise the breathing rate increases to provide extra oxygen. At rest the average adult breathes around 12 times a minute, taking in and out about 6 litres of air. In fit athletes this can be increased to between 45 and 80 litres in females and 80 and 100 litres in males during exercise. Regular exercise results in increased lung volume and efficiency, and an increase in the capillary blood supply to the lungs.

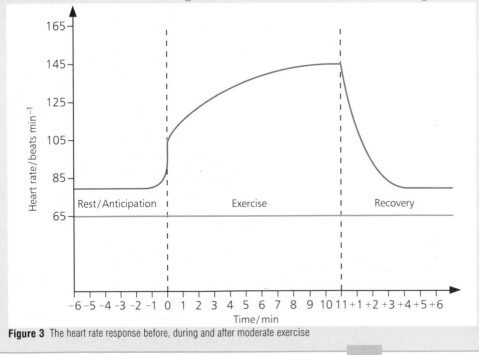

Figure 3 The heart rate response before, during and after moderate exercise

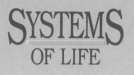

Although the breathing rate increases with the onset of exercise, it cannot respond quickly enough to the demand for more oxygen. In the first stage of exercise the muscle cells respire **anaerobically** (without oxygen), so that glucose and glycogen are not completely broken down. During this period an **oxygen deficit** builds up, when the oxygen requirement of the body is not fully met. At the end of the period of exercise the oxygen deficit has to be paid off as the **oxygen debt**, and a raised breathing rate continues until this has happened and all the lactic acid produced in anaerobic respiration has been metabolised. The size of the oxygen deficit and the time it takes to make up the oxygen debt are measures of the fitness of an individual. With increasing fitness the respiratory system responds more efficiently to exercise and so the deficit is smaller and is paid off more rapidly.

Exercise and the skeletal system

The complex system of bones and joints that make up the skeleton, along with the muscles needed to move them, are of vital importance in exercise. Vigorous exercise carried out without preparation can damage the skeletal system, but in general exercise is beneficial. The bones of the body are built up and dissolved away continuously. They are built up where stresses on them indicate that more bone mass is needed. When the bones are rarely stressed they are dissolved away. Plenty of exercise therefore tends to increase the overall bone mass. This is of great importance, particularly for women. With age, the bone mass is gradually reduced. If there is insufficient material to begin with, this erosion of the bone may result in crippling disease. Regular exercise helps to prevent this and also, when not carried out to excess, keeps the joints flexible and helps to maintain a full range of movement.

Muscles are used to move the skeleton in exercise. Different types of exercise use different muscles. When muscles are used regularly they get larger and stronger which has the effect of 'toning' the body. The development of the muscles (taken to extremes in figure 6) involves an

Figure 6 Regular exercising of specific muscles can produce spectacular results. Whilst not everyone wants a physique like this, exercise is beneficial to everyone in maintaining a flexible and supple body.

increase in the cross-sectional areas of the muscle fibres, rather than an increase in the numbers of fibres. As the proportion of the body mass made up of muscle increases, the

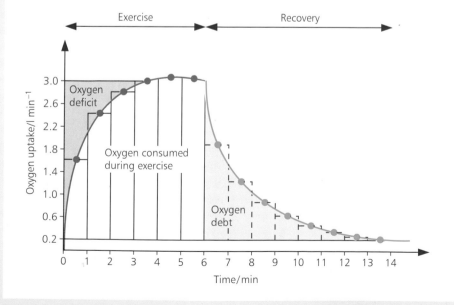

Figure 5 The oxygen debt on the graph translates into breathless agony for the athlete at the end of a race.

amount of energy used in the basic maintenance of the body increases too. Muscle cells use up more energy than fat cells. This in turn helps to reduce obesity.

Exercise and energy

In 532 BC Milo of Croton won the wrestling competition at the Olympic Games seven years in succession. It was recorded that he trained on a diet of 20 lb of bread, 20 lb of meat and 18 pints of wine each day. Thus sportsmen attempting to improve their performance with special diets has a very long history. The link between diet and exercise is relatively simple, and certainly for the majority of people attempting to improve their physical fitness the most important factor is that exercise uses up energy.

When the human body is completely at rest, lying still on a bed, it uses about 4.3 kJ of energy per hour for every kilogram of body weight. Sitting up reading a book uses marginally more – 4.45 kJ kg^{-1} h^{-1}! Many people spend large parts of the day effectively sitting still or sleeping, and spend part of the rest of the time eating. In many cases the energy taken in in the food is greater than the energy expended on staying alive. The excess is stored as fat, and increasingly large proportions of the population are carrying around with them considerable loads of adipose tissue. What is more, a substantial proportion of the population at any one time is attempting to shed this fat, usually by restricting food intake as part of a calorie-controlled diet.

Obesity is a major factor to poor health. It contributes to the development of heart disease, high blood pressure and diabetes and is a known risk factor for kidney disease, gall bladder disease and joint disease. Many other people, whilst not obese to the point that their health is at risk, would nevertheless like to carry a little less fat for reasons of vanity. Very often, weight-loss programmes are doomed to fail because they concentrate solely on the food intake, rather than on the amount of energy which is expended.

Compare the energy requirements of a sedentary person with those when exercise is undertaken (see table 1) and it is easy to see the difference. An hour on the tennis court or a brisk hour's walk demands around 25 kJ of energy for every kilogram of body weight. Steady running uses 36.5 kJ per hour and a hard game of squash around 50 kJ kg^{-1} h^{-1}. It is therefore easy to see that the development of a programme of regular exercise has benefits not only for the fitness of the cardiovascular systems and the skeletal system, but also in the loss of weight and the maintenance of that weight loss.

Is exercise a good thing?

Too much or inappropriate exercise can be very bad for you – several people die each year because they attempt to run marathons without the required training beforehand. Sporting injuries devastate the lives of some individuals each year. But in spite of this the overwhelming body of evidence suggests that moderate exercise, carried out sensibly and with a gentle build-up to fitness, carries many benefits. Several of the major body systems which will be studied in some detail in this section function better in a fit individual than in an unfit one. It seems that exercise, far from wearing us out, can actually prolong our active lives.

Activity	Energy/kJ h^{-1} kg^{-1}
Lying in bed	4.3
Sitting down, reading	4.45
Standing	5.2
Bowling	12.6
Canoeing	25.2
Cycling for pleasure	18.3
Aerobics	33.6
Golf	21.4
Squash	50.0
Football	42.0
Brisk walking	24.4

Table 1 The approximate energy costs of a variety of activities

2.1 Systems in plants: Roots

As organisms increase in complexity, becoming multicellular, evolving distinct methods of providing for their energy needs, perhaps moving around, then the levels of organisation within the organism must also become more complex. In unicellular organisms the single cell must perform all the functions necessary for life. In multicellular organisms there is scope for specialisation of function. Groups of cells become organised into systems dealing with only one or two particular aspects of life – perhaps obtaining energy, perhaps reproducing the species – whilst at the same time becoming dependent on different systems to provide their other needs. These **systems of life** will be considered in this section.

PLANT ORGANISATION

The living world can be divided in many ways, but two major groups are the **plants** and the **animals**. Plants are **autotrophic** (self-feeding). They can utilise the energy of the Sun to drive reactions involving water and carbon dioxide, with glucose as the immediate end product. The glucose can then be broken down by the process of **cellular respiration**, producing ATP as an energy supply for the other reactions of the cell, or used for the synthesis of amino acids and lipids. This process of using the Sun's energy to synthesise sugars and so provide an energy supply for the cell is known as **photosynthesis**, and is described in detail in section 3.2. Most organisation in plants is related in one way or another to this feature of plant life.

Figure 2.1.1 Multicellular organisms need oxygen and food, must get rid of waste, and have to reproduce themselves. They have evolved specialised systems made up of highly modified cells to carry out these functions.

The needs of plants

When considering the systems of any living organism it is important to look at the demands made by the way of life of that organism. When looking at a group as extensive as the plants, generalisations inevitably have to be made, but these do hold true for the majority of the group. Most of the points below are particularly relevant to the multicellular plants.

(1) Plants need to be anchored in some way to give a base for the photosynthetic part of the plant and to allow the plant to take advantage of a particular habitat.

(2) Plants must be able to capture as much sunlight as possible for photosynthesis.

(3) Plants must be able to obtain all the water, carbon dioxide and minerals that they need, and also get rid of any toxic waste products.

(4) For sexual reproduction to occur, male and female sex cells must be brought together and the seeds protected and nurtured as they develop and then dispersed to give the maximum possible chance for the species to continue.

Most multicellular plants show some differentiation into at least two, and often three, distinct permanent regions. One area is involved in anchorage and often also in the uptake of water and minerals. Another region may act as a support for the third, which consists of the major photosynthetic organs. These divisions are more commonly referred to as **roots**, **stems** and **leaves**, shown in figure 2.1.2. Many plants also develop further specialised reproductive areas (flowers, spore capsules, fruit, etc.) at certain times of year. These will be considered in more detail in section 5.

THE ROOT SYSTEM

The need for roots

Plants need to trap as much energy from the Sun as possible to make food by photosynthesis. In order to do this they generally have a very large surface area. This in turn makes them very vulnerable to water loss by evaporation. To reduce this water loss the plant is protected from dehydration by a waxy covering. This means that water cannot be absorbed through the exposed surfaces of the plant and so alternative methods of obtaining water have evolved – most commonly, the **roots**. These are structures which contain **vascular** (transport) tissue.

Some plant groups such as the liverworts and mosses (see section 6.3) do not have a waterproofing layer of wax over their leaves. This means that they can absorb water over the entire plant surface and do not have a true vascular root system. Instead they have a system of **rhizoids**, simple root-like structures whose major role is that of anchoring the plant. If you have ever tried to prise moss off a wall, you will know that they are very successful at this. This lack of vascular tissue restricts the distribution of liverworts and mosses to moist or wet habitats.

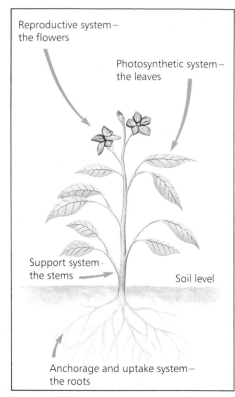

Figure 2.1.2 The main systems in a generalised plant. They are all interlinking and interdependent to give a coordinated and successful organism.

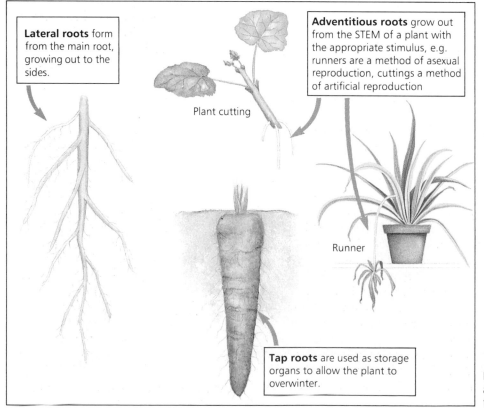

Lateral roots form from the main root, growing out to the sides.

Plant cutting

Adventitious roots grow out from the STEM of a plant with the appropriate stimulus, e.g. runners are a method of asexual reproduction, cuttings a method of artificial reproduction

Runner

Tap roots are used as storage organs to allow the plant to overwinter.

Figure 2.1.3 Root systems come in a variety of shapes and forms to enable them to perform their variety of functions.

True root systems

Most land plants have a **true root system**. This contains cells specialised for the uptake and transport of water and minerals from the soil. The root system of a plant is usually very large – often at least as extensive as the visible plant above it. The more extensive the root system of a plant, the larger will be the surface area available for the uptake of water and minerals from the soil. Large root systems are also very effective at anchoring plants. Figure 2.1.3 shows the different types of roots.

Roots are very important to the structure of the soil. The extensive network of plant roots traps soil, binding it together and helping to prevent erosion by wind or rain.

STRUCTURE OF THE ROOT SYSTEM

The cells which make up true root systems are specially adapted for their functions. One of their most striking features is that, under normal conditions, they do not contain chloroplasts or the green pigment chlorophyll which is usually associated with plant cells. As little or no light penetrates the soil, photosynthetic pigments would be of no use to root cells. There are several different types of cells within the root system, each adapted to its particular function. To enable them to work effectively they are arranged in a distinctive way.

The root system involves a complex, branching network to give a large surface area, and to maximise this surface area the roots are covered in microscopic **root hairs**, extensions of the outer cells. Plants cannot move around to find minerals in the soil, but the growth of these roots and root hairs allows them to exploit the soil minerals over a wide area. They are the equivalent of a 'foraging strategy' for plants. The outer layers of cells are permeable to water. Within the roots the tissues can be divided into two main types. Much of the root – indeed, much of the plant – is made up of supportive packing tissue. This largely consists of the most common type of plant cells, known as **parenchyma** cells. These cells do not need particular specialisation to carry out their supportive role, but they can be modified in a variety of ways to make them suitable for other functions such as storage and photosynthesis.

The remainder of the root consists of vascular tissues, the **phloem** and **xylem**. The phloem is living tissue which carries the dissolved products of photosynthesis around the plant. Root cells need energy to grow, divide and live. Glucose is made in the leaves, converted to sucrose and transported by the phloem to areas of the plant where it is needed. The flow within phloem tissues can go either up or down the plant. The xylem carries water and dissolved minerals from the soil to the photosynthetic parts of the plant and the movement within xylem tissues is always upwards under normal circumstances. Xylem starts off as living tissue, but increasing amounts of **lignin** (woody material) are laid down in the cell walls to make the tissue stronger and more supportive, and the living contents of the cells die. This means that most of the functioning xylem in a plant is dead. A simple picture of the structure of a typical root is given in figure 2.1.4. Figure 2.1.5 builds on this basic picture of the root to show us exactly where the different tissues are found, what they look like and what they do.

The flowering plants contain two groups – **monocotyledons**, in which the embryo has one seed leaf, and **dicotyledons**, in which it has two. This will be considered in more detail in section 5.3. Figure 2.1.5 shows the structure of a dicotyledonous root.

The functions of roots

The cells that make up a root system have to be able to carry out a variety of functions, listed below, and several different types of cells have evolved for this purpose.

(1) Roots take up water from the soil.

(2) Roots move mineral salts from the soil water into the plant cells, often against a concentration gradient.

(3) Roots anchor and stabilise the plant to allow for the effective functioning of the photosynthetic systems.

(4) Roots need to be flexible and able to withstand the strong pulling forces exerted when the wind blows without snapping and destroying the plant.

(5) In some plants, roots form the means by which the plant stores the food it needs over winter.

(6) In some plants, roots form a means by which the plant reproduces.

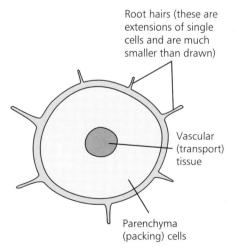

Root hairs (these are extensions of single cells and are much smaller than drawn)

Vascular (transport) tissue

Parenchyma (packing) cells

Figure 2.1.4 The arrangement of tissues in the root system

Epidermis

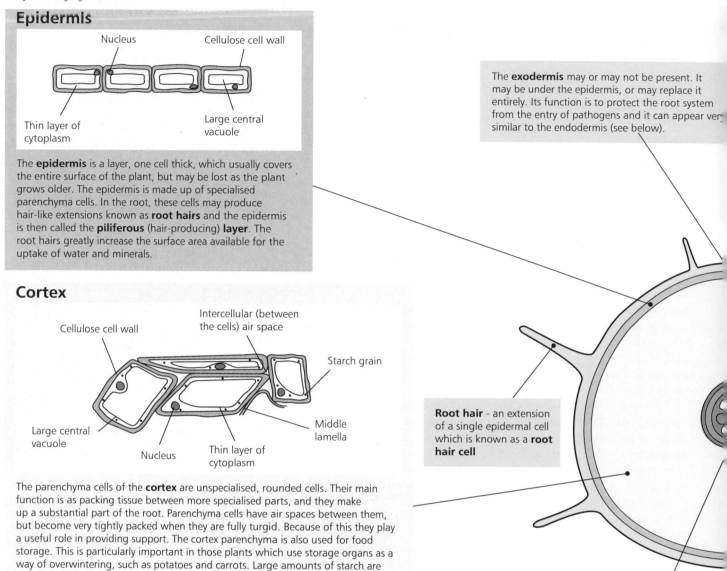

The **epidermis** is a layer, one cell thick, which usually covers the entire surface of the plant, but may be lost as the plant grows older. The epidermis is made up of specialised parenchyma cells. In the root, these cells may produce hair-like extensions known as **root hairs** and the epidermis is then called the **piliferous** (hair-producing) **layer**. The root hairs greatly increase the surface area available for the uptake of water and minerals.

The **exodermis** may or may not be present. It may be under the epidermis, or may replace it entirely. Its function is to protect the root system from the entry of pathogens and it can appear very similar to the endodermis (see below).

Cortex

The parenchyma cells of the **cortex** are unspecialised, rounded cells. Their main function is as packing tissue between more specialised parts, and they make up a substantial part of the root. Parenchyma cells have air spaces between them, but become very tightly packed when they are fully turgid. Because of this they play a useful role in providing support. The cortex parenchyma is also used for food storage. This is particularly important in those plants which use storage organs as a way of overwintering, such as potatoes and carrots. Large amounts of starch are then packed away in organelles called **amyloplasts**. This ability to store food has been utilised by humans for thousands of years to provide a valuable food source.

Root hair - an extension of a single epidermal cell which is known as a **root hair cell**

Figure 2.1.5 This section through a young (primary) dicotyledonous root shows the main elements of a typical root system. The details may vary in monocotyledons or in older plants, but the main types of cells found in roots, along with the modifications they show for their particular function, are displayed here.

Endodermis

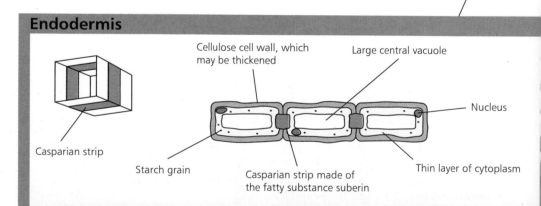

The **endodermis** is a layer of cells surrounding the vascular tissue of the roots. It is particularly noticeable in roots because of the **Casparian strip** which develops. The Casparian strip is a band of fatty material called **suberin** which runs around each cell, and its possible role in the regulation of the movement of water and minerals across the root is discussed on page 98. The starch grains found in the endodermis cells may be important in the way that roots respond to gravity (**geotropism**).

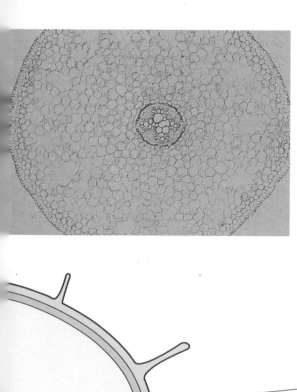

Xylem

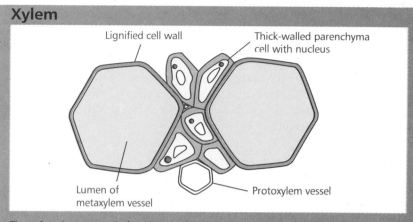

Lignified cell wall

Thick-walled parenchyma cell with nucleus

Lumen of metaxylem vessel

Protoxylem vessel

The **xylem** has two main functions in the plant – the transport of water and mineral ions, and support. In roots it is the transport function which is the most important. Xylem is made up of several different types of cells, most of which are dead when they are functioning within the root. The xylem vessels are long tubular structures made by several cells fusing end to end by the breakdown of their end walls. There is no equivalent of the phloem sieve plates – xylem vessels are just hollow tubes.
The first xylem to form is called **protoxylem**. It is capable of stretching and growing because its walls are not fully lignified. However, once growth has stopped **metaxylem** is formed. The walls are fully lignified and the cell contents die. Water moves out of the xylem into the surrounding cells either through unlignified areas or through specialised **pits**. In the root the xylem is positioned centrally, allowing it to perform a supportive role by helping the plant cope with the tugging strains imposed on it by the leaves and the stems as they bend in the wind. The only living cells found in xylem tissue are **xylem parenchyma** which stores food, holds deposits of tannins and various crystals and is involved in the transport of various substances. There are two other dead elements found with the xylem – **tracheids**, elongated lignified single cells more common in primitive plants, and **fibres**, important for additional mechanical strength.

The cells of the **pericycle** are again parenchyma cells, similar to the cells of the cortex, acting as packing tissue between the endodermis and the vascular tissue. The cells of the pericycle can divide to form new lateral (side) roots if given the appropriate stimulus.

Phloem

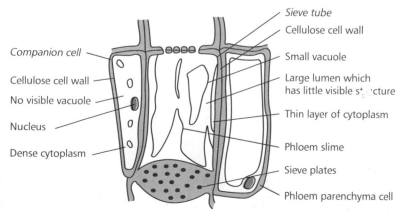

Companion cell

Cellulose cell wall

No visible vacuole

Nucleus

Dense cytoplasm

Sieve tube

Cellulose cell wall

Small vacuole

Large lumen which has little visible structure

Thin layer of cytoplasm

Phloem slime

Sieve plates

Phloem parenchyma cell

The **phloem** makes up a system of tubes which transport organic solutes such as sucrose around the plant. It supplies the root system with the raw material for cellular respiration. Phloem tissue is basically made up of **sieve tubes**, cells which have fused end to end at **sieve plates** to form continuous tubes. As sieve tubes have no nucleus when mature, they are supported and kept alive by **companion cells**. Sieve tubes and companion cells form the functional unit of the phloem. The phloem is supported by **phloem parenchyma cells** – these look very like other parenchyma cells but are elongated to fit alongside the phloem cells. Also giving mechanical support and protection are **fibres** and spherical **sclereids**. These are dead because their walls are heavily lignified.

The root system of a plant can thus be seen to be a collection of a variety of cells, many of which have evolved to carry out very specialised functions. As you have seen, the large quantities of packing cells and the arrangement of the vascular tissue in the centre of the root are particularly important in the role of support. But how do the various elements of the system come together to perform the other major function of the roots – the uptake of water and mineral ions from the soil?

THE ROOT SYSTEM AT WORK

The uptake of water

An adequate supply of water is of prime importance to a plant. It is needed for photosynthesis. It is also important to maintain turgor in the cells, ensuring support of the tissues by the parenchyma cells, and to prevent wilting so that the leaves are positioned correctly for photosynthesis. (Note that wilting is *not* simply loss of turgor – when a plant wilts whole cells shrink. However, unless the wilting continues for a long time, no permanent damage is done.) Water is also important for the transport of organic solutes and mineral salts around the body of the plant.

There is a thin film of water containing minute traces of dissolved minerals around soil particles. This water is absorbed mainly by the younger parts of the roots, where the majority of the root hairs are found as the epidermis is still present. These microscopic hairs increase the surface area of the roots enormously, and they also come into very close contact with the soil particles. At its simplest, uptake of water by the roots depends on the water potential gradient (see section 1.4, page 54) across the root from the soil water to the xylem. Water moves from the soil into a root hair cell along a water potential gradient by osmosis. This makes the root hair cell contents more dilute than those of its neighbour, and water again moves by osmosis and so on across the root, as figure 2.1.6 shows.

The picture is in fact rather more complicated than this simple model suggests. The water potential gradient across the root from the piliferous layer to the cells closest to the xylem is the result of two factors – firstly, water is continually moved up the xylem, lowering the water potential of the xylem contents, and secondly, the xylem contents have a more negative solute potential than the soil water. The water does not simply flow from one cell to another – there appear to be three alternative routes which may be taken, as shown in figure 2.1.7. The **vacuolar pathway** allows water to move by osmosis

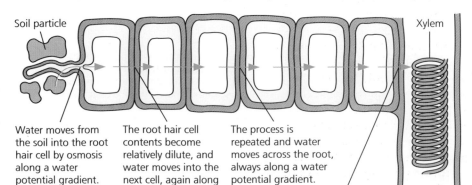

Soil particle | Xylem

Water moves from the soil into the root hair cell by osmosis along a water potential gradient.

The root hair cell contents become relatively dilute, and water moves into the next cell, again along a water potential gradient.

The process is repeated and water moves across the root, always along a water potential gradient.

The movement of water from the cells into the xylem is not fully understood. It could result from active transport or could be simply the result of physical processes such as evaporation from the leaf surface.

Figure 2.1.6 A simple model of the movement of water into the root of a plant

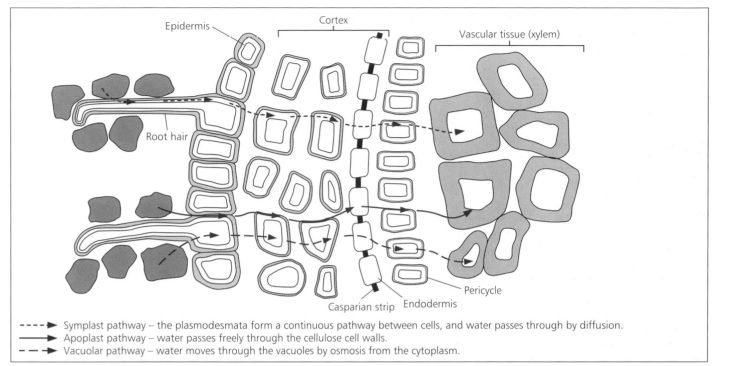

Symplast pathway – the plasmodesmata form a continuous pathway between cells, and water passes through by diffusion.
Apoplast pathway – water passes freely through the cellulose cell walls.
Vacuolar pathway – water moves through the vacuoles by osmosis from the cytoplasm.

Figure 2.1.7 The cells of the roots are organised into a system which is very efficient at taking up water from the soil.

along a water potential gradient from the soil water to the xylem across the vacuoles of the cells of the root system. In the **symplast pathway** the water moves along the same water potential gradient, but this time across the interconnected cytoplasm (**symplast**) of the cells of the root system. The **apoplast pathway** is the movement of water through adjacent cell walls (the **apoplast**) from the root hair cell to the xylem. The movement of water through this pathway is largely the result of the transpiration pull, explained in section 2.3.

A closer look at water movement in the root

To understand the pathways for the movement of water through the root system needs a closer examination of the cellular situation.

The vacuolar pathway

The soil water has a higher water potential (it is a more dilute solution) than the cells of the piliferous layer. As a result water moves across the cell wall and cytoplasm into the vacuole of the root hair cell by osmosis. It then moves from vacuole to vacuole across the root until it reaches the cells next to the xylem, moving down a water potential gradient which is set up as follows. As water moves up the xylem it is replaced. This replacement water is drawn from the neighbouring parenchyma cells. It seems that water must be moved from the cells next to the xylem into the xylem by active means, or it may just be by the suction created by evaporation at the top of the plant. As yet no one is sure exactly how this movement of water into the xylem vessels is brought about. As water leaves these cells, their water potential becomes more negative and so water enters them from the vacuole of neighbouring cells, which in turn develop a more negative water potential and so moves water from *their* neighbours. This continues right across the root system, and as a result, a continuous movement of water is seen from the soil solution to the xylem across the vacuoles of the root cells.

The symplast pathway

The cytoplasm of adjacent plant cells is connected by strands called **plasmodesmata** which go through pores in the cellulose cell walls. Figure 2.1.8 shows these connections. Water moves from the soil across the cytoplasm (the **symplast**) of adjacent cells along the water potential gradient from the root hair cells to the xylem.

The apoplast pathway

Because of the loose, open-network structure of cellulose, up to half of the volume of the cell wall can be filled with water molecules. As water is drawn into the xylem, attraction between the molecules ensures that more water is drawn across from the adjacent cell wall, and so on. As water is taken into the root hair from the soil, the mineral ions dissolved in it move too. The water and minerals move across the cells of the root in the cell walls until they reaches the endodermis. The Casparian strips of suberin stop any further progress through the cell walls. Any water and minerals that are to continue the journey have to enter the cytoplasm temporarily to avoid the Casparian strip. As minerals often need to enter the cytoplasm against a concentration gradient, active transport must be involved. This seems to be a method of controlling the amounts of water and minerals which move from the soil into the xylem. In spite of the barrier of the Casparian strip, the end result of the apoplast pathway is again a continuous stream of water across the root to the xylem.

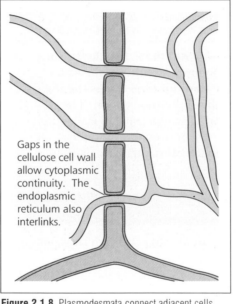

Gaps in the cellulose cell wall allow cytoplasmic continuity. The endoplasmic reticulum also interlinks.

Figure 2.1.8 Plasmodesmata connect adjacent cells through the cell wall.

The uptake of minerals

Along with its functions of support, anchorage and water uptake, the root system has also developed to absorb minerals from the soil. Although plants can synthesise their own carbohydrates by photosynthesis, they also need other molecules such as proteins and fats. To produce these they require certain minerals such as nitrates, and these they must extract from the soil.

Minerals dissolved in the soil water are carried in an unselective manner in the apoplast pathway of water uptake and move through adjacent cell walls. This means that the parenchyma cells of the root cortex are bathed in a very dilute solution of mineral ions because they are surrounded by the soil water held in the cellulose cell walls. Any ions which are needed may be moved across the cell membrane by diffusion if the membrane is freely permeable to the ion, or by facilitated diffusion or active transport.

On reaching the impermeable Casparian strip, mineral ions can no longer move through the cellulose wall. They may pass into the cytoplasm of the cells either by diffusion along a concentration gradient, or by active transport if they are being moved against a concentration gradient. In these ways the root system takes up mineral ions, including those which are required for the functioning of the plant as a whole. The water and minerals are then supplied to the other major systems of the plant.

SUMMARY

- As organisms become larger and more complex, they develop specialised systems to carry out particular functions.

- Multicellular plants need systems to carry out the functions of: anchorage; capture of sunlight for photosynthesis; taking in the water, carbon dioxide and minerals they need; sexual reproduction.

- Land plants commonly consist of stems, roots and leaves. **Roots** carry out the functions of anchorage and obtaining water and mineral salts from the soil. They may also store food for overwintering and provide a means of reproduction.

- The overall root structure is an outer layer of **epidermis**, **parenchyma** (packing) cells in the **cortex**, and central **vascular** (transport) tissue.

- **Epidermal** cells are specialised parenchyma cells, some of which contain hair-like extensions called **root hairs**.

- The **exodermis** may or may not be present; it protects the root from pathogens.

- The **cortex** parenchyma cells are unspecialised cells with intercellular spaces which disappear when the cells are fully turgid, playing a role in support. In some plants these packing cells may contain **amyloplasts** to store starch for overwintering.

- **Endodermal** cells are specialised parenchyma cells with a band of fatty **suberin** around them called the **Casparian strip**, involved in regulating the movement of water and minerals across the root.

- The **pericycle** contains unspecialised parenchyma cells between the endoderm and the vascular tissue of the phloem and xylem.

- **Phloem** vessels transport organic materials around the plant. **Sieve tubes**, long cells which fuse at **sieve plates**, have no nuclei and are supported and kept alive by **companion cells**. Phloem parenchyma cells which are elongated to fit alongside the sieve tubes support the phloem vessels.

- **Xylem** vessels transport water and mineral ions up the plant and help support the plant. They are long tubular vessels in the centre of the root, called **protoxylem** when first formed which mature into **metaxylem**. Metaxylem is **lignified** (lignin is impermeable) so its cells are dead. Living xylem parenchyma cells store food and hold deposits.

- Water passes into the root hair cells from the soil along a water potential gradient. It passes to the xylem in the centre of the root by three pathways:
  the **vacuolar pathway**, by osmosis along a water potential gradient, through the vacuoles of the root cells

the **symplast pathway**, by osmosis through the interconnected cytoplasm of the root cells

the **apoplast pathway**, by attraction between the water molecules, through the cell walls of the root cells, controlled by the Casparian strip of the endodermis.

● Minerals in the soil water are also taken up and are held in solution in the water in the plant cell walls so are available to cells and can enter by diffusion, facilitated diffusion or active transport, depending on the mechanisms available for each ion.

QUESTIONS

1 a Describe the main functions of a plant.

  b With reference to your answer to **a**, explain the need for cells to be arranged into tissues and organs in multicellular organisms.

2 The structure of a root is closely related to its functions.

  a What are the main functions of a plant root?

  b Describe the structure of a typical plant root and explain how this structure may be modified for the root to perform a variety of functions.

3 Explain the role of plant root cells in the uptake of:

  a water

  b minerals.

The root system of a plant is highly specialised to perform its particular functions. The other major plant systems, the stem and the leaf, show similar levels of organisation.

THE STEM SYSTEM

Functions of the stem

The needs of plants were discussed on page 91. The primary function of a stem system in fulfilling these needs is support. Stems hold the leaves above the ground and support them in the best position for obtaining sunlight for photosynthesis. They also support flowers to maximise the likelihood of pollination and thus sexual reproduction – sometimes this means drooping stems, sometimes stems are erect. The stem system must be flexible in order to withstand buffeting by wind and rain, and yet have the strength to remain upright. In extremes of heat plants wilt, which helps prevent excessive water loss by reducing the surface area of the leaves in direct sunlight. However, they need to be able to revive rapidly as soon as conditions are suitable again. Thus the support system has to fulfil a variety of complex and in some cases conflicting needs.

The other major function of stems is the movement of materials about the plant. The products of photosynthesis are carried from the leaves where they are formed via stems to the buds, flowers, fruit and roots where they are needed. Water moves through the stems in a steady stream from the roots up to the leaves, and the mineral ions needed for the synthesis of complex chemicals are transported in this stream.

Most stems are green – they contain chlorophyll. They carry out a small amount of photosynthesis, but this is not a major function.

Not all plants have stems. The liverworts have a simple thallus structure and the mosses have leaves which arise directly from a pad of rhizoids (see section 6.3 for more details). Neither of these groups has specialised transport tissues, and both grow close to the ground. However, the majority of land plants possess stems.

The functions of stems

As in the root system, the cells of the stem carry out a variety of functions and have become specialised.

(1) Stems provide support for the main body of the plant.

(2) Stems support leaves in positions such that they capture as much sunlight as possible, and reproductive organs in positions to maximise pollination and aid seed dispersal.

(3) Stems must be flexible and able to withstand the pulling and bending forces exerted by the wind.

(4) Stems allow the transport of water, sugars and other dissolved substances around the plant.

(5) In some plants stems form the means by which the plant stores food for overwintering.

(6) In some plants stems can form adventitious roots.

(7) In some plants stems form a means by which the plant reproduces.

(8) In some plants, such as the cactus, the stem forms the main photosynthetic organ and also stores water. The leaves have been reduced to spines to cut down water loss.

Figure 2.2.1 Stems link the leaves, the roots and the reproductive systems and provide support.

STRUCTURE OF THE STEM SYSTEM

The cells which make up a stem system are adapted for their various functions. Many of the cells have structures similar to those of the root, but their arrangement is different. Unlike the root, the outer layers of the stem should not be permeable to water, in order to prevent unwanted water loss. A waxy, waterproof layer is produced over the surface of the epidermis which is known as the **cuticle**. The outer layers of the stem may contain some chloroplasts in modified parenchyma cells, ensuring that as much food as possible is produced. The bulk of the stem is made up of supporting and packing (parenchyma) cells with the transport tissue arranged around the edge in **vascular bundles** to provide both support and flexibility in the young stems which need it.

Figure 2.2.2 gives a simple picture of the main regions of a typical stem, whilst figure 2.2.3 shows in detail the types of cells and their functions. The example used is the stem of a young, dicotyledonous plant. Although the details of arrangement vary in monocotyledons and older plants, the basic cell types and principles of arrangement of stem systems can be seen here.

Epidermis

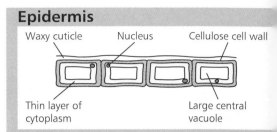

The **epidermis** is similar to that of a root but it remains throughout the plant's life. The cells secrete **cutin**, a waxy substance which helps to prevent water loss from the stem surface and to protect against the entry of pathogens. Epidermal hairs may be formed, sometimes as an extension of a single epidermal cell as in root hairs and sometimes involving several modified epidermal cells. These hairs perform a variety of functions. Some form a layer trapping moist air and helping to prevent water loss. Others are hooked and help climbing plants grip their supports. Others are protective – stiff and bristly, or loaded with irritant chemicals such as on stinging nettles.

Cortex

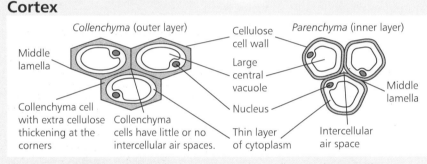

The **cortex parenchyma** cells in the stem act as packing cells, as in the root. The **collenchyma** consists of cells similar to parenchyma, but more specialised in mechanical support. The increased amounts of cellulose, particularly at the corners of the cells, and the lack of air spaces between cells, means that they are effective in providing support regardless of their level of turgor. The outer cells of this layer may contain chloroplasts so that photosynthesis can take place.

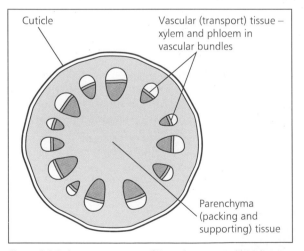

Figure 2.2.2 A stem appears very different from a root, although many of the same tissues are involved. The different distribution of the tissues allows the stem to carry out its supportive functions and yet resist the effects of the weather.

The **endodermis** is a single-celled layer that is difficult to identify unless the section is specially stained to show up the starch grains which are stored within it.

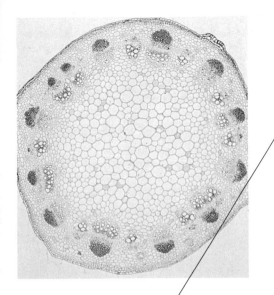

The centre of the stem is made up of **pith**. The cells are unspecialised parenchyma. They may store starch or fatty material if part of the stem forms a storage organ for overwintering, e.g. in crocus corms.

Vascular bundles

Pericycle made up of sclerenchyma fibres

Phloem – made up of sieve tubes, companion cells, phloem parenchyma and fibres

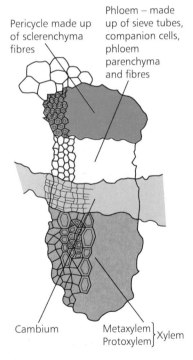

Cambium

Metaxylem
Protoxylem } Xylem

The **vascular bundles** contain the transport tissue of the stem. The xylem and phloem have the same structure and function as in the root system, with xylem carrying water and dissolved minerals up the stem and phloem moving dissolved organic substances around the plant. In contrast to the root, the transport tissue in the stem system is arranged around the outside in young dicotyledons. This contributes greatly to the support function of the stem. The xylem itself is lignified and the phloem contains strengthening fibres. In addition, the **sclerenchyma** present has a purely mechanical function. The walls are greatly thickened with deposits of lignin. This kills the cells, but gives them both tensile and compressive strength – they can resist stretching and squashing by the wind without breaking or buckling. The **cambium** is a band of unspecialised cells which divide, giving rise to more specialised cells which form both the xylem and the phloem.

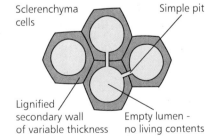

Sclerenchyma cells

Simple pit

Lignified secondary wall of variable thickness

Empty lumen - no living contents

Figure 2.2.3 The main elements of a stem system. The arrangement of tissues enables the functions of support and transport to be carried out.

THE STEM SYSTEM AT WORK

The relationship between structure and function

The arrangement of the tissues in the stem system, with large amounts of parenchyma for packing and support and the additional mechanical strength supplied by collenchyma and sclerenchyma, enables the stem to support the body of the plant. At the same time it maintains the flexibility needed to change the orientation of the leaves and flowers to follow the Sun, and to withstand the stresses imposed by the wind.

In monocotyledonous plants the vascular bundles are not arranged around the edge of the stem but are scattered more generally through the parenchyma. They still provide effective support. Plants which live for a very long time or grow particularly large may add to their support systems with **secondary thickening**, involving the laying down of **lignified** or woody tissue which gives a great deal of additional support and strength.

The presence of the vascular tissue within the stem system provides for the transport of materials. The way in which the movement of water and solutes is brought about will be considered in more detail in section 2.3.

THE LEAF SYSTEM

The photosynthetic system

All life depends on the photosynthetic system of plants. Without the oxygen- and sugar-producing capabilities of plants, animal life as we know it could not exist. Leaves have evolved to become photosynthetic factories. They contain specialised cells and have a high level of organisation which allows them to take full advantage of the prevailing conditions in order to synthesise food. In this section we shall consider how leaves provide a system arranged to make available the raw materials for photosynthesis, and to provide the 'chemical plant' required for the process itself. The biochemistry of photosynthesis will be considered later.

Of the needs of a plant already discussed, the leaf system's primary function is to capture sunlight and also take in carbon dioxide in order to carry out photosynthesis. The photosynthetic organs of the plant are highly organised to ensure that the needs for photosynthesis are met.

STRUCTURE OF THE LEAF SYSTEM

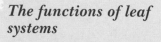

The functions of leaf systems

(1) Leaves capture light energy.
(2) Leaves contain chlorophyll which is involved in the reactions of photosynthesis, and are positioned so as to maximise their exposure to light.
(3) Leaves supply water and carbon dioxide to the site of photosynthesis.
(4) Leaves have a transport system to remove the products of photosynthesis so that they can be carried around the plant.

Figure 2.2.4 Leaves show a wide variety of adaptations for different extremes of environment. In spite of this variety, the underlying organisation of most leaves is the same.

In general, the leaves of plants offer as large a surface area as possible to the Sun. They tend to be relatively thin, so that sunlight can reach all the photosynthetic tissue. Transport tissue supplies water to the cells of the leaf and removes the products of photosynthesis. This transport tissue frequently has another role – it supports the thin leaf and thus helps to expose it to the sunlight. These features can be seen clearly by considering both the external and the detailed internal structure of a leaf, in figures 2.2.5 and 2.2.6.

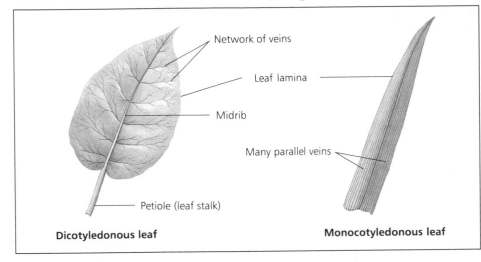

Figure 2.2.5 The external appearance of a leaf shows the overall arrangement with the transport tissues supporting the rest.

THE LEAF SYSTEM AT WORK

The relationship of structure and function

The structure of the leaf in figure 2.2.6 shows how the various tissues are arranged to allow them to function with maximum efficiency. The photosynthetic cells are closely packed near the upper surface of the leaf, closest to the Sun. The details of their functioning and the biochemistry of photosynthesis will be considered later. But the remainder of the leaf has other important functions including the transport of water, minerals and the products of photosynthesis in the xylem and phloem, as well as the support of the photosynthetic tissue of the leaf lamina. An important transport role is gaseous exchange between the cells of the leaf and the external atmosphere.

Gaseous exchange in the leaf

The cells of the spongy mesophyll are irregularly shaped and arranged with large air spaces between them. The surfaces of these cells are moist. This and the large exposed surface area means that gaseous exchange can occur freely between the cells of the leaf and the air spaces by a simple process of diffusion. During the day carbon dioxide (required for photosynthesis) moves by diffusion *into* the cells and oxygen (the main waste product of photosynthesis) moves *out*. Water passes by evaporation from the cells into the air spaces.

The inside of the leaf and the air spaces there are not a closed system. The impermeable waxy cuticle on the surface of the leaf acts as a barrier to gaseous diffusion, and particularly to the evaporation of water. A particularly thick waxy cuticle can virtually eliminate evaporation. Gases move into and out of the leaf through the **stomata**, specialised pores found mainly in the epidermis on the underside of the leaf, shown in figure 2.2.7. The pores open to allow the uptake of carbon dioxide by the leaf. The intercellular spaces are saturated with water vapour so evaporation of water takes place while the stomata are open.

Vascular bundles

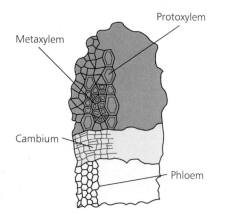

Metaxylem
Protoxylem
Cambium
Phloem

The **midrib** contains the vascular bundles and is the main vessel of transport. A large system of smaller vessels branch out from the midrib. Water and mineral salts are conducted to the leaf in the xylem and the products of photosynthesis are removed in the phloem. The vessels, and in particular the midrib, are also important as a support for the thin lamina of the leaf.

Epidermis

The **epidermis** of a leaf is made up of the same type of simple, unspecialised, flattened cells seen in root and stem epidermis. Those on the top surface of the leaf are referred to as **upper epidermis** and those on the underneath surface as **lower epidermis**. They secrete cutin, forming a waxy cuticle which helps prevent water loss through the leaves. However, gases such as carbon dioxide, oxygen and water vapour need to both enter and leave the leaf for photosynthesis and respiration. **Stomata** allow this to happen. These are specialised pores with **guard cells** on either side of the opening. Stomata are found particularly in the lower epidermis, and in some plants stomata also occur in the epidermis of the stems. Stomata can be either open or closed, depending on the prevailing conditions, and they therefore help control water loss from a plant.

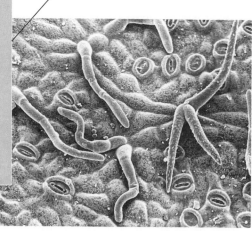

Palisade mesophyll

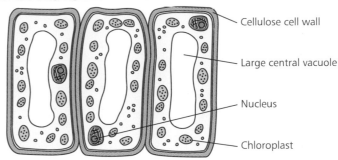

Cellulose cell wall

Large central vacuole

Nucleus

Chloroplast

Palisade mesophyll is the main photosynthetic tissue of the plant. The cells are frequently columnar and contain large numbers of chloroplasts, tightly packed together. Within the palisade mesophyll cells the chloroplasts can move about, probably by cytoplasmic streaming (mass movement of the cytoplasm). In strong sunlight chloroplasts are fairly evenly distributed throughout the cytoplasm, as they can all receive enough light to photosynthesise effectively. At lower light levels, chloroplasts move closer together at the top of the cell to obtain maximum radiation. This maximises the chances of the chlorophyll pigment receiving enough light for photosynthesis to occur. Palisade mesophyll cells are a very specialised form of parenchyma cell which has been modified through evolution to carry out photosynthesis.

Spongy mesophyll

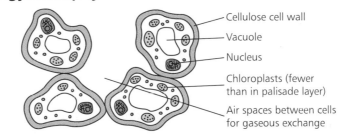

Cellulose cell wall

Vacuole

Nucleus

Chloroplasts (fewer than in palisade layer)

Air spaces between cells for gaseous exchange

Spongy mesophyll is the second main photosynthetic tissue after the palisade mesophyll. It is lower down in the structure of the leaf and so less likely to receive enough light to photosynthesise effectively. There are fewer chloroplasts in spongy mesophyll cells than in palisade ones. Spongy mesophyll cells are again modified parenchyma, are smaller and have more irregular shapes than palisade mesophyll cells.

Large **intercellular air spaces** in the spongy mesophyll, particularly around the stomata, are of great importance for effective gaseous exchange – in photosynthesis the uptake of carbon dioxide and the removal of oxygen, and also the evaporation of water from the leaf. They therefore have an important role in water balance and the transport of materials round the plant.

Collenchyma cells may be found in the leaf, sometimes above and often as part of the **midrib** (main vein). With their areas of cellulose thickening they give mechanical strength and support.

Figure 2.2.6 The specialised cells and their functional arrangement can be clearly seen in a section through a dicotyledonous leaf.

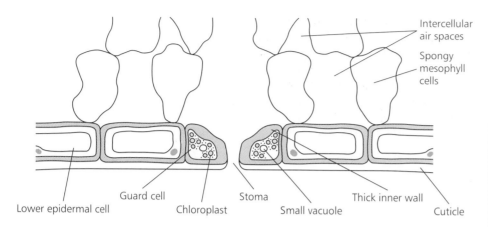

Lower epidermal cell   Guard cell   Chloroplast   Stoma   Small vacuole   Thick inner wall   Cuticle

Figure 2.2.7 Stomata provide a route for gases in and out of the leaf, connecting the intercellular spaces with the outside world.

Open or closed?

There is a constant state of conflict in the needs of a plant. For photosynthesis to occur successfully at its maximum rate carbon dioxide must move into the leaf and oxygen out of it, so stomata must be open. A steady flow of water and minerals up from the soil into the aerial parts of the plant then results from the evaporation of water from the leaf cells. On the other hand, water is often in relatively short supply. In dry climates or in drying conditions such as strong wind an enormous amount of water would be lost from the leaves if free evaporation were possible, so that the plant would suffer damage or death. How is this conflict resolved?

Stomata are not simple pores. The pore is bordered by two **guard cells**, specialised epidermal cells. They are sausage shaped and, like other epidermal cells, contain a sap vacuole. Unlike other epidermal cells they contain chloroplasts, and the cellulose of their walls is unevenly distributed. Figure 2.2.8 shows how stomata open as a result of turgor changes.

It appears that as water moves into the guard cells from the surrounding epidermal cells, the turgidity of the guard cells increases. The stomatal pore opens as a result of the uneven bending which takes place because of the unevenly thickened cellulose walls and the cellulose hoops around the guard cells. But what causes the osmotic situation of the cells to change so that water moves into the guard cells? As yet there is no definitive answer, although there are several hypotheses.

The best current theory on stomatal opening suggests that the osmotic movement of water is the result of some active ion transport. Levels of potassium ions have been observed to increase in the light to a degree which would affect water movement into the cells. It can be shown that the potassium ions are brought into the cell by active transport because metabolic poisons

Photosynthetic opening

The stomata tend to open during the day and close at night. They contain chloroplasts and can therefore photosynthesise. A theory was developed suggesting that as photosynthesis occurs in the guard cells in daylight, levels of sucrose (produced as a result of photosynthesis) rise. This in turn affects the water potential of the guard cells, causing water to move in along a concentration gradient, increasing turgor and causing the stomata to open.

Unfortunately, the theory comes unstuck when examined closely. The build-up of sucrose in the guard cells as a result of photosynthesis appears to be too slow to explain the very rapid response of the stomata to light. Neither has it been shown that the levels of sucrose build up enough to cause this level of movement of water. Some guard cells do not contain chloroplasts and yet still open and close, whilst others open at night and close during the day. So although at first glance the 'photosynthetic hypothesis' appeared to fit the observed facts well, further work showed that a different hypothesis was needed.

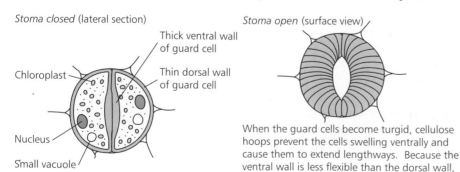

Stoma closed (lateral section)

Chloroplast

Nucleus

Small vacuole

Thick ventral wall of guard cell

Thin dorsal wall of guard cell

Stoma open (surface view)

When the guard cells become turgid, cellulose hoops prevent the cells swelling ventrally and cause them to extend lengthways. Because the ventral wall is less flexible than the dorsal wall, the cells become semicircular and the stoma opens.

Figure 2.2.8 The mechanics of stomatal opening and closing

stop not just stomatal opening but also the accumulation of potassium ions. If this active transport stops in the dark, closure of the stomata would be brought about by the rapid diffusion of the potassium ions out of the cell along a very steep concentration gradient, followed by water movements due to osmosis. This theory provides an explanation for both the opening and the closing of stomata, is possible in all plants whether there are chloroplasts in the guard cells or not, and has some experimental evidence to support it. Until a better explanation is put forward, it seems that potassium ions are the key to the control of gaseous exchange in plants.

SUMMARY

- The **stem** system fulfils the role of support of the leaves and flowers of a plant, and also the transport of water, mineral ions and organic substances around the plant. It may also store food for overwintering, form adventitious roots, provide a means of reproduction or form the photosynthetic organ of the plant.

- The stem consists of an external **cuticle**, **vascular** tissue arranged in bundles, and central **packing** tissue. In dicotyledons the vascular bundles are arranged around the edge of the stem; in monocotyledons they are scattered through the parenchyma.

- **Epidermal** cells secrete cutin to form the cuticle in order to cut down water loss from the stem surface. They sometimes have epidermal hairs for various specialised functions.

- **Cortex** parenchyma cells act as packing cells. The cortex also contains **collenchyma** cells, which are tightly packed and strengthened with thickened cellulose at their corners, and provide support. Some photosynthesis occurs in the stem cortex.

- The **endodermis** is a single-celled layer with stored starch grains.

- Vascular bundles contain an outer **pericycle** made of strengthened **sclerenchyma** cells for support. The **phloem**, which transports organic solutes, and **xylem**, which brings water and mineral ions, are also strengthened. The cells of the **cambium** divide and differentiate to form xylem or phloem cells.

- The central **pith** contains unspecialised parenchyma cells.

- The **leaf** system fulfils the function of photosynthesis. Leaves capture sunlight and also supply carbon dioxide and other requirements to photosynthetic cells. The products of photosynthesis are transported from the leaves to the stem and around the plant.

- The overall structure of the leaf includes a **lamina** which is usually thin with a large surface area, and **vascular** tissue organised in a network of veins in dicotyledons and parallel veins in monocotyledons. Dicotyledons also have a **petiole** (leaf stalk) and central **midrib**.

- The **epidermal** cells of a leaf secrete cutin to form a cuticle. The lower epidermis particularly contains pores called **stomata**, surrounded by **guard cells**, which can open or close to allow gaseous exchange.

- The **palisade mesophyll** cells are columnar and tightly packed. They contain many chloroplasts and are the main photosynthetic cells of the leaf.

- The **spongy mesophyll** cells are smaller and more irregularly shaped and have large intercellular spaces between them, important in gaseous exchange. These connect with the stomata. Some photosynthesis takes place in these cells.

- **Collenchyma** cells provide support to the vascular tissue, particularly around the midrib.

- **Vascular bundles** in the midrib and veins contain **xylem**, **phloem** and **cambium** as in the stem, with a sheath of **parenchyma** or **sclerenchyma**.

- Stomata open to allow carbon dioxide into the leaf, and as a result water vapour passes out of the leaf. The guard cells elongate and open the stomata as a result of increased turgor, probably brought about by active transport of potassium ions.

QUESTIONS

1 a Describe the structure of a plant stem and how it is adapted to its function of transport.

  b Suggest how stems may be modified to perform functions other than the major function of transport.

2 a How is the structure of a dicotyledonous leaf related to its photosynthetic function?

  b How do stomata control gaseous exchange within the leaf?

3 Plants are found in a wide variety of habitats. Explain how adaptations of the stems and leaves can help plants to survive in a variety of adverse situations.

2.3 Transport systems in plants

Within a living organism substances need to be moved from one place to another. A chemical may be synthesised in one place for use somewhere else; materials need to be taken in from the external environment; waste products of reactions must be removed. Between an individual cell and the external environment, or a cell and its close neighbours, there is a variety of methods by which transport may occur. These include diffusion, osmosis, active transport and endo- and exocytosis.

For unicellular organisms, and multicellular organisms with a large surface area:volume ratio, the transport of substances in these ways is quite adequate. However, as organisms become larger and more complex in shape and design, the distances to be travelled become too large for these simple methods of transport. Specialised transport systems have evolved to carry substances rapidly around larger organisms.

TRANSPORT SYSTEMS

Within any large organism, plant or animal, a transport system will be organised to carry out its functions as effectively as possible. Most major transport systems have certain features in common:

1 A system of vessels in which substances are carried. These are usually tubes, sometimes following a very specific route, sometimes widespread and branching.

2 A means of ensuring that substances are moved in the right direction.

3 A means of moving materials fast enough to supply the needs of the organism. This may involve a manipulation of physical processes (for example, actively changing concentrations of solutes to ensure a diffusion gradient) or a mechanical method (for example, the pumping of the heart).

4 A suitable transport medium.

5 Mass flow may be involved – bulk transport brought about by a pressure difference of some sort. In a mass flow system all the moving material travels at the same rate.

Both plants and animals need transport systems. Plants produce food in the leaves and must then transport the products of photosynthesis around the entire organism. They obtain water and minerals via the roots, and these too are needed by all the cells. Animals obtain food and oxygen in very specific systems within their bodies. They produce carbon dioxide in every cell. Transport systems are needed in animals to distribute food and oxygen to the active cells and to remove all the waste products.

TRANSPORT SYSTEMS IN PLANTS

Routes for transport

Carbon dioxide enters plants and moves into the cells by diffusion. It is used up in photosynthesis, so a constant concentration gradient is maintained between the inside of the cells and their environment. Thus diffusion occurs continuously and a more specific transport route is not needed. The oxygen produced by photosynthesis is similarly moved out of the plant if it is not used in respiration.

When it comes to the transport of water, minerals and the products of photosynthesis, specific transport routes are required. An integral part of plant roots, stems and leaves is the vascular system, with xylem and phloem, shown in figure 2.3.1, making up the transport routes. In general the xylem carries water and minerals up from the roots to the other areas of the plant, whilst the phloem carries the soluble products of photosynthesis from the leaves to the other cells. The xylem generally transports substances upwards, while substances can move both up and down in the phloem. The structures of these transport tissues were discussed earlier.

Evidence for the movement of water through the xylem

How do we know that water moves up the plant through the xylem? Evidence can be obtained in several ways:

1 If the cut end of a shoot is placed in a solution of eosin dye, the dye is carried to the vascular tissue of the leaves and appears in the xylem vessels, as figure 2.3.2 shows.

2 Ringing experiments involve killing with a steam jet a complete outer ring of bark, including the phloem vessels, but leaving the xylem intact. The upward movement of water through the plant is unaffected.

3 Radioactive isotopes can be introduced in minerals into the water available to the plant. Their path through the xylem, transported by water, can then be followed using autoradiography (see the box below).

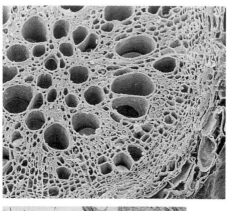

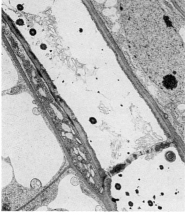

Figure 2.3.1 The structures of xylem (top) and phloem are well suited for their transport functions.

Autoradiography

Autoradiography is a very useful technique for following the movements of various substances around plants. It involves the following steps:

(1) The plant is exposed to a **labelled** substance – a substance containing a radioactive isotope (for example carbon 14 in the carbon dioxide supply to investigate the movement of the products of photosynthesis).

(2) The radioactive label is taken up as the plant metabolises.

(3) The movement of substances can then be traced in several ways. An autoradiograph is produced when the plant is left against photographic film. The labelled substance causes the photographic film to shadow, and so the areas of the plant in which it has accumulated can be seen.

Figure 2.3.3 An autoradiograph of a geranium leaf showing labelled material in the veins

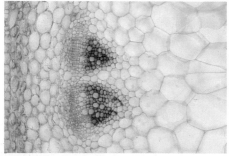

Figure 2.3.2 When sections of the stem and leaves are examined under the light microscope, eosin dye can be seen in the xylem vessels only.

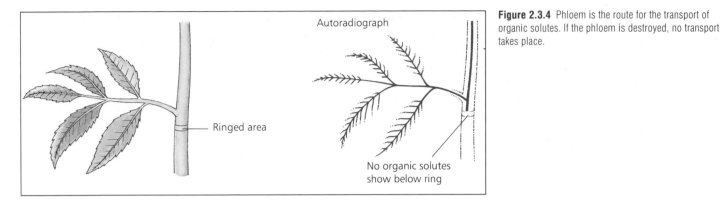

Figure 2.3.4 Phloem is the route for the transport of organic solutes. If the phloem is destroyed, no transport takes place.

Evidence for the movement of solutes in the phloem

Similarly, evidence can be obtained for the movement of solutes in the phloem:

1 The movement of solutes in the phloem can be observed by exposing the plant to carbon dioxide containing carbon 14. This labels the sugars produced by photosynthesis. Autoradiography can be used to follow the route of labelled sucrose around the plant. If the experiment is repeated in a plant with a ring of dead tissue, so that the phloem cells are no longer functioning, no radioactively labelled sucrose appears below the damaged area, as figure 2.3.4 shows.

2 A remarkably simple but very effective way of showing that organic solutes are carried by the phloem involves the use of aphids such as greenfly and is shown in figure 2.3.5.

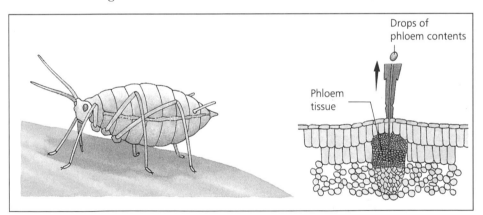

Figure 2.3.5 Aphids penetrate plant stems with their pointed mouthparts called stylets. Examination of the penetrated stem with a microscope shows that the stylet enters the phloem cells. T.E. Mittler in California realised that if the insect is removed from the mouthparts, a liquid oozes out of the end of the stylet. This is some of the contents of the phloem, and when analysed can be shown to contain sugars and other organic solutes.

Movement of substances in plants

The movement of substances around plants is usually referred to as **translocation**. Plants do not have mechanical systems to force materials along the narrow tubes of the xylem and phloem, but have evolved ways of manipulating a variety of physical processes to their advantage.

The movement of substances up from the roots to the leaves is well understood on the whole, though the mechanism by which organic solutes are transported in the phloem is much less clear.

TRANSPORT IN THE XYLEM

Translocation of water

The vessels of the xylem system are very narrow, dead tubes. They have a lumen diameter of 0.01–0.2 mm, and so have a great resistance to movement through them. Yet water has been shown to move up through the xylem vessels at speeds of 1–8 m h^{-1}, and to heights of up to 100 m above the ground in trees such as the giant redwood. How is this achieved?

Transpiration

The movement of water in the xylem of plants depends on **transpiration**. Transpiration is the loss of water vapour from the surface of the plant, mainly from the leaves, as figure 2.3.6 shows. This loss of water has the effect of 'pulling' water up through the xylem, as will be explained later.

When a liquid turns into a gas, it absorbs energy. The energy needed to turn a liquid into a gas is known as the **latent heat of vaporisation** and is high for water. When water evaporates from the surfaces of a plant, the latent heat of vaporisation is supplied by the Sun. Thus the energy for the transport of water and minerals in a plant comes from the Sun, without involving photosynthesis.

The amounts of water lost by a plant due to transpiration can be surprisingly large. A sunflower may transpire 1–2 dm^3 in a day, whilst a large oak tree can lose up to 600 dm^3 per day. Water moves from the xylem cells into the mesophyll cells of the stem and leaf along the vacuolar, symplast and apoplast pathways, with most using the apoplast route. (These routes were described when considering the movement of water across the root *into* the xylem.) Evaporation of water takes place from the cellulose walls of the spongy mesophyll cells into the air spaces. Water vapour then moves through open stomata into the external air along a diffusion gradient. Even on a windy day, each leaf has a layer of still air around it. The thickness of this layer varies with the wind speed. The water vapour diffuses through this still layer before it is swept away by the mass of moving air. Figure 2.3.7 illustrates the effect of wind speed on transpiration.

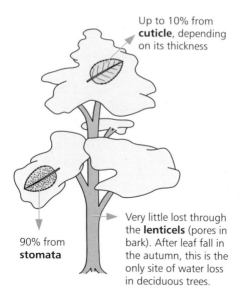

Up to 10% from **cuticle**, depending on its thickness

90% from **stomata**

Very little lost through the **lenticels** (pores in bark). After leaf fall in the autumn, this is the only site of water loss in deciduous trees.

Figure 2.3.6 The main sites of transpiration from the surface of a plant

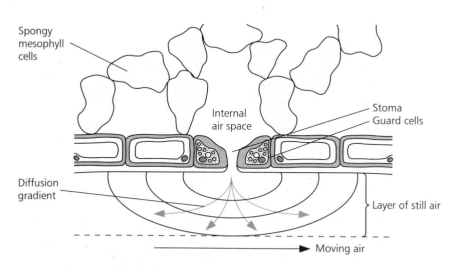

Spongy mesophyll cells

Internal air space

Stoma
Guard cells

Diffusion gradient

Layer of still air

Moving air

Figure 2.3.7 The thickness of the layer of still air around a leaf affects the rate of transpiration. The thinner the layer, the more rapidly is water lost, because the diffusion gradient is greater.

Factors affecting transpiration

Water loss from a plant occurs mainly as a result of the opening of the stomata for gaseous exchange, to provide the carbon dioxide needed for photosynthesis. The result of transpiration is that water is moved through the plant, carrying minerals with it. Evaporation of water from the leaf surfaces may also help to cool the plant. Yet in spite of these advantages, it appears that water loss is a major problem for plants, because there is rarely an abundance of available water.

Demonstrating transpiration

Loss of water from the surfaces of a plant can be demonstrated by enclosing the pot of a potted plant in a sealed plastic bag. This prevents the evaporation of water from the soil surface interfering with the experiment. The plant is then sealed in a bell jar. As transpiration proceeds, a colourless liquid collects on the inside of the bell jar. Cobalt chloride or anhydrous copper(II) sulphate paper can be used to show that this liquid contains water.

It is not easy to measure amounts of water transpired from a plant. We can more readily measure the uptake of water by a plant, which gives us a measure of transpiration because most of the water taken up by a plant under normal conditions is transpired, with a minute fraction being used for other purposes. Uptake of water is demonstrated using a **potometer**, as shown in figure 2.3.8.

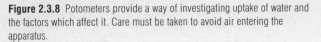

Figure 2.3.8 Potometers provide a way of investigating uptake of water and the factors which affect it. Care must be taken to avoid air entering the apparatus.

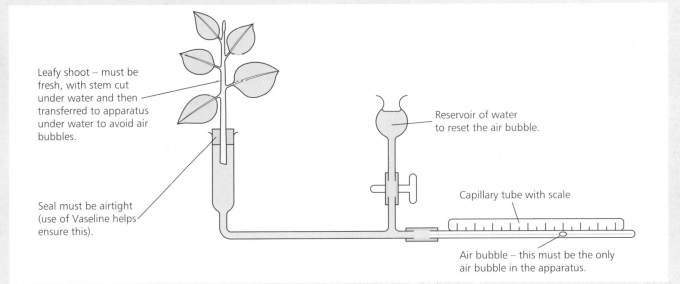

Leafy shoot – must be fresh, with stem cut under water and then transferred to apparatus under water to avoid air bubbles.

Reservoir of water to reset the air bubble.

Seal must be airtight (use of Vaseline helps ensure this).

Capillary tube with scale

Air bubble – this must be the only air bubble in the apparatus.

The rate of water loss from the leaves of a plant may be affected by a variety of factors. In many instances plants have features to minimise water loss, such as curled, hairy and grooved leaves which trap layers of still, moist air around the stomata.

The following factors affect transpiration:

1 Light has a major effect on transpiration, as figure 2.3.9 shows. Stomata usually open in the light for photosynthetic gas exchange, and close in the dark. Thus transpiration rates increase with light intensity until all the stomata are open and transpiration is at a maximum.

2 Temperature is the next factor affecting transpiration after light. At a given light intensity, an increase in temperature increases the amount of evaporation from the spongy mesophyll cells, and also increases the amount of water vapour the air can take before it becomes saturated. Both of these factors increase the water potential gradient between the air inside and outside the leaf, increasing the rate of transpiration.

3 Air movement or wind increases the rate of transpiration because it reduces the shell of still air around the stomata and so increases the water potential gradient between the leaf and the air.

4 Air humidity is the concentration of water vapour in the air. A high air humidity lowers the rate of transpiration because of the reduced water

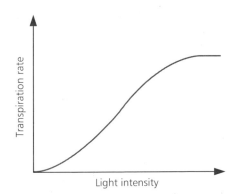

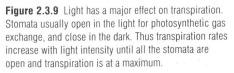

Figure 2.3.9 Light has a major effect on transpiration. Stomata usually open in the light for photosynthetic gas exchange, and close in the dark. Thus transpiration rates increase with light intensity until all the stomata are open and transpiration is at a maximum.

potential gradient between the leaf and the air. Very dry air – low humidity – has the opposite effect.

5 The availability of water from the soil has an effect too. If there is little soil water, the plant is under water stress and transpiration is reduced.

The transpiration stream

When water is lost by transpiration from the leaves of a plant, this affects the contents of the xylem, a continuous column of water. Water evaporating from the surface of one spongy mesophyll cell results in the osmotic movement of water across from the next-door cell, and so on to the xylem itself.

When molecules of water leave the xylem to enter a cell by osmosis, this creates tension in the column of water in the xylem, which is transmitted all the way down to the roots. This is due to the **cohesion** of water molecules. Because of their dipolar nature, water molecules are held together by hydrogen bonds. As a result the column of water has high **tensile strength** – it is unlikely to break.

The water molecules also **adhere** strongly to the walls of the narrow xylem vessel and (more importantly, it is thought) to the millions of tiny channels and pores within the cellulose cell walls of the apoplast pathway through the leaf. **Adhesion** is the attraction between unlike molecules. The adhesive force is sufficient to support the entire column of water. Thus the combination of adhesive and cohesive forces allows the whole column of water to be pulled upwards. More water is continuously moved into the roots from the soil to replace that lost from the leaves by transpiration. Figure 2.3.11 shows the transpiration stream.

Water is lost from the surface of the leaves by evaporation.

Water moves across the cells of the leaf by osmosis along water potential gradients, mainly along the apoplast pathway. Strong adhesive forces form between the pores of the cellulose cell walls and the water molecules.

As water molecules are lost by evaporation and moved out of the xylem, cohesion between the water molecules means that the whole column of water in the xylem is pulled upwards.

Water moves into the root hair from the soil by osmosis.

Water is moved across the root by osmosis to maintain the continuous column in the xylem.

Figure 2.3.10 A village pump relies on the pressure of atmospheric air to raise the water. The maximum column of water that can be raised in this way is about 10 m. The giant redwood tree regularly raises water to over 30 m.

◀ **Figure 2.3.11** The transpiration stream is set up as a result of physical processes, and a pressure of around 4 000 kPa, moving the water upwards, can result. This is enough to supply water to the tops of the tallest trees.

Root pressure

Transpiration is not the only method by which water is moved through the xylem. Transpiration seems to be a passive process, yet there are aspects of water transport which are affected by metabolic inhibitors and lack of oxygen. During the night, when transpiration rates are extremely low, drops of water may be forced out of the leaves of some plants in a process known as **guttation**. If a plant is cut off from the root, root sap will continue to ooze from the root xylem. These observations are thought to be the results of root pressure.

Artificial transpiration

A vacuum pump can cause a column of water to rise 11 m or a column of mercury to rise 760 mm. Yet trees regularly pull columns of water up 20 m and more. Figure 2.3.12 shows an experiment which demonstrates neatly the effect of evaporation on a column of liquid, carried out by Josef Bohm in 1893.

Adhesive forces between the water molecules and the pores of the porous pot are strong enough to support this enormous column of water, and cohesive forces between the water molecules stop the column breaking under the strain. This gives us our best model so far of the transpiration stream.

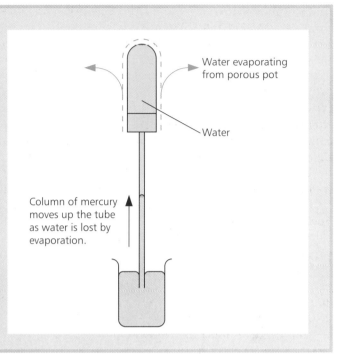

Figure 2.3.12 Drawn by evaporation, the column of mercury rises to over 1 m. It has been calculated that if there was water in the system, the column could be pulled to a height of more than 1 km – far greater than the height of any living plant.

Root pressure seems to be produced by the active secretion of salts into the root cells, increasing the movement of water by osmosis. The pressure generated is about 100–200 kPa. This contributes to the movement of water in the xylem of many plants, particularly in situations where the transpiration rate is low.

Translocation of ions

The route by which mineral ions are taken in from the soil and moved into the xylem involves partially permeable membranes and active transport against concentration gradients. Once the ions reach the xylem, they are carried along in the transpiration stream in a mass flow system, being delivered by the extensive vascular network to the tissues where they are needed. They may be moved out into the cells by either diffusion or active transport, depending on the permeability of the cell membrane and the relative concentrations of the solutes.

TRANSPORT IN THE PHLOEM

Translocation of the products of photosynthesis

The leaves of a plant produce large amounts of glucose. This is needed throughout the entire organism. Experiments such as those mentioned on page 113 show that the phloem is involved in the movement of organic solutes, and that the translocation is active. Sucrose makes up 90% of the solutes found in the phloem fluid, the rest being amino acids, organic phosphates and nitrates, etc. Sucrose is the form in which the products of photosynthesis are transported around the body of the plant. When it reaches the cells it may be converted back to glucose for use in respiration, or to starch for storage.

Whilst the process of transport in the xylem of plants is relatively well understood, the way in which solutes are moved in the phloem is much less certain. The flow rates are known to be about 0.2–6 m h^{-1}. In many species the distances moved are large, up to 100 m. There is relatively little phloem compared with the often extensive xylem vessels. The tubes of the phloem are very narrow, with a high resistance to movement which is increased by the presence of sieve plates. However, large quantities of solute can be moved. A large tree may transport 250 kg of sucrose down its trunk in a year.

The mechanism of phloem transport

The soluble products of photosynthesis seem to enter the sieve tubes by an active process. The sucrose content of most plant cell sap is only 0.5%, yet sucrose makes up 20–30% of the phloem sap content. Active transport must therefore be involved to move the sucrose against this sort of concentration gradient. Specialised parenchyma cells known as **transfer cells**, shown in figure 2.3.13, appear to be involved in transporting the sucrose into the sieve tubes.

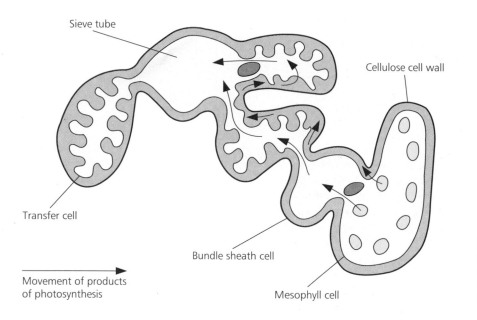

Sieve tube

Cellulose cell wall

Transfer cell

Bundle sheath cell

Mesophyll cell

Movement of products of photosynthesis

Figure 2.3.13 Regular infoldings of the cell wall and membrane give transfer cells a large surface area for active transport. They contain many mitochondria and so have the supply of energy needed to be metabolically active.

Transfer cells are also found in roots, storage organs and growing shoots, where they are involved in helping to move solutes such as nitrates from the sieve tubes into the cells that need them.

The mechanism for the movement of substances along the sieve tubes themselves is still very much open to debate. Experiments have shown that killing the phloem prevents the movement of solutes, and that sugars and amino acids travel through the system too quickly to be relying on diffusion alone, so it seems reasonable to assume that an active process is involved. More evidence for this is the observation that translocation is reduced by lack of oxygen and by respiratory inhibitors. The companion cells with their mitochondria can supply the energy needed for this active process. However, there is no entirely satisfactory explanation of how the translocation of solutes in phloem occurs.

A mass flow system of some sort seems to be involved, but the fact that different solutes move at different speeds and sometimes in different directions mean this cannot be the whole story.

Plants, although they can be very large, have relatively slow metabolisms. As a result, transport systems relying mainly on physical processes are sufficient to supply all their needs. In animals, with their more active lifestyles and greater energy demands, the situation is not quite so simple, as section 2.5 will show.

Phloem transport theories

Münch's mass flow theory

In 1930 a purely physical explanation for the mass flow of substances through sieve tubes was proposed by Münch. The model in figure 2.3.14 shows the principle behind it.

Initially, water moves into both containers by osmosis. As X contains a much more concentrated solute solution than Z, water will move into X more rapidly and so there will be a flow of solution from X to Z. The hydrostatic pressure that this creates forces water out of Z. The flow will continue until the concentrations of the solutions in X and Z are the same.

The pressure flow hypothesis

The modern application of Münch's model is known as the **pressure flow hypothesis**. In a plant, X is the phloem in the leaves where the sucrose concentration is high due to photosynthesis and the transfer cells actively load the sieve tubes. Z is the area of phloem where the sucrose is unloaded and used by the cells. Water can move into the phloem by osmosis at any point, and the return route for the water to the cells is through the xylem. Unlike the physical model, the flow can be continuous because sucrose is continually being added at one end and removed at the other.

A variety of ideas has been put forward to try and extend this model, none of which are supported by much evidence. The pressure flow hypothesis certainly explains much of what is seen, but there is doubtless more to be discovered in this area. Some of the most recent and exciting research suggests that there are proteins present in the phloem cells closely related to the actin and myosin which make up muscles (see section 2.6, page 158). It has been proposed that the contraction of these proteins may be involved in translocation in the phloem, but much more work is needed before the hypothesis can be accepted.

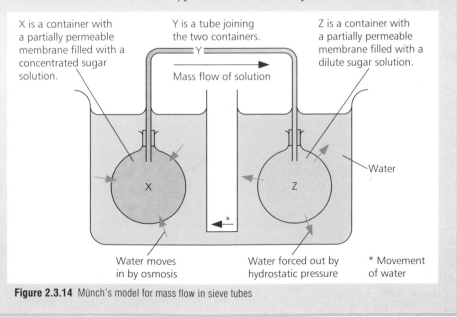

X is a container with a partially permeable membrane filled with a concentrated sugar solution.

Y is a tube joining the two containers.

Z is a container with a partially permeable membrane filled with a dilute sugar solution.

Mass flow of solution

Water

Water moves in by osmosis

Water forced out by hydrostatic pressure

* Movement of water

Figure 2.3.14 Münch's model for mass flow in sieve tubes

SUMMARY

- **Transport systems** have evolved to carry substances rapidly around multicellular organisms. They generally consist of a system of vessels, a means of ensuring that transport occurs in the right direction, a means of moving substances quickly and a transport medium. The mass flow mechanism may be involved.

- In plants, specific transport systems are the **xylem**, which transports water and mineral ions up from the roots, and **phloem**, which transports the soluble products of photosynthesis from the leaves around the plant.

- Evidence for the movement of water and ions in the xylem is given by the uptake of eosin dye, ringing experiments and autoradiography using labelled mineral salts.

- Evidence for the movement of organic compounds in the phloem includes autoradiography on plants exposed to labelled carbon dioxide combined with ringing, and the use of aphid stylets to penetrate the phloem and subsequent analysis of the liquid carried there.

- Movement of water in the xylem depends on **transpiration**, the loss of water vapour from the leaves through stomata which open to allow carbon dioxide to diffuse in for photosynthesis.

- Transpiration is affected by light, temperature, air movement, air humidity and the availability of water from the soil. Plants may develop specialised features to reduce the rate of transpiration, by increasing the size of the envelope of still air around the leaf.

- The **transpiration stream** is the continual movement of water up through the xylem vessels caused by the **cohesion** of water molecules and the **adhesion** between water molecules and pores in the cellulose cell walls of the apoplast pathway. The entire column of water moves upwards, pulled by transpiration from the leaves.

- **Root pressure** contributes to transpiration under conditions when the transpiration rate is low, and involves the active secretion of salts into the root cells.

- Mineral ions are transported by mass flow in the xylem and are carried into and out of the plant cells by diffusion, facilitated diffusion and active transport, depending on the mechanism available for that ion.

- The sugar products of photosynthesis are transported as sucrose in the phloem, with small amounts of amino acids and other organic compounds.

- Transport in the phloem involves active transport, with **transfer cells** loading sucrose into the sieve tubes against a concentration gradient. **Münch's mass flow theory** and the **pressure flow hypothesis** provide models for the mechanism of transport in the phloem.

QUESTIONS

1 a Produce a table comparing the structure and functions of xylem and phloem tissues within a plant.

  b Explain the value of radioisotopes in determining the functions of these two transport systems.

2 a Describe an experiment which could be used to measure transpiration.

  b Comment on any precautions which need to be observed in the setting up of the apparatus and on the limitations of the apparatus for measuring transpiration.

  c How and why would **i** lowered light intensity and **ii** increased air movement affect the results of the experiment?

3 Trace the path followed by a molecule of water from the soil through the root, stem and leaf of a vascular plant until it enters the atmosphere, naming the tissues and cells along the route.

2.4 Respiratory systems

As we have seen, complex multicellular organisms have systems within them specialised to fulfil particular needs. **Respiratory systems** have evolved to enable organisms to obtain oxygen and to remove carbon dioxide from their bodies in a process known as **gaseous exchange**. Oxygen is vital for almost all living organisms as it is needed for efficient cellular respiration, the process by which energy is released from foods. Cellular respiration is described in detail later, but can be summarised by the equation:

$$\textbf{Glucose + oxygen} \rightarrow \textbf{energy + carbon dioxide + water}$$

Carbon dioxide is formed as a waste product of this process. Carbon dioxide is toxic and so cannot be allowed to accumulate in cells or tissues – it has to be removed.

For unicellular and simple multicellular organisms, obtaining oxygen and getting rid of carbon dioxide happens by diffusion along concentration gradients. As organisms become larger their surface area to volume ratio gets smaller, and simple diffusion through the body surface is no longer sufficient to supply all cells with their requirements.

FEATURES OF RESPIRATORY SYSTEMS

A respiratory system is an area specialised for gaseous exchange between the organism and the environment. Oxygen and carbon dioxide are transported to and from the respiring cells by a transport system. Gaseous exchange takes place by diffusion at the **respiratory surface**. For effective diffusion to take place, all respiratory surfaces have certain features in common:

1   A large surface area gives sufficient gaseous exchange to supply all the needs of the organism. The surface area of the respiratory surface has to compensate for the relatively small surface area:volume ratio of the organism as a whole.

2   Permeable surfaces allow the free passage of the respiratory gases.

3   Thin surface structures allow effective diffusion from one side of the surface to the other.

4   Moist surfaces ensure that the gases are in solution and so can diffuse rapidly.

5   In many animals, a rich blood supply to the respiratory surfaces allows the transport of the respiratory gases between the respiring cells and the site of gaseous exchange.

RESPIRATORY SYSTEMS IN PLANTS

As plants get larger, their surface area:volume ratio gets smaller. However, there are no specialised systems within large and complex multicellular plants for obtaining oxygen and getting rid of carbon dioxide. The stomata connected with the large air spaces in the spongy mesophyll layer in the leaf, along with lenticels (slit-like openings in woody stems), ensure that any necessary gas exchanges occur smoothly. The large surface area of cells in the

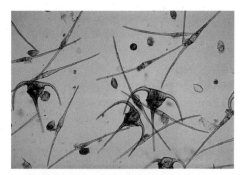

Figure 2.4.1 A cheetah needs much more oxygen to satisfy its needs than a dinoflagellate. The cheetah's surface area : volume ratio means that diffusion through its skin cannot supply all the oxygen it needs, so it has a specialised respiratory system.

spongy mesophyll along with their moist surfaces do meet some of the requirements for a respiratory system as they are well adapted for gaseous exchange. But large multicellular plants do not need specialised respiratory systems in the way many animals do.

There are several reasons for this. One of the most important is that plants use up carbon dioxide in photosynthesis, thus removing much of this toxic waste product of respiration. Oxygen is a waste product of photosynthesis and is thus immediately available for cellular respiration. Another important factor is that the oxygen demand of plant tissues is relatively low. Most multicellular plants do not move around and the energy requirements of their tissues are not high. This means that the demand for oxygen for cellular respiration is small, and relatively low levels of carbon dioxide are produced by the process. Diffusion is sufficient to meet the needs for gaseous exchange for both cellular respiration and photosynthesis.

THE NEED FOR RESPIRATORY SYSTEMS IN ANIMALS

Animals do not photosynthesise, so they have no internal system of generating oxygen or removing carbon dioxide. Animals are generally active for at least part of the day – plant eaters must find sufficient food to eat and avoid predators whilst meat eaters need to capture their prey or find carrion, and both must carry out digestion. They also need to carry out all the normal cellular processes common to plants and animals. Thus the oxygen demands of animals for cellular respiration are relatively high, and they consequently produce large quantities of carbon dioxide. As a result of the surface area:volume ratio, diffusion through the outer surface alone is insufficient for most animals except the very small, and so specialised respiratory systems have evolved.

Respiratory systems in land animals

In terrestrial life there is a perpetual conflict between the need for oxygen and the need for water. The conditions which favour the diffusion of oxygen into an organism also favour the diffusion of water out. Animals need a large, moist surface area for gaseous exchange, and yet they need to limit the water loss from this same surface as much as possible. Air-breathing vertebrates have solved the problem by developing **lungs**. Some lungs are quite simple and merely add to the area for gaseous exchange already provided by the body surface, for example in frogs. Other animals cannot use their outer surface for respiration at all and so are much more dependent on efficient lungs. Mammals have one of the most complex respiratory systems, which demonstrates the general features of most respiratory systems. We shall here look at the example of the human respiratory system in detail.

THE STRUCTURE OF THE HUMAN RESPIRATORY SYSTEM

Most of the human respiratory system is found within the protective walls of the chest. The system is linked with the outside world by a tube which may be entered through either the mouth or the nose. Figure 2.4.3 (pages 124–5) shows the human respiratory system.

THE HUMAN RESPIRATORY SYSTEM AT WORK

In looking at the human respiratory system we shall consider the following questions in turn:

Animals without a respiratory system

We have already mentioned single-celled organisms, but some multicellular animals also manage without a specialised respiratory system. These animals include the coelenterates, the flatworms and a large proportion of the annelid worms. Many of these are slow moving and so have low oxygen demands. Others have increased their effective surface area for gaseous exchange without investing in a specialised respiratory system. Examples are shown in figure 2.4.2.

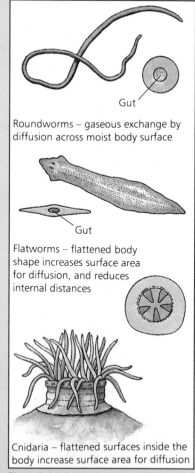

Roundworms – gaseous exchange by diffusion across moist body surface

Flatworms – flattened body shape increases surface area for diffusion, and reduces internal distances

Cnidaria – flattened surfaces inside the body increase surface area for diffusion

Figure 2.4.2 Organisms such as these can obtain sufficient oxygen and get rid of their carbon dioxide waste without specialised respiratory systems.

1 How is air brought into the respiratory system, to reach the respiratory surfaces?

2 How is gaseous exchange brought about at the respiratory surfaces?

3 How is the process of breathing controlled and regulated to meet the varying demands placed upon it?

(1) Breathing – bringing the air in

The lungs are the site of gaseous exchange, but they play only a passive part in getting the gases there. The chest cavity is effectively a sealed unit, with only one way in or out – through the trachea. **Breathing** is the way in which air is moved in and out of the lungs, passing through the trachea, bronchi and bronchioles.

What does breathing involve?

Simple observation tells us that there are two parts to the process of breathing – taking air into the chest (**inhalation**) and breathing air out again (**exhalation**). Breathing involves a series of pressure changes in the chest cavity which in turn bring about movements of the air.

Inhalation is an active, energy-using process. The diaphragm, normally dome shaped, contracts and as a result is lowered and flattened. The external intercostal muscles contract to raise the rib cage upwards and outwards. This results in the *volume* of the chest cavity increasing, which in turn means the *pressure* in the cavity is lowered. As the pressure within the chest cavity is now lower than the atmospheric pressure of the air outside, air moves through the trachea, bronchi and bronchioles into the lungs to equalise the pressures inside and out.

Normal exhalation is a largely passive process. The muscles of the diaphragm relax so that it moves up into its resting dome shape. The external intercostal muscles relax and the internal intercostal muscles contract so that the ribs move down and in. The elastic fibres around the alveoli of the lungs return to their normal length. As a result of all this the volume of the chest cavity decreases, causing an increase in pressure. As the pressure in the chest cavity is now greater than that of the outside air, air moves out of the lungs, travelling along the same passageways as on the inward journey. Figure 2.4.4 shows the mechanism of breathing.

There are times when passive exhalation is not enough. On the occasions when we want to force the air out of our lungs more rapidly than passive exhalation allows, the muscles of the abdomen contract. This increases the pressure in the abdomen, forcing the diaphragm upwards further and so increasing the pressure in the chest cavity. This is known as **forced exhalation**.

Boyle's law

Boyle's law states that the pressure of a fixed mass of gas is inversely proportional to the volume in which it is contained, or:

$$P \propto 1/V$$

Thus as volume increases, pressure is lowered. Conversely, as the volume decreases, the pressure in the system increases.

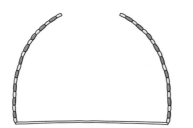

Inhalation:
Ribs lift up and out as external intercostal muscles contract.
Diaphragm contracts and flattens.
Volume increases, pressure falls.

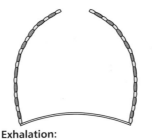

Exhalation:
Ribs drop down and in as internal intercostal muscles contract.
Diaphragm relaxes and arches.
Volume falls, pressure increases.

Figure 2.4.4 These movements of the ribs and diaphragm result in pressure changes which in turn cause the movement of air in and out of our lungs.

Figure 2.4.3 The human respiratory system and the associated organs are well adapted for gaseous exchange, for ventilating the respiratory surfaces and for preventing excess water loss.

Nasal cavity
The nasal passages have a relatively large surface area, but no gaseous exchange takes place here. The passages have a good blood supply, and the lining secretes mucus and is covered in hairs. The hairs and mucus filter out much of the dust and small particles such as bacteria that we breathe in. The moist surfaces mean the level of water vapour in the inspired air is increased and the rich blood supply raises the temperature of cold air. The entry of air into the lungs therefore has as little effect as possible on the internal environment there.

Mouth
Air can enter the respiratory system via the mouth, but does not gain the cleaning, warming and moistening effects of the nasal route.

Epiglottis and Glottis
The epiglottis closes over the glottis in a reflex action when food is swallowed. This stops food entering the respiratory system, preventing infection and damage to the lungs.

{ Epiglottis
{ Glottis

Larynx
The larynx makes use of the flow of air in and out of the respiratory system to produce sounds which in humans are developed into the spoken language.

Bronchioles

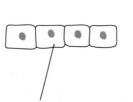 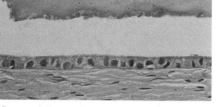

Flattened cuboidal epithelium – the flattened shape gives an increased surface area and a smaller distance for gases to diffuse.

Bronchioles are much smaller than the bronchi and there are many more of them. Larger bronchioles have cartilage rings, unlike those with a diameter of 1 mm or less. These small bronchioles collapse quite easily. Their main function is as an airway, but a little gaseous exchange may occur here. As the bronchioles get smaller the lining epithelium changes from columnar to **flattened cuboidal**, making diffusion of gases more likely.

Pleural membranes
The pleural membranes surround the lungs and line the chest cavity.

Pleural cavity
The pleural cavity is the space between the pleural membranes. It contains lubricating fluid which allows the membranes to slide easily during the breathing movements.

Ribs
The ribs form a protective bony cage around the respiratory system.

Diaphragm
The diaphragm is a broad sheet of muscle which forms the floor of the chest cavity. Movements of the diaphragm are important in the physical process of breathing.

Heart

Trachea

Columnar epithelial cell – these line the trachea and the bronchi.

Cilia – these all beat *away* from the lungs.

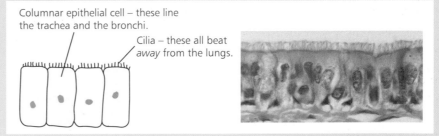

The trachea is the major airway leading down into the chest cavity. It is lined with **columnar epithelial cells**. In the layers below the epithelium are **mucus-secreting cells**. The cilia of the columnar epithelial cells beat to move the mucus and any trapped microorganisms and dust away from the lungs. The inhalation of tobacco smoke stops the cilia beating.

Incomplete rings of cartilage

These cartilage rings support both the trachea and the bronchi and prevent them collapsing. The rings are incomplete to allow the easy passage of food down the oesophagus which runs behind the trachea.

Bronchus

Within the chest cavity the trachea divides to give two bronchi, one leading to the left lung and one to the right. The bronchi are very similar in structure to the trachea, but are slightly narrower. The left bronchus divides into two, the right bronchus into three.

Alveoli

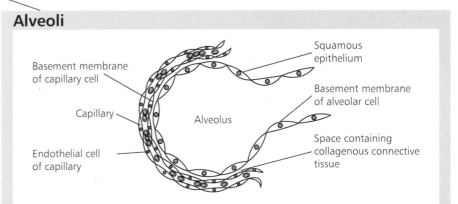

Basement membrane of capillary cell

Squamous epithelium

Capillary

Basement membrane of alveolar cell

Alveolus

Endothelial cell of capillary

Space containing collagenous connective tissue

The alveoli are the main respiratory surfaces of the lungs where most of the gaseous exchange takes place. Alveoli are made up of **squamous epithelial cells** which facilitate diffusion as they have a large surface area and are thin, reducing the distance for gases to travel. The capillaries which run close to the alveoli also have a wall which is only one cell thick, creating the best possible conditions for gas exchange. Between the capillary and the alveolus is a layer of elastic connective tissue which holds them together. The elastic elements in this tissue help to force air out of the stretched lungs. This is known as the **elastic recoil** of the lungs.

Intercostal muscles

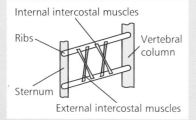

Internal intercostal muscles

Ribs

Vertebral column

Sternum

External intercostal muscles

The intercostal muscles are found beween the ribs. The **external intercostal** muscles contract to raise the rib cage upwards and outwards during inspiration, while the **internal intercostal** muscles contract to bring the ribs closer together and lower the rib cage during expiration.

Components of lung volume

A certain amount of air is always present in the respiratory system, simply filling up the spaces when no air is flowing. Other than this, the volume of air which is drawn in and out of the respiratory system can be very variable. There are different components of lung volume which have the following specific names for ease of reference:

(1) The **tidal volume** (V_T) is the volume of air that enters and leaves the lungs at each natural resting breath.

(2) The **inspiratory reserve volume** (**IRV**) is the volume of air that can be taken in by a maximum inspiratory effort, over and above the normal inspired tidal volume. In other words, this is the extra air that you can take in when you breathe in as deeply as possible after a normal exhalation.

(3) The **expiratory reserve volume** (**ERV**) is the volume of air that can be expelled by the most powerful expiratory effort, over and above the normal expired tidal volume. This is the extra air breathed out when you force the air out of your lungs as hard as possible after a normal expiration.

(4) The **vital capacity** (**VC**) is the sum of the tidal volume and the inspiratory and expiratory reserves. It is the volume of air which can be breathed out by the most vigorous possible expiratory effort following the deepest possible inspiration.

(5) The **residual volume** (**RV**) is the volume of air left in the lungs after the strongest possible exhalation. It has to be measured indirectly.

(6) The **total lung capacity** (**TLC**) is the sum of the vital capacity and the residual volume.

(7) The **inspiratory capacity** (**IC**) is the volume that can be inspired from the end of a normal exhalation – in other words, the sum of the tidal volume and the inspiratory reserve volume.

Figure 2.4.5 illustrates these volumes.

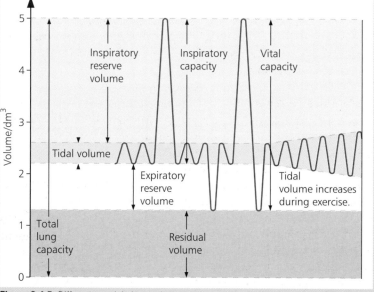

Figure 2.4.5 Different people's lung volumes vary – men usually have greater lung volumes than women, and athletes have larger lung volumes than non-athletes due to their training. This diagram gives the *average* figures.

Breathing rhythms

The body constantly needs oxygen to be delivered and carbon dioxide to be removed, so breathing continues throughout our lives. The pattern of our breathing will alter under differing conditions of exercise, stress, fitness, etc., but in normal circumstances a rhythm of some kind will be maintained.

The **tidal volume**, the amount of air naturally breathed in or out, is usually about 500 cm^3 in a normal person at rest, or about 15% of the vital capacity of the lungs. The rate of breathing can be expressed as the **ventilation rate**, and this is a measure of the volume of air breathed in a minute:

Ventilation rate = tidal volume × frequency of inspiration (per minute)

The ventilation rate is a useful measurement which is affected by two things – the amount of air taken into the lungs at each breath and the number of breaths per minute. For example, the tidal volume can increase from 15% to 50% of the vital capacity during heavy exercise – a great deal more air is then passing through the system. The frequency of inspiration shows similar increases during exercise – you can easily measure this yourself.

The ventilation of the lungs is regulated to bring in sufficient air to allow

Measuring the volume of inhaled or exhaled air

A piece of apparatus known as a **spirometer** is used to find out information such as the vital capacity of a person's lungs, or to measure the inspiratory or expiratory reserve volume. Spirometers come in a wide variety of shapes and sizes, but they all work in the same way – that is, as a gasometer.

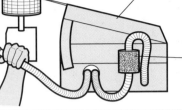

The subject of the experiment breathes in and out of the air-tight chamber, making it move up and down, until all the oxygen is used up.

Revolving drum on which a trace is drawn out as the lid moves up and down

Air-tight chamber – in this case a perspex lid floating on water. The chamber is filled with oxygen at the beginning of the experiment. Attached to the lid of the chamber is an arm with a pen on the end.

Canister of soda lime to remove carbon dioxide from the exhaled air. Carbon dioxide levels affect the rate of breathing and would affect the investigation if allowed to accumulate.

Figure 2.4.6 The volume of gas inhaled and exhaled under a variety of conditions can be measured using a spirometer.

effective gaseous exchange. This exchange of gases must supply all the oxygen required by the tissues of the body appropriate to their level of internal respiration, and must also remove all the carbon dioxide which is formed as a waste product.

(2) Gaseous exchange

Gaseous exchange takes place in the alveoli. The air in the alveoli does not have the same composition as the atmospheric air which is breathed in. In fact, the levels of oxygen and carbon dioxide in atmospheric air would be toxic to most body cells with prolonged exposure.

There is always a substantial volume of air in the respiratory tract. The incoming air mixes with this, changing the relative proportions of the gases to levels which suit the cells much better. In normal quiet breathing, about $500\,cm^3$ of air are drawn in with each breath. Of this, about $350\,cm^3$ reach the alveoli and mix with the air that is already there – usually about $2500\,cm^3$ ($2.5\,dm^3$). This mixture or **alveolar gas** is involved in gaseous exchange with the blood. Table 2.4.1 shows the relative proportions of the main respiratory gases at various points of the respiratory tract.

Gas	Inspired air	Alveolar air	Expired air
Oxygen	20.8	13.1	15.3
Carbon dioxide	0.04	5.2	4.2

Table 2.4.1 The percentages of the main respiratory gases found in inspired air, alveolar air and expired air

The alveoli provide the enormous surface area needed for the exchange of gases in the human body. They fulfil the requirements for gaseous exchange – a large surface area, thin walls (single flattened cuboidal epithelial cells), moist surfaces and close proximity to the blood. The number of alveoli increases tenfold from birth to adulthood, giving an increased surface area to supply the increased body size. Within an adult human lung there are around 300 million alveoli, supplied by 280 million capillaries. The two lungs together give a surface area of between 60 and 80 m² for gaseous exchange – equivalent to the surface area of a tennis court.

Keeping the enormous surface for gaseous exchange folded into the controlled environment of the lungs is vitally important in the control of water loss. The alveolar air is saturated with water vapour, so evaporation from the moist alveolar epithelium is reduced. If this area of moist respiratory surface was on the outside of the body, death by desiccation would occur very rapidly due to an enormous water loss by evaporation.

Partial pressures of gases

The relative proportions of the respiratory gases in table 2.4.1 have been shown as their percentages of the total constituents of the air. A more scientific way of comparing the gases is to consider their **partial pressures**.

The partial pressure of a gas in a mixture of gases in a fixed volume is the pressure that the individual gas would exert if it alone occupied the container. Kinetic theory says that gas molecules exert a pressure by colliding with the sides of the container. The partial pressure of a gas in a mixture is related to the number of molecules of that gas present in the mixture – in other words, the mole fraction of the gas.

Partial pressures are measured in kilopascals (kPa). In the case of the respiratory gases in the air, the percentages and the partial pressures give very similar proportions. The percentage of oxygen in the inspired air is 20.8%, while the partial pressure of oxygen is 21.1 kPa. The percentage of carbon dioxide in inspired air is 0.04% and the partial pressure is 0.04 kPa.

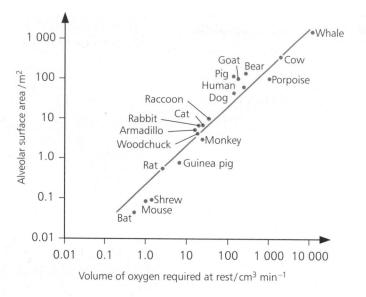

Figure 2.4.7 The alveoli of the lungs are well adapted to supply the oxygen demands of a wide range of mammals, as this graph shows. (*After S.M. Tenney and J.E. Remmes, Comparative quantitative morphology of the mammalian lung: diffusing area. Nature 197: 54, 1963.*)

Gaseous exchange occurs by a process of simple diffusion between the alveolar air and the deoxygenated blood in the capillaries. This blood has a relatively low oxygen content and a relatively high carbon dioxide content. The gases are exchanged between the blood and the air so that the blood leaving the alveolus has similar proportions of oxygen and carbon dioxide as the expired air, as figure 2.4.8 illustrates.

The alveoli are basically minute bubbles of gas. They naturally have a tendency to collapse, but this is prevented by a chemical known as **lung surfactant**. This is a phospholipid which is first produced in any quantity in the lungs of the fetus from about the 28th week of pregnancy and it helps to ensure that the lungs do not collapse after every breath. The absence of lung surfactant is one of the major causes of problems in babies born prematurely, as they have to struggle much harder to breathe than a full-term infant.

(3) Control and regulation of breathing

The respiratory system must be tightly controlled. If too little oxygen is taken in the tissues work less efficiently – and eventually death results. Too little carbon dioxide in the blood results in the pH of the body fluids rising which causes, surprisingly, lack of oxygen. Too much oxygen means the body is

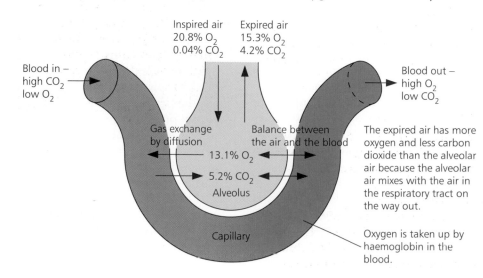

Figure 2.4.8 Diffusion across the alveolar surface provides a very successful means of obtaining oxygen and getting rid of carbon dioxide.

Lung surfactant – the theory

As small gas bubbles we would expect alveoli to have a tendency to collapse, which will depend largely on the surface tension of the alveolar lining. The pressure P exerted by the surface tension of a bubble in a liquid is expressed as:

$$P = 2T/r$$

where T is the surface tension and r is the radius of the bubble.

In breathing out the radius of the alveolus will get smaller. As you can see, this will increase the pressure resulting from the surface tension and so make it more likely that the alveolus will collapse. Lung surfactant ensures that this does not happen.

Molecules of lung surfactant are well spread out as the alveolus expands during inhalation. But during exhalation they become more closely packed and have the effect of lowering the surface tension (T). This in turn counteracts the effect of the decrease in radius, and so the tendency of the alveoli to collapse is reduced.

The breath of life

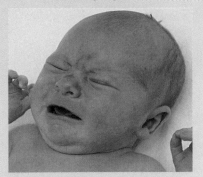

Figure 2.4.9 The first breath is the most important and probably the most difficult of your life.

The lungs of a fetus are fluid filled and have a relatively small volume. Breathing movements are 'practised' by the fetus whilst it is still in the uterus, but of course no air can be taken in. Very soon after birth the first breath must be taken. This first inspiratory effort is brought about by a very powerful contraction of the diaphragm. To successfully accomplish this breath the newborn infant has to exert a force 15 to 20 times greater than that needed for a normal inhalation. As a result the lungs are greatly stretched and the elastic tissue they contain never again returns to its original length. Assuming that plenty of lung surfactant is present, the infant will then establish a breathing rhythm which will continue for the rest of its life.

---

doing unnecessary work and wasting energy, whilst too much carbon dioxide can lead to death.

The oxygen needs of the body can change very rapidly from the relatively low levels at rest to the high levels during strenuous exercise, and the amount of carbon dioxide to be removed changes similarly. The ventilation rate must be able to adjust, by increasing both the tidal volume and the frequency of inspiration. How is this control brought about?

The basic stimulus to inhale and exhale is given by an area of the hindbrain known as the **respiratory centre**. This gives a basic deep, slow breathing rhythm. Inputs from stretch receptors in the bronchi, other receptors sensitive to the carbon dioxide levels of the blood and the higher centres of the brain, all interact with the basic rhythm to give a finely tuned respiratory response to most situations. This control system will be considered in more depth in section 4.

The human respiratory system has evolved to cope with the problems of gaseous exchange for a large, land-dwelling animal. All mammals and most land vertebrates have developed a somewhat similar system. But internal lungs are not the only way to solve these problems, as a look at the respiratory systems of insects shows.

THE INSECT RESPIRATORY SYSTEM

Many insects are very active during parts of their life cycles. They are therefore in a similar position to the mammals, being complex and largely land-dwelling animals with relatively high oxygen requirements and an external surface through which little or no gaseous exchange can take place.

Most animals have an internal respiratory system linked to a transport system, with oxygen diffusing into the blood or a similar fluid and being carried around the body to the individual cells in this way. Figure 2.4.10 shows an example of the respiratory system of insects, which has evolved to deliver oxygen directly to the cells and to remove carbon dioxide in the same way.

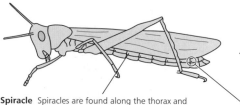

Spiracle Spiracles are found along the thorax and abdomen of most insects. They are the site of the entry and exit of the respiratory gases. In many insects they can be opened or closed by sphincters, which is of great value in the control of water loss.

Watery fluid in the tracheoles. Water can be withdrawn from the tracheoles by osmosis when extra oxygen is needed, giving a form of control over the level of respiration.

Trachea Tracheae are the largest tubes of the insect respiratory system, carrying the air into the body. They may be up to 1 mm in diameter. They run both into and along the body of the insect. The tubes are lined by spirals of chitin which keep them open if they are bent or pressed. Chitin is the material which makes up the cuticle. The tracheae are therefore relatively impermeable to gases and little gaseous exchange takes place there. The tracheae branch to form narrower and narrower tubes until they break up into the tracheoles.

Tracheoles Tracheoles are minute tubes of diameter 0.6–0.8 μm. Each is a single greatly elongated cell, with no chitin lining, so they are freely permeable to gases. Because of their very small size they spread throughout the tissues of the insect, running between individual cells. Most of the gaseous exchange takes place between the air and the respiring cells in the tracheoles.

Figure 2.4.10 The respiratory system of an insect has to fulfil the same requirements as the human respiratory system. In spite of its very different design, there are many similar features.

How the insect respiratory system works

Air enters the system through the spiracles. To minimise the amount of water lost the spiracle sphincters are kept closed as much as possible. For example, an adult flea which has sphincters is much more resistant to desiccation than a larval flea which has no sphincters. Figure 2.4.11 shows the effect of opening the spiracles on water loss in an insect.

When an insect is inactive and its oxygen demands are very low, the spiracles will all be closed. The occasional opening of just one or two pairs brings in enough air for gaseous exchange. When the oxygen demand is higher or the carbon dioxide levels build up, more spiracles open.

Air moves along the tracheae and tracheoles and reaches all the tissues by diffusion alone. The vast numbers of tiny tracheoles, even penetrating into the cells themselves, give a very large surface area for gaseous exchange. The tracheoles contain a watery fluid towards the end of their length. This limits the penetration of the gases for diffusion. However, when oxygen demands build up – when the insect is flying, for example – lactic acid in the tissues causes water to be withdrawn from the tracheoles by osmosis and exposes additional surface area for gaseous exchange.

All the oxygen needed by the insect's cells is supplied to them by the respiratory system. However, up to 25% of the carbon dioxide produced by the cells is lost directly through the cuticle.

The extent of respiration in most insects is controlled by the opening and closing of the spiracles. There are respiratory centres in both the ganglia of the nerve cord and the brain. They are stimulated by increasing carbon dioxide levels and by the lactic acid which builds up in active tissues when there is a lack of oxygen. It seems that a combination of lack of oxygen and carbon dioxide build-up work together to provide the insect with a flexible and responsive respiratory system.

RESPIRATORY SYSTEMS IN FISH

As we have seen, animals living on the land have to overcome the problems of water loss from their respiratory surfaces. For animals which live in, and obtain oxygen from, water this is not a problem – but there are other difficulties to overcome.

Water is 1000 times denser than air. It is 100 times more viscous and has a much lower oxygen content. To cope with the viscosity of water and the slow rate of oxygen diffusion, aquatic animals such as fish have evolved very specialised

Very active insects

The type of respiratory system described so far works well for small insects and for large but slow ones. Those insects which have more active lifestyles, for example larger beetles, locusts and grasshoppers, bees, wasps and flies, have much higher energy demands. To supply the extra oxygen needed, alternative methods of increasing the level of gaseous exchange are used.

(1) Some form of mechanical ventilation of the tracheal system may be introduced – in other words, air is actively pumped into the system. This is usually brought about by increased opening of the spiracles along with muscular pumping movements of the thorax and/or abdomen.

(2) Some active insects have collapsible tracheae or **air sacs** which act as air reservoirs and are used to increase the amount of air moved through the respiratory system. They are usually inflated and deflated by the ventilating movements of the thorax and abdomen. In some insects they can be ventilated by the general body movements, for example, when a locust is in flight the air sacs within the muscles are automatically inflated.

respiratory systems. Moving water in and out of lung-like respiratory organs under water would use up an enormous amount of energy. Moving water in one direction only is much simpler and more economical in energy terms.

How the fish respiratory system works

The bony fish, such as cod, salmon and sticklebacks, cannot undergo gaseous exchange through their scaly external covering, but they have evolved a respiratory system which works very well in the water. **Gills** incorporate the large surface area, large blood supply and thin walls needed for efficient gaseous exchange. In the bony fish they are contained in a gill cavity covered by a protective **operculum** which is also active in maintaining a flow of water over the gills. Figure 2.4.12 shows the respiratory system of a bony fish.

When looking at the respiratory system of the fish, there are two main questions we shall consider:

1 How is a flow of water maintained over the gills to allow continuous gaseous exchange?

2 How is gaseous exchange carried out as effectively as possible in a medium where diffusion tends to be slow?

(1) Water flow over the gills

While fish are swimming it is easy to see that they could keep a flow of water over their gills simply by opening the mouth and operculum. Problems arise when the fish stops moving. The more primitive cartilaginous fish such as the sharks often rely on continual movement to ventilate the gills. But the majority of the bony fish have evolved a more sophisticated system involving the operculum, shown in figure 2.4.13.

(2) Effective gaseous exchange

Gills, like other respiratory surfaces, have a large surface area, a rich blood supply and thin walls. There are two other aspects of the gills which help to ensure that effective gaseous exchange occurs.

The tips of adjacent gill filaments overlap. This increases the resistance to the flow of water, slowing down the movement of water over the gill surfaces. The result is more time for the exchange of gases to take place.

The blood in the gill filaments and the water moving over the gills flow in different directions. Diffusion occurs down a concentration gradient, and the

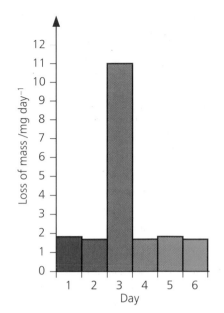

Figure 2.4.11 The effect of opening the spiracles on water loss can be seen clearly from this graph. An adult *Rhodnius* (a blood-sucking bug) was kept in dry air for 6 days. It was not fed, to keep it relatively inactive throughout. On day 3 the bug was exposed to a raised level of carbon dioxide in the air which made the spiracles open. The air was returned to normal on day 4.

Figure 2.4.12 The gills make up the respiratory surface of the fish. They have many features in common with both mammalian and insect respiratory surfaces, but are particularly adapted to their environment.

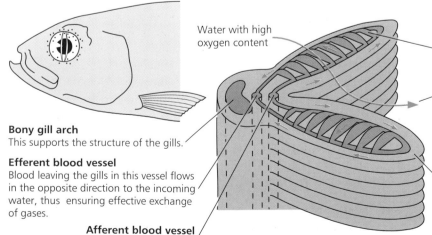
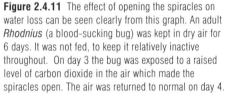

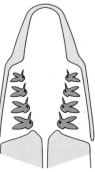

Bony gill arch
This supports the structure of the gills.

Efferent blood vessel
Blood leaving the gills in this vessel flows in the opposite direction to the incoming water, thus ensuring effective exchange of gases.

Afferent blood vessel
This brings blood into the system.

Water with high oxygen content

Gill lamellae
These are the main site of gaseous exchange in the fish. They have a very rich blood supply and give the gill filaments their large surface area.

Water with low oxygen content

Gill filaments
The fragile gill filaments occur in large stacks. They need water to keep them apart and so to expose the large surface area needed for gaseous exchange. A fish does not last long out of water because these stacks of filaments collapse onto each other and insufficient surface area is exposed to supply the respiratory needs.

Arrangement of the gill stacks

steeper the concentration gradient, the more effective is diffusion. By having this **countercurrent exchange system**, steeper concentration gradients are maintained and so more gaseous exchange can take place than if the blood and water flowed in the same direction (in a **parallel system**). Figure 2.4.14 illustrates this. The cartilaginous fish – the sharks and rays – have parallel systems and extract only about 50% of the oxygen from the water. The bony fish, with their countercurrent systems, remove about 80%.

From these examples it can be seen that respiratory systems are very highly developed to enable larger organisms to carry out gaseous exchange. Mammalian lungs, insect tracheal systems and the gills of fish are not the only ways of carrying out gaseous exchange. However, they demonstrate the type of features which enable an animal with an unfavourable surface area:volume ratio to obtain the oxygen it needs for life.

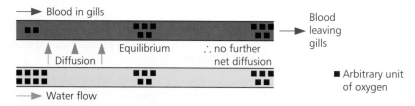

If the blood in the gills travelled in the same direction as the water flowing over the gills there would be a steep concentration gradient between the two at the beginning. Diffusion would take place until the blood and water were in equilibrium, after which no net movement of oxygen into the blood or carbon dioxide out of the blood would occur.

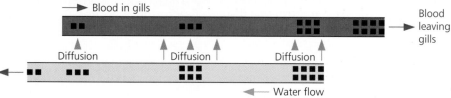

In a countercurrent system a concentration gradient between the water and the blood is maintained along the gill. As a result, a far higher saturation of the blood with oxygen is possible, and larger amounts of carbon dioxide can be removed.

The buccal cavity (mouth) is first expanded as the floor is lowered. This causes the pressure to drop in the mouth and water moves in through the mouth opening. The operculum is shut and the opercular cavity containing the gills expands. This lowers the pressure in the opercular cavity containing the gills. The floor of the buccal cavity moves up, increasing the pressure there and so water moves from the buccal cavity over the gills.

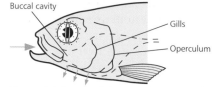

The mouth is then closed, the operculum opens and the sides of the opercular cavity move inwards, increasing the pressure and forcing water out of the operculum. The floor of the buccal cavity continues to be raised, maintaining a flow of water over the gills.

Figure 2.4.13 This continuous process ensures that water is flowing over the gills at all times.

Figure 2.4.14 The advantages of a countercurrent exchange system. These systems occur in a variety of different roles throughout the animal kingdom.

SUMMARY

- **Respiratory systems** are a means of effecting gaseous exchange in multicellular organisms, necessitated by the process of cellular respiration which uses oxygen and produces carbon dioxide.

- Gaseous exchange takes place at **respiratory surfaces**, which have a large surface area and are permeable, thin, moist and in many animals have a rich blood supply.

- In plants the gaseous exchange needed for both photosynthesis and cellular respiration takes place through the stomata.

- The land vertebrates have **lungs** to provide a respiratory surface while limiting water loss.

- The **human respiratory system** consists of the **trachea** (windpipe) connected to the nose and mouth and covered by the **epiglottis** during swallowing. The trachea is lined with **columnar epithelial** cells with **cilia** which move mucus and trapped microorganisms and dust away from the lungs. The **mucus-secreting** cells are below the epithelium. The trachea splits to form two **bronchi** which like the trachea are lined with C-shaped **cartilage rings**.

- The bronchi are further divided to form **bronchioles**. The larger ones are lined with columnar epithelium and supported with cartilage rings, while the smaller ones are lined with **flattened cuboidal epithelium**. Some gaseous exchange takes place in these bronchioles.
- **Alveoli** are sacs at the ends of the smallest bronchioles, lined with **squamous epithelium**. Capillaries with walls one cell thick run alongside the alveoli and gaseous exchange occurs between the alveolar air and the blood in the capillaries.
- Two **pleural membranes** surround the lungs, between which is the **pleural cavity**.
- The **ribs** protect the lungs and their movement is brought about by the **intercostal muscles**. The **diaphragm** forms the floor of the chest cavity.
- **Inhalation** is brought about by contraction of the **external** intercostal muscles pulling the rib cage up and out, and the contraction of the diaphragm which moves it down. This increase in volume decreases the pressure in the lungs so air moves in.
- During **exhalation**, the **internal** intercostal muscles contract, pulling the rib cage down and in, while the diaphragm relaxes and moves up, so forcing air out of the lungs.
- The rate of breathing is expressed as the **ventilation rate**, which is the **tidal volume** (the volume breathed in) × the **frequency of inspiration**.
- The ventilation rate is controlled by signals from the respiratory centre in the brain, from stretch receptors in the bronchi and from receptors sensitive to carbon dioxide levels in the blood.
- The **insect respiratory system** consists of **spiracles** which can open and close, leading to tubes called **tracheae**, lined with chitin for support and so impermeable. These lead to **tracheoles**, single elongated cells permeable to gaseous exchange, which run between the insect body cells. Watery fluid can be withdrawn from the tracheoles when necessary to increase the surface area for gaseous exchange.
- The **respiratory system of a bony fish** consists of **gills** within a protective **operculum**. A bony **gill arch** supports stacks of **gill filaments** which contain **gill lamellae**, the site of gaseous exchange, which are well supplied with blood.
- Water is constantly forced over the gills by the combined action of the buccal cavity and the operculum. The overlapping gill filaments slow down the flow of water, and the blood runs in the opposite direction to the water, so **countercurrent exchange** takes place.

QUESTIONS

1 a Why are respiratory systems necessary?
  b What are the main features of successful respiratory systems?
  c Why are active respiratory systems not found in plants?
  d Sketch the human respiratory system and annotate the main regions.

2 a How is the rate of breathing in human beings determined experimentally?
  b What factors are most likely to affect the breathing rate?
  c Describe how the rate of breathing is controlled.

3 Draw up a table to compare the respiratory systems of a mammal, an insect and a fish.

2.5 Transport systems in animals

The features of transport systems were discussed in section 2.3. Most of the larger animal groups possess a specific system for the transport and distribution of materials around the organism. This usually takes the form of a **cardiovascular system**.

THE CARDIOVASCULAR SYSTEM

A cardiovascular system is made up of a series of vessels with a pump (the heart) to move blood through the vessels. The heart and blood vessels form a transport system with the blood as the transport medium. The system delivers the materials needed by the cells of the body, and carries away the waste products of their metabolism. The cardiovascular system often carries out other functions as well, such as carrying hormones (chemical messages) from one part of the body to another, transporting the defence system of the body and distributing heat. Transport systems of this type are found in a wide range of animals, but we shall concentrate on the human cardiovascular system. This has been widely researched and is a good representative of systems of this type.

The human cardiovascular system

The human cardiovascular system includes various structures with different functions – the heart, the arteries, the arterioles, the capillaries, the venules and the veins. Mammals and birds possess the most complex type of transport system, known as a **double circulation** which involves two systems, shown in figure 2.5.1. One, the **systemic circulation**, carries oxygen-rich blood from the heart to the cells of the body and deoxygenated blood back to the heart. The other, the **pulmonary circulation**, carries deoxygenated blood to the lungs and oxygenated blood back to the heart.

The blood is contained within the closed circulation system, making a continuous journey out to the most distant parts of the body and back to the heart. The ancient Chinese had a good understanding of the circulation of the blood around 100 BC, but it was not until William Harvey, in his writings of 1628, showed how the circulation worked that the idea was accepted in Europe.

THE BLOOD VESSELS

Transport routes

The various types of blood vessels which make up the circulatory system show differences in their structures which are closely related to their functions. The largest vessels are the major named arteries and veins, such as the pulmonary vein coming away from the lungs or the mesenteric artery supplying the small intestine. The blood flows quickly through these vessels, which then divide into smaller and smaller vessels until they are linked by the vast, branching and spreading capillary network. As the diameter of the vessels decreases, so their resistance to the flow of blood increases. It is interesting to look at the approximate numbers of some of these linked vessels shown in table 2.5.1 and figure 2.5.2.

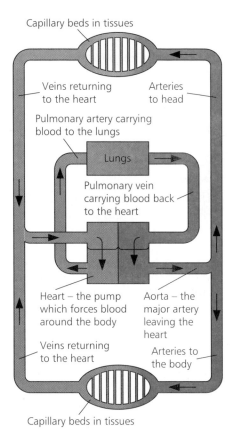

Figure 2.5.1 A double circulation of this type ensures that the blood sent to the active cells of the body is carrying as much oxygen as possible.

Type of blood vessel	Total number
Mesenteric artery	1
Main branches of mesenteric artery	15
Short and long branches to intestine	1 899
Branches to villi	328 500
Arteries of villi	1 051 000
Capillaries of villi	47 300 000
Veins at base of villi	2 102 400
Branches from villi	131 400
Long and short branches from intestine	1 899
Branches of mesenteric vein	15
Mesenteric vein	1

Table 2.5.1

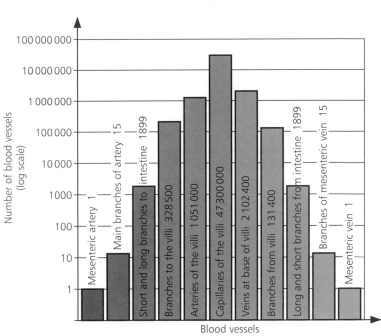

Figure 2.5.2 The numbers of different blood vessels within the mesenteric system of a dog. The mesenteric system carries blood to and from the very active intestinal region.

When considering the blood vessels we shall look at each type separately, considering the relationship between structure and the function. It must be remembered, however, that the vessels do not exist as separate structures – they are interlinked within the complexities of the whole circulatory system.

Arteries

The **arteries** carry blood *away from* the heart. The structure of an artery is shown in figure 2.5.3. Arteries almost all carry **oxygenated** or oxygen-rich blood towards the cells of the body. The only exceptions are the pulmonary artery, which carries deoxygenated blood from the heart to the lungs, and the umbilical artery, which carries deoxygenated blood from the fetus to the placenta. As arteries get further from the heart they branch, and the diameter of the lumen gets smaller. The very smallest branches of the arterial system are referred to as **arterioles**.

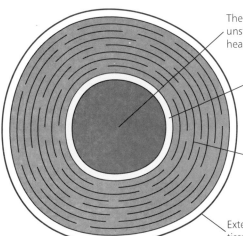

The lumen is small when the artery is unstretched by the flow of blood from the heart.

Smooth lining of squamous epithelium allows a smooth flow of blood.

The middle layers of the artery structure are made up of varying proportions of elastic fibres and smooth muscle. The arteries nearest the heart have a lot of elastic fibres, those further from the heart have a greater proportion of muscle tissue.

External layer of connective tissue, mainly collagen

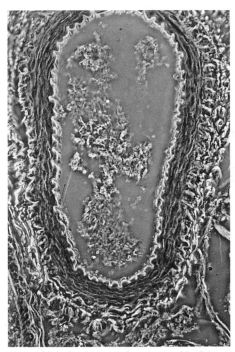

Figure 2.5.3 The structure of an artery enables it to cope with the surging of the blood as the heart pumps.

Blood is pumped out from the heart in a regular rhythm, about 70 times a minute. The major arteries close to the heart have to be able to withstand the pressure of these spurts of blood, and stretch to accommodate the increased volume. They also 'even out' the blood pumped from the heart to give a continuous, if pulsing, flow. The large proportion of elastic fibres in artery walls helps them fulfil these functions, allowing them to stretch without being damaged when blood is pumped out from the heart. Then, when the heart relaxes and no further blood is being forced into the arteries, the elastic fibres return to their original length, squeezing the blood and so moving it along in a continuous flow. The pulse which can be felt in an artery is the effect of the surge each time the heart beats, which the arteries do not completely eliminate.

In the more peripheral arteries, contracting or relaxing the muscle fibres in the artery walls can be used to change the size of the lumen. The smaller the lumen, the harder it is for blood to flow through the vessel. Thus the muscular walls of the arteries supplying blood to the various organs of the body can control the amount of blood that flows into these organs, and so affect the level of their activity.

The blood pressure in all arteries is relatively high, but drops with distance away from the heart. Figure 2.5.4 compares the structures of different sized arterial vessels.

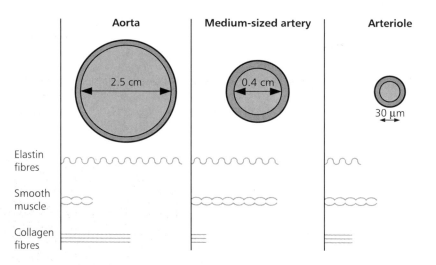

Figure 2.5.4 The differences in function of these three different types of artery are reflected in the proportions of the tissues making up their walls.

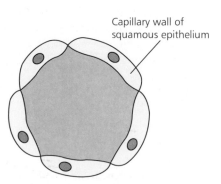

Capillaries

Linking the arterial and venous system is the network of **capillaries**. These minute vessels spread throughout the tissues of the entire body. They are found between cells, and no cell is far from a capillary to supply its needs. The capillary system provides an enormous surface area for the diffusion of substances into and out of the blood. Also, because the diameter of each individual capillary is very thin, the blood travels relatively slowly through them, again giving more opportunity for diffusion.

Capillaries have a very simple structure well suited to their function. Their walls contain no elastic fibres, smooth muscle or collagen as this would interfere both with the ability of the capillaries to penetrate between individual cells and with efficient diffusion. The walls consist of **squamous epithelium**, which is extremely thin in cross-section. Oxygen is removed from the blood as it travels through the capillaries, and carbon dioxide is loaded into it, so the capillaries are said to carry mixed blood – oxygenated when it enters the capillary network but deoxygenated by the time it leaves.

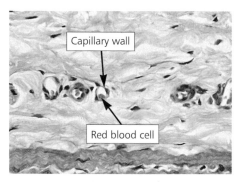

Figure 2.5.5 Capillary walls are extremely thin to allow for diffusion of oxygen, carbon dioxide and food molecules.

The lymphatic system

Capillary walls are permeable to everything in the blood apart from the erythrocytes (red blood cells) and the plasma proteins. As blood flows through the capillaries under pressure from the arterial system, fluid is squeezed out of the vessels. This fluid fills the spaces between the cells and is known as **intercellular fluid** or **tissue fluid**. It is through this fluid that diffusion between the blood and the cells takes place.

Much intercellular fluid is eventually returned to the blood. Some of it returns by osmosis and diffusion at the venous end of the capillaries. Most of it drains into a system of blind-ended tubes called **lymph capillaries** and once in these vessels the fluid is called **lymph**. The lymph capillaries join up to form larger and larger vessels. The fluid is transported through them by the squeezing effect of muscular movements and backflow is prevented by a system of valves. The lymph is returned to the blood in the neck area, into the left and right subclavian veins (underneath the collar bone or clavicle).

Situated along the lymph vessels are **lymph glands**. Lymphocytes accumulate in these glands and produce antibodies which are then emptied into the blood. The glands also filter out bacteria and other foreign particles from the lymph to be ingested by phagocytes. Thus the lymphatic system, shown in figure 2.5.6, plays a major role in the defence mechanisms of the body.

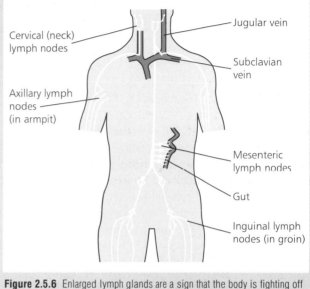

Figure 2.5.6 Enlarged lymph glands are a sign that the body is fighting off an invading pathogen. You can see why doctors often examine the neck, armpits, stomach and groin of patients – these are the main sites of the lymph glands.

Veins

The **veins** carry blood *towards* the heart. This means that most of them carry blood which is **deoxygenated** – it has given up its oxygen to the cells. There are two exceptions to this. The pulmonary vein carries oxygen-rich blood from the lungs back to the heart for circulation around the body, and the umbilical vein of the fetus carries oxygenated blood from the placenta into the fetus.

From the capillary network in the tissues, blood enters tiny **venules** which merge into larger and larger vessels. Eventually only two veins carry the returning blood to the heart – the inferior vena cava from the lower parts of the body and the superior vena cava from the upper parts of the body.

Veins can hold a large volume of blood – in fact, more than half of the total body volume of blood is in the venous system at any one time. The veins act as a blood reservoir. The blood pressure in the veins is relatively low – the pressure surges from the heart have been eliminated as the blood passes through the capillary beds. Figure 2.5.7 shows how the structure of veins differs from that of arteries.

The venous system has two main methods of ensuring that blood is returned to the heart:

1 There are one-way valves at frequent intervals throughout the venous system called **semilunar valves**. These are formed from infoldings of the inner wall of the vein, shown in figure 2.5.8. Blood can pass through in the direction of the heart, but if it tends to flow backwards the valves close, preventing this.

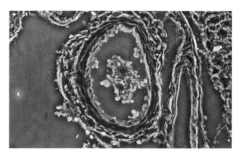

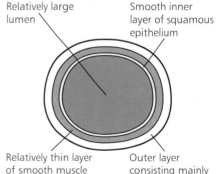

Figure 2.5.7 Veins do not have to withstand the high pressures of the arterial system and this is reflected in their structure.

2 Many of the larger veins are situated between the large muscle blocks of the body, particularly in the arms and legs. When the muscles contract during physical activity they squeeze these veins. As the valves make sure that the blood can only travel towards the heart, this squeezing helps to force the returning blood on its way.

THE HEART

The heart is the organ that moves blood around the body. In some animal groups it is no more than a simple muscular tube. In mammals the heart is a complex, four-chambered muscular bag. It is found in the chest, protected by the ribs and sternum. An effectively beating heart is vital to the successful functioning of the human transport system and to the overall health of the body.

The structure of the human heart

The heart is one of the first organs to be completed as the embryo develops. It consists of two pumps joined together and working synchronously. The right-hand side of the heart receives blood from the body and pumps it to the lungs. The left-hand side of the heart receives blood from the lungs and pumps it to the body. The blood from the two sides of the heart does not mix. The whole heart is surrounded by the inelastic **pericardial membranes**, one attached to the heart itself. Fluid is secreted between these membranes which allows them to move easily over each other. They help prevent the heart from over-distending with blood. The structure of the heart is shown in figure 2.5.9.

How the heart works

Relating the structures of the heart to their functions as in figure 2.5.9 gives a good picture of how the heart works. In order to do this a particular volume of blood is followed on its journey round the heart. However, it is important to remember that the heart works as a unit, with both sides contracting at the same time.

Months before birth the heart has begun its regular contraction and relaxation which will continue throughout life. The contraction of the heart is called **systole**. First the atria contract, closely followed by the ventricles. Systole forces blood out of the heart to the lungs and general body circulation. Between contractions the heart relaxes and fills with blood. This relaxation stage is called **diastole**. Diastole and systole together make up a single heartbeat, which lasts about 0.8 s in humans and is known as the **cardiac cycle**. This cycle is made up of the following sequence of events:

1 Blood enters the right atrium from the body and the left atrium from the lungs. The atria fill and then the atrioventricular valves open under pressure so allowing the ventricles to fill as well. This is **diastole**.

2 The two atria contract together, forcing more blood into the ventricles which in turn both contract at the same time. During this **systole** blood is forced out of the heart into the pulmonary artery and the aorta, as figure 2.5.10 shows.

The heartbeat

The beating of the heart produces sounds known as the **heartbeat**. The heartbeat can be heard by putting an ear to the chest wall, but an instrument called a **stethoscope** makes the heartbeat clearer. The sounds of the heartbeat are made not by the contracting of the heart muscle, but by blood hitting the

Blood moving in the direction of the heart forces the valve open, allowing the blood to flow through.

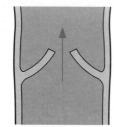

A backflow of blood will close the valve, ensuring that blood cannot flow away from the heart.

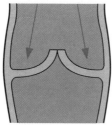

Figure 2.5.8 An extensive system of valves in the venous system makes sure that the blood only flows towards the heart.

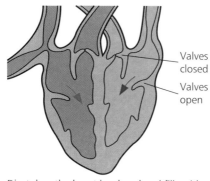

Valves closed

Valves open

Diastole – the heart is relaxed and fills with blood.

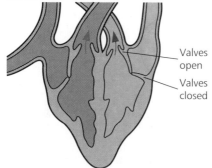

Valves open

Valves closed

Systole – the heart (atria followed by ventricles) contracts and forces blood out to the lungs and round the body.

Figure 2.5.10 The sequence of events in the cardiac cycle

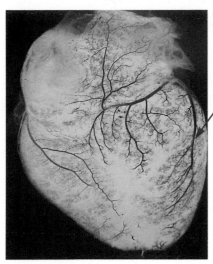

Outside view of
the heart

Coronary arteries
supply heart
muscles with
food and oxygen.

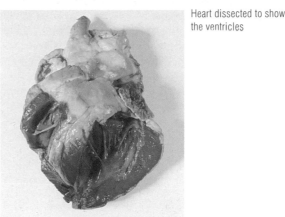

Heart dissected to show
the ventricles

Pulmonary artery carries
deoxygenated blood from the
heart to the lungs.

Carotid arteries
to the head

Aorta: This is the major artery of the body, carrying
oxygenated blood away from the heart. It forms the
aortic arch and other major arteries branch off from it.

The **great veins** receive
deoxygenated blood from the
body and deliver it to the heart.

Superior
vena cava

Semilunar valves prevent the blood flowing back
into the ventricles.

Pulmonary veins: carry oxygenated blood back
from the lungs to the left side of the heart at
relatively low pressure as the blood has travelled
through the extensive capillaries of the lungs.

Inferior vena
cava

Right atrium: This chamber receives the
deoxygenated blood from the great veins at
relatively low pressure. It retains the blood until
the pressure builds up and opens the tricuspid
valve so that the ventricle can fill with blood.
When both atrium and ventricle are filled with
blood the atrium contracts, forcing all the
blood into the ventricle. The atrium has thin
muscular walls as the blood it receives is at low
pressure and it needs to exert relatively little
pressure to move it into the ventricle. One-way
valves at the entrance to the atrium stop a
backflow of blood into the great veins.

Left atrium: This is another thin-walled chamber
which performs the same function as the right
atrium, contracting to force blood into the left
ventricle. Backflow is again prevented by one-way
valves at the atrial entrance.

Tendinous cords

Bicuspid valve

The **bicuspid valve** is made up of two flaps. It is
also an atrioventricular valve. It functions in the
same way as the tricuspid valve, allowing blood
through from the atrium to the ventricle but closing
as the ventricle contracts to prevent a backflow of
blood.

The **tricuspid valve** is made up of three flaps.
It is also known as an atrioventricular valve as
it separates the atrium and ventricle. The valve
allows blood to move from the atrium to the
ventricle, but not in the other direction. The
tendinous cords make sure the valve is not
turned inside out by the great pressures
exerted when the ventricles contract.

Tricuspid
valve

Tendinous
cords

The heart is made up of **cardiac muscle**. This
muscle has special properties which enable the
heart to carry out its regular contractions without
fatigue. These properties will be considered in
more detail in section 2.6.

Left ventricle
Very thick muscular wall

Right ventricle: The right ventricle is filled with
deoxygenated blood by contraction of the right
atrium. Shortly afterwards the ventricle also
contracts, its thicker muscular walls allowing it to
force the blood out of the heart into the
pulmonary artery and on to the capillary beds of
the lungs.

Septum: The inner dividing wall of the heart is not
complete until after birth. In the fetus the blood is
oxygenated in the placenta, not in the lungs. As a
result all the blood in the heart is very similar and so
mixes freely. In the days after birth the gap in the
septum closes to separate the deoxygenated and
oxygenated bloods. Any gap remaining in the
septum after the first few weeks of life is referred to
as a 'hole in the heart', and if large, can lead to
severe health problems.

The **left ventricle** is filled with oxygenated blood
as the left atrium contracts. When the left ventricle
contracts it forces the blood out of the heart and
into the aorta to be carried around the body. The
muscular wall of the left side of the heart is much
thicker than that on the right because the lungs are
relatively close to the heart, but the left side has to
produce sufficient force to move the blood under
pressure to all the extremities of the body.

Figure 2.5.9 The structure of the heart is closely related to its function. In an average lifetime it will beat more than
2.5×10^9 times, and each ventricle will pump over 150 million litres of blood.

heart valves. The two sounds of a heartbeat are described as 'lub-dub'. The first sound is made when blood is forced back against the atrioventricular valves as the ventricles contract, and the second when a backflow of blood hits the semilunar valves in the aorta and pulmonary artery as the ventricles relax. Some of the changes in the heart which cause the heartbeat are shown in figure 2.5.11.

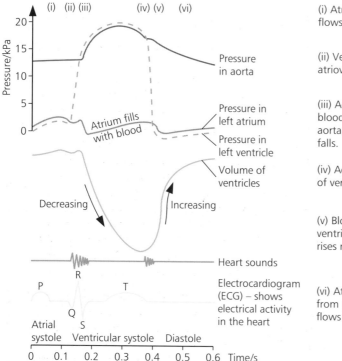

(i) Atrium contracts, blood flows into ventricle.

(ii) Ventricle starts to contract, atrioventricular valve closes.

(iii) Aortic valve forced open, blood flows from ventricle into aorta, and volume of ventricle falls.

(iv) Aortic valve closes, volume of ventricle starts to rise.

(v) Blood flows from atrium to ventricle, volume of ventricle rises rapidly.

(vi) Atrium fills with blood from pulmonary vein, blood flows from atrium to ventricle.

Figure 2.5.11 The cardiac cycle and causes of the heartbeat. The first and second heart sounds are due to the closure of the atrioventricular and aortic valves. Look at how the differences in pressure cause these closures – the atrioventricular valve closes when the ventricular pressure rises above the atrial pressure (ii), and the aortic valve closes when the ventricular pressure falls below the aortic pressure (iv). (*After Winton and Bayliss,* Human Physiology, *Churchill*)

Control of the heart

The heart beats continually throughout life, with an average of about 70 beats per minute, although in small children the heart rate is much higher. The heart can respond to the varying needs of the body – during physical exercise when the tissues need more oxygen, it beats faster to supply more oxygenated, glucose-carrying blood to the tissues and to remove the increased waste products. Stress can raise the heart rate, whilst rest and relaxation can lower it.

How is the heart controlled? In the very early embryo, cells which are destined to become the heart begin contracting rhythmically long before the organ forms. They have **intrinsic rhythmicity**. An adult heart removed from the body will continue to contract as long as it is bathed in a suitable oxygen-rich fluid. The intrinsic rhythm of the heart, that is, the rate at which it beats when isolated from the nervous and hormonal control of the body, is around 60 beats per minute. This rhythm is maintained by a wave of electrical excitation similar to a nerve impulse which spreads through special tissue in the heart muscle. Figure 2.5.12 shows the spread of excitation through the heart.

The intrinsic rhythm of the heart does not explain how the heart is able to respond to changes in the body's requirements. This sensitivity is supplied by nervous control of the heart. A nerve from the sympathetic nervous system speeds up the heart rate and the vagus nerve from the parasympathetic system slows it down. More details about the nervous system will be found in section 4. Other factors such as hormones, pH changes caused by carbon dioxide levels and temperature all have an effect on heart rate.

The **sinoatrial node (SAN)** or **pacemaker** sets up a wave of electrical excitation which causes the atria to start contracting and also spreads to an adjacent area of similar tissue. This is the P wave in the ECG in figure 2.5.11.

The **atrioventricular node (AVN)** is excited as a result of the SAN's excitation and from here the excitation passes into...

the **Purkyne tissue**. Formerly called the Purkinje tissue, this penetrates through the septum between the ventricles. As the excitation travels through the tissue it sets off the contraction of the ventricles, starting at the bottom and so squeezing blood out. This is the Q,R,S and T waves of the ECG. The speed at which the excitation spreads makes sure that the atria have stopped contracting before the ventricles start.

Figure 2.5.12 The area of the heart with the fastest intrinsic rhythm is a group of cells in the right atrium known as the **sinoatrial node**, and this acts as the heart's own natural pacemaker.

The heart can respond to an increased demand for glucose and oxygen by the body in two ways. The rate at which the heart beats can be increased, as mentioned above. Also, the volume of blood pumped at each heartbeat, called the **cardiac volume**, can be increased by a more efficient contraction of the ventricles. The combination of these two factors gives a measure called the **cardiac output**:

$$\text{Cardiac output (dm}^3 \text{ min}^{-1}) = \text{cardiac volume} \times \text{heart rate}$$

Blood pressure

The blood travels through the arterial system at pressures which vary as the heart beats. The blood pressure is also affected by the diameter of the blood vessels themselves. Narrowing the arteries is one way in which the body affects and controls local blood pressure, but permanent changes in the artery can cause severe health problems.

Most people have their blood pressure taken at some point in their lives. The blood pressure reading is expressed as two figures, the first higher than the second. The most common way to take the blood pressure uses a **sphygmomanometer**. A cuff is connected to a mercury manometer (an instrument which measures pressure using the height of a column of mercury). The cuff is placed around the upper arm and inflated until the blood supply to the lower arm is completely cut off.

A stethoscope is positioned over the blood vessels at the elbow. Air is slowly let out of the cuff. The pressure in the cuff at which blood sounds first reappear is recorded. The first blood to get through the vessels under the cuff is that under the highest pressure – in other words, when the heart is contracting strongly. The height of the mercury at this point gives the **systolic blood pressure**. The blood sounds return to normal when blood at the lowest pressure during diastole can get through the vessels under the cuff. The pressure in the cuff at this point gives the **diastolic blood pressure**. A systolic reading of 120 mm Hg and a diastolic reading of 80 mm Hg is regarded as 'normal'. The blood pressure is expressed as '120 over 80' or '120/80'.

Blood pressure is used as an indicator of the health of both the heart and the blood vessels. A weakened heart may produce a low blood pressure, whereas damaged blood vessels which are closing up or becoming less elastic will give a raised blood pressure.

THE BLOOD

The components of blood

Blood is the transport medium of the body. It is an extremely complex substance carrying a wide variety of cells and substances to all areas of the body. Figure 2.5.13 shows the components of blood.

Functions of the blood

The blood carries out a wide variety of functions, summarised as follows:

1 transport of digested food products from the villi of the small intestine to all the areas of the body where they are needed for either immediate use or storage
2 transport of food molecules from storage areas to the cells which need them
3 transport of excretory products from the cells where they have been formed to the areas where they will be excreted
4 transport of chemical messages (hormones) from the glands where they are made around the body
5 helping to maintain a steady body temperature by distributing heat from deep-seated or very active tissues around the system
6 transport of oxygen from the lungs to all cells for respiration

Blood component	Main features
Plasma	This straw-coloured liquid is the main component of blood, and consists largely of water. Plasma contains **fibrinogen**, a protein vital for the clotting of the blood. The removal of fibrinogen from plasma results in **serum,** which contains a wide range of dissolved substances to be transported.
Erythrocytes (red blood cells)	Erythrocytes are biconcave discs. There are approximately 5 million erythrocytes per mm^3 of blood. The cells contain **haemoglobin**, the red oxygen-carrying pigment which gives them their colour.
	Erythrocytes are formed in the red bone marrow of the short bones. Mature erythrocytes do not contain a nucleus and have a limited life of about 120 days.
Leucocytes (white blood cells) Granulocytes Neutrophils (engulf bacteria) Eosinophils (antihistamine properties) Basophils (produce histamine and heparin) Agranulocytes Monocytes (engulf bacteria) Lymphocytes (produce antibodies)	Leucocytes are much larger than erythrocytes, but can squeeze through narrow gaps by changing their shape. There are approximately 7000 leucocytes per mm^3 of blood. They all contain a nucleus and have colourless cytoplasm. Most are formed in the white bone marrow of the long bones, although some lymphocytes are formed in the lymph glands and spleen. Their main function is in defence against pathogens (see section 1.6).
Platelets	Platelets are tiny fragments of large cells called **megakaryocytes** which occur in the bone marrow. There are about 0.25 million platelets per mm^3 of blood. They are involved in the clotting of the blood.

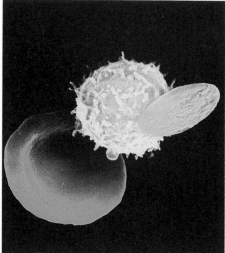

Figure 2.5.13 The main components of the blood. The plasma and the erythrocytes are involved in transport of oxygen and food molecules while the leucocytes defend the body against pathogens and the platelets, along with blood proteins, enable the blood to clot in case of damage to the circulatory systems.

False-colour scanning electron micrograph showing an erythrocyte (left), a leucocyte (centre) and a platelet (right)

7 transport of carbon dioxide from respiring cells to the lungs

8 clotting of the blood, to prevent excessive blood loss and the entry of pathogens

9 providing immunity through the lymphocytes

10 engulfing and digesting pathogens by the phagocytotic action of the granulocytes

11 acting as a buffer to pH changes.

The first six of these functions are carried out by the plasma. Substances move into and out of the plasma by diffusion or active transport, and are carried around the body in a mass flow system as the blood is pumped by the heart.

The defence functions of the blood discussed in section 1.6 are separate from its role as a transport system, though they take advantage of the rapid transport the blood offers to all areas of the body.

The erythrocytes are specialised transport cells used only for carrying oxygen and carbon dioxide.

The carriage of oxygen

The erythrocytes are ideally adapted for the carriage of oxygen. They contain the red pigment haemoglobin, which is largely responsible for oxygen transport. The shape of the cells means that they have a large surface area:volume ratio for

The blood as a damage limitation system

In theory a minor cut or scrape, or even a hard bang on the body surface, could endanger life due to loss of blood as the torn blood vessels bleed. In normal circumstances this does not happen because the body has a highly efficient damage limitation system in the **clotting mechanism** of the blood.

When a blood vessel is damaged a semi-solid mass called a **clot** forms and seals the wound as the result of a complex sequence of events:

- Blood plasma, blood cells and platelets flow from a cut vessel.
- Contact between the platelets and some of the tissue components (for example collagen fibres) causes the platelets to break open in large numbers. They release several substances, two of which are particularly important. **Serotonin** causes the smooth muscle of the arterioles to contract and so narrows the vessels, cutting off the blood flow to the damaged area. **Thromboplastin** is an enzyme which sets in progress a cascade of events that lead to the formation of a clot.
- In the presence of sufficient levels of calcium ions, thromboplastin catalyses a large-scale conversion of the globular plasma protein **prothrombin** into the enzyme **thrombin**.
- Thrombin acts on another plasma protein called **fibrinogen**, converting it to **fibrin**. Fibrin consists of fibrous strands.
- Further platelets and blood cells pouring from the wound get trapped and tangled within the fibrin meshwork, forming a clot.
- Contractile proteins in the cytoskeleton of the platelets contract, pulling the clot into a tighter and tougher configuration.

These events are an example of a cascade system where a relatively small event is **amplified** through a series of steps, as shown in figure 2.5.14. The speed and efficiency of clotting is effective in preventing blood loss. The difficulties of life without it are seen in the genetic condition of **haemophilia**, where normal life is impossible without regular injections of clotting factors.

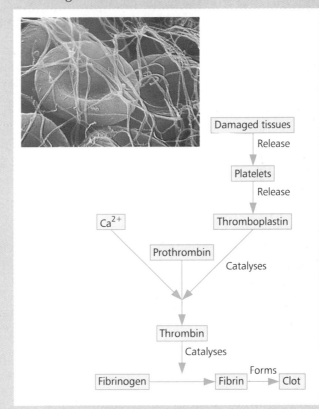

Figure 2.5.14 A cascade of events which result in a life-saving clot. This seals the blood vessels and prevents excessive blood loss, and also protects the delicate new skin growing underneath.

the diffusion of gases, and the lack of a nucleus means that the maximum amount of space is available to pack in haemoglobin molecules. In fact, each red blood cell contains around 250 million molecules of haemoglobin, giving it the capacity to carry 1000 million molecules of oxygen.

Haemoglobin

Haemoglobin is the key to oxygen uptake. It is a very large globular protein molecule made up of four globin polypeptide chains. Each chain has a prosthetic haem group which contains iron and gives the molecule its red colour. Haemoglobin (Hb) has a high affinity for oxygen. The oxygen is bound quite loosely to the haem groups to form **oxyhaemoglobin** (HbO_8):

$$\mathbf{Hb + 4O_2 \rightarrow HbO_8}$$

The first oxygen molecule to be bound to haemoglobin alters the shape of the molecule in such a way that it is easier for the next oxygen to be taken on.

This in turn alters the shape and makes it easier for the next oxygen to be taken up, until the fourth and final oxygen molecule combines with the haemoglobin several hundred times more rapidly than the first. The same process happens in reverse when oxygen dissociates from haemoglobin – it gets progressively harder to remove the oxygen.

This has very important implications for the way in which oxygen is taken up in the lungs and released in the respiring cells. Figure 2.5.15 shows that a very small change in the proportion of oxygen in the air (represented by the partial pressure) makes a big difference to the saturation of the blood with oxygen. This means that as the blood enters the lungs it is rapidly loaded with oxygen. Equally, a relatively small drop in the oxygen levels of respiring tissues will initiate the release of oxygen from the blood.

An additional factor is the effect of carbon dioxide. The way in which haemoglobin takes up and releases oxygen is affected by the proportion of carbon dioxide in the tissues, as figure 2.5.16 demonstrates.

The effect of the Bohr shift seen in figure 2.5.16 is that in higher partial pressures of carbon dioxide, haemoglobin needs higher levels of oxygen to become saturated. More importantly, it gives up oxygen more easily. Thus in active tissues with high carbon dioxide levels, haemoglobin releases the oxygen needed very readily. On the other hand, carbon dioxide levels in the lung capillaries are relatively low, and so it is easier for oxygen to bind to the haem groups.

Other respiratory pigments

Two respiratory pigments other than haemoglobin are found in mammals. One is another form of haemoglobin known as **fetal haemoglobin**. The fetus in the uterus is dependent on the mother to supply it with oxygen. Oxygenated blood from the mother runs through the placenta close to the deoxygenated fetal blood. If the blood of the fetus had the same affinity for oxygen as the blood of the mother, transfer of oxygen would occur with difficulty. Fetal haemoglobin has a higher affinity for oxygen than the mother's haemoglobin, and so can take up oxygen from the maternal blood.

Myoglobin is structurally like a single haemoglobin chain. It is found mainly in the skeletal muscle, and has a much higher affinity for oxygen than haemoglobin, thus readily becoming saturated with oxygen. It does not give up oxygen easily, however, and so acts as an oxygen store. When the oxygen levels in very active muscle tissue get extremely low, and the carbon dioxide levels are correspondingly high, then myoglobin releases its store of oxygen.

Figure 2.5.17 compares the oxygen dissociation curves for haemoglobin, fetal haemoglobin and myoglobin.

The carriage of carbon dioxide

Carbon dioxide can be carried in the blood in three different ways. A small amount – about 5% – is carried in solution in the plasma. A further 10–20% combines with amino groups in the polypeptide chains of haemoglobin molecules to form **carbaminohaemoglobin**. But the bulk of the carbon dioxide is transported in the blood as **hydrogencarbonate ions**. This is explained in figure 2.5.18.

The reactions shown in figure 2.5.18 take place in the tissues, where there is a relatively high concentration of carbon dioxide and a low concentration of oxygen. In the lungs there are high oxygen concentrations and low carbon dioxide concentrations. The reactions go in reverse, releasing carbon dioxide into the lungs and freeing the haemoglobin to pick up more oxygen.

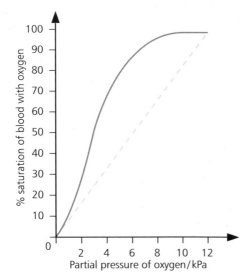

Figure 2.5.15 Oxygen dissociation curve for human haemoglobin. The dotted line shows what the curve would look like without haemoglobin's particular affinity for oxygen.

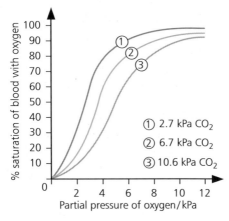

Figure 2.5.16 As the proportion of carbon dioxide increases, the haemoglobin curves move downwards and to the right. This is known as the **Bohr shift**.

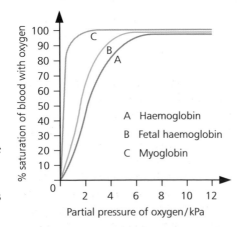

Figure 2.5.17 Both fetal haemoglobin and myoglobin have higher affinities for oxygen than does haemoglobin, so they can take up oxygen from haemoglobin.

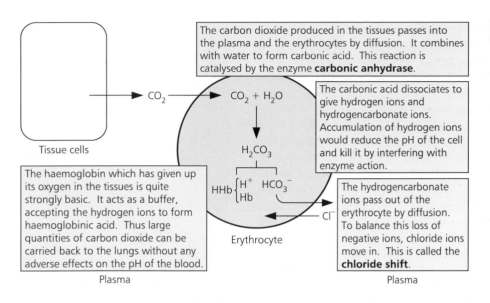

Tissue cells

The carbon dioxide produced in the tissues passes into the plasma and the erythrocytes by diffusion. It combines with water to form carbonic acid. This reaction is catalysed by the enzyme **carbonic anhydrase**.

CO_2

$CO_2 + H_2O$

H_2CO_3

$HHb \begin{cases} H^+ \\ Hb \end{cases} HCO_3^-$

Cl^-

The carbonic acid dissociates to give hydrogen ions and hydrogencarbonate ions. Accumulation of hydrogen ions would reduce the pH of the cell and kill it by interfering with enzyme action.

The haemoglobin which has given up its oxygen in the tissues is quite strongly basic. It acts as a buffer, accepting the hydrogen ions to form haemoglobinic acid. Thus large quantities of carbon dioxide can be carried back to the lungs without any adverse effects on the pH of the blood.

The hydrogencarbonate ions pass out of the erythrocyte by diffusion. To balance this loss of negative ions, chloride ions move in. This is called the **chloride shift**.

Erythrocyte

Plasma

Plasma

Figure 2.5.18 The carriage of carbon dioxide by the blood

ADAPTATIONS OF TRANSPORT SYSTEMS

The transport vessels, the pumping heart and the blood of a mammal are all highly specialised and adapted for their functions. They have evolved to suit the way of life of the majority of mammals – land-dwellers at or around sea-level. But some species live in the oceans, and others at high altitudes. Humans attempt to survive in almost all the available environments. How do the cardiovascular and respiratory systems cope with some of these adverse conditions?

The diving mammals

The seals and the whales are two of the best-known groups of water-living mammals. They are very successful aquatic animals, in spite of their need to return to the surface at regular intervals to breathe air into the lungs. Land-dwellers of comparable size need to breathe many times in a minute. If the diving mammals had to breathe as often as this they would not be able to stay underwater long enough to catch any food.

There are several major adaptations of the respiratory and cardiovascular systems of the diving mammals. The lungs are large, so that extra air can be taken in to oxygenate the blood as fully as possible before diving. There is an increased blood volume, so that large amounts of oxygen can be loaded into the blood. Once a dive begins, the capillary beds in many of the non-vital organs are shut off, so that blood flows mainly to the brain and the swimming muscles. This minimises the oxygen consumption of the tissues. Also, the heart rate slows dramatically to only a few beats per minute. This is known as **bradycardia**. Again this greatly reduces the oxygen consumption of the tissues. There is increased tolerance of oxygen debt, which means that the tissues can cope with relatively high levels of lactic acid without damage. These features enable diving mammals to dive for considerable periods of time, as figure 2.5.19 shows.

Interestingly, humans who make a habit of prolonged diving without additional air, such as pearl divers, begin to develop some of these adaptations. The lung volume increases and a degree of bradycardia occurs, with the heart rate slowing noticeably. Although these adaptations are not to the same extent as in true diving mammals, they make possible dives of much longer duration than would normally be expected.

Figure 2.5.19 Dives of up to 20 minutes are not uncommon in some seals and the great whales.

Living at altitude

With increased height above sea level, both the atmospheric pressure and the partial pressure of oxygen in the air are reduced. At heights of more than about 3650 m above sea level there is a noticeable oxygen lack, and the highest permanent human habitations are at about 5500 m above sea level. That is well below the summit of Everest at 8848 m above sea level.

For most people, moving to a high altitude area is an unpleasant experience. Symptoms resulting from lack of oxygen include headaches, dizziness, breathlessness even at rest, sweating, dim vision and reduced hearing, even loss of consciousness on physical exertion. These symptoms are known as **mountain sickness**. Much of the problem is the result of the fact that lack of oxygen causes an increased breathing rate, but this removes too much carbon dioxide. Thus the normal stimulus to breathe is lost, resulting in very abnormal breathing patterns.

After some time, people experience **acclimatisation**. This includes a sustained increase in the ventilation of the alveoli, an increase in the oxygen-carrying capacity of the blood with the formation of extra erythrocytes, and an increase in cardiac output. However, someone who has moved to a high altitude after years at sea level can never truly acclimatise and may suffer severe mountain sickness at any time. Those who are born at high altitude show a very different level of acclimatisation, although even they may be struck with mountain sickness at times.

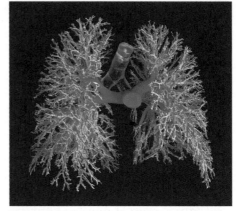

Figure 2.5.20 People who live high in the mountains of the Andes and the Himalayas are born and live at high altitudes. They develop an increased lung volume by increasing the numbers of both alveoli and capillaries. The number of red blood cells rises to a level where the blood becomes measurably thicker. The acid/base balance in the system is altered to affect the control of breathing. With these adaptations the people of the mountains can carry out tasks which would daunt most of us even at sea level, as they maintain an existence in a most inhospitable environment.

SUMMARY

- Transport systems in larger animals take the form of **cardiovascular systems** consisting of vessels carrying blood, the transport medium, pumped around the animal by the heart.

- The **human cardiovascular system** is a **double circulation** consisting of the **systemic circulation** to the body and the **pulmonary circulation** to the lungs.

- **Arteries** are vessels which carry blood away from the heart. They are lined with **squamous epithelium** and have elastic walls to allow them to stretch and absorb the pulsing flow of the blood pumped from the heart. There is an external layer of connective tissue, mainly collagen, to prevent overstretching.

- Arteries divide to form smaller branches called **arterioles** which have a lower proportion of elastin in their walls than larger arteries. Arterioles subdivide further to form the **capillaries**.

- Capillary walls consist of a layer of squamous epithelium thus allowing diffusion to take place. As blood in the capillaries passes through the tissues, oxygen is removed and carbon dioxide loaded in.

- Fluid is forced out of the capillary vessels, forming the **intercellular** or **tissue fluid**. Diffusion takes place between the tissue fluid and the cells of the tissues. Tissue fluid is collected into **lymph capillaries**, vessels with valves to prevent backflow, and this forms the **lymph**. **Lymph glands** produce antibodies and filter out foreign particles from the lymph, which is returned to the blood into the subclavian veins.

- Capillaries join to form **venules** which lead to **veins** – vessels which carry blood back to the heart. Veins have a large lumen, are lined with squamous epithelium and have thin walls with few elastic fibres. Their outer layer is made up mainly of collagen.

- Veins have one-way **semilunar valves** which prevent the backflow of blood. Many larger veins are between big muscle blocks, contraction of which forces blood along the veins.

- The **heart** is a four-chambered organ surrounded by the **pericardial membranes**. The left-hand side pumps blood from the lungs to the body, while the right-hand side pumps blood from the body to the lungs.

- The **right atrium** receives deoxygenated blood from the great veins. Blood fills the atrium and flows into the **right ventricle** until both atrium and ventricle are full. The right atrium then contracts, forcing all the blood into the right ventricle, which then also contracts to force the blood into the pulmonary artery. The **tricuspid atrioventricular valve** prevents flow from the ventricle back into the atrium, while similar one-way valves prevent backflow from the right atrium to the great veins.

- The **left atrium** receives oxygenated blood from the pulmonary veins, which passes down into the **left ventricle**. Contraction of the left atrium is followed by contraction of the left ventricle, forcing blood into the aorta. Again, valves in the aorta and the **bicuspid atrioventricular valve** prevent backflow. The left ventricle is more muscular than the right ventricle as it has to force blood to the capillary beds of the extremities of the body.

- The **cardiac cycle** describes the chain of events that makes up a heartbeat: the relaxed atria fill and then the ventricles (**diastole**), then both atria contract forcing the blood into the ventricles which then also contract (**systole**), forcing the blood into the pulmonary artery and aorta. The sounds of the **heartbeat** are caused by the closing of the aortic and atrioventricular valves.

- **Cardiac muscle** cells have an intrinsic rhythm of contraction maintained by a wave of electrical excitation spreading from the **sinoatrial node** or **pacemaker** to the **atrioventricular node** and then down into the **Purkyne tissue**. The heart rate is also under the control of the sympathetic and parasympathetic nervous systems, hormones, pH changes and temperature changes.

- **Cardiac output** is given by the **cardiac volume** (the volume of blood pumped per heartbeat) multiplied by the **heart rate**.

- **Blood** consists of liquid **plasma**, **erythrocytes** (red blood cells), **leucocytes** (white blood cells) and **platelets**. Plasma contains **fibrinogen**, a protein involved in blood clotting, and **serum** is plasma with fibrinogen removed. Erythrocytes are biconcave discs with no nuclei, containing **haemoglobin**. They are formed in the red bone marrow. Leucocytes include **lymphocytes** and **monocytes** (the **agranulocytes**) and **neutrophils**, **eosinophils** and **basophils** (the **granulocytes**) mainly formed in the white bone marrow. They are involved in defence against pathogens. Platelets are involved in the clotting of blood.

- The blood transports oxygen, carbon dioxide, food molecules, excretory products and hormones. It distributes heat around the body and provides immunity, engulfs pathogens and buffers pH changes.

- Erythrocytes have a large surface area:volume ratio and carry large amounts of haemoglobin, a globular protein made up of four polypeptide chains each with a prosthetic iron-containing **haem group**. Oxygen can bind to each haem group forming **oxyhaemoglobin**.

- The binding of one oxygen molecule to a haem group causes successive haem groups to bind more easily, and vice versa, giving a characteristic **oxygen uptake curve** for haemoglobin. This curve is shifted by higher carbon dioxide concentrations so that oxygen is given up more easily (the **Bohr shift**).

- **Fetal haemoglobin** and **myoglobin** both have higher affinities for oxygen than does haemoglobin, to enable them to take up oxygen from the blood.

- Carbon dioxide is carried in the blood in solution, as carbaminohaemoglobin, and (mainly) as hydrogencarbonate ions. **Carbonic anhydrase** catalyses the formation of carbonic acid, and the hydrogen ions from this acid are accepted by haemoglobin to form **haemoglobinic acid**. The hydrogencarbonate ions pass out of the erythrocyte and are replaced by chloride ions. These reaction are reversed in the lungs to give up carbon dioxide.

- The diving mammals have large lungs and an increased blood volume. The blood supply to capillary beds in non-vital organs is shut off during a dive, and the heart rate slows (**bradycardia**).

- People at high altitudes can suffer **mountain sickness**, where lack of oxygen brings about an increased breathing rate, while too much carbon dioxide is removed so interfering with the normal breathing stimulus. **Acclimatisation** takes place as the ventilation rate of the alveoli is increased, more erythrocytes are formed so more oxygen can be carried in the blood, and cardiac output is increased.

QUESTIONS

1 a What is the role of diffusion in the distribution of gases and nutrients to the individual cells of:
  i single-celled animals
  ii large multicellular animals?
  b What is the function of a circulatory system?

2 a Visualise the journey of a red blood cell from the extremity of a toe around the body and back to the toe. Describe the route taken by the red blood cell.

b Produce a table to summarise the structure and functions of the main types of blood vessels – arteries, capillaries and veins.
c Blood is moved around the circulatory system by the heart. How is the heart rate controlled?

3 Summarise the main functions of mammalian blood. Select two of these functions and explain them fully.

2.6 Support systems and movement in animals

Almost all animals and plants have some kind of support system. In general, as organisms get larger, they need increasingly complex systems of support. We have already considered in section 2.2 how the cellulose walls, the turgor of the cells, the vascular xylem and phloem tissues and lignification in woody plants support even the giant redwood trees of America, the largest living organisms on Earth today. We are now concerned with support systems in animals, which usually take the form of a skeleton of some type.

FUNCTIONS OF SUPPORT SYSTEMS

Support systems provide a framework for the body of the organism and often help to determine its shape. Land animals in particular need skeletons to support their weight against gravity. Many organs are attached to the skeleton for support and stability.

Another major function of many support systems is protection. Whether inside or outside the body, a hard skeleton can protect particularly important or delicate organs.

Strong, relatively rigid skeletons would fulfil these functions of support and protection. Indeed, in large plants the rigid wooden trunks of trees support the great mass of photosynthesising leaves and growing fruits. But almost all animals need to locomote – that is, move their whole bodies around in order to feed, escape, find a mate, etc. Most animals have evolved to use their support systems to enable locomotion to occur, with muscles acting against and often joining to the skeleton via tendons.

Figure 2.6.1 Over 100 million years ago bony skeletons were already performing their prime functions of support, protection and movement, as this fossil triceratops shows.

Types of support systems

Three main types of support system have evolved in animals. In each case there is a close relationship between the skeleton which forms the support system and the muscles which act upon it to bring about locomotion.

(1) Hydrostatic skeletons

Hydrostatic skeletons are found in soft-bodied creatures such as sea anemones, flatworms and annelids (segmented worms). These animals have a fluid-filled body cavity surrounded by tubes of muscles. The muscles contract against the fluid. The contraction of the muscles and the fluid pressure together maintain the shape of the animal and bring about movement. There are usually two layers of muscles, one circular and one longitudinal, as shown in figure 2.6.2, which contract and relax alternately to cause changes in shape and locomotion.

Figure 2.6.2 In an animal like an earthworm, the contraction or relaxation of circular and longitudinal muscles can have a localised effect on small groups of segments. In unsegmented animals the fluid is moved around the whole tube.

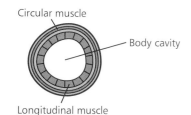

Circular muscle

Body cavity

Longitudinal muscle

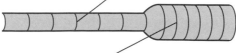

Circular muscles contracted, longitudinal relaxed - body lengthened and thinned

Longitudinal muscles contracted, circular relaxed - body shortened and thickened

(2) Exoskeletons

Exoskeletons are found exclusively in the arthropods. This group includes the insects (for example bees, water boatmen and ladybirds) and the crustaceans (such as crabs, lobsters and woodlice). As the name suggests, an exoskeleton is on the outside of the body. To allow movement an external skeleton is made of many jointed pieces, similar to a suit of armour, with strong solid sections protecting delicate organs.

The basic material of exoskeletons is chitin. This is secreted by the epidermal cells of the arthropod and is tough, light and flexible. In most areas it is impregnated with proteins which make it harder and less flexible, so improving support and protection. In crustaceans calcium carbonate is used to strengthen the chitin. At the joints where movement is necessary, the chitin is less rigid and has flexibility. The muscles which bring about movement are attached inside the skeleton, as shown in figure 2.6.3.

Growth within an exoskeleton is obviously limited, so the skeleton is shed at regular intervals. This moulting or **ecdysis** results in periods of great vulnerability, and it is also very expensive in terms of resources. A skeleton which covers the entire external surface employs a great deal of material, and body resources must be used up every time it is replaced. This is one of the main factors limiting the ultimate size to which arthropods grow.

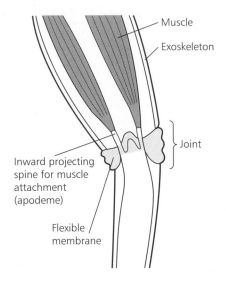

Figure 2.6.3 Joints and the contraction of the internal muscles within an insect exoskeleton allows a rigid protective box to move about.

(3) Endoskeletons

The cuttlefish bone seen in budgerigar cages is one unusual example of an **endoskeleton**. Endoskeletons are most typically found in vertebrates, although they do occur in other groups such as sponges and some molluscs. An endoskeleton is an internal skeleton, encased within the body tissues. Vertebrate skeletons are made of living material, either cartilage or bone, which can grow as needed with the organism. A variety of different types of joints make a whole range of movements possible. The functioning of an endoskeleton will be considered by looking in detail at the skeleton of a mammal.

THE MAMMALIAN SKELETON

Structure of the mammalian skeleton

Mammalian skeletons are typical of vertebrate endoskeletons. They are composed largely of **bone**, with varying amounts of **cartilage**. **Ligaments** hold bones together at joints and **tendons** join muscles to bones to allow movement. The precise arrangements of the different bones vary from species to species

---

Endoskeletons and size limits

The size of the arthropods is limited to a large extent by the resources needed to repeatedly replace the entire body surface. But what limits the size of animals possessing an internal skeleton capable of growing with them? It is the properties of bone which put a brake on the growth of vertebrates.

The strength of a bone under compression is mainly affected by its cross-sectional area. As an animal doubles in size, the length and width of the leg bones will double too. This increases the cross-sectional area by a factor of 4 (2^2). The volume of the animal, and so its mass, will increase by a factor of 8 (2^3). A point is reached where the limb bones can no longer withstand the compressive forces in an enormous animal, and would snap or crumble as soon as the weight is shifted as the creature moves.

Organisms can get bigger under various circumstances, such as living in water to gain extra support or evolving areas of relatively light tissue or air-filled cavities. However, eventually limits of support by the endoskeleton are reached.

depending on their adaptations for a particular way of life. However, the overall pattern of mammalian skeletons is similar to that of humans, in spite of our unusual habit of walking on two legs. The human skeleton and its associated ligaments and tendons are described in figure 2.6.4, pages 152–3.

The endoskeleton and movement

The way in which the internal skeleton of mammals is used to give support and protection is explained by looking at the arrangement and properties of the tissues, particularly in the strong bones and protective cages. However, the use of the internal skeleton to bring about locomotion is not so immediately obvious. It involves the structure of the joints within the skeleton, and also the interaction with the muscles.

The structure of the joints

The **joints** are points of weakness in the support system of the skeleton, though they are vital to allow movement and locomotion. As you can see in figure 2.6.5, bones are held in position in the joints by ligament capsules which allow movement of the bones over, but not away from, each other. Anyone who has experienced the pain of a dislocated joint will know that keeping the bones lined up correctly is necessary to the working of a joint.

The ends of the bones at a joint are shaped to move smoothly over each other, and the way in which the two bones meet varies according to the type of movement required. The **ball and socket joints** found at the hip and shoulder give very free movement, whereas the **hinge joints** of the fingers and knees are much more restrictive.

Two solid masses moving over each other whilst subjected to quite severe compressive forces would soon wear each other away. To prevent the bones from being eroded in this way the joint is lined with a replaceable layer of rubbery cartilage which allows the joint to articulate smoothly. The most mobile joints produce a liquid lubricant known as **synovial fluid** which fills the joint cavity and ensures easy, friction-free movement, as shown in figure 2.6.5.

How is movement brought about?

Movement is brought about by the action of muscles on bones. Each of the skeletal muscles of a vertebrate is attached by tendons to at least two different bones, spanning at least one joint. The attachment nearest to the heart is known as the **origin** of the muscle and this bone moves very little when the muscle contracts. The attachment furthest away from the heart is called the **insertion** and this is the bone which moves when the muscle contracts.

Figure 2.6.5 The adaptations of the skeleton normally allow smooth, easy, pain-free movement. But bones can become damaged and replacements do not come cheap – £1500 to buy and fit an artificial hip.

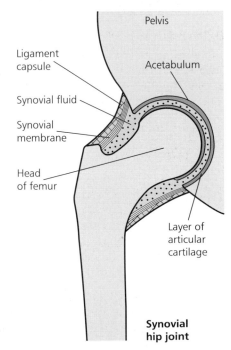

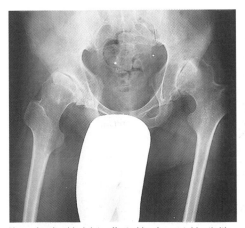

X-ray showing hip joints affected by rheumatoid arthritis

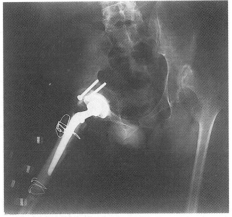

X-ray showing replacement hip joint

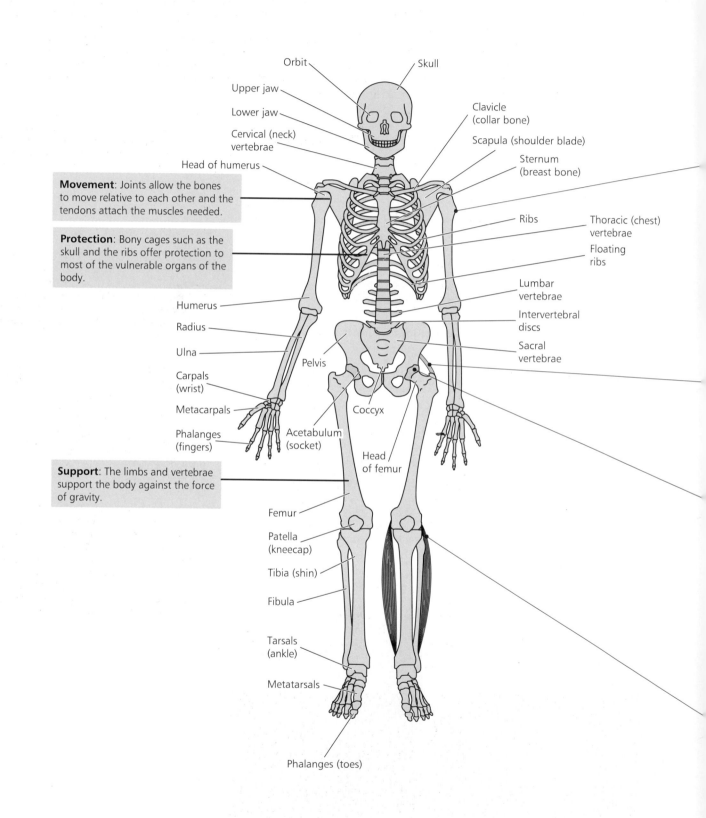

Movement: Joints allow the bones to move relative to each other and the tendons attach the muscles needed.

Protection: Bony cages such as the skull and the ribs offer protection to most of the vulnerable organs of the body.

Support: The limbs and vertebrae support the body against the force of gravity.

Orbit

Upper jaw

Lower jaw

Cervical (neck) vertebrae

Head of humerus

Humerus

Radius

Ulna

Carpals (wrist)

Metacarpals

Phalanges (fingers)

Pelvis

Acetabulum (socket)

Coccyx

Femur

Patella (kneecap)

Tibia (shin)

Fibula

Tarsals (ankle)

Metatarsals

Phalanges (toes)

Skull

Clavicle (collar bone)

Scapula (shoulder blade)

Sternum (breast bone)

Ribs

Thoracic (chest) vertebrae

Floating ribs

Lumbar vertebrae

Intervertebral discs

Sacral vertebrae

Head of femur

Figure 2.6.4 The human endoskeleton shows specialist adaptations for support, protection and movement.

Bone

Compact bone

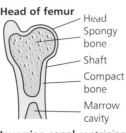

Head of femur

- Head
- Spongy bone
- Shaft
- Compact bone
- Marrow cavity

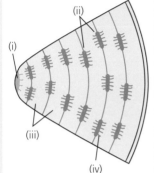

(i) **Haversian canal** containing an artery and a vein. Capillaries run from these to the lacunae and so to the bone cells. Haversian canals also contain a lymph vessel and nerve fibres.

(ii) **Lacuna** containing the inactive bone cells or **osteocytes**

(iii) Bony lamellae or layers

(iv) **Canaliculi,** thin links of cytoplasm ramifying through the structure maintaining a communicating network

Bone is strong and hard. It is a composite material, which means it is made up of more than one substance. The bone cells or **osteocytes** are embedded in a **matrix** which consists of about 30% organic **collagen fibrils** impregnated with **calcium salts** which make up the remaining 70% of the matrix. Bone is particularly strong under compressive (squashing) forces, which means it stands up well to the stresses of supporting and moving an animal. It is relatively weak under tensile or pulling forces, but is exposed to these far less frequently.

Bone needs to be strong and hard, and as light as possible to reduce the weight an animal must move about. There are two types of bone. **Compact bone** is very dense and heavy, but very strong. It is found in areas such as the shafts of the long bones of the body. **Spongy bone** has a much more open structure and is lighter. It is found in growing regions and in large masses of bone such as the head of the femur.

Bones contain **marrow**. The long bones contain white bone marrow where the white blood cells are made. The short bones such as the ribs contain red bone marrow where the red blood cells are made.

We tend to think of bones as rigid stable structures. In fact bone is constantly laid down or removed in response to the different stresses imposed upon it. New activities can cause the bones to be altered slightly. This change is brought about by the bone cells. The **osteocytes** are inactive bone cells. If more bone is needed, they are rapidly stimulated and form **osteoblasts** which lay down new matrix. If bone needs to be removed, **osteoclasts** result which reabsorb the matrix.

Cartilage

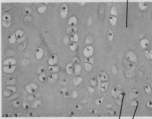

Hyaline cartilage

- Hyaline matrix
- Chondrocytes

Cartilage is a hard but flexible tissue made up of cells called **chondrocytes** within an organic **matrix** which consists of varying amounts of **collagen fibrils**. The matrix is produced by the chondrocytes. Cartilage is elastic and able to withstand compressive forces. It is a very good shock absorber and is frequently found between bones such as the vertebrae and in the joints. A few vertebrates, mainly the cartilaginous fishes, have skeletons made up entirely of cartilage.

There are three main types of cartilage:

(1) **Hyaline cartilage** found at the ends of bones and in the nose, air passageways and parts of the ear.

(2) **Yellow elastic cartilage** which has a high proportion of elastic fibres so that the tissue quickly recovers its shape after distortion. It is not usually part of the skeleton.

(3) **White fibrous cartilage** has bundles of densely packed collagen in the matrix. It has great tensile strength but is less flexible than the other forms of cartilage. It forms the intervertebral discs and is found between the bones in joints.

Ligaments

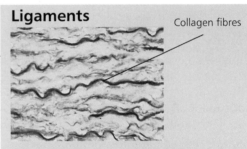

- Collagen fibres

The **ligaments** form capsules around the joints, holding the bones in place. However, they also need to be elastic to allow the bones of the joint to move when necessary. **Yellow elastic tissue** gives a combination of strength with elasticity. By varying the amounts of collagen and incorporating some white fibrous tissue if needed, the ligament capsules can be relatively loose or very tight as required by the body at different joints.

Tendons

Collagen fibres

White fibrous tissue
Tendons are made up almost entirely of **white fibrous tissue**. This consists of bundles of collagen fibres and gives a tissue which is strong but relatively inelastic. This makes it ideal for joining the muscles to the bones. If the tendons stretched, much of the work done by the muscles would be wasted as it would not move the bones.

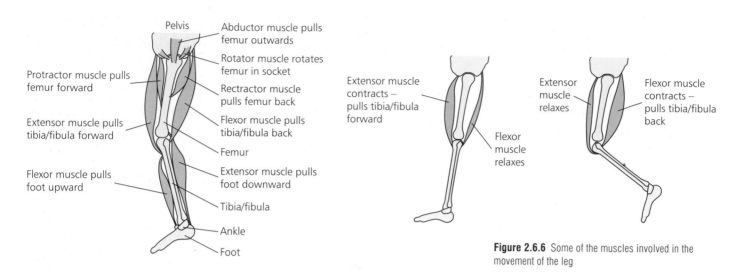

Pelvis

Abductor muscle pulls femur outwards

Rotator muscle rotates femur in socket

Protractor muscle pulls femur forward

Rectractor muscle pulls femur back

Flexor muscle pulls tibia/fibula back

Extensor muscle pulls tibia/fibula forward

Femur

Flexor muscle pulls foot upward

Extensor muscle pulls foot downward

Tibia/fibula

Ankle

Foot

Extensor muscle contracts – pulls tibia/fibula forward

Flexor muscle relaxes

Extensor muscle relaxes

Flexor muscle contracts – pulls tibia/fibula back

Figure 2.6.6 Some of the muscles involved in the movement of the leg

When muscles contract they exert a pull on a bone and so it moves relative to another bone. However, when muscles relax they do not exert a corresponding push – they simply stop contracting, and become capable of being pulled back to their original shape. Thus the muscles of the skeleton are usually found in pairs. One pulls the bone in one direction, the other pulls it back to its original position. Because they work in direct opposition to each other these muscle pairs are known as **antagonistic pairs**. A clearer picture of how movement is brought about can be gained from figure 2.6.6.

Stability

When an object is stationary it will remain standing upright as long as all the forces acting on it are balanced. For four-legged animals, the four legs act like the legs of a table with the centre of gravity somewhere in the middle, so quadrupeds are very unlikely to fall over. Once only three legs are in contact with the ground, during locomotion, the animal must shift its weight so its centre of gravity falls within the tripod of legs left in contact with the ground. By judicious shifting of the weight animals can remain upright with two, one or even no legs on the ground.

The complexity of this becomes apparent when watching a newborn animal such as a calf or a foal trying out different gaits in the first few days of life – and not always getting the centre of gravity in the right place. Remaining stable with only two legs is even more difficult. For example, if a human being takes in too much alcohol the brain cannot correctly interpret the information it receives, and remaining upright becomes a major problem.

LOCOMOTION

Locomotion is important for animals. Whether the animal depends upon a hydrostatic skeleton, an exoskeleton or an endoskeleton there are three possible locations in which they may move – on land, through water or through the air.

Locomotion on land

Locomotion on land involves the animal moving through air, which offers relatively little resistance to movement. Contraction of the muscles causes the limbs to act as levers, pressing downwards and backwards into the ground. As a result of this a force acts upwards and forwards on the animal, as figure 2.6.7

shows. This force brings about locomotion. By altering the angle at which the limb presses into the ground, the balance between the forward force and the upward force can be altered, allowing for leaps into the air and low, fast running.

Locomotion through water

Water presents a very different medium for locomotion. It is much denser than air, and also much more viscous (thicker). This means that it offers much more resistance to movement, but also provides a great deal of support and purchase. Animals that always move through water – the fish in particular – can grow very large with comparatively small skeletons, since much of the supporting role of the skeleton is taken over by the water. Animals that move at speed through water also tend to have a very streamlined shape, minimising the resistance to movement of the water.

In many fish, locomotion through the water is brought about by lashing movements of the vertebral column. These may involve the whole animal, as in eels (**anguilliform locomotion**). The method used by many fish involves the rear part of the fish and particularly the tail (**carangiform locomotion**). Finally there is the relatively rare **ostraciform locomotion** where just the tail lashes from side to side. In each case the movements are brought about by the alternate contraction and relaxation of sets of antagonistic muscles along the spine, shown in figure 2.6.8. The movements of the fish's body exert a backwards and sideways pressure on the water. Just as on the land, an equal and opposite force results pushing the fish forwards and sideways. The sideways element is balanced, at least partly, by the resistance to movement of the water on the head and dorsal fins of the fish, so that the fish swims forwards.

Most of the fins of a fish are not actively involved in locomotion, but give stability to the animal. Different sets of fins are used to prevent the fish from rolling around in the water, which would interfere with efficient forward movement.

Locomotion through the air

Vast numbers and varieties of animals successfully move through the water and over the land. In contrast, only three groups – birds, insects and bats – have mastered the art of flying. Air offers virtually no support and no purchase for limbs to press against.

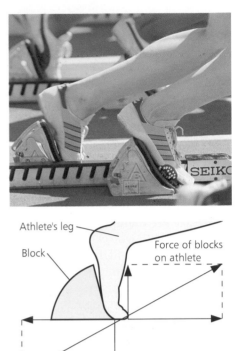

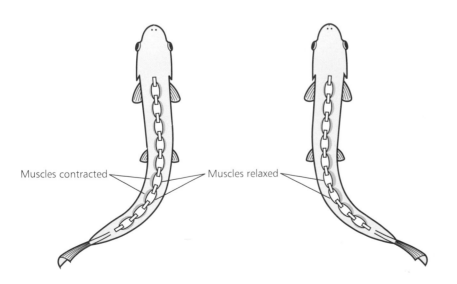

Figure 2.6.7 Newton's third law states that for every action there is an equal and opposite reaction. Land animals use this to propel themselves along.

Figure 2.6.8 The antagonistic muscles on the two sides of the body contract alternately, bending the vertebral column from side to side.

Muscles contracted — Muscles relaxed

Wings are the feature which all flying animals have in common, and they are vital for successful flight. However, many other adaptations are necessary too. The animal must be as light as possible. Insects are so small that this is not a major problem for most of them but birds and bats have evolved hollow bones for lightness with internal cross-bracing for strength. Flight muscles need to be enormously well developed – in birds they may make up to a third of the body weight. There are well-adapted areas of the skeleton for the attachment of the muscles – in birds the sternum is usually deep and keel-like. Even the body chemistry of flying animals is adapted. Flying uses up extremely high levels of energy, so an ability to convert food to usable energy very quickly (a high metabolic rate) is needed.

Figure 2.6.9 A bird needs to produce enough lift to overcome its natural tendency to fall out of the sky.

Flight in birds

There are two main types of flight in birds – powered and unpowered flight. Powered flight involves flapping the wings to cause a movement of air over them. Unpowered flight involves gliding and soaring, using natural air currents. Both powered and unpowered flight depend on the shape of the wing.

The wing of a bird acts as an aerofoil – that is, a smooth surface moving through the air at an angle to the airstream. The air above the wing has to move further than the air below the wing, resulting in a greater air pressure under the wing than above it. The shape of the wing and the feathers make it very effective, with the air flow over the wing resulting in lift, as figure 2.6.9 shows.

Some birds, particularly larger ones, rely heavily on gliding flight. Birds such as eagles live high up and have large wings to make use of thermals and upcurrents. Many smaller birds rely much more on active flight, flapping their wings relatively rapidly with short glides between bursts of flapping. The flapping of wings is not a simple up and down motion, as this would force birds up and then down again. The wing edge moves through a figure-of-eight pattern and the wing angle is altered to maximise the lift which can be obtained.

Flight in insects

Insect flight relies on lift produced by a flow of air over the wings in just the same way as bird flight does. However, most insects are too small for gliding flight – they generally have to beat their wings very fast to become and remain airborne. The naked eye can often see the beating of a bird's wing in flight, but when we look at most flying insects the wings are a blur. The wings of the housefly beat 200 times a second, those of the mosquito 600 times a second and the wings of some midges beat at over a thousand times per second.

Figure 2.6.10 Insects constantly change the shape of their thorax to move their wings for flight.

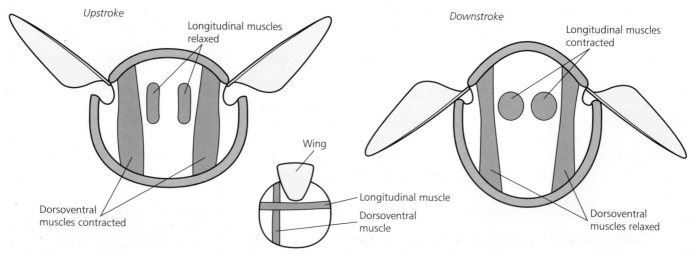

Insect wings beat in a very different way from birds' wings. Although insect flight involves muscle contractions, the main flight muscles are not connected to the wings themselves. The wings are attached to the outside of the thorax of the insect. The flight muscles are attached to the inside of the thorax, and are called **indirect flight muscles**. There are two sets of these flight muscles which contract and relax alternately. They cause changes in the shape of the thorax and this in turn raises and lowers the wings, as shown in figure 2.6.10. There are also **direct flight muscles** attached to the wings, which alter the angle of movement through the air to create the maximum lift and the minimum downwards force.

The speed at which some insects beat their wings cannot be controlled by nervous stimulation of the muscles alone because nervous transmission is not fast enough. The structure of the thorax makes these high speeds possible. The contraction of the indirect flight muscles causes deformation of the thorax. This causes the wings to move up or down. Then tension on the contracting muscle is eased, but the antagonistic muscle is stretched. This stimulates it to contract and another deformation is set up in the other direction. This results in rapid alternating contractions of the muscles, with the speed of beating being largely determined by the resonating properties of the thorax wall (which can be altered by other muscles). The nervous system starts the process up and then sends an occasional impulse to keep the muscles activated.

HOW MUSCLES WORK

Types of muscle

Throughout this section we have referred to the interaction of muscles with skeletal systems to bring about movement and locomotion. But what *is* muscle? It is a very specialised tissue which is remarkably similar throughout the animal kingdom, so we will concentrate mainly on mammalian muscle. Muscles are largely made up of protein and they consist of large numbers of very long cells known as **muscle fibres** bound together by connective tissue. They can contract (shorten) and relax, and have a good blood supply to provide them with the glucose and oxygen they need when they are working and remove the waste products which result.

In mammals the muscle tissues can make up as much as 40% of the body weight. There are three main types of muscle, each specialised to perform a particular function.

- **Skeletal muscle** is also known as striated muscle or voluntary muscle. This is the muscle which is attached to the skeleton and so is involved in locomotion. It is under the control of the voluntary nervous system, and its appearance under the microscope is striated or stripy. It contracts rapidly but also fatigues or tires relatively quickly.

- **Smooth muscle** is also known as involuntary muscle. It does not appear striated under the microscope and is under the control of the involuntary nervous system. It is found in the gut where it is involved in moving the food along, and in the blood vessels. It both contracts and fatigues slowly.

- **Cardiac muscle** is found exclusively in the heart. It is striated and the fibres are joined by cross-connections. It contracts spontaneously and does not fatigue.

Skeletal muscle

There are two types of skeletal muscle in mammals. **Slow-twitch muscles** are adapted for steady action over a period of time. They can stay in tetanus and

Properties of skeletal muscle

If an isolated calf muscle (gastrocnemius) from a frog is given a variety of different electrical stimuli the effects on the length of the muscle can be recorded on a revolving drum (kymograph). The results tell us several things about the way in which muscles work. Experiments from a single muscle fibre give the clearest results – in a whole muscle, different fibres work together to give more confusing results.

A single stimulus causes a single contraction or twitch of the muscle fibre. It is an 'all-or-nothing' response. If the stimulus is below a certain level, nothing happens. If it is above the **threshold** level, the muscle fibre twitches. However big the stimulus above the threshold level, the size of the twitch is always the same. Single twitches are relatively rare.

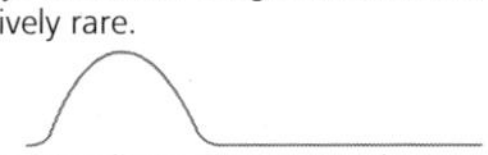

If two stimuli are given quite close together, the muscle fibre will contract again before it is fully relaxed and so it gets shorter than with a single twitch.

If the two stimuli are close enough together the two contractions are so close that there is no relaxing and lengthening of the muscle fibre between them. This gives the appearance of a single large contraction and is called **summation**.

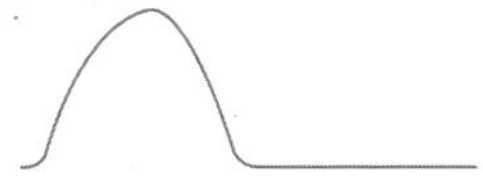

When a series of rapid stimuli are given the muscle fibre becomes fully contracted (as short as possible), and stays like this. This is known as **tetanus** and is the normal situation in a muscle when you are lifting an object or standing up and maintaining your posture against gravity.

A muscle cannot remain in a tetanic contraction continuously. Eventually it **fatigues** and cannot contract any longer.

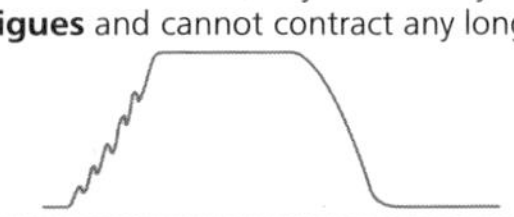

Figure 2.6.11 Kymographic traces from experiments on a muscle fibre taken from a frog gastrocnemius

- Bone consists of bone cells or **osteocytes** in a matrix of collagen fibrils and calcium salts. **Compact bone** is dense and strong, **spongy bone** has a more open, lighter structure. In the matrix, the **Haversian canal** contains an artery and vein with capillaries running to the osteocytes within spaces called **lacunae**. Osteocytes may form **osteoblasts** to lay down new bone, or **osteoclasts** to reabsorb the matrix.

- Cartilage is made up of cells called **chondrocytes** in a matrix of collagen fibrils. It is elastic but hard. The three types are **hyaline cartilage** found at the ends of bones, **yellow elastic cartilage** and **white fibrous cartilage** with densely packed collagen in the matrix which forms the intervertebral discs.

- Tendons are made up of **white fibrous tissue** – bundles of collagen fibrils with cells called **fibrocytes**. They are strong but inelastic. Ligaments are made up of **yellow elastic tissue** containing collagen fibres for strength and elasticity.

- Bones are held in position at joints by ligaments. **Ball and socket** joints allow free movement in most directions while **hinge** joints are more restrictive. A **synovial joint** such as the hip joint consists of a ligament capsule enclosing the head of a bone in a socket, both of which are lined with cartilage. **Synovial fluid** lubricates the joint.

- Muscles act on bones to bring about movement. The **origin** is the point of attachment on a bone nearest the heart and the **insertion** that on another bone further from the heart. Contraction of a muscle brings about movement of the bone at the insertion. Muscles act as **antagonistic pairs**, one moving a bone one way and the other reversing this movement.

- Locomotion on land depends on the reaction of the ground to a force exerted by the animal.

- Water provides more support but more resistance to locomotion. Aquatic creatures are often **streamlined**. They locomote by lashing movements of the vertebral column. **Anguilliform** locomotion involves movements of the whole animal, **carangiform** locomotion the rear part and the tail and **ostraciform** locomotion just the tail. Fins cancel the sideways and rolling motions produced so the fish moves forwards.

- Flying animals have light hollow bones, highly developed flight muscles and bones for their attachment and a high metabolic rate.

- Birds have wings shaped to act as an **aerofoil** to provide lift, flapped in a motion to provide an angle that causes upward movement. Birds glide between flapping movements.

- Insects flap their wings very quickly to remain airborne and do not glide. **Indirect flight muscles** inside the thorax contract and relax to change the shape of the thorax and flap the wings. **Direct flight muscles** alter the angle of the wings in the air.

- Muscles may be **skeletal** (striated or voluntary, involved in locomotion), **smooth** (involuntary) or **cardiac** (heart muscle).

- Skeletal muscle includes **slow-twitch muscles** for steady prolonged action and **fast-twitch muscles** for rapid activity.

- Skeletal muscle fibres are made up of **myofibrils**, themselves consisting of units called **sarcomeres**. **Sarcoplasm** makes up the cytoplasm, which contains many mitochondria and a network of membranes called the **sarcoplasmic reticulum**. **T tubules** which store and release calcium ions run between the myofibrils.

- **Actin** (thin) and **myosin** (thick) filaments interlock in the sarcomere. The **I band** contains actin filaments only and the **H zone** contains myosin filaments only. The overlap or **A band** contains both actin and myosin filaments.

- When a myofibril contracts, the actin filaments slide over the myosin filaments, increasing the overlap or A band. **Actomyosin bridges** form between the filaments to move the actin filaments along. The formation and breaking of the bridges depends on calcium ions and ATP.

QUESTIONS

1 a What are the main functions of a skeletal system?
  b What are the differences between an endoskeleton and an exoskeleton?
  c Taking an insect as your example, demonstrate how an exoskeleton can perform all the functions of a skeletal system.

2 a Name and compare the major tissues which make up the mammalian skeletal system.

  b Using a named joint, illustrate how these tissues interact to allow movement to occur.

3 a Describe the structure of striated muscle.
  b How does a muscle fibre contract? (Include reference to ATP, calcium, troponin, actin, myosin and tropomyosin in your answer.)

Questions

1 State the significant chemical features and then explain the biological significance of haemoglobin. **(6 marks)**

2 a Describe the structure of each of the following tissues, indicating in each case how structure is related to function:
  i parenchyma
  ii collenchyma
  iii sclerenchyma. **(12 marks)**
  b Compare the distribution of tissues in a dicotyledonous stem and root in relation to the mechanical functions of the stem and root. **(8 marks)**
  (ULEAC January 1991)

3 a Describe the structure of striated (skeletal) muscle tissue. **(7 marks)**
  b Explain the mechanism of contraction in striated muscle tissue. **(10 marks)**
  (ULEAC January 1991)

4 Write a comparative account of the mechanisms used for transport in plants and animals.
  (ULEAC January 1991)

5 a Relate the structure of a leaf to its activities of photosynthesis and transpiration. **(12 marks)**
  b Explain how variation of **i** light and **ii** temperature can affect these functions. **(8 marks)**
  (NEAB June 1991)

6 a What are the essential features of the mammalian blood vascular system? **(4 marks)**
  b Comment on the absence of a blood vascular system in animals such as protozoa and coelenterates. **(4 marks)**
  c Describe the formation and function of lymph. **(6 marks)**
  d How is oxygen carried in the blood? **(6 marks)**

7 Discuss the concept of tissues as structural and functional units in living organisms. Include reference to named examples of both plant and animal tissues. **(20 marks)**
  (ULEAC June 1990)

8 'Exercise stimulates the circulatory, respiratory and muscular systems.' Comment on this statement with reference to the physiological mechanisms involved. **(20 marks)**
  (ULEAC June 1990)

9 a Describe the mode of action of the mammalian heart. **(8 marks)**
  b Describe the mechanisms which regulate heart rate during and after vigorous exercise. **(8 marks)**
  (NEAB June 1989)

10 Describe the sequence of events which occurs when resting striated muscle is stimulated, contracts and then relaxes. **(14 marks)**
  (ULEAC June 1991)

11 Diagram X below shows part of the respiratory tract of a mammal. Diagram Y is an enlargement of the indicated region of diagram X.

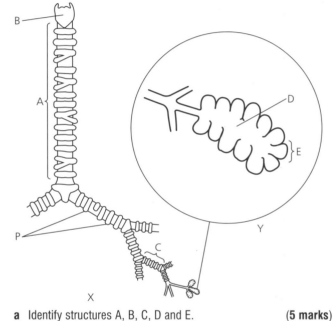

a Identify structures A, B, C, D and E. **(5 marks)**
b i Name the tissue found in structure P. **(1 mark)**
  ii Explain the importance of this tissue in the functioning of the respiratory tract. **(2 marks)**
c i Name the type of epithelium lining E. **(1 mark)**
  ii How is the structure of this type of epithelium suited to its function in E? **(2 marks)**
  (Total 11 marks)
  (ULEAC January 1991)

12 The diagram below shows a section through a human heart with associated blood vessels.

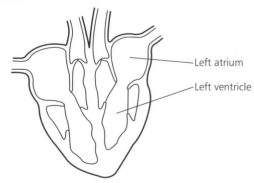

Left atrium

Left ventricle

a On a copy of the diagram label with the appropriate letter the part which fits each of the following descriptions.

   A A structure which prevents the backflow of *oxygenated* blood when the pressure in the ventricles falls

   B A structure which prevents the backflow of *deoxygenated* blood when the pressure in the ventricles rises

   C The vessel from which the coronary arteries branch

   D A vessel bearing deoxygenated blood at low pressure

   E The chamber in which pressure rises first when the sino-atrial node ('pacemaker') discharges **(5 marks)**

b Some babies are born with an incomplete septum between the two atria of the heart.

   i Suggest what effect this would have had during fetal life. **(2 marks)**

   ii Suggest what effect this would have after birth. **(2 marks)**

                                  (Total 9 marks)

(ULEAC January 1992)

13 The diagram below shows part of a dicotyledonous stem. The section enclosed in the box labelled X is shown in the second diagram as it would appear under high power of a light microscope.

a **i** Name the tissues labelled A and B. **(2 marks)**

   ii State *two* functions of the tissue labelled A. **(2 marks)**

b Explain how the functioning of the vascular cambium is important in the life of a flowering plant. **(4 marks)**

c **i** Where would the tissues labelled C be located in a root? **(1 mark)**

   ii Explain how this difference in position is related to the adaptations of roots and stems to their mechanical functions. **(2 marks)**

   iii State *two* ways in which tissue A would differ in a young root. **(2 marks)**

                                  (Total 13 marks)

(ULEAC January 1992)

14 **a** Give *one* location in a mammal of each of the following tissues, and indicate the relationship between the structure of the tissue and its function at that location:

   i pavement epithelium **(3 marks)**

   ii stratified epithelium. **(3 marks)**

b **i** Make a fully labelled diagram to show the structure of compact bone as seen in transverse section. **(4 marks)**

   ii Select *three* of the structures you have labelled and state their functions in compact bone. **(3 marks)**

                                  (Total 13 marks)

(ULEAC January 1992)

15 Water loss from the leaves of two different species of plants, *Species A* and *Species B*, was compared using two experimental methods.

Experiment 1

Leafy shoots were taken, one from each plant. They were weighed, hung in air and re-weighed at 15-minute intervals. The results are shown in the table (— means that a result is not available).

Time after start of experiment/minutes	Mass of leafy shoot/g	
	Species A	**Species B**
0	210.0	240.0
15	195.3	—
30	184.4	—
45	176.4	—
60	170.1	—
75	166.3	213.1

a **i** Using graph paper, plot a graph of mass against time for *Species A*.

   ii Account for the shape of the curve for *Species A*. **(5 marks)**

b **i** Calculate the percentage change in mass for each species after 75 minutes.

   ii Explain which of these two species you would expect to be more successful in an arid habitat. **(3 marks)**

Experiment 2

Different leafy shoots from each species were placed in a simple potometer, as shown below.

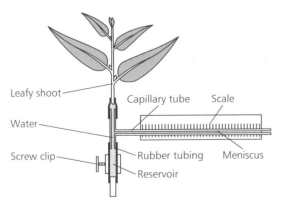

c **i** State *two* precautions that should be taken when using a potometer, to ensure that the results obtained are comparable for the two different species.

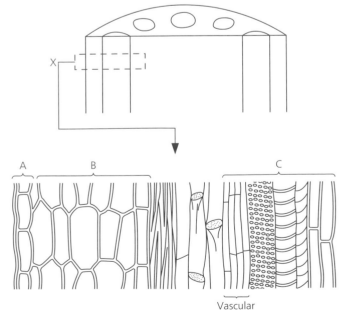

Vascular cambium

ii What *three* measurements are required to calculate the rate of water taken up in cm³ min⁻¹ g⁻¹ in *Experiment 2*?

iii Give *one* reason why the values obtained for water loss in *Experiment 2* were likely to have been higher than those from *Experiment 1* for the same species. **(6 marks)**
(Total 14 marks)
(NEAB June 1990)

16 The diagram below shows some of the structures visible in a ventral view of a section through a mammalian heart.

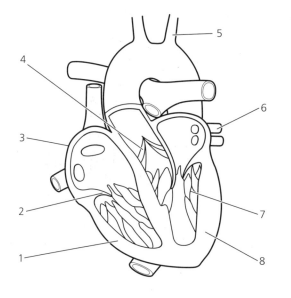

a Name the parts numbered 2, 3, 4, 5, 6 and 7. **(3 marks)**
b Explain the significance of the difference in thickness between:
  i parts 1 and 3
  ii parts 1 and 8. **(4 marks)**
c Explain how the heart beat is:
  i initiated and propagated
  ii controlled. **(6 marks)**
d List *five* contrasting features of the structure and function of the blood and lymph systems. **(5 marks)**
(Total 18 marks)
(O & C June 1992)

17 Write an essay on the range of structure and function in animal skeletons. **(20 marks)**
(ULEAC January 1992)

18 The respiratory surfaces of teleost fish consist of numerous gill lamellae.
a Explain how the following features of gill lamellae make gas exchange more efficient.
  i the lamellae being folded into many gill plates **(2 marks)**
  ii blood flowing through the plates in the opposite direction to water flowing over the plates. **(2 marks)**
b The following table summarises some features of the gill lamellae of three species of teleost fish.
Comment on these figures, given that herring are very active swimmers, trout are moderately active, and eels are relatively inactive. **(4 marks)**

Species of fish	Thickness of lamellae in μm	Distance between lamellae in μm	Distance between blood and surrounding water in μm
Eel	26	30	6
Trout	12	35	3
Herring	7	20	1

c Suggest *two* reasons why gill lamellae would not provide an efficient respiratory surface on land. **(2 marks)**
(Total 10 marks)
(ULEAC June 1991)

19 The diagram below shows some of the structures in the hind limb of a rabbit.

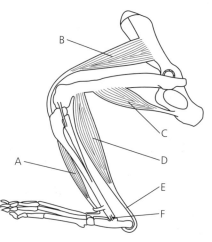

a Which of the muscles labelled A, B, C or D are responsible for the following?
  i straightening the knee **(1 mark)**
  ii lifting the heel **(1 mark)**
b For each of structures E and F, state *one* function and the property which enables this function to occur. **(4 marks)**
c State *two* ways in which friction is reduced at joints. **(2 marks)**
(Total 8 marks)
(ULEAC June 1991)

20 The graph below shows dissociation curves of oxyhaemoglobin from two mammals, the elephant and the mouse.

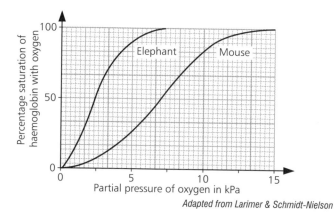

Adapted from Larimer & Schmidt-Nielson

a State *three* properties important for any respiratory pigment such as haemoglobin. **(3 marks)**

b i Describe the effect of a decrease in partial pressure of oxygen on the percentage saturation of haemoglobin in the elephant. **(3 marks)**

ii Describe the differences between the dissociation curves for the elephant and the mouse. **(3 marks)**

iii How might these differences be related to the sizes of the two animals? **(3 marks)**

c The graph below shows the effect of acid on the oxygen dissociation curve for the mouse (the Bohr effect).

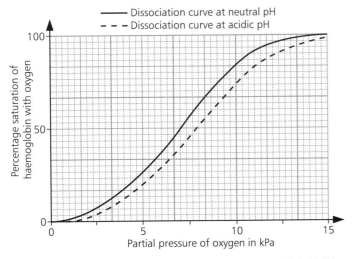

Adapted from Larimer & Schmidt-Nielson

i Comment on the effect of increase in acidity on the curve for the mouse. **(2 marks)**

ii Why might the pH change in respiring tissues? **(2 marks)**

iii Explain the physiological advantage of the Bohr effect. **(4 marks)**
(Total 20 marks)
(ULEAC January 1991)

21 a The apparatus shown is often used to measure the rate of transpiration. To be more precise, what does it measure? **(1 mark)**

b Give *two* reasons why movement of the meniscus may not give a measure of transpiration rate. **(2 marks)**

c Give *four* precautions you would adopt to avoid failure or inaccuracies when using the apparatus. **(4 marks)**

d In which tissue does water travel from the cut surface to the leaves? **(1 mark)**

e Explain how water travels up the stem once it has reached the xylem in the root. **(5 marks)**

f Name *three* climatic conditions which tend to accelerate the rate of water loss from the plant. **(3 marks)**

g Name *two* possible advantages of transpiration for the plant. **(2 marks)**
(Total 18 marks)
(O & C June 1991)

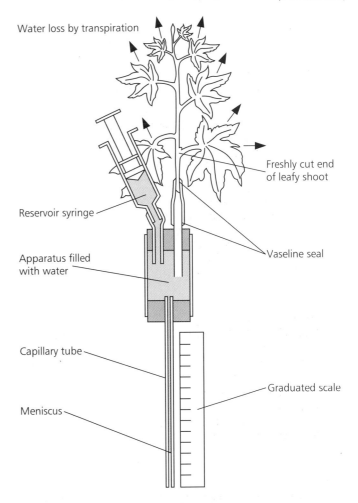

ENERGY SYSTEMS

Biomass – the energy of the future?

Throughout the twentieth century the world has faced an ever increasing energy crisis. 80% of the world's energy resources are consumed by 25% of the world population – those people living in the developed world. Cheap oil has been used to finance rapid industrial development in North America, the countries which made up the old USSR, Europe, Australia, Japan and South Africa. Now at the end of the century the supplies of fossil fuels such as coal, oil and natural gas are dwindling rapidly. Nuclear energy could be used as a replacement, but as a result of several accidents and near accidents along with media scare stories public confidence in this source of energy is low. New and renewable energy sources are needed, and soon.

The 75% of the world who cannot afford oil, coal, gas or the electricity produced from them have a constant need for cheap, renewable and readily available sources of energy for cooking and warmth. For example, at present most of the energy needs of the African countries are provided by firewood. The increasing demands of a growing population mean that wood is becoming a scarcer and scarcer resource. There is as yet no replanting programme so women spend most of the day looking for wood. Moreover, the removal of the wood that is left exposes the soil to erosion.

What is biomass?

Concern about energy supplies has led to much research into energy sources for the future. Wind power, tidal power and solar power are all well known possibilities. Perhaps less well known is the concept of using biomass. **Biomass** is all organic material produced by living organisms. Developing ways of utilising the energy held in the molecules of living things may perhaps lead to the ultimate energy supply. Not only is biomass readily available, but the source of the energy in living organisms is (directly or indirectly) the Sun. The Sun's energy is free to everyone, and as an added bonus it is not due to run out for some considerable time!

At the moment about 15% of the world's energy is provided from biomass. Most of it – wood, straw and animal dung – is used as a simple fuel, being burned in rural areas of the developing world. This simple combustion is an inefficient and dirty use of biomass. Much of the energy is wasted and the process is smoky and often smelly. The task for scientists and biotechnologists is to develop ways of using biomass which are more efficient but not too expensive for the developing world. Similar technology may be useful in the developed world as fossil fuels are increasingly depleted.

Figure 1 Whether for the basic necessities of life or for more frivolous activities, a readily available source of energy is needed by people all over the world.

Table 1 Some of the questions which need to be asked before a new energy source is introduced into a community

1	How abundant is the energy source, and is it readily available?
2	Can it be stored easily?
3	Can it be used easily for a variety of purposes?
4	Is it a constantly renewable source of energy?
5	How much pollution does it cause?
6	Is it safe?
7	What is the relationship between the energy put into the system to make it work and the energy gained from it (the output:input ratio)?
8	How expensive is it compared with other forms of energy?
9	Is technology sufficiently advanced to allow its use on a large scale?
10	Can the members of the community afford the energy?
11	Does the energy source place conflicting demands on a scarce resource, such as land?

Where does biomass come from?

Energy radiates from the Sun at an enormous rate – around 10^{31} kJ each year. The energy results from multiple collisions of protons in the central core of the Sun at temperatures of around 20 million degrees Celsius. However, only a small proportion of the radiation from the Sun reaches the Earth – about 3×10^{21} kJ – and on arriving at the Earth's atmosphere much more is absorbed or otherwise prevented from reaching the surface of the Earth. Light is absorbed by the ozone layer, scattered by dust and water particles or reflected back into space by the clouds and water at the surface of the Earth. Even when the light does reach the surface, not all is available to living organisms as much falls on relatively barren areas such as oceans, mountains and deserts. However, in the region of 10^{19} kJ of the Sun's energy is captured each year by plants – around 0.33% of the total energy that reaches the Earth.

3

The energy from the Sun is captured by plants using the green pigment chlorophyll. Solar energy is then transferred into chemical energy in the process of photosynthesis. It is stored in the chemical bonds of the glucose molecules made from carbon dioxide and water. This energy is then transferred into other organisms – animals, fungi and microbes – by processes of digestion and assimilation which are not efficient. For example, only about 10% of the biomass eaten by a cow is converted into new cow, and only about 10% of the cow eaten by a person gets turned into new person. All along the process, biomass and therefore potential energy is lost.

Using biomass as a fuel

Part of the energy within a plant or animal is in the form of ATP, which is used to supply energy for all the cells' metabolic reactions. Metabolism results in heat production, and the energy transferred to heat is effectively lost for further use. However, considerable energy remains in the tissues of a plant, and animals pass out large quantities of energy-containing material in their faeces. Using biomass as a fuel takes advantage of these sources of energy which would otherwise be wasted.

As we have already seen, both plant material (wood and straw) and animal dung are already used as fuel, but generally for inefficient combustion – only about 10% of the energy in the biomass is used. However, cleaner and more efficient ways of using biomass are being developed. Some, such as a process known as gasification, are complex and require expensive technology. Other more simple techniques rely on the actions of microorganisms to produce either methane gas (biogas) or ethanol. The anaerobic digestion by bacteria of waste from cattle to

biogas is 60% efficient. An added advantage of producing biogas from dung is that the slurry left at the end is a valuable fertiliser for fields as all the nitrates, phosphates and other minerals remain intact. This is another advantage over burning the dung, when the resource disappears in flames and a much-needed fertiliser is lost completely.

Dung digestion and biogas production

Cattle dung is a potentially enormous energy resource. It can be added to by the dung of pigs, sheep and poultry and, of course, human excrement. The only difficulty with using human waste is that in some cultures there are religious and social taboos against it. The successful conversion of manure into biogas depends on manipulating the digestive and respiratory processes of microorganisms. By the process of respiration, the energy in molecules produced by photosynthesis or digested by animals is transferred into molecules of ATP, the energy currency of the cell. When respiration takes place in the presence of oxygen, carbon dioxide and water are the waste products of the process. But in some bacteria, when respiration occurs in the absence of air or oxygen (that is, anaerobically), the gas produced is a mixture of methane and carbon dioxide in the ratio of about 2 volumes of methane for every 1 volume of carbon dioxide. After the carbon dioxide is removed, the methane provides a clean and efficient fuel and the residual sludge is a good fertiliser.

A mixed population of bacteria is needed to digest the dung, and the process is similar to that which occurs in the guts of ruminants like the cow. Small family biogas producers are currently used in both China and India. In China waste vegetables, animal dung and human waste is used and the digesters produce excellent fertiliser but relatively low quality gas. In India

Figure 2 You may not realise that if you eat beef, all the energy you are taking in is derived originally from the Sun. What is more, the waste biomass from the cow and from your body could be used to cook more beef.

Figure 3 A regular supply of dung from animals or people can supply a clean and efficient energy source for cooking and heating when processed in a biomass digester.

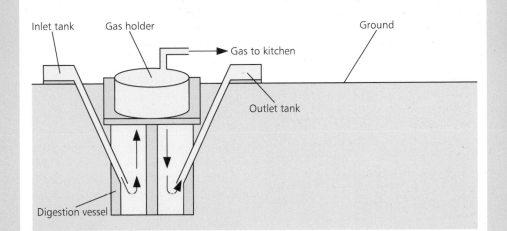

Inlet tank   Gas holder   Ground

→ Gas to kitchen

Outlet tank

Digestion vessel

Drying dung for fuel in Pakistan

only cattle and buffalo dung is used and the generators are used to produce high quality gas but less fertiliser.

Although the main use of biogas generators has so far been in the developing world, more and more countries are experimenting with them on a larger scale to deal with the problems of municipal sewage and rubbish tips, and with a view to reducing dependence on fossil and nuclear fuels. The efficiency of biogas generators is increased when the fuel is readily available. In India, for example, many people do not own cows and therefore cannot use a generator, as religious beliefs prevent the use of human excrement. Even those people who do own cows often only manage to find and collect 30–40% of the dung produced. In intensive rearing units in the developed world, almost 100% of the dung produced can be collected and used, making the production of biogas more efficient.

Alcohol – another fuel of the future?

As well as the production of biogas from dung, energy may be obtained from biomass by the fermentation of carbohydrate-rich crops such as sugar cane, sweet potato and maize into ethanol. This alcohol can then be used as a fuel for cars instead of petrol. Not only does this avoid dependence on rapidly disappearing oil resources, but also the combustion of the ethanol is far less polluting to the atmosphere.

Table 2 Anaerobic biogas generators convert the energy of the Sun into energy for cooking and heating through a complicated pathway which includes plants, photosynthesis, animals, digestion and egestion, microorganisms and respiration. This table sums up some of the pros and cons of biogas as a fuel of the future.

Advantages of anaerobic biogas generators	Disadvantages of anaerobic biogas generators
Relatively cheap	Best adapted for warm climates, although underground tanks and the exothermic nature of the reactions do make it feasible elsewhere
Versatile – a whole range of waste and raw materials can be used	The chemical processes are slow and it may take up to four weeks to produce the gas from a load of dung
The process can be carried out on either a small or a large scale	Carbon dioxide has to be removed from the gas mixture
Works at relatively low temperatures of 30–35 °C so energy input is not high	
pH is not critical – works within a fairly wide pH range	
Can be run continuously or in batches when fuel is available	

To produce alcohol from the biomass involves yeasts in a process of anaerobic respiration known as **fermentation**. The sugars are broken down incompletely to form ethanol and water. To make this process feasible in terms of output:input ratios it is important that there is as little pretreatment of the biomass as possible. As a result, only countries such as Brazil which have an enormous capability for growing crops such as sugar cane have so far been able to make much use of this process. However, in Brazil oil imports have been reduced by about 20% and car engines have been converted to run on a mixture of petrol and ethanol. Most new cars in Brazil will run on ethanol alone. The disadvantage of this development is that land valuable for growing food crops has been taken over to produce the enormous quantities of sugar cane needed to replace oil.

The major stumbling block in the conversion of biomass to ethanol as a major world-wide proposition is that we need to be able to utilise the cellulose element of the biomass which makes up much of agricultural waste (straw) and dung. At the moment it can be used only after expensive shredding, chemical or enzyme treatments. For ethanol to become the motor fuel of the future a major breakthrough is required in the development of a low-cost technology for preparing cellulose for fermentation. It may be that genetic engineering will produce enzymes capable of doing this cheaply and easily, so that waste biomass can be used to produce fuel instead of specific crops being cultivated specially.

What does the future hold?

No one can gaze into a crystal ball and see what will happen in world energy production in the future. We do know that changes in our energy sources will be forced upon us in the next century or two by the depletion of stocks of fossil fuels. Energy from biomass, taking advantage of photosynthesis to trap the energy of the Sun and respiration to release suitable fuel molecules, must surely have a role to play in replacing fossil fuels. Whether this will involve biogas or ethanol production, or some new biotechnology as yet undiscovered, we must wait and find out. Within this section you will discover more about energy in biological systems, and consider the processes of photosynthesis and respiration by which energy is obtained and released in living things.

Figure 4 So far, only countries like Brazil with vast natural resources can produce useful amounts of ethanol from biomass, but the development of new cellulose-digesting processes could see alcohol-powered cars for everyone in the future.

3.1 Energy

ENERGY FLOW IN THE BIOSPHERE

Where does energy come from?

Living organisms need energy to grow, to move, to reproduce – energy is needed for life. There is a massive flow of energy in the biosphere. Organisms can be classed according to their different energy sources. **Autotrophic** organisms make organic compounds from carbon dioxide. Most autotrophic organisms do this by **photosynthesis** – they trap solar energy, which is transferred to chemical energy within the bonds of glucose and other organic products. There are a few autotrophic organisms, the **chemosynthetic** bacteria, which synthesise organic compounds from carbon dioxide using chemical, rather than solar, energy. **Heterotrophic** organisms take in organic compounds by feeding on autotrophs or other heterotrophs. They use the chemical energy in the products of photosynthesis to make the structural molecules they need, and as fuels to supply energy for a wide variety of activities. The Sun is thus the ultimate source of energy for almost all organisms (the exceptions being the chemotrophs).

Where does energy go?

Energy is not cycled in the biosphere – there is a one-way flow, with all energy eventually being dissipated. Solar energy is captured by photosynthetic cells and transferred to chemical energy in the products of photosynthesis. It is then used by both heterotrophs and autotrophs to carry out cell work – synthesis of new materials for growth and reproduction, contraction of muscles for movement, etc. In the course of these reactions much of the energy is transferred to relatively useless forms from a biological point of view. For example, much is transferred into heat energy, which may warm the animal on a temporary basis but is then simply dissipated in the environment. (It is important to remember that energy is not destroyed – all the energy that is captured by photosynthesis still exists, but is eventually converted to a less useful form.)

The flow of energy in the biosphere, shown in figure 3.1.1, is of enormous proportions. Around 10^{19} kJ of solar energy is used each year to convert carbon dioxide into biological material (**biomass**) in photosynthetic organisms. This biological energy flow is some 20 times greater than the flow of energy through all the machines that people have created in the world.

Identifying energy sources

In everyday life we often refer to energy, for example in terms of how energetic we feel, or of saving electrical energy. Much of the energy we know about comes from particular sources. The electrical energy we use is produced either from the release of heat energy when fossil fuels are burned or during a controlled nuclear reaction, or from the conversion of potential to kinetic energy as water moves downhill, or from the kinetic energy as wind turbines are turned. The chemical energy stored in the molecules of gas, coal or other fuels may be transferred to heat energy for heating, cooking or electricity

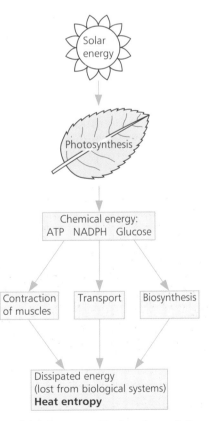

Figure 3.1.1 A summary of the flow of energy in the biosphere

production. The chemical energy present in the molecules of petrol may be transferred to kinetic energy to move a car.

In the same way, the energy used by living organisms for their daily functioning has very specific sources. As we have already mentioned, the source of energy for almost all living things is ultimately the Sun. The Sun is the direct source of energy for those organisms that are capable of trapping it. (The process of photosynthesis by which the energy of the Sun is captured and transferred to chemical energy in organic molecules within the plant is considered later.) For heterotrophs, such as animals, the direct source of energy is the food they eat. This food may be plants for **herbivores**, other animals for **carnivores** or both plants and animals for **omnivores**. The chemical energy stored in the bonds of the food molecules, particularly the carbohydrates and fats, is used for all the biochemical reactions of life.

Energy stores

Most living things need to store energy. Plants cannot photosynthesise at night and animals do not spend all their time feeding, yet the demands of the cells for energy are continual. Thus energy must be stored, and then produced when needed in a readily usable form. In plants, some products of photosynthesis are immediately used to create new plant tissue or to provide energy for their synthesis. However, any products not used in this way are converted into starch (in dicotyledons) or sucrose (in monocotyledons). These carbohydrate molecules are stored in the cells until needed, when they are broken down into glucose.

In animals, some digested food which is not immediately required may be converted into a carbohydrate store of glycogen in the liver, muscles and brain. The larger part is converted into fat and stored in special fat cells. The way these stores may be used by an animal is shown in figure 3.1.2.

For the individual cells of any living organism, energy has to be available in a much more accessible form than fat or starch, ready for use in any one of a multitude of different reactions immediately it is needed. One molecule is believed to be the universal energy supplier in cells. It is found in all living organisms in exactly the same form. Anything which interferes with its production or breakdown is fatal to the cell and, ultimately, the organism. This remarkable compound is called **adenosine triphosphate**, more commonly referred to as **ATP**.

Figure 3.1.2 Animals that hibernate eat copious quantities of food before the winter. This is converted into large fat stores. During the time of hibernation this fat is slowly used up to supply energy to the cells, with an amount saved for the final surge of energy needed to bring the creature out of hibernation at the end of the winter. The weight loss can be quite dramatic.

Adenosine triphosphate (ATP)

A chemical energy store

Figure 3.1.3 shows the structure of ATP. When energy is needed, the third phosphate bond can be broken by a hydrolysis reaction catalysed by the enzyme **ATPase**. The result of this hydrolysis is **adenosine diphosphate** or

Figure 3.1.3 ATP is a nucleotide with three phosphate groups attached. It is the chemical energy stored in the phosphate bonds, particularly the last one, which is made available to cells.

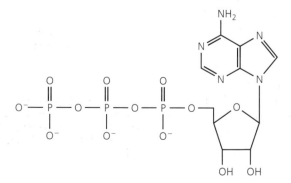

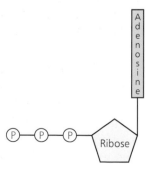

ADP, and a free inorganic phosphate group (**P**ᵢ). About 34 kJ of energy are released per mole of ATP hydrolysed. Some of this energy is lost as heat and wasted, but some is used for any energy-requiring biological activity.

The breakdown of ATP into ADP and phosphate is a reversible reaction. Catabolic reactions, for example the breakdown of glucose in respiration in both plants and animals, can be coupled to the synthesis of molecules of ATP from ADP and phosphate groups. This reaction is also catalysed by ATPase. Energy is thus stored in the ATP molecule, ready for use when required. Figure 3.1.4 shows the cycle between ADP and ATP, and figure 3.1.5 illustrates how the cycle is used in living organisms.

ATP is the single energy-providing and energy-storing molecule for all processes in all cells.

SOURCES OF ENERGY FOR *ATP* SYNTHESIS

We have already stated that ATP is made as a result of catabolic or breakdown reactions. Most of the ATP synthesised in a cell is produced as a result of the breakdown of food molecules in the process known as cellular respiration, which takes place in the mitochondria. Much of the detail of this process will be considered later. However, ATP synthesis also takes place in chloroplasts and in chemosynthetic bacteria. We shall consider the way ATP is produced during respiration as a basic model, and look at this in isolation. We can then later insert this model of ATP synthesis into large biochemical pathways when needed.

When sugars are broken down in respiration, energy for the production of ATP from ADP and inorganic phosphate is made available in two ways:

1 Catabolic reactions are exothermic – they release energy. This energy may be sufficient to drive the production of a molecule of ATP. Relatively little ATP is actually formed this way.

2 The other main source of energy for ATP synthesis results from the removal of hydrogen atoms from several intermediates in a metabolic pathway. This removal of hydrogen atoms brings about a series of **redox reactions** in a chain of compounds called the **electron transfer chain**. The term 'redox' is explained in the box below. The redox reactions in the electron transfer chain each release a small amount of energy, which is incorporated into an ATP molecule.

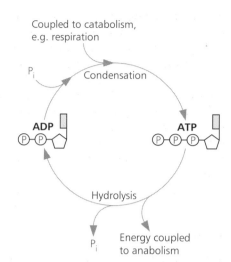

Figure 3.1.4 The storing and release of energy in ATP

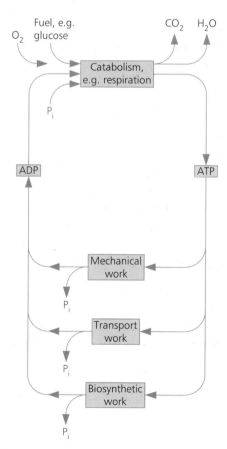

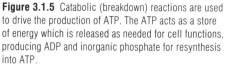

Figure 3.1.5 Catabolic (breakdown) reactions are used to drive the production of ATP. The ATP acts as a store of energy which is released as needed for cell functions, producing ADP and inorganic phosphate for resynthesis into ATP.

Redox reactions

Reduction and **oxidation** are two important chemical concepts which help us to understand the terminology of the electron transfer chain.

Reduction is the addition of electrons to a substance. In biological systems this addition of electrons is usually brought about by the addition of hydrogen or the removal of oxygen. Any compound which has oxygen removed, or hydrogen or electrons added, is said to be **reduced**.

Oxidation is the removal of electrons from a substance. In biological systems this removal of electrons is usually brought about by the addition of oxygen or the removal of hydrogen. Any compound which has oxygen added, or hydrogen or electrons removed, is said to be **oxidised**.

In the electron transfer chain, electrons are passed from one member of the chain to the next. The components of the chain are therefore reduced when they receive the electrons, and oxidised again when they pass them on.

In a **redox reaction**, one compound will be oxidised while another is simultaneously reduced.

The electron transfer chain

Figure 3.1.6 shows the electron transfer chain. The various elements of the chain are described below. As you can see, two hydrogens are removed from a compound (which is therefore oxidised) and are picked up by NAD, the first **hydrogen carrier** or **acceptor**. The acceptor is therefore reduced. The hydrogens are then passed to the next acceptor and then along the molecules of the electron transfer chain, so a series of redox reactions takes place.

Although it is hydrogen atoms which are removed from the compounds of the metabolic pathways, and hydrogen atoms which eventually join up with oxygen atoms at the end of the chain to form water, it is in fact electrons which are passed along the carrier system. Hydrogen atoms split into protons and electrons and it is the electrons that are passed from one carrier to the next. This is why the system is most accurately known as the electron transfer chain rather than the hydrogen carrier chain. The various elements of the chain are at different energy levels, the first member of the chain being at the highest level with subsequent steps down. Thus the electron is passed down from one energy level to another, powering the production of ATP.

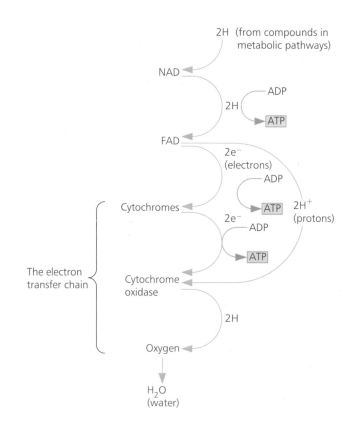

Figure 3.1.6 These are the main known components of the electron transfer chain. As they become reduced and then oxidised again, sufficient energy is released to drive the production of molecules of ATP.

Hydrogen acceptors

The hydrogen acceptors include:

1   **NAD** or **nicotinamide adenine dinucleotide**. When this accepts hydrogen atoms from a metabolic pathway it becomes reduced to **NADH$_2$**.

2   **FAD** or **flavine adenine dinucleotide**. This is synthesised from vitamin B2, riboflavin. It accepts hydrogens from NADH$_2$, which is therefore oxidised again to NAD. The FAD is reduced to **FADH$_2$** at the same time. A molecule of ATP is produced at this stage.

The electron transfer chain

1   **Cytochromes** are protein pigments with an iron group rather like haemoglobin. Cytochromes are reduced by electrons from FADH$_2$ which is oxidised again to FAD. A molecule of ATP is produced at this stage.

2   **Cytochrome oxidase** is an enzyme which receives the electrons from the cytochromes and is reduced as they are oxidised. A molecule of ATP is produced at this stage.

3   **Oxygen** is the final electron acceptor in the chain. The reduction of oxygen forms water at the end of the chain.

As a result of the electrons from each molecule of hydrogen passing along the electron transfer chain, sufficient energy is released to make three molecules of ATP. The process is also known as **oxidative phosphorylation** – ADP is phosphorylated in a process which depends on the presence of oxygen.

THE SITE OF *ATP* SYNTHESIS

The main site of ATP synthesis in the cell is the mitochondria. The main reactions of respiration appear to take place in the matrix, while the reactions of the electron transfer chain and so the production of ATP seem to take place on the inner membrane of the mitochondria. This is folded up to form cristae, which in turn are covered with closely packed stalked particles, shown in figure 3.1.7.

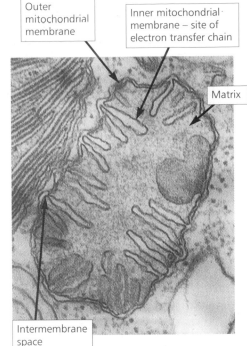

Figure 3.1.7 The site of most ATP production in cells appears to be the inner membrane of the mitochondria, with the ATPase sited in the stalked particles.

Evidence for the site of ATP production

There are several strands of evidence that ATP synthesis takes place on the inner membrane of the mitochondria.

(1) It is possible to break open cells and centrifuge the contents to obtain a fraction containing just mitochondria. If these are kept supplied with pyruvate (see pages 216–17) and oxygen they will produce ATP.

(2) High-powered electron micrographs show the surface of the inner membrane of the mitochondrion to be covered in closely packed stalked particles. These give a greatly increased surface area for enzymes to work.

(3) These stalked particles and the bits of membrane associated with them can be separated from the rest of the mitochondrial structure. It can then be demonstrated that they alone out of the contents of the mitochondrion are capable of ATP synthesis.

The chemiosmotic theory of ATP production

In the 1960s Peter Mitchell put forward a mechanism for the production of ATP in the mitochondria. Other scientists were sceptical at first but by 1978 Mitchell's **chemiosmotic theory** was widely accepted, and he won the Nobel prize.

The theory is very elegant. The members of the electron transfer chain are found within the inner mitochondrial membrane, and electrons are passed along them as we have seen. What happens to the hydrogen ions (protons) that are left?

Figure 3.1.8 shows the mechanism. The protons are actively transported into the space between the inner and outer mitochondrial membranes. The inner mitochondrial membrane is impermeable to protons. This means that as a result of the active transport of the protons there are different hydrogen ion concentrations on the two sides of the inner membrane. The intermembrane space has a higher concentration of hydrogen ions than the matrix, so there is a concentration gradient across the membrane. As a result of the different hydrogen ion concentrations there is also a pH gradient. Because positive hydrogen ions are concentrated in the membrane space, there is also an electrochemical gradient.

All this means that there is a tendency for the hydrogen ions to move back into the matrix. However, the membrane is generally impermeable to hydrogen ions. The only way they can move back into the matrix is through special pores involving the stalked particles where the electron transfer chain is situated. The movement of the hydrogen ions along their electrical, concentration and pH gradients is thereby linked to an ATPase enzyme. The energy from the gradients is used to drive the synthesis of ATP. Thus the universal energy carrier is produced in a universal process, found in all living things.

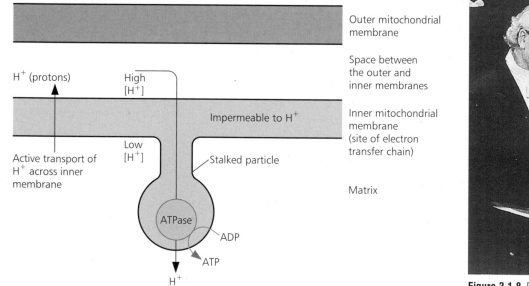

H^+ (protons) leave the space between the membranes along the only route available to them – through the stalked particles. They move along concentration, pH and electrochemical gradients, driving the production of ATP.

Figure 3.1.8 Peter Mitchell, seen here receiving a Nobel prize for his work, developed the chemiosmotic theory which explains the formation of ATP not just in the mitochondria but also in chloroplasts and elsewhere. The hydrogen ion gradient explains all the current observations about the process.

SUMMARY

- Organisms may be classed according to their energy sources. **Autotrophs** make organic compounds from carbon dioxide, mainly by photosynthesis using solar energy. The exceptions are **chemotrophs** which synthesise organic compounds using chemical energy. **Heterotrophs** take in organic compounds by feeding on autotrophs or other heterotrophs.

- Energy is transferred within organisms to different forms and is eventually dissipated.

- Energy is stored in chemicals within organisms – in plants as starch or sucrose and in animals as glycogen or fat.

- At a cellular level energy is made available for biochemical reactions in the form of **adenosine triphosphate** or **ATP**. The terminal phosphate group can be hydrolysed from this molecule forming **adenosine diphosphate** or **ADP**, and energy is released which can be used to drive other reactions. The reaction is reversible and so ATP can act as an energy store. The reaction is catalysed by **ATPase**.

- ATP may be produced as a result of exothermic catabolic reactions, though this is not very common.

- Most ATP is produced as electrons pass down the **electron transfer chain** during the process of **oxidative phosphorylation**. The process results in the reduction of oxygen and the oxidation of hydrogen (from metabolic intermediates) to form water. As electrons move down the different energy levels in the chain, energy is made available to synthesise ATP.

- **Hydrogen acceptors** that take the hydrogen from other molecules and pass it on to the electron transfer chain include **nicotinamide adenine dinucleotide** or **NAD**, which is reduced to **NADH$_2$**, and **flavine adenine dinucleotide** or **FAD**, which is reduced to **FADH$_2$**. This reaction results in the production of a molecule of ATP.

- Hydrogens split into protons and electrons and the electrons pass from FADH$_2$ to the **cytochromes** and then to **cytochrome oxidase**, both of which reduction reactions result in a molecule of ATP. The electrons finally reduce oxygen, and water is formed at the end of the chain.

- The **chemiosmotic theory** says that the protons, once separated from the electrons, are actively transported to the space between the inner and outer mitochondrial membranes. The concentration of protons in the intermembrane space is higher than that in the matrix, resulting in a chemical, electrochemical and pH gradient. Protons can only move back into the matrix at the stalked particles, where their movement is linked to an ATPase enzyme and drives the synthesis of ATP.

QUESTIONS

1 a What are the main energy sources in the biosphere?
  b Where is energy stored within the biosphere?
  c Describe the flow of energy through the biosphere.

2 a What is the main function of ATP within a cell?
  b Describe the structure of an ATP molecule and explain how its structure is related to its function.
  c Explain where ATP production takes place and describe the evidence which supports the hypothesis.

3 With reference to the structure of the mitochondrion, describe how Mitchell's chemiosmotic pump works and how ATP is produced.

3.2 Photosynthesis

WHAT IS PHOTOSYNTHESIS?

Photosynthesis is the process by which living organisms, particularly plants, capture solar energy and use it to convert carbon dioxide and water into simple sugars. The process is dependent on pigments, particularly chlorophyll, present in the chloroplasts of the plant cells. These pigments absorb light energy and transfer it to chemical energy. Photosynthesis is summarised by the equation:

$$\text{Carbon dioxide} \quad + \quad \text{water} \quad \xrightarrow[\text{chlorophyll}]{\text{light energy}} \quad \text{glucose} \quad + \quad \text{oxygen}$$

$$6CO_2 \quad + \quad 6H_2O \quad \xrightarrow[\text{chlorophyll}]{\text{light energy}} \quad C_6H_{12}O_6 \quad + \quad 6O_2$$

The whole structure of the plant body has evolved around the process of photosynthesis. As we have seen in section 2 the stems, roots and leaves of plants are adapted both for obtaining the raw materials carbon dioxide and water, and for trapping the energy of the Sun.

As already mentioned, organisms capable of synthesising their own food are known as **autotrophs**, which can be roughly translated as 'self-feeding'. Of the autotrophs, the **phototrophs** or 'light feeders' rely on the Sun as their source of energy, and the **chemotrophs** rely on energy from breaking chemical bonds to synthesise their food. The most important autotrophs are the plants, which are one of the most successful groups on Earth. Their ability to synthesise new biological material (roughly 3.5×10^{16} kg of carbon dioxide are fixed into sugars each year), along with the resulting waste product oxygen, means that phototrophs are of vital importance to the survival of all species on Earth.

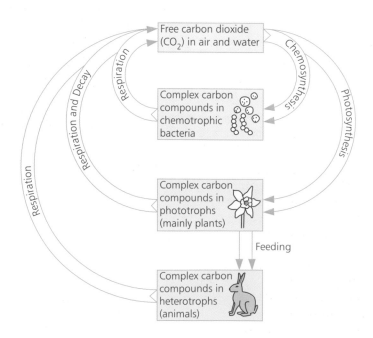

Figure 3.2.1 The carbon cycle shows the way in which carbon is cycled in nature, and also demonstrates the interdependence of animals and plants.

FACTORS AFFECTING PHOTOSYNTHESIS

As shown in the equation for photosynthesis, carbon dioxide and water are needed for the process, along with a supply of light energy and the means to capture that energy. In order to show that certain factors are needed for photosynthesis, or have an effect on its rate, we need a way of demonstrating that photosynthesis has in fact taken place. The simplest way to do this is to look at the end products of the process.

Demonstrating the products of photosynthesis

The immediate products of photosynthesis are glucose and oxygen. As glucose is osmotically very active, it is rapidly converted to the large polysaccharide molecule starch, which has little effect on the movement of water in cells. Thus we have two simple ways of demonstrating that photosynthesis is taking or has taken place – the production of oxygen gas and the formation of starch.

The production of oxygen by a photosynthesising plant can be shown very simply by collecting the gas over water. But as is shown in figure 3.2.2 this basic technique can be relatively easily adapted into a much more sophisticated experimental device. By measuring the amount of oxygen gas given off in a period of time a measure of the rate of photosynthesis can be found. By changing the conditions such as light intensity or temperature the influence of these factors on the rate of photosynthesis can be seen.

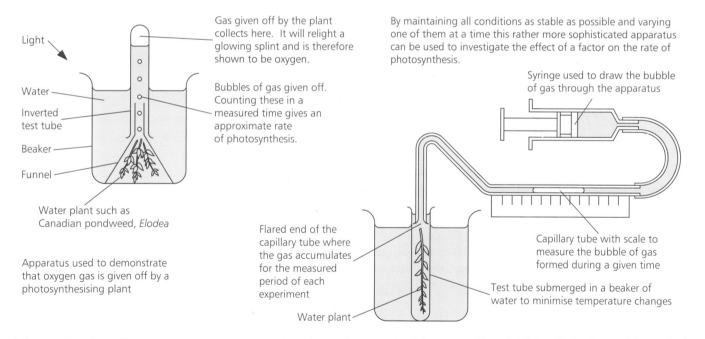

Light

Water

Inverted test tube

Beaker

Funnel

Gas given off by the plant collects here. It will relight a glowing splint and is therefore shown to be oxygen.

Bubbles of gas given off. Counting these in a measured time gives an approximate rate of photosynthesis.

Water plant such as Canadian pondweed, *Elodea*

Apparatus used to demonstrate that oxygen gas is given off by a photosynthesising plant

By maintaining all conditions as stable as possible and varying one of them at a time this rather more sophisticated apparatus can be used to investigate the effect of a factor on the rate of photosynthesis.

Syringe used to draw the bubble of gas through the apparatus

Capillary tube with scale to measure the bubble of gas formed during a given time

Test tube submerged in a beaker of water to minimise temperature changes

Flared end of the capillary tube where the gas accumulates for the measured period of each experiment

Water plant

Figure 3.2.2 Investigating the rate of photosynthesis

The production of oxygen gas as a waste product by a photosynthesising plant is relatively easy to demonstrate in water plants, as shown in figure 3.2.2, but considerably less so in land plants. Thus for land plants the most useful indicator of the occurrence of photosynthesis is whether starch has been produced. If a plant is deprived of light for about 48 hours, then the stores of starch in the leaves are substantially depleted. The plant is said to be **destarched**. Any subsequent production of starch is then the result of new photosynthesis occurring since the destarching. Testing a leaf for the presence of starch using iodine solution is a familiar procedure, summarised in the box below. Destarched plants can be subjected to different conditions and the effects of various factors on photosynthesis in land plants thus investigated.

Testing a leaf for starch

A common way of demonstrating whether a plant has carried out photosynthesis is to test a leaf for the presence of starch. The test reagent used is iodine in potassium iodide solution, which turns from yellowish-red to blue-black in the presence of starch.

Simply applying the iodine solution to the surface of a leaf is not effective as the leaf is covered by a waterproof layer – the waxy cuticle. In order for the test to show clearly whether starch has been produced, a series of experimental steps must be followed:

(1) Remove a leaf from the plant, plunge it into boiling water and continue to boil briefly. This serves two main purposes – it stops all the biochemical processes by killing the leaf, and it breaks open the cells making them more accessible to the removal of chlorophyll and to the iodine solution.

(2) Boil the leaf again in methanol for several minutes until all the colour is removed. This must be carried out in a water bath and great care taken as methanol is very flammable. The removal of the green pigment from the leaf enables any colour changes in the test reagent to be more clearly seen.

(3) Wash the leaf once more in hot water. Methanol makes the leaf brittle, and this softens it.

(4) Spread the leaf out on a white tile, again to make colour changes more obvious, and apply the iodine solution.

Figure 3.2.3 The striking colour change in iodine solution when starch is present demonstrates photosynthesis very clearly.

Requirements for photosynthesis

Light

Light is needed for some, but not all, of the reactions of photosynthesis. There is a **light-dependent stage** of photosynthesis, and also a **light-independent stage** in which the reactions occur in the absence of light.

For most plants the source of light energy for the reactions of the light-dependent stage is the Sun, although artificial light of appropriate wavelengths can be used. If plants are deprived of light for any substantial amount of time they will die, because once the stores of starch have been used up they are not replaced and so there is no energy available for the metabolic reactions of the cells.

The simplest way of demonstrating the requirement of a plant for light is to cover either a whole leaf or part of a leaf of a destarched plant with black paper or foil. This prevents light from reaching the covered area. The plant is left in the light for several hours and the covered leaf tested for the presence of starch and compared with an uncovered leaf. The difference is plainly visible, as figure 3.2.4 demonstrates.

Carbon dioxide

A source of carbon is needed for the synthesis of sugars. There are numerous carbon-containing chemicals in existence, but carbon dioxide from the air, or in solution or as hydrogencarbonate ions (HCO_3^-) in water, is the only form which plants can use in photosynthesis. Carbon dioxide is found more or less everywhere and is produced by plants as a result of their metabolic processes.

Figure 3.2.4 Only the 'L' on the lower leaf was exposed to light, and the resulting absence of starch can be clearly seen.

However, although there is always sufficient carbon dioxide available for some photosynthesis to take place, there are circumstances when the levels are too low for plants to take full advantage of the light available.

Demonstrating the absolute requirement of a plant for carbon dioxide is not easy. Carbon dioxide can readily be removed from the air surrounding a leaf or a plant using potassium hydroxide solution, which absorbs carbon dioxide, as figure 3.2.5 shows. However, the cells produce carbon dioxide as they respire and so it is almost impossible to entirely deprive a plant of the gas. A more valid approach is to change the levels of carbon dioxide in the air surrounding a plant in high-intensity light, and measure the changes in the rate of photosynthesis. As the carbon dioxide level increases, the rate of photosynthesis rises.

Water

Carbon dioxide alone is not sufficient to produce carbohydrates. Hydrogen ions are needed too, and water is the only source of hydrogen ions that plants can make use of. As a result of metabolic processes and the transpiration stream, there is always an adequate supply of water for photosynthesis.

Water is vital to all the functions of a plant. We cannot demonstrate that water is required for photosynthesis just by depriving the plant of it – the plant would die long before any effect of water lack on photosynthesis could be seen. To show that water is needed for photosynthesis, the plant can be supplied with water containing the 'heavy' isotope oxygen-18. This experiment shows that water is needed for photosynthesis, and also makes clear its role. We shall look at this role in more detail later, on pages 183–5.

Chlorophyll

The final requirement for photosynthesis is a means of capturing the energy from the Sun. The photosynthetic pigment **chlorophyll** fulfils this role, although it has been estimated that of all the light which reaches the Earth from the Sun, only about 0.33% is used in photosynthesis.

The simplest way to demonstrate that chlorophyll is required for photosynthesis is to consider the leaves of a variegated plant. Variegated leaves have areas which contain chlorophyll and areas which do not. The chlorophyll-free regions are usually yellow or creamy-white in colour. If a destarched variegated plant is exposed to light for several hours and one of the leaves tested for the presence of starch, the iodine solution changes colour only in those regions of the leaf which were green. This shows that without chlorophyll, photosynthesis does not take place.

Photosynthetic pigments

The green colouring in plants that we have so far loosely referred to as chlorophyll is not produced by a single pigment, but by a group of five closely related ones. These are **chlorophyll a** (blue-green), **chlorophyll b** (yellow-green), **carotene** (orange) and **xanthophyll** (yellow) along with a grey pigment **phaeophytin**. This has been considered for many years to be a breakdown product of the other photosynthetic pigments. Recent reports suggest that it may after all play a part in the process of photosynthesis.

Chlorophyll *a* is found in all photosynthesing plants and is the most abundant of the five. The other pigments are found in varying proportions, and it is these differences which give the leaves of plants their variety of shades of green.

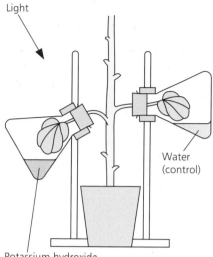

Light

Water (control)

Potassium hydroxide solution absorbs carbon dioxide from the air.

Figure 3.2.5 Although rather crude and not entirely successful, this experiment can be used to show that when carbon dioxide levels are very low, photosynthesis is substantially reduced.

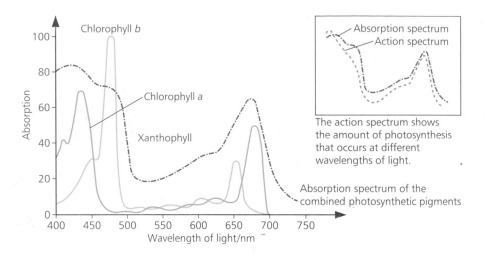

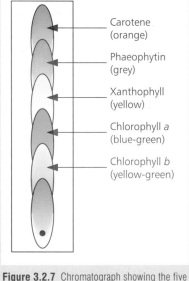

Figure 3.2.6 The different photosynthetic pigments absorb light at a variety of different wavelengths, making more of the light available for use by the plant.

Why is there this variety of photosynthetic pigments? Figure 3.2.6 shows that each of the pigments absorbs light well from particular areas of the spectrum. As a result, far more of the light falling on the plant can be absorbed than if only one pigment was present. It is also interesting to note that none of the pigments absorb well in the green/yellow areas of the spectrum, around 500–550 nm. This light, not being absorbed, is reflected, which is why plants appear green.

Explaining the absorption spectrum of the photosynthetic pigments

Figure 3.2.6 shows that the photosynthetic pigments tend to absorb light most readily at the blue or shorter wavelength end of the spectrum. Is this just coincidence, or is there a good reason for it?

There are two models of light. One is the **wave model**, which we use to look at the light absorbed by photosynthetic pigments. Light of different wavelengths is shone into solutions of the photosynthetic pigments and the amount of absorption measured.

The other model of light is **particulate**. In this model light behaves as a series of particles called **photons**. Each photon of light contains a fixed amount of energy called a **quantum**. The size of the quantum varies with the wavelength of the light – the shorter the wavelength, the larger the quantum. This is the model of light we use when we are explaining the events in photosynthesis.

By absorbing strongly in the blue end of the spectrum, plants are using photons of light with larger quanta and are therefore obtaining more energy.

Evidence for photosynthetic pigments

Plants look green. We can extract the pigments from a plant by grinding leaves with acetone and filtering. The resulting filtrate looks green. So how do we know that there are five different pigments? The answer is paper chromatography. With a suitable solvent the pigments travel up the paper at different speeds and are readily separated.

Figure 3.2.7 Chromatograph showing the five photosynthetic pigments

THE BIOCHEMISTRY OF PHOTOSYNTHESIS

The equation for photosynthesis shown earlier gives the impression that photosynthesis is a simple, one-step process by which carbon dioxide and water are converted into simple sugars and oxygen. In fact, photosynthesis is a complex series of reactions making up a biochemical pathway fundamental to the existence of all life on Earth.

As we have seen, photosynthesis is a two-stage process. The light-dependent reactions produce materials which are then used in the light-independent stages. The whole process takes place all the time during the hours of daylight, but only the light-independent reactions can occur in the dark. The light-independent reactions of photosynthesis are sometimes referred to as the **dark reactions**. This phrase can be misleading, implying that these reactions take place only in the dark, whereas in fact they occur continuously. We shall begin by looking at the events of the light-dependent stage of photosynthesis, and then see how the products of these reactions are used in the light-independent stage.

Evidence for two stages of photosynthesis

There are several strands of evidence for the two stages of photosynthesis.

(1) A chemical reaction or series of reactions which are dependent on light should be completely independent of the temperature of the surroundings. However, as figure 3.2.8 shows, the rate of photosynthesis is not independent of temperature. Temperature has a very distinct effect on the rate of photosynthesis, particularly at higher light intensities.

(2) A plant which is exposed to rapidly alternating periods of dark and light will form more carbohydrate than a plant which is exposed to continuous light. This suggests that the light-dependent reactions result in a product which is then fed into the light-independent stage. Continuous light would cause a build-up of this product which might then inhibit further reactions, whereas a period of darkness would ensure that all the product of the light stage was converted into the end product thus making the system more efficient.

(3) More recent techniques have allowed areas of the chloroplast to be isolated. The reactions occurring on the grana have been shown to depend on the presence of light, but those of the stroma do not.

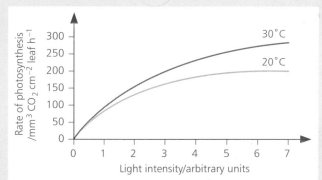

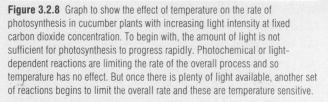

Figure 3.2.8 Graph to show the effect of temperature on the rate of photosynthesis in cucumber plants with increasing light intensity at fixed carbon dioxide concentration. To begin with, the amount of light is not sufficient for photosynthesis to progress rapidly. Photochemical or light-dependent reactions are limiting the rate of the overall process and so temperature has no effect. But once there is plenty of light available, another set of reactions begins to limit the overall rate and these are temperature sensitive.

The light-dependent stage

The light-dependent stage of photosynthesis has two main functions. Water molecules are split in a photochemical reaction. This provides hydrogen ions which can then be used to reduce fixed carbon dioxide and so produce carbohydrates. Also ATP is made, which supplies the energy for the synthesis of carbohydrates. How does it work?

An understanding of the light stage of photosynthesis depends on the following idea. When a photon of light hits a chlorophyll molecule, the quantum of energy is transferred to the electrons of that molecule. The electrons are **excited** – they are raised to higher energy levels. One may be raised to a sufficiently high energy level to leave the chlorophyll completely. If this happens the excited electron will be picked up by a carrier molecule, and this can result in the synthesis of ATP by one of two processes – **cyclic** or **non-cyclic photophosphorylation**.

Cyclic photophosphorylation

The light-excited electron may be passed along an electron transfer chain (see section 3.1), with each member of the chain at a lower energy level, until it is returned to the chlorophyll molecule that it left. As the electron moves along the chain, down the energy levels, ATP is produced by the phosphorylation of ADP. The electron leaves the chlorophyll and returns to it, so may then be excited in exactly the same way again. This is known as **cyclic photophosphorylation** (the phosphorylation of ADP in a cyclical process which depends on light). The process is illustrated in figure 3.2.9.

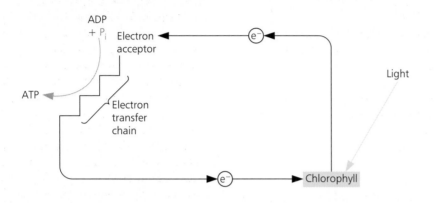

Figure 3.2.9 Cyclic photophosphorylation – the electron passes down the electron transfer chain back to the chlorophyll molecule.

Non-cyclic photophosphorylation

The excited electron may instead be used to provide the reducing power needed in the second, light-independent stage of the photosynthetic process. Water dissociates spontaneously into hydrogen (H^+) ions and hydroxide (OH^-) ions. As a result there are always plenty of these ions present in the cell, including in the interior of the chloroplasts. Interactions between these ions and chlorophyll molecules bring about the process of **non-cyclic photophosphorylation**, described below and illustrated in figure 3.2.10.

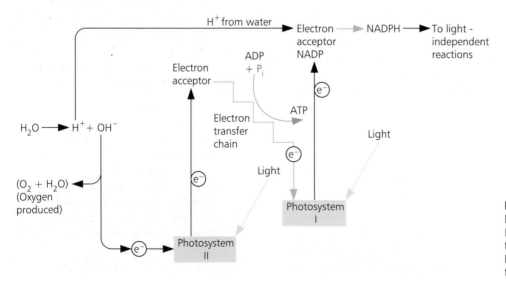

Figure 3.2.10 Non-cyclic photophosphorylation. NADPH and ATP are produced, which power the light-independent reactions. The biochemistry of the two photosystems was worked out by R.Hill and R.Bendall at Cambridge and is often referred to as the Z scheme because of its shape.

There are two distinct complexes or forms of chlorophyll known as **photosystem I** and **photosystem II**. An excited electron from photosystem II passes to an electron acceptor and down an electron transfer chain to photosystem I, which is at a lower energy level than photosystem II. This loss of energy allows the synthesis of a molecule of ATP. Light energy can then excite an electron from photosystem I, and this excited electron passes to another electron acceptor – **nicotinamide adenine dinucleotide phosphate**

(**NADP**). NADP also takes up a hydrogen ion from water and is thus reduced, forming NADPH. The NADPH is a source of reducing power for the light-independent reactions.

So photosystem I receives electrons via the electron transfer chain from photosystem II. This leaves photosystem II electron deficient. The electron from photosystem II is replaced by an electron from a hydroxide ion:

$$4OH^- - 4e^- \rightarrow O_2 + 2H_2O$$

The hydroxide ions are 'left behind' from the hydrogen ions taken up in the reduction of NADP to NADPH. The removal of the electrons from hydroxide ions by the photosystem II results in the by-product oxygen.

Thus the reactions of the light-dependent stage of photosynthesis provide a source of reducing power (NADPH) and the universal energy-supplying molecule ATP, with oxygen gas given off as a waste product. To find out how the NADPH and ATP are used to make carbohydrates we must move on and consider the reactions of the light-independent stage.

The light-independent stage

The light-independent reactions are known as the **Calvin cycle**. This is a cyclic reaction consisting of a series of small steps resulting in the reduction of carbon dioxide to bring about the synthesis of carbohydrates. NADPH and ATP from the light-dependent reactions provide the reducing power and the energy needed for the various steps. The stages of the cycle are controlled by enzymes and are independent of light. Figure 3.2.11 shows the reactions of the Calvin cycle.

Carbon dioxide from the air combines with ribulose bisphosphate (RuBP), a 5-carbon compound which **fixes** the carbon dioxide by accepting it and making it part of the photosynthetic reactions. The enzyme ribulose bisphosphate carboxylase is necessary for this step. The result is a theoretical highly unstable 6-carbon compound which immediately splits to give two molecules of glycerate 3-phosphate (GP), a 3-carbon compound. This is reduced to give glyceraldehyde 3-phosphate (GALP), a 3-carbon sugar. The hydrogen for the reduction comes from NADPH and the energy required from ATP, both produced in the light-dependent stage. Some of the glyceraldehyde 3-phosphate is synthesised into the 6-carbon sugar glucose,

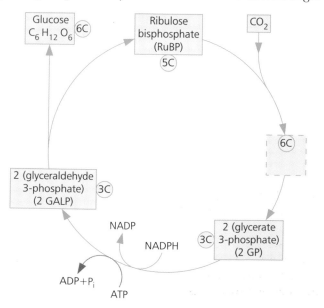

Compensation point

In the dark, a plant takes in oxygen and releases carbon dioxide due to respiration. Under dark conditions only the light-independent reactions of photosynthesis can occur. As the light intensity increases, so the amount of oxygen taken in and carbon dioxide given out by the plant becomes less. With the increase in light the rate of photosynthesis increases, thus more oxygen is produced as a waste product of photosynthesis which can be used for respiration. The carbon dioxide produced in cellular respiration is used up in photosynthesis.

As the light intensity continues to increase, a point is reached at which there is no net exchange of gases. All the oxygen produced by photosynthesis is used up in respiration, and all the carbon dioxide produced in respiration is used for photosynthesis. This is known as the **compensation point**.

With any further increase in light intensity, oxygen is produced by the plant and carbon dioxide taken in. More oxygen is produced by photosynthesis than can be used up in respiration, and insufficient carbon dioxide is produced in respiration to supply the rapid photosynthesis taking place.

Figure 3.2.11 The Calvin cycle – here the products of the light-dependent stage are used in a continuous cycle, the end result being new carbohydrates.

which is supplied to the cells or converted to starch for storage. However, much of the glyceraldehyde 3-phosphate is passed through a series of steps to replace the ribulose bisphosphate, without which further carbon dioxide cannot enter the cycle.

The products of photosynthesis, although initially carbohydrates, are rapidly fed into other biochemical pathways to produce amino acids and lipids for the requirements of the cells of the plant.

Melvin Calvin and the Calvin cycle

Melvin Calvin worked at the University of California. He came up with a novel, simple method of investigating the reactions which occur in photosynthesis. He produced a thin, transparent vessel known as a 'lollipop' because of its shape. Into this was placed a suspension of photosynthetic protoctists called *Chlorella* which were supplied with radioactively labelled carbon dioxide containing carbon-14. Light was shone through the suspension of organisms and they were allowed to photosynthesise.

The experiment was repeated time after time, with the *Chlorella* being killed at intervals ranging from a few seconds to a few minutes after the start of photosynthesis. They were killed in boiling ethanol, which stopped all enzyme-controlled reactions immediately. The radioactive compounds formed were then extracted, separated by paper chromatography and identified. In this way the biochemical pathway which we now call the Calvin cycle was built up.

Summary of photosynthesis

Figure 3.2.12 shows how the light-dependent stages of cyclic and non-cyclic photophosphorylation and the light-independent Calvin cycle interact.

Figure 3.2.12 The process of photosynthesis, which occurs continuously in plants that are exposed to light

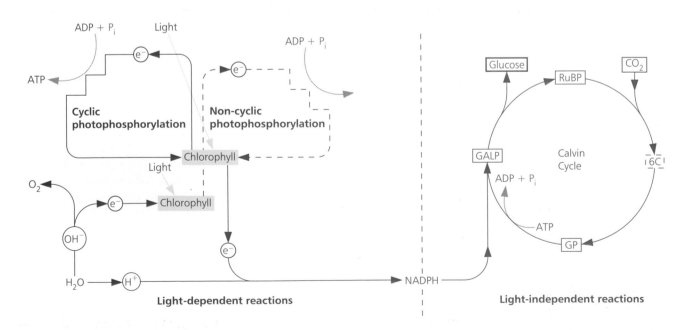

C_3 and C_4 plants

The Calvin cycle is the pathway by which carbohydrates are made in all plants. But not all plants fix carbon dioxide directly into the Calvin cycle using ribulose bisphosphate. The ones that do – and they are the majority of the plant kingdom – are known as **C_3 plants** because the product of the carbon fixation is the three-carbon (C_3) compound glycerate 3-phosphate.

We already know that in order to obtain carbon dioxide for photosynthesis, plants must open their stomata. This in turn increases their water loss. In hot dry conditions C_3 plants are at extreme risk from desiccation with open stomata and so the stomata are closed. This prevents gaseous exchange taking place, and so levels of oxygen build up and carbon dioxide is depleted. As a result of the raised oxygen levels, the enzyme ribulose bisphosphate carboxylase joins oxygen to ribulose bisphosphate instead of to carbon dioxide as the two compete for the active site. Phosphoglycerate is metabolised and carbon dioxide is lost. This wasteful process, whereby no carbohydrate is formed and energy is used, is known as **photorespiration**.

Some plants in these areas, for example sugar cane and maize (sweetcorn), have overcome the problems of photorespiration by evolving a different pathway for fixing carbon dioxide, shown in figure 3.2.13. The ribulose bisphosphate carboxylase is concentrated deep in bundle sheath cells, insulated from fluctuations in oxygen levels, and the fixed carbon dioxide is transported to the bundle sheaths. The product of the carbon fixation is a four-carbon (C_4) compound, oxaloacetic acid, and so these plants are known as **C_4 plants**.

The carbon dioxide acceptor in C_4 plants is phosphoenolpyruvate (PEP) and the reaction is catalysed by the enzyme PEP carboxylase. The importance of this is that the carbon dioxide is fixed extremely rapidly, even at very low concentrations. As a result, sufficient carbon dioxide can be fixed to supply the plant for a day by having the stomata open for a very short time. The oxaloacetate is converted into malate, and this donates carbon dioxide to ribulose bisphosphate in the Calvin cycle. The result of the removal of carbon dioxide from malate is pyruvate, which is recycled into PEP using energy from ATP. The pathway was worked out by Hal Hatch and Roger Slack and is known as the **Hatch-Slack pathway**. In C_4 plants, carbohydrate is synthesised in the Calvin cycle as usual.

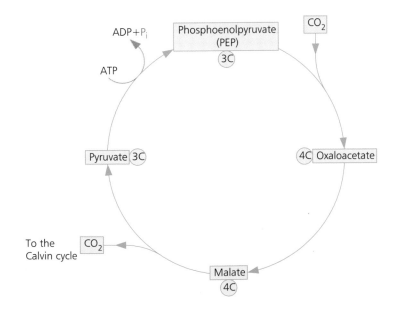

Figure 3.2.13 The Hatch-Slack pathway of carbon dioxide fixation in C_4 plants

THE SITE OF PHOTOSYNTHESIS

Any part of a plant that is green contains chlorophyll and so is capable, at least in theory, of carrying out photosynthesis. In practice, most photosynthesis takes place in the leaves. As we have seen earlier, the leaves are the organs of the plant which are specially adapted for photosynthesis. Carbon dioxide and water are readily available to the cells. The palisade mesophyll cells in particular contain a vast number of chloroplasts which can be moved around the cells depending on the light intensity, enabling the plant to trap the maximum amount of solar energy. Further chloroplasts in the spongy mesophyll layer enable extra photosynthesis to take place.

The chloroplasts are the site of photosynthesis. Chloroplasts are large organelles containing the chlorophyll and other pigment molecules and the enzymes associated with photosynthesis. Their internal structure, shown in figure 3.2.14, consists of **thylakoids**, discs made up of pairs of membranes with a small gap between. Stacks of thylakoids make up the grana, and elongated thylakoids the intergrana. Arranged on these membranes are the molecules of chlorophyll, held in the best possible position for trapping light energy. Electron micrographs have shown that the membranes of the grana are covered with particles which seem to be involved in ATP synthesis. Isolated fragments of the granal membranes will split water and release oxygen. The picture which results from all these threads of evidence is that the grana are the site of both the chlorophyll and the enzymes involved in the light-dependent stage of photosynthesis. The enzymes associated with the Calvin cycle are found in the stroma of the chloroplasts.

Grana – stacks of membranes or thylakoids. The chlorophyll molecules and the enzymes of the light-dependent reactions are sited here. On average a chloroplast contains about 60 grana, and each granum is made up of about 50 thylakoids, so the resultant surface area available for the light-dependent reactions of photosynthesis is substantial.

Stroma – the enzymes of the light-independent stage are found here.

Chloroplast envelope

Starch grain   Intergrana

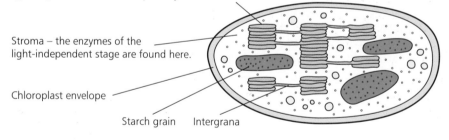

Figure 3.2.14 A chloroplast – site of photosynthesis

UNDERSTANDING LIMITING FACTORS IN PHOTOSYNTHESIS

Light

With an understanding of the process of photosynthesis, it is easy to see how certain factors might affect the ability of a plant to photosynthesise. In a biochemical process which depends on or is affected by a number of factors, it is a matter of common sense that the process will be limited by the factor which is nearest to its minimum value. For example, the amount of light available affects the amount of chlorophyll which has excited electrons and therefore the amount of NADPH and ADP produced in the light-dependent stage. If there is a low level of light then insufficient NADPH and ATP will be produced to allow the reactions of the light-independent stage to progress at their maximum rate. In this situation light is said to be the **limiting factor** for the process.

Carbon dioxide

Carbon dioxide levels are very important in photosynthesis – insufficient carbon dioxide available for fixing in the Calvin cycle means that the reactions cannot proceed at the maximum rate. When this is the case, carbon dioxide is the limiting factor. In the natural situation of plants it is most often carbon dioxide which is the limiting factor. Changes in the level of carbon dioxide have a clear effect on the rate of photosynthesis, as figure 3.2.15 shows. Commercial growers of some fruits and vegetables make use of this effect to increase their production – tomatoes, for example, may be grown in greenhouses with a carbon dioxide-enriched atmosphere.

Temperature

The other main factor which limits the rate of photosynthesis is temperature. All the Calvin cycle reactions and many of the light-dependent reactions of photosynthesis are controlled by enzymes and are therefore sensitive to temperature. This means that even when the light and carbon dioxide levels are suitable for a very high rate of photosynthesis, unless the temperature is also satisfactory the plant will be unable to take advantage of the conditions.

The rate of photosynthesis in a wild plant is often determined by a combination of factors, some or all of them limiting the process to an extent.

Limiting factors in action

Photosynthesis and its limiting factors are relatively easy to investigate in the laboratory – but what about the real world? The way in which plants grow, and the ecosystems which develop, are basically governed by competition between plants for those factors which can limit photosynthesis and growth.

Carbon dioxide levels do not generally vary very much, but plants compete for situations with suitable conditions of light and warmth, as well as for soil nutrients which enable the carbohydrate produced by photosynthesis to be converted to proteins and fats. Growth in height, spreading of leaves into a mosaic pattern, climbing and developing large leaves are all ways in which plants endeavour to obtain as much light as possible so that photosynthesis is not limited. Methods of seed dispersal have also evolved to reduce competition as much as possible by ensuring that seedlings do not develop in the shade of their parents. Figure 3.2.16 shows the effect of light as a limiting factor in a laboratory investigation – limiting factors also play their part in the real world outside.

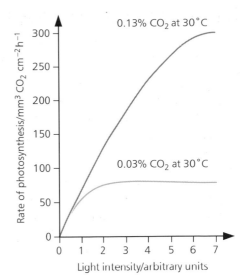

Figure 3.2.15 The effect of carbon dioxide as a limiting factor in photosynthesis

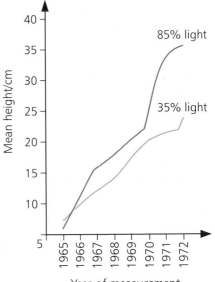

Figure 3.2.16 Growth of oak tree seedlings grown in 85% light and 35% light. The effect of growing in an environment where light is a continually limiting factor can be seen clearly.

SUMMARY

- **Photosynthesis** is the process by which **phototrophs** convert carbon dioxide and water into simple carbohydrates and oxygen in the presence of chlorophyll, using sunlight energy.

- The effects of various factors on the rate of photosynthesis can be followed by collecting oxygen gas from an aquatic plant or using iodine solution to detect the presence of starch in a previously destarched plant. Such experiments can help demonstrate that light, carbon dioxide, water and chlorophyll are needed for photosynthesis.

- Photosynthetic pigments include **chlorophyll *a*** (blue-green), **chlorophyll *b*** (yellow-green), **carotene** (orange), **xanthophyll** (yellow) and **phaeophytin** (grey). None of these pigments absorbs light well in the green area of the spectrum, so plants look green.

- Photosynthesis has a **light-dependent stage** which provides ATP and NADPH for the **light-independent stage** which uses carbon dioxide to produce glucose.

- The light-dependent stage involves the excitation by light energy of an electron in chlorophyll, which is picked up by an electron acceptor and passed to the electron transfer chain, and brings about the synthesis of ATP by either **cyclic** or **non-cyclic photophosphorylation**.

- In **cyclic photophosphorylation** the excited electron is passed along the electron transfer chain, moving down the energy levels. This is coupled to the production of ATP. The electron returns to the chlorophyll.

- In **non-cyclic photophosphorylation**, an excited electron from **photosystem II** passes down an electron transfer chain, again coupled to the production of ATP, to **photosystem I**, which is at a lower energy level than photosystem II. An excited electron from photosystem I then teams up with a hydrogen ion from the splitting of a water molecule to reduce NADP to NADPH. The hydroxide ion remaining from the splitting of the water molecule provides an electron to photosystem II, and produces oxygen gas as a by-product.

- The ATP and NADPH produced by the light-dependent reactions are used to drive the light-independent reaction of the **Calvin cycle**. Carbon dioxide is fixed by combination with ribulose bisphosphate to form an unstable 6-carbon intermediate which splits to form two molecules of **glycerate 3-phosphate** (**GP**). These are reduced to **glyceraldehyde 3-phosphate** (**GALP**) using NADPH and ATP. The GALP molecules are then either used to form glucose, or allow the cycle to continue by reforming ribulose bisphosphate.

- **C_3 plants** fix carbon dioxide to form the 3-carbon molecule GP. **C_4 plants** in hot dry areas have a more efficient method of fixing carbon dioxide prior to the Calvin cycle which involves stomata being open for a shorter time. In the **Hatch-Slack pathway** carbon dioxide is accepted by the 4-carbon compound **phosphoenolpyruvate**, which is converted to **oxaloacetate** and then **malate**. Malate supplies carbon dioxide to the Calvin cycle and is converted to **pyruvate**, which can be used to regenerate phosphoenolpyruvate using ATP.

- The site of photosynthesis in plants is the chloroplasts. Stacks of membranes called **thylakoids** make up the grana, and these membranes hold the chlorophyll molecules. The light-dependent reactions take place here, while the light-independent reactions take place in the **stroma**.

- **Limiting factors** for photosynthesis include light levels, carbon dioxide levels and temperature. Plants have evolved to compete for these limiting factors.

QUESTIONS

1 Photosynthesis involves two sets of reactions – those which are light dependent and those which are independent of light.
   a Where do the two stages of photosynthesis take place?
   b Describe the light-dependent events, explaining clearly the difference between cyclic and non-cyclic photophosphorylation.

2 Describe the structure of chlorophyll and relate this to the role of the molecule in the process of photosynthesis.

3 What is meant by the term 'limiting factors' in photosynthesis? Give a brief explanation of how limiting factors might be investigated experimentally. Summarise the effects of limiting factors on the growth of plants in their natural habitats.

3.3 Heterotrophic nutrition

HETEROTROPHS

As we saw in section 3.1, all living organisms which cannot provide their own energy supply by either photosynthesis or chemosynthesis are known as **heterotrophs**. 'Heterotrophic' means 'feeding on others'. The carbohydrates, proteins and fats originally made by autotrophs are broken down and reassembled by heterotrophs in a wide variety of ways.

The most obvious heterotrophic organisms are animals, but many bacteria, some protoctists (unicells), a few flowering plants and all fungi also use this method of nutrition. In this section we shall consider mainly mammals, particularly humans, although other groups with interesting specialisations will be mentioned as appropriate.

Types of heterotrophic nutrition

There are three main types of heterotrophic nutrition:

1 **Holozoic nutrition** means feeding on solid organic material from the bodies – living or dead – of other organisms, which may be either plant or animal. This method of nutrition is usually seen in animals, and carnivorous plants and some protoctists also feed in this way.

2 **Parasitic nutrition** is found in most groups of organisms – animals, plants, protoctists, fungi and bacteria. A **parasite** feeds on organic material, often but not always soluble, from the body of another living organism known as the **host**.

3 **Saprotrophic nutrition** means feeding on soluble organic material from dead animals or plants. It occurs mainly in protoctists, bacteria and fungi, although there are a few saprotrophic animals. Saprotrophic nutrition is of great importance because it plays a significant role in decomposing biological material and returning nutrients to the soil and the atmosphere.

So what does heterotrophic nutrition involve? A heterotrophic organism has to be able to obtain, digest and absorb food.

Obtaining food

Adaptations designed to make obtaining food easier or more effective appear to have been the driving force behind much evolutionary development.

Those heterotrophs which rely on eating plants, the **herbivores**, do not have to catch their food. However, because of the cellulose walls of plant cells, digesting the food efficiently is not a straightforward process. The enzymes needed to digest cellulose are present in very few organisms. Also, plants are constantly evolving defences against being eaten, via either digestibility reducers (celluloses, tannins, silicates) or specific toxins. Herbivorous mammals can *only* digest plant material with the help of microbes. Herbivores frequently need to eat very large quantities of plant material to get enough energy to survive. As a result much of their time is taken up with eating, leaving them open to attack as they graze. They also have less time for other activities such as sleeping and searching for a mate. Great demands are put on the teeth of herbivores too, with special adaptations needed to cope with the enormous amounts of chewing, as figure 3.3.2 illustrates.

Figure 3.3.1 There are many ways in which heterotrophs can obtain their nutrition – herbivores eat plants, carnivores eat other animals and omnivores such as ourselves eat either. Parasites get their nutrition from a living host whilst saprotrophs live on dead and decaying tissues. Whatever their food source, all heterotrophs ultimately depend on energy from the Sun via autotrophs.

Heterotrophs which kill and eat other animals, the **carnivores**, have food that is easier to digest as it contains more accessible protein, but it is frequently much harder to obtain. Carnivores have a variety of adaptations which allow them to successfully capture their prey. These range from the burst of speed of a hunting cheetah to the paralysing poison of a spider, from the worm-like snapping turtles' tongue to the communication system between the members of a wolf pack, allowing them to hunt as a team. They include sonar systems in bats and whales along with stings and many kinds of complex traps. The teeth of carnivorous mammals are specialised for killing prey and for tearing flesh, and the jaws are frequently very strong, as figure 3.3.3 shows.

Parasites, heterotrophs which feed off a living host, need adaptations which will enable them to survive either within or on their host without being attacked or rejected. Saprotrophs which gain their food from decaying animal or plant material have a food supply that is finite, and they need to be able to find more decaying material when one rotting organism has been used up.

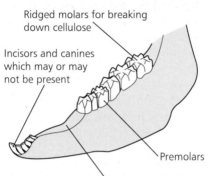

Ridged molars for breaking down cellulose

Incisors and canines which may or may not be present

Premolars

Diastema - gap to manipulate the food onto the molars and keep chewed and unchewed material separate

Figure 3.3.2 Many herbivores live in groups or herds to minimise the risk of attack whilst feeding and also to make finding a mate a less time-consuming activity. They have a variety of front teeth which may be adapted to nibbling, gnawing, tearing or biting. The back teeth or molars almost always show ridges of alternating dentine and enamel which help to break open the cellulose cell walls of plants. Another common adaptation of herbivores is to have many sets of teeth or continuous growth of the teeth to cope with the almost continuous wear.

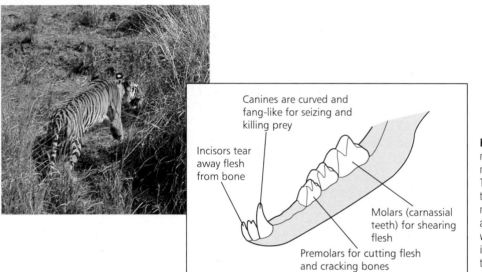

Canines are curved and fang-like for seizing and killing prey

Incisors tear away flesh from bone

Molars (carnassial teeth) for shearing flesh

Premolars for cutting flesh and cracking bones

Figure 3.3.3 Carnivores sometimes live and hunt in groups, but are more likely than herbivores to lead quite solitary lives. They eat relatively infrequently, often gorging at a meal and then sleeping for long periods of time to digest it. They have more 'spare time' to search for a mate. Their teeth are specialised for holding and killing animals and removing flesh from the bone.

Digestion of food

All heterotrophs have to digest food to a greater or lesser extent, but particularly those which, like the mammals, feed holozoically. The food is usually taken into the body in relatively large pieces. The molecules present in these food pieces, be they plant or animal, are large, complex and frequently insoluble. Thus the majority of heterotrophs must break down these large complex molecules into simple, soluble ones that can later be absorbed. We call this process of breaking down large molecules **digestion**. The food of parasites and saprotrophs is often partially digested, but some form of digestion is frequently still needed.

Digestion is catalysed by the digestive enzymes. Digestion may be intracellular, occurring within the cell, as in the protoctists (see figure 3.3.4) and other small organisms which survive by heterotrophic nutrition. However, in the larger heterotrophic organisms, including the mammals, digestion is mainly extracellular, with the digestive enzymes working in a specialised environment known as the **alimentary canal** or **gut**.

How do heterotrophs take in food?

Most heterotrophic organisms – herbivores, carnivores, parasites and saprotrophs – deal with their food in similar ways once it is inside the system. However, there are three main forms in which the food may be taken into the body. These are as big chunks, small chunks and liquids.

Mammals and most larger animals take in their food in relatively large chunks – they use teeth, mandibles, tentacles or other specialised organs to bite or tear portions off their food which then enter the gut. This is known as **macrophagous feeding**.

Other heterotrophs take in their food in the form of much smaller particles relative to their body sizes. Examples include filter feeders such as many of the shellfish, and protoctists – organisms such as *Amoeba* which use a pseudopodium (see figure 3.3.4) and *Paramecium* which relies on cilia. These are known as **microphagous feeders**.

Some heterotrophic organisms are capable of taking in only liquid food, for example, the aphids met in section 2.3. They are known as **fluid feeders**.

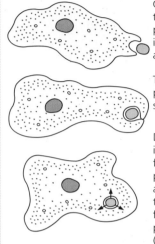

Cytoplasmic streaming leads to the formation of a pseudopodium. If this comes into contact with a food particle a 'food cup' is formed.

The cup flows round the food particle until it forms a complete sealed **food vacuole**.

Digestive enzymes are secreted into the food vacuole and the food particle is digested. The products of digestion are absorbed into the cytoplasm through the membrane of the food vacuole. The whole process takes place intracellularly (within the single cell).

Figure 3.3.4 Intracellular digestion is demonstrated within the cell of a protoctist such as *Amoeba*.

Absorption of food

The products of digestion have to be absorbed into the body of the organism and distributed around its cells before they can be used to provide the energy needed by the cells for contraction, transport and biosynthesis. Thus heterotrophs have many specialised features to enable the products of digestion to be absorbed into the body quickly and effectively.

HETEROTROPHIC NUTRITION IN HUMANS

We shall develop an understanding of heterotrophic nutrition by looking in detail at a fairly representative animal – the human. We can then consider some of the different specialisations found elsewhere in the living world.

People are holozoic, omnivorous, macrophagous feeders – that is to say, we take in large chunks of organic material from plants and other animals which are then broken down into simple soluble molecules to be absorbed and used by our cells. The same is true of all mammals, except that not all are omnivorous – the majority are either carnivores or herbivores.

Holozoic nutrition can be considered as a series of steps. These are:

1 **Ingestion** – the taking in of complex organic food by the organism.
2 **Digestion** – the breakdown of large, complex molecules into simpler, soluble ones by enzyme action.
3 **Absorption** – the uptake of the soluble products of digestion into the main system of the organism, either directly into the cells, or into the bloodstream and then into the cells.
4 **Assimilation** – the ultimate use of the products of digestion, either as an energy source or to be incorporated into new biological material.
5 **Egestion** – the removal from the organism of any undigested food materials.

In the human being, these processes are closely associated with the alimentary canal or gut. The gut is a continuous muscular tube which runs from the mouth to the anus. It is an integral part of the body and yet in a way is not 'inside' the body at all. The cavity of the gut is a contained part of the outside world, although the structures relating to it and supplying it are part of the body proper.

THE HUMAN DIGESTIVE SYSTEM

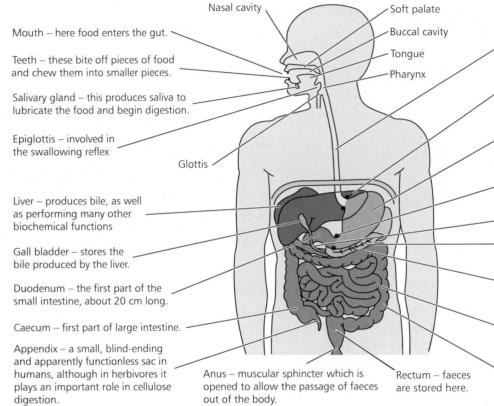

Nasal cavity

Soft palate

Mouth – here food enters the gut.

Buccal cavity

Tongue

Teeth – these bite off pieces of food and chew them into smaller pieces.

Pharynx

Salivary gland – this produces saliva to lubricate the food and begin digestion.

Epiglottis – involved in the swallowing reflex

Glottis

Liver – produces bile, as well as performing many other biochemical functions

Gall bladder – stores the bile produced by the liver.

Duodenum – the first part of the small intestine, about 20 cm long.

Caecum – first part of large intestine.

Appendix – a small, blind-ending and apparently functionless sac in humans, although in herbivores it plays an important role in cellulose digestion.

Anus – muscular sphincter which is opened to allow the passage of faeces out of the body.

Rectum – faeces are stored here.

Oesophagus – a tube about 25 cm long, which carries food from the pharynx to the stomach.

Cardiac sphincter – a thickened ring of circular muscle which opens to allow food into the stomach.

Stomach – a muscular bag which churns food up and produces enzymes, hydrochloric acid and mucus.

Pyloric sphincter – a thickened ring of circular muscle which opens to allow food out of the stomach.

Bile duct – carries the bile into the gut.

Pancreas – produces digestive enzymes (and hormones).

Pancreatic duct – carries the digestive enzymes produced by the pancreas into the gut.

Ileum – the main part of the small intestine, about 5 m long and the major site of digestion and absorption in the gut.

Colon – main part of the large intestine in which much reabsorption of water into the blood takes place.

The human digestive system shown in figure 3.3.5 is a long tube with a series of areas and compartments specialised for particular functions. The bulk of the food we eat, whether from a plant or an animal source, consists largely of complex proteins, carbohydrates and fats, along with indigestible material known as fibre or roughage. There are also minute amounts of various minerals and vitamins which are vital to our health but need no processing as they are already in a form which is soluble and can be absorbed into the body. How does the gut bring about the breakdown of the food? If we consider each area of the alimentary canal in turn we can see how it works as a coordinated whole to carry out both physical and chemical digestion.

Figure 3.3.5 Up to 9 m of gut can look relatively simple when drawn, but a glance at the real state of affairs seen during abdominal surgery shows the complexity of the gut. Each part of the alimentary canal is basically a muscular tube, adapted to carry out a specific function.

The mouth or buccal cavity

The process of digestion begins in the mouth. Food is taken in and subjected to both physical and chemical breakdown before it is passed on.

Physical digestion – the teeth

Physical digestion begins as pieces of food are bitten off by the teeth and then chewed. The main reason for chewing is to break down the food into smaller pieces, making it easier for them to be swallowed, and also increasing the surface area available for enzymes to act on.

Teeth have evolved to be very strong and are important to mammals in the obtaining and initial digestion of their food. The structure of human teeth is shown in figure 3.3.6. In the wild, it is frequently the loss of effective teeth in older animals which heralds the end of their lives, as they can no longer feed properly. This is not true for humans as we can prepare and cook food to soften it, and can even replace worn out, damaged or decayed teeth with artificial ones.

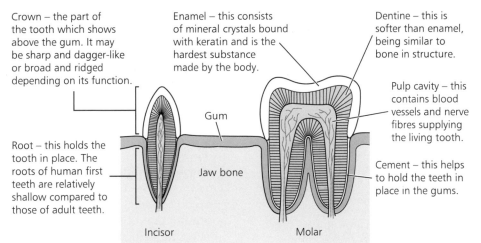

Crown – the part of the tooth which shows above the gum. It may be sharp and dagger-like or broad and ridged depending on its function.

Enamel – this consists of mineral crystals bound with keratin and is the hardest substance made by the body.

Dentine – this is softer than enamel, being similar to bone in structure.

Pulp cavity – this contains blood vessels and nerve fibres supplying the living tooth.

Root – this holds the tooth in place. The roots of human first teeth are relatively shallow compared to those of adult teeth.

Gum

Jaw bone

Cement – this helps to hold the teeth in place in the gums.

Incisor

Molar

Figure 3.3.6 The structure of teeth makes them well suited to their function. Problems arise when the protective and relatively impenetrable layer of enamel on the surface of the tooth is damaged. This is commonly due to a build-up of **plaque** – sugary food residues along with the bacteria which feed on them. The bacterial waste products are acidic and eat into the enamel and destroy it, leaving an entry hole for the bacteria to attack the inside structure of the tooth.

Different teeth are specialised for different functions, as shown in figure 3.3.7. In humans, the canines are not particularly long and sharp as we have long since replaced them with more effective weapons. If you look back at the herbivore and carnivore teeth shown in figures 3.3.2 and 3.3.3, you can see how the dentition structure is adapted to its function.

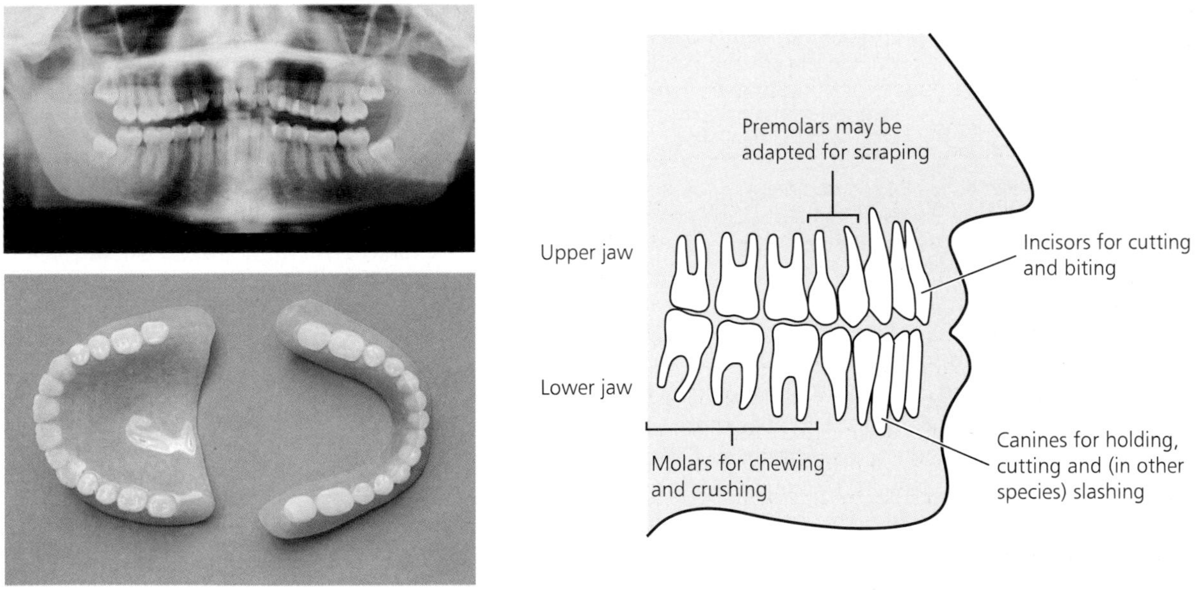

Premolars may be adapted for scraping

Upper jaw

Lower jaw

Incisors for cutting and biting

Canines for holding, cutting and (in other species) slashing

Molars for chewing and crushing

Figure 3.3.7 The different types of teeth in the human mouth

Saliva

As food is chewed in the mouth, it is coated with **saliva** secreted by the **salivary glands**. Saliva production and secretion increases in response to the taste, smell or thought of food as a result of nervous stimulation by the parasympathetic nervous system (see section 4.2). Saliva performs a variety of functions, although digestion can occur without it.

Saliva moistens and lubricates the food, making swallowing easier. Eating several dry crackers in succession will help you to see these effects of saliva. Saliva also dissolves some of the food, allowing it to be tasted. This in turn stimulates enzyme production in other parts of the gut. The food begins to be diluted, reducing the osmotic disruption which could result from the ingestion of, for example, large amounts of sweet or salty food.

Saliva is often very slightly alkaline. This makes it effective in helping to prevent dental caries – it neutralises the effects of acidic foods or acids produced by mouth bacteria. Saliva also contains an enzyme, **salivary amylase**. This begins the process of carbohydrate digestion by hydrolysing the polysaccharide starch into disaccharide units of maltose. Once the food has been chewed and coated in saliva it is swallowed (see figure 3.3.8) and so moved into the next part of the gut, the oesophagus.

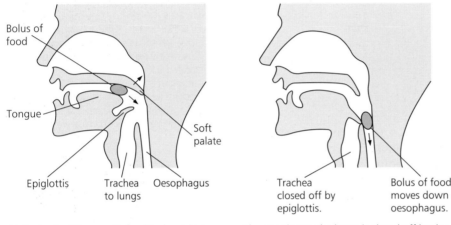

At the back of the mouth food is shaped into a bolus or ball by the tongue. The pressure of this against the soft palate triggers the swallowing reflex. The upward movement of the soft palate closes off the nasal cavity so that food does not return down the nose.

The opening to the lungs is closed off by the epiglottis, so the food has a clear route down into the oesophagus. For the brief moment of the reflex breathing is inhibited, so that in theory at least it is impossible to swallow and breathe at the same time.

Figure 3.3.8 It is important that food does not enter the respiratory system during swallowing, as the risk of blocked airways or infection would be high. The swallowing reflex ensures the smooth passage of the food to the gut, and the violent choking response ensures that any food entering the trachea is returned to its rightful route.

The oesophagus

The oesophagus is a tube about 25 cm long which links the buccal cavity with the stomach. The inner lining is folded, which allows it to stretch and accommodate the bolus of food as it is swallowed. Waves of muscular activity known as **peristalsis** move the food down the oesophagus. This peristaltic activity occurs throughout the gut. We are usually unaware of this gut movement apart from during vomiting, when the squeezing actions of the gut muscles become obvious. The structure of the oesophagus, as seen in figure 3.3.9, exemplifies the basic structure of the entire gut.

The stomach

At the end of the oesophagus, the entrance to the stomach is marked by a greatly thickened ring of circular muscle known as the **cardiac sphincter**. Most of the time this sphincter is closed, keeping the food in the stomach. It is relaxed and opened only to allow food into the stomach from the oesophagus, and during vomiting.

Pavlov's dogs and the control of digestive juices

In the late nineteenth and early twentieth centuries Ivan Pavlov performed a series of classic experiments on the control of the secretion of digestive juices. He was awarded a Nobel prize for his work. He implanted tubes in dogs in order to allow the secretions of various regions of the digestive tract to be collected and measured. Pavlov demonstrated clearly the important role of the nervous system, including the sight and smell of food, in the stimulation of the secretions of the gut, particularly the saliva and the gastric juices.

He also showed how the increased production of saliva could become linked not to the sight or smell of food but to some other stimulus which was linked to feeding. In his most famous experiment, a bell was rung shortly before his dogs were fed. In a short time, the ringing of the bell alone, without any other stimulus, was sufficient to cause the saliva of the dogs to flow in anticipation of the food to come. This is known as a **conditioned reflex**. Doubtless in schools and factories throughout the country the sound of the break bell or hooter has the same effect on the salivary glands of the students and workers.

Circular muscle
Longitudinal muscle

Coordinated movements of these two layers of muscle bring about peristalsis, waves of muscular activity responsible for moving the food along the gut.

Submucosa - a layer of connective tissue, blood vessels and nerves which helps to hold the structure together.

Lumen - the space through which the food travels

Mucosa made up of **stratified squamous epithelium**. These simple epithelial cells are regularly replaced. They are interspersed with mucus-producing cells.

Mucus glands - The function of the mucus is to lubricate the food as it moves through the oesophagus and protect the oesophageal lining from damage.

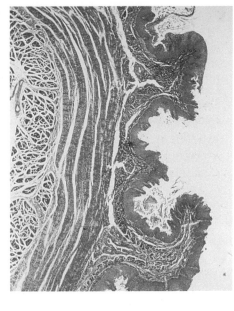

Figure 3.3.9 The structure of the oesophagus is typical of that of the digestive system, with specific modifications in different regions.

The stomach is a muscular bag, which can hold about 5 dm³ of food when fully distended. It is rarely stretched to this limit – imagine the discomfort of the equivalent of about eight pints of milk in your stomach at any one time! Both physical and chemical digestion is brought about by the stomach. Physical digestion is caused by the continual churning movements of the muscular walls which mix the food and gastric juice thoroughly and also help reduce the food to a uniform creamy paste known as **chyme**. Chemical digestion is brought about by the enzymes and acid secreted by the cells of the gastric glands, shown in figure 3.3.10.

The liquid in the stomach is a powerful protein-digesting mixture. **Pepsinogen** is an **enzyme precursor** – an inactive form of an enzyme which must be activated before it can catalyse a reaction. Hydrochloric acid brings about the conversion of pepsinogen to **pepsin**, an enzyme which catalyses the hydrolysis of proteins into polypeptides. It also provides the optimum pH (around pH2) for the enzyme pepsin to work. The acidic environment also helps destroy any microorganisms which might be taken in with the food.

Both acid and pepsinogen are released largely when food enters the stomach. This is partly as a result of the nervous stimulation which results from the sight, smell and taste of food, and partly as a result of the physical distending of the stomach. The presence of food in the stomach also causes the release of the hormone **gastrin**. This is a polypeptide molecule released by the lining of the pyloric (lower) region of the stomach into the blood. It travels back to the stomach in the blood and stimulates the production of acidic gastric juice for about 4 hours.

Rennin is of much less general importance, but in young mammals it is needed to coagulate the soluble protein in milk which is known as casein. This coagulation curdles the milk, forming solid curds and liquid whey. The curds stay in the stomach much longer than the liquid milk, making digestion more effective.

Food stays in the stomach on average for about 4 hours, although this depends on the type of food taken in. Liquid is retained for a much shorter time, while a large, protein-rich meal might be held for longer. Once the stomach has turned all the food to chyme, it releases the paste a little at a time into the duodenum by the opening and closing of the **pyloric sphincter**.

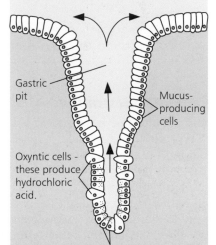

Gastric pit

Mucus-producing cells

Oxyntic cells - these produce hydrochloric acid.

Chief cells - these secrete pepsinogen. They are also known as peptic or zymogen cells.

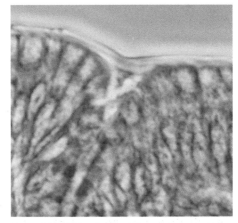

Figure 3.3.10 The digestive juices of the stomach are produced by the cells of the gastric glands. The main secretions are pepsinogen and hydrochloric acid, whilst in babies rennin is also formed.

Protecting the gut from autodigestion

A brief consideration of the human gut shows that it is a potentially dangerous system, capable of causing self-destruction by autodigestion. Protein- and lipid-digesting enzymes along with areas of high hydrochloric acid concentration are far from ideal conditions for cells. Autodigestion is prevented by several strategies.

(1) The lining of the gut is often thicker than that of other systems and is constantly and rapidly replaced.

(2) Large quantities of mucus are produced by many areas of the gut. This has a double function. Physical damage to the cells of the gut by the passage of food is greatly reduced as the mucus acts as a lubricant. Chemical damage is reduced as the mucus forms a protective barrier. Mucus can eventually be digested, reabsorbed and used again.

(3) Enzymes are only produced or released once food arrives in a particular area of the gut. This coordination is brought about by nervous control and by the production of hormones by some regions of the gut, for example gastrin in the stomach.

(4) Several of the protein-digesting enzymes which are most likely to inflict damage on the cells of the gut are produced as inactive precursors, for example pepsinogen and trypsinogen, which are then converted into the active, protein-digesting form only when the food is present.

If some of these protective mechanisms go wrong, then autodigestion can occur. Both gastric (stomach) ulcers and duodenal ulcers result from the very acidic stomach contents breaking down the wall of the digestive tract. These complaints cause great pain, and the gut wall can be completely digested so that the contents of the gut are released into the body cavity, with potentially fatal results. The most used pharmaceutical drug worldwide is for the treatment of stomach ulcers.

The small intestine

The small intestine is a long, coiled tube about 5–6 m long with a rich supply of blood and lymph vessels. It performs two major functions. Firstly it brings about most of the chemical breakdown of large food molecules into simpler and more soluble ones. Much of this function is carried out in the first few centimetres of the small intestine known as the **duodenum**. Secondly, the small intestine brings about the absorption of digested food into the blood supply so that it can be transported to the active cells of the body. This takes place along the length of the **ileum**. The structure of the small intestine is very closely related to its function, as can be seen in figure 3.3.11.

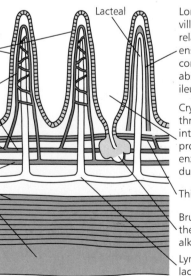

Villi – found throughout the small intestine, these finger-like extensions are frequently covered with microvilli only clearly visible with the electron microscope.

Capillary plexus
Arteriole

Rich blood supply to villi – arterioles bring the blood and venules carry it away to the hepatic portal vein.

Venule

Circular muscle
Longitudinal muscle

The muscles contract and relax to move the villi within the food, mixing the enzymes and moving the food along.

Lacteal

Longitudinal muscle in the villus – contraction and relaxation moves the villus ensuring that there is a constant contact with food for absorption, particularly in the ileum.

Crypt of Lieberkuhn, found throughout the small intestine – the cells at the base produce many digestive enzymes, particularly in the duodenum.

Thin layer of muscle

Brunner's glands, found only in the duodenum – these produce alkaline secretions and mucus.

Lymph vessel into which the lacteals drain

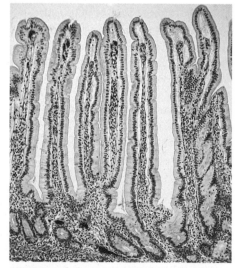

Figure 3.3.11 The duodenum and the ileum are very similar in structure, with differences due to the different emphasis of their functions.

The structure of the small intestine

The lining of the small intestine forms millions of finger-like projections known as **villi**. These increase the surface area available both for enzyme action and, more importantly, for the absorption of digested foodstuffs. The outer layer of cells of the villi are columnar epithelial cells, shown in figure 3.3.12, which have a brush border of slender, closely packed finger-shaped projections or microvilli, about 1700 per cell. It has been estimated that this brush border increases the surface area of the lining of the small intestine 15–40 times. Add to this the 600-fold increase in surface area given by the villi themselves, and the total surface area of the small intestine available for action is in the region of 200 m² – about the area of a tennis court.

The villi are well supplied with both blood and lymph vessels. These ensure that good diffusion gradients are maintained and also carry away the dissolved food molecules. A normal meal consisting of a mixture of protein, carbohydrate and fat is absorbed to a large extent within the first metre or so of the small intestine – the remainder acts as a reserve.

Between the villi are the **crypts of Lieberkuhn**. New columnar epithelial cells are made here, but their most important role is in the production of digestive enzymes. Some of these enzymes such as **amylase**, **maltase** and **lactase** work in association with the membrane of the villus. Others are released into the lumen of the intestine.

In the duodenum, **Brunner's glands** secrete into the crypts and so into the lumen. They produce mucus and an alkaline fluid. These secretions both protect the lining of the duodenum from damage by the acidic chyme which leaves the stomach and also, by neutralising the acid, provide the optimum pH (around 8) for the duodenal enzymes to work in.

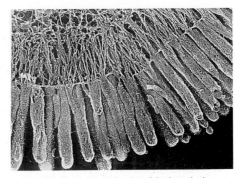

Figure 3.3.12 The brush border of the intestinal columnar epithelial cells increases the surface area available for the absorption of digested food, as well as being the site of much enzyme activity. The mitochondria present in these cells are indicative of the level of activity in the cytoplasm, both in producing enzymes and in active transport.

Digestion in the small intestine

Physical digestion plays a relatively small part in the functioning of the small intestine, though peristalsis moves the food along. Chemical digestion is brought about by a variety of enzymes, some produced by the small intestine, some by the pancreas. Other chemicals are also produced which affect digestion.

Bile from the gall bladder is released into the ileum. This contains sodium hydrogencarbonate which helps to neutralise the stomach acid and maintain the correct pH for the intestinal enzymes to function effectively. It also contains the **bile salts**, sodium taurocholate and sodium glycocholate. These have a physical effect on large fat droplets, emulsifying them into many smaller droplets and so greatly increasing the surface area available for digestion by enzymes. Alkaline juice from the pancreas is also of value in the maintenance of an optimum pH.

The activity of the various components of the intestinal juices is summarised in table 3.3.1.

The table shows the wide variety of chemicals, mainly enzymes, which are produced by the small intestine and the closely associated pancreas and gall bladder. The chemical breakdown of the larger food molecules, particularly the carbohydrates and proteins, often occurs in two stages. One enzyme breaks up the large molecule into smaller pieces (polysaccharides to disaccharides, proteins to polypeptides and peptides). Another enzyme then acts to complete the process (disaccharides to monosaccharides, peptides to amino acids). As a result of the action of the enzymes of the small intestine, a watery fluid known as **chyle** is produced. This contains all the products of digestion which are then absorbed into the blood.

Secretion	Source	Function
Bile – contains no enzymes	Made in liver, stored in gall bladder, released into duodenum	Neutralises gut contents, emulsifies fats
Amylase	Made in pancreas, released into duodenum. Also made in crypts of Lieberkuhn and associated with brush border of villi in ileum	Breaks down *starch* to *maltose*
Trypsin - secreted as trypsinogen	Trypsinogen is made in the pancreas and secreted into the duodenum. Activated by enterokinase	Breaks down *proteins* to *polypeptides*
Lipase	Made in the pancreas and secreted into the duodenum	Breaks down *fats* to *fatty acids* and *glycerol*
Endopeptidase, e.g. chymotrypsin, trypsin	Made in the pancreas and secreted into the duodenum	Break down *proteins* to *polypeptides* and *polypeptides* to short *peptides*
Exopeptidases, e.g. aminopeptidase, carboxypeptidase	Made in the pancreas and secreted into the duodenum	Break down *peptides* to *amino acids*
Nuclease	Made in the pancreas and secreted into the duodenum	Breaks down *nucleic acids* (DNA, RNA) into *nucleotides*
Enterokinase	Secreted by the lining of the small intestine	Converts inactive *trypsinogen* to active *trypsin*
Maltase	Secreted by the lining of the small intestine	Breaks down *maltose* to *glucose*
Sucrase	Secreted by the lining of the small intestine	Breaks down *sucrose* to *glucose* and *fructose*
Lactase	Secreted by the lining of the small intestine	Breaks down *lactose* (milk sugar) to *glucose* and *galactose*
Nucleotidases	Secreted by the lining of the small intestine	Breaks down *nucleotides* to *pentose sugars* + *phosphate* + *organic base*

Table 3.3.1 The main secretions associated with the small intestine

Endopeptidases and exopeptidases

As can be seen in table 3.3.1, the breakdown of proteins is brought about by two different groups of enzymes known as **endopeptidases** and **exopeptidases**. What are they, and what is their purpose?

To put it simply, endopeptidases break bonds within the polypeptide chains producing shorter peptides, whilst exopeptidases act only on the bonds at the very end of the peptide chain, thus releasing individual amino acids. Figure 3.3.13 shows this.

The value to the organism of having these two different types of protein-digesting enzymes is simple. Exopeptidases on their own would take a relatively long time to break down proteins, working on one amino acid at a time from the two ends of the polypeptide chain. Endopeptidases speed up the process enormously, because the result of their action is to provide many short peptide chains with terminal amino and carboxyl ends for exopeptidases to attack, effectively increasing the substrate concentration and so the rate of reaction.

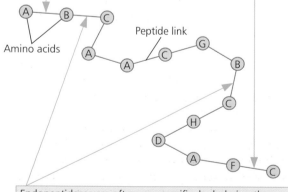

Exopeptidases act only on the terminal peptide links in a peptide chain – for example, aminopeptidase will only hydrolyse the final link at the amino end of the molecule and carboxypeptidase will only hydrolyse the link at the carboxyl end. As a result the amino acids are released from the chains one at a time.

Amino acids

Peptide link

Endopeptidases are often very specific, hydrolysing the peptide links only between particular amino acids. They break these particular bonds wherever they occur in the polypeptide molecule. As a result, endopeptidases such as trypsin produce many short peptide chains.

Figure 3.3.13 The action of endopeptidases and exopeptidases

Control of secretions in the small intestine

Table 3.3.1 shows the large number of secretions which are produced by the small intestine and its associated organs. For the digestive functions to work properly it is important that the right substances are present only at the appropriate time. How is this control brought about?

As with other areas of the gut, the sight, smell and taste of food cause increases in the secretions of both bile from the liver and pancreatic juices. This is the result of nervous reflexes. In addition, the acidic chyme entering the duodenum from the stomach has a direct effect on the cells lining the small intestine. It stimulates them to produce enzymes, and also to produce hormones. These hormones are carried in the blood and in turn stimulate the liver, gall bladder and pancreas. The main hormones are **secretin** and **cholecystokinin** (also known as cholecystokinin-pancreozymin or CCK-PZ). The action of these hormones is described in figure 3.3.14.

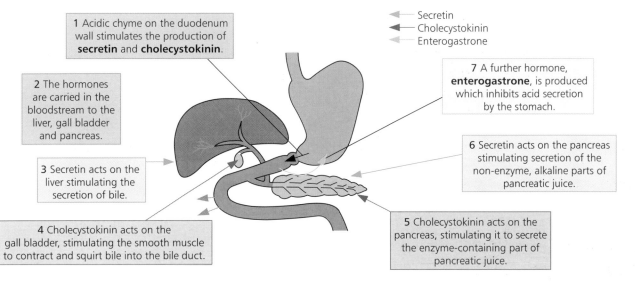

1 Acidic chyme on the duodenum wall stimulates the production of **secretin** and **cholecystokinin**.

2 The hormones are carried in the bloodstream to the liver, gall bladder and pancreas.

3 Secretin acts on the liver stimulating the secretion of bile.

4 Cholecystokinin acts on the gall bladder, stimulating the smooth muscle to contract and squirt bile into the bile duct.

Secretin
Cholecystokinin
Enterogastrone

7 A further hormone, **enterogastrone**, is produced which inhibits acid secretion by the stomach.

6 Secretin acts on the pancreas stimulating secretion of the non-enzyme, alkaline parts of pancreatic juice.

5 Cholecystokinin acts on the pancreas, stimulating it to secrete the enzyme-containing part of pancreatic juice.

Figure 3.3.14 A complex control system by hormones in addition to nervous control means that the digestive functions of the small intestine work only as and when they are needed.

Absorption in the small intestine

One of the major roles of the small intestine is the chemical digestion of food. However, the products of digestion must be taken into the bloodstream for distribution to the cells of the body, and this is the other major role of the small intestine.

We have already discussed the enormous surface area of the small intestine available for the absorption of digested food which results from the villi and the brush border cells. The villi also have an excellent blood supply which ensures a good diffusion gradient. The products of the digestion of carbohydrates and proteins – that is, monosaccharides, disaccharides and amino acids – pass through the epithelial cells to the blood capillaries of the villi. They are moved mainly by simple diffusion and active transport, although facilitated diffusion may play some part.

The results of fat digestion (fatty acids and glycerol) may also be taken directly into the blood. However, there is an alternative route. Many years ago it was observed that after a fatty meal the fluid in the lymph vessels appeared milky-white. In fact, this is why the tiny lymph vessels of the villi are known as 'lacteals' – the word means 'milky'. The milky appearance is due to tiny lipoprotein droplets appearing in the lymph.

The fatty acids and glycerol, once absorbed, can be recombined in the lining cells of the villi forming minute droplets of fat. These pass into the lacteals and are coated in protein to stop them forming much larger droplets. These lipoproteins are known as **chylomicrons** and they travel in the lymphatic system until they reach the thoracic duct where they are returned to the blood system. They are then converted back into fatty acids and glycerol before being taken up by the cells.

Blood is brought to the small intestine by the **mesenteric artery**, but it is not then returned in the veins directly to the heart in the usual way. The blood which leaves the ileum is rich in digested food products. It is carried in the **hepatic portal vein** to the liver, which is a site of major biochemical activity in the body. Here some of the digested products are stored or converted into other forms, whilst others continue to the heart via the **hepatic vein** to be distributed to the cells. Figure 3.3.15 shows the blood supply to and from the ileum.

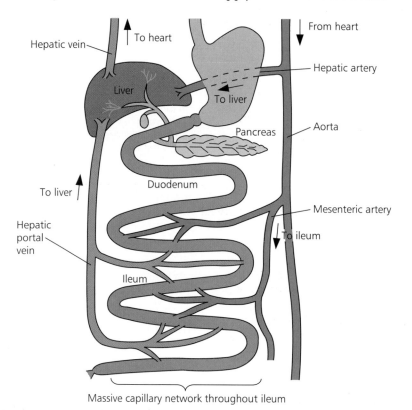

Massive capillary network throughout ileum

Figure 3.3.15 The products of digestion are carried to the liver for processing before continuing in the bloodstream to be supplied to all the body cells.

The large intestine

By the end of the small intestine, all the digested food products and the minerals and vitamins that are useful to the body should have been removed from the watery contents. The remaining fluid consists of indigestible or undigested food, bacteria, dead cells sloughed off from the digestive tract, mucus, bile and a large amount of water, much of which has been secreted into the gut by the body in the form of digestive juices. The function of the large intestine is to reabsorb as much of this as is useful and to remove the remainder from the body in the process of egestion.

The large intestine is thin-walled and wider in lumen than the small intestine, but much shorter in length. The main function of the large intestine is to reabsorb water into the blood. The remaining waste material is known as **faeces**. It accumulates in the colon and the rectum until that area of the digestive tract is stretched sufficiently to trigger the opening of the muscular sphincter known as the anus.

The length of time spent by the indigestible food in the large intestine determines the amount of water which will be reabsorbed and so the state of the faecal material. For example, when the lining of the colon is irritated by a viral or bacterial infection, the food is passed through very quickly and little reabsorption of water takes place. This results in diarrhoea, which can in turn cause rapid dehydration of the system if the water lost is not replaced. This sequence of events – infection of the gut, water loss through diarrhoea and subsequent death from dehydration – kills more children worldwide than anything else, clearly demonstrating the importance of this final stage of the alimentary canal.

THE HUMAN DIET

Any heterotrophic organism must take in sufficient raw materials of the right type for the synthesis of new biological material, and enough energy-providing foods. To maintain the human body a balance of foods is needed, which must include carbohydrates, proteins and fats along with mineral salts, vitamins, water and fibre. We shall consider each in turn.

Carbohydrates

Carbohydrates provide energy – they are used largely in cellular respiration in the production of ATP. Some carbohydrate is stored as glycogen, a storage carbohydrate found in the liver, muscles and brain. Any excess carbohydrate is converted to fat for storage.

Fats

Fats are also used to provide energy, and they are stored in readiness for times of food shortage.

Proteins

Proteins are used for body building. They are broken down in digestion to their constituent amino acids and these are then rebuilt by the process of protein synthesis to form the appropriate proteins for that particular individual. Certain amino acids are vital in the diet because they cannot be synthesised. Mainly found in animal protein, these are called the **essential amino acids**.

Mineral salts

Mineral salts are in general needed in minute amounts, but lack of them in the diet can lead to a variety of adverse conditions, as table 3.3.2 shows.

Table 3.3.2 Some of the main minerals needed in the diet and the deficiency diseases associated with a lack of them

Mineral	Functions	Deficiency symptoms	Source
Calcium (Ca^{2+})	Activates some enzymes. Needed for skeleton and teeth formation, muscular contraction and blood clotting	Poor bone growth, rickets (soft bones), muscle spasms, delayed blood clotting	Milk, cheese, fish
Iron (Fe^{2+})	Part of cytochromes. Activates the enzyme catalase. Needed for haemoglobin and myoglobin	Anaemia	Liver, red meat, eggs, apricots, cocoa powder
Iodine (I^-)	Constituent of thyroxine	Goitre (swollen thyroid gland)	Sea-fish, shellfish, drinking water, added to salt
Nitrogen (N)	Constituent of proteins	Kwashiorkor – swelling, stunted growth and weakness	Protein foods – milk, eggs, meat, pulses
Phosphorus (P)	In phospholipids, proteins, ATP, nucleic acids. Also needed for skeletal growth	—	Most foods

Vitamins

Vitamins are similarly required in very small amounts. They are usually complex organic substances which can nevertheless be absorbed directly into the bloodstream from the gut. If any particular vitamin is lacking from the diet in the long term, it will result in a deficiency disease, shown in table 3.3.3. These deficiency diseases can be avoided or remedied using vitamin supplements if the dietary intake remains inadequate.

Table 3.3.3 The main vitamins needed in the human diet and their associated deficiency diseases

Vitamin letter	Name	Function	Principal sources	Deficiency diseases
Fat-soluble vitamins				
A	Retinol	Involved in photochemical reaction in rods in retina of eye	Liver, carotenoid pigments in vegetables, particularly carrots	Poor dark adaptation, xerophthalmia (drying and degeneration of cornea)
D	Calciferol	Calcification and hardening of bone and teeth	Fish liver oil	Softening of bones – rickets in children, osteomalacia in adults
K	Phylloquinone	Required for synthesis of certain blood-clotting factors	Cabbage, spinach, pig's liver	Prolonged blood clotting time
Water-soluble vitamins				
B_1	Thiamine	Coenzyme for decarboxylation of pyruvic acid to acetyl CoA in respiratory pathway	Yeast, cereals	Beri-beri – wasting of muscles, gastric upsets, circulatory failure and paralysis
B_2	Riboflavine	Forms flavine coenzymes (FAD etc.) – electron carriers in cell respiration	Leafy vegetables, fish, eggs	Sore mouth, ulcerations
PP	Nicotinic acid	Forms coenzymes NAD and NADP – hydrogen acceptors in cell respiration	Meat, fish, wheat	Pellagra – diarrhoea, dermatitis and mental disorder; pigmentation of neck (Casal's necklace)
M or Bc	Folic acid	Required for formation of erythrocytes	Leafy vegetables, liver, kidney	Anaemia
C	Ascorbic acid	Required for formation of intercellular material	Citrus fruits and green vegetables	Scurvy

Evidence for the importance of vitamins

Over the years an enormous amount of evidence for the importance of vitamins in the diet has been built up. Modern techniques can show the positions in biochemical pathways where vitamins are used. Work done at Cambridge in the early years of this century by Sir Frederick Gowland Hopkins remains a classic piece of evidence. He took two sets of eight young rats and fed both sets a diet consisting of purified casein, starch, sucrose, mineral salts, lard and water. In addition, one set only was given 3 cm³ of milk every day for the first 18 days of the experiment. At this point the milk supplement was stopped and given to the other set of young rats until the end of the experiment.

Figure 3.3.16 These results show that a diet consisting of carbohydrate, protein, fat, minerals and water is not enough for the long-term health and growth of young rats. The vitamins present in the milk were also needed.

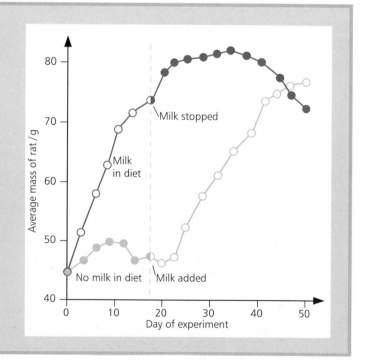

Water

The importance of water in biological systems has already been considered. Suffice it to say that, whereas the average person can survive with little or no food for days if not weeks, complete lack of water will bring about death in two to four days, depending on other conditions such as temperature.

Fibre

Roughage or fibre cannot be digested in the human gut, yet it is an essential part of the diet because it provides bulk for the intestinal muscles to work on and also holds water. In a diet low in roughage peristalsis is sluggish and the food moves through the gut relatively slowly. This can lead to minor ailments such as constipation and haemorrhoids, and is implicated in more serious conditions such as cancer of the bowel.

A balanced diet

The balance of food required by different types of animals will obviously vary, but for any species the right balance of food is of enormous importance to the overall health and well-being of the animal. If too little food is eaten (undernutrition) then the organism will suffer from **malnutrition**. Too much food may also be eaten (overnutrition). This is well illustrated by the human animal.

One of the most important factors of a balanced diet is that enough food is eaten to supply the energy needs of the organism. For vast areas of the world, particularly the developing world, this is a major problem. There is simply not enough food available to supply the energy needs of the people. As a result much of the population is seriously underweight, with shortened lifespans and reduced resistance to disease. Hand in hand with insufficient food go insufficiencies of essential amino acids, minerals and vitamins and so the deficiency diseases seen in tables 3.3.2 and 3.3.3 are also found.

However, it is also important that too much food is not consumed. The energy requirements of each individual vary depending on age, sex and levels of activity. If more energy is taken in than is required, the excess is stored as fat and obesity may result. Frequently a problem in the developed world, up to a third of the population of the USA is thought to be seriously overweight, mainly due to eating a diet rich in high-energy fat. This causes coronary heart disease, high blood pressure and other disorders which reduce life expectancy. A high proportion of sugar in the diet leads to the formation of dental caries, as bacteria digest the sugar in the mouth, producing acid which in turn attacks the enamel of the teeth. Many people have lost their teeth as a result of their diet and poor dental hygiene.

Control of body fat

The most obvious effect of overeating is an increase in body fat, eventually resulting in obesity. The long-term effects of overeating are a shortened lifespan and poor health in old age. Longer, fitter lives result from eating just sufficient to supply the needs of the body. However, it is more the fashion-driven desire to look younger and sexier that has resulted in a large proportion of the adult populations in countries such as Britain, France and the USA spending at least part of their life on a weight-loss diet.

Sadly for these well-meaning efforts, recent evidence suggests that the weight of an organism is strongly governed by factors other than the amount of food eaten. There appears to be a **set point** rather like a thermostat, a genetically determined level for each individual's body weight and in particular the amount of fat stored. In most people tested and many other mammals, a large increase in calories results in little weight gain over the set point – the excess calories are burned off and lost as body heat. Similarly, a lowering of the food intake did not result in a large weight loss below the set point – the metabolism simply functioned more efficiently. The only way of adjusting the set point thermostat seems to be to change the level of physical activity.

In humans we can observe aberrations of the normal appetite and feeding controls in conditions known as **anorexia nervosa** and **bulimia**. In the former there is minimal food intake over a prolonged period of time. Bulimia is another form of the same condition, where huge amounts of food are taken in and then vomited out again. These conditions are usually linked to deep psychological problems, and the ways in which food, love and control can become distorted in the human mind. Sadly, in spite of treatment, they can result in death.

Figure 3.3.17 Reducing food intake and increasing the levels of exercise can result in quite dramatic weight loss, but for most adults with fairly stable eating and exercise habits their genetic set point will largely determine their levels of body fat.

The Western diet also contains many added chemicals. Strangely, the diet of the average inhabitant of the developed world, whilst full of fat and refined carbohydrate, may be low in vitamins as relatively little fresh fruit and vegetables are consumed. Roughage is also frequently missing in any quantity, causing not only constipation but also many diseases of the gut and bowel, including, it would seem, some bowel cancers. The saying 'You are what you eat' may be truer than many of us would like to think.

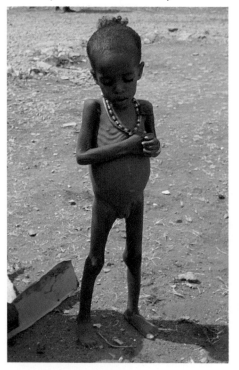

Figure 3.3.18 Whether you are forced by circumstances to eat too little or choose to eat too much, the wrong type of nutrition can seriously damage your health.

VARIATIONS ON A THEME

Carnivores and herbivores

The human digestive system gives us a good picture of the major adaptations and mechanisms necessary for heterotrophic nutrition. Other mammalian groups add to the picture showing specialisations that apply to different types of organisms as well.

Carnivores eat a diet which is largely made up of protein – the largest component of the skin, muscle and many of the internal organ systems of their prey. Protein is relatively easy to digest. The breakdown begins in the stomach and is completed by the endo- and exopeptidases of the small intestine. Therefore the guts of carnivores tend to be shorter than those of omnivorous or herbivorous animals of the same size, as figure 3.3.19 shows. Also, relatively little undigested material is egested by carnivores.

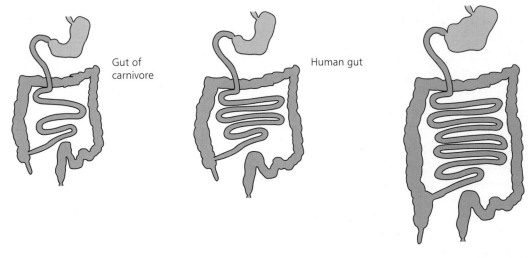

Gut of carnivore

Human gut

Gut of herbivore – often either the stomach (rumen) or the caecum and appendix are enlarged.

Figure 3.3.19 The modifications observed in the guts of different animals reflect the varying digestive demands of their diet.

In the case of herbivores, most animals cannot digest cellulose, so there is no food value in the cellulose cell walls of the plants they eat. The nutrition, in the cytoplasm, is trapped inside an indigestible box. We have already considered the adaptation of herbivore teeth for the chewing needed to break open the cell walls. Alongside this, most herbivores have extremely long guts to maximise the opportunities for nutrients to be extracted, and also to hold the large amounts of food that must be taken in.

Another adaptation is that some herbivores (for example ruminants such as the cow and some insects such as the termites) have set up **symbiotic** relationships with microorganisms. These are mainly bacteria which manufacture the enzyme cellulase, and which can therefore digest cellulose. This means that less food has to be taken in to supply the herbivore's needs, although the more effective digestive process often takes some time. The cellulose digestion takes place in specialised areas. In ruminants this is the **rumen**, shown in figure 3.3.19, but in other herbivores the much-enlarged appendix is the home of the cellulose-digesting bacteria. Ruminants in particular 'chew the cud' – they chew their food once, swallow it for some preliminary digestion, regurgitate it for another prolonged chew and then carry on digesting it again. Other species such as rabbits allow the food to pass through the gut once, eat the faeces produced and then redigest the food to obtain as much value as possible from it. In spite of measures such as these, herbivores in general egest large quantities of undigested plant material as faeces.

Other feeding relationships

There are many other variations on the theme of digestion. Parasites have particular adaptations. Internal parasites, particularly those found in the gut such as many roundworms and the tapeworms, have little digestion to perform. They absorb the ready-digested material present in the gut of their host. However, they must protect themselves from being digested in turn. Even if they do not live within the gut, they still have to avoid destruction by the host's immune system. External parasites too have many difficulties to overcome before they begin digestion. Hosts must be found and attacked without damage to the parasite. As a result of these obstacles many parasitic feeders have evolved extremely complex ways of life and immune systems which in turn make them a major problem for the human race. We will meet them again in more detail in section 6.

Symbiosis – parasitism, mutualism and commensalism

Symbiosis literally means 'living together'. More specifically it is used to mean 'the living together of two or more organisms of different species in close association with each other'. For organisms from two different species to be commonly found together, it is probable that one at least benefits from the arrangement. The more specific terms **parasitism**, **mutualism** and **commensalism** give us more detail about symbiotic relationships.

Parasitism describes a way of life which benefits one partner (the parasite) and harms the other (the host). The parasite gains food and frequently shelter from the host and inflicts damage which can range from very slight to causing the death of the host. However, a parasite which kills its host is obviously not well adapted.

Mutualism describes a way of life where the association benefits both organisms. Examples include the microorganisms in the guts of ruminants, where the microbes live in a warm and protected environment and are supplied with food. The larger animal receives the benefit of the cellulose-digesting ability of the microbe.

Commensalism describes a way of life where one partner benefits and the other is unaffected by it. One of the best known examples is the relationship between a type of hermit crab and the colonial hydroid *Hydractinia echinata*. The hydroid gains a way of moving about and a ready supply of food scraps. As far as anyone has been able to discover, the crab gains nothing by the relationship, but neither is it disadvantaged.

SUMMARY

- **Heterotrophs** obtain their food by eating autotrophs or other heterotrophs. **Holozoans** feed on the bodies of other organisms. **Parasites** feed on organic material from a host. **Saprotrophs** feed on soluble organic material from dead animals and plants.

- Heterotrophs have to obtain, digest and absorb their food. **Herbivores** expend little energy obtaining food but cannot digest cellulose so have to eat large quantities to obtain enough energy. They have ridged molars adapted to break down cellulose cell walls. **Carnivores** expend more energy obtaining their food and have teeth adapted for trapping and tearing flesh, but they can digest their food more easily. **Parasites** are adapted to prevent rejection by the host and **saprotrophs** to find new food sources.

- **Digestion** of food is the breakdown of large complex food molecules into simple soluble molecules that can be absorbed into the body and distributed. Digestion is catalysed by enzymes and takes place in the **alimentary canal** or **gut**.

- Humans are holozoic feeders. Holozoic nutrition consists of **ingestion** (taking food in), **digestion**, **absorption** of the digested products into the cells, **assimilation** (use of the products of digestion by the cells) and **egestion** (removal of undigested food material).

- Food is ingested in the mouth and chewed to break it down (mechanical digestion). Chemical digestion starts in the mouth by the action of **saliva**, which also moistens and lubricates the food. A **bolus** of food is swallowed and passes into the oesophagus.

- The **oesophagus** has a folded inner lining typical of the alimentary canal. The **mucosa** is made up of **stratified squamous epithelium** interspersed with **mucus-producing cells**. Mucus protects the oesophageal lining and lubricates the food. The **submucosa** is a layer of connective tissue, nerves and blood vessels. **Peristalsis** is brought about by the coordinated movement of the **circular** and **longitudinal muscles**.

- The food passes from the oesophagus through the **cardiac sphincter** to the **stomach**, a muscular organ which churns the food and continues chemical digestion. The stomach lining contains **gastric pits** which have **mucus-producing cells**, **oxyntic cells** which produce hydrochloric acid and **chief cells** which secrete pepsinogen. The **chyme** which results from the action of the stomach passes through the **pyloric sphincter** to the duodenum.

- The **duodenum** is the first part of the small intestine where digestion is more or less completed by secretions from the duodenum wall, the pancreas and the liver. The later part of the small intestine is the **ileum** where absorption of digested food takes place. The small intestine has finger-like **villi** which are frequently covered with a **brush border** of **microvilli** to increase the surface area for absorption. Villi contain a **lacteal** (lymph vessel) and a capillary. **Crypts of Lieberkuhn** in the duodenum wall produce digestive enzymes while **Brunner's glands** produce alkaline secretions and mucus. Circular and longitudinal muscles in the wall of the small intestine move the villi.

Food molecule	Converted to	By (enzyme)	Secreted by	Acts in (organ)
Polysaccharides e.g. starch	Disaccharides e.g. maltose	Salivary amylase	Salivary glands	Mouth
		Amylase	Pancreas and crypts of Lieberkuhn in duodenum	Duodenum
Maltose	Glucose	Maltase	Lining of small intestine	Small intestine
Sucrose	Glucose + fructose	Sucrase	Lining of small intestine	Small intestine
Lactose	Glucose + galactose	Lactase	Lining of small intestine	Small intestine
Protein	Polypeptides	Pepsin	Stomach as pepsinogen, activated by hydrochloric acid	Stomach
		Trypsin	Pancreas as trypsinogen, activated by enterokinase which is secreted by the lining of the small intestine	Small intestine
	Polypeptides and short peptides	Endopeptidases e.g. chymotrypsin	Pancreas	Duodenum
Peptides	Amino acids	Exopeptidases e.g. aminopeptidase	Pancreas	Duodenum
Casein (soluble milk protein)	Coagulated protein	Rennin	Stomach of young mammals	Stomach
Fats	Fatty acids and glycerol	Lipase	Pancreas	Duodenum
Nucleic acids	Nucleotides	Nuclease	Pancreas	Duodenum

Table 3.3.4 Chemical digestion in the human alimentary canal

- Chemical digestion is summarised in table 3.3.4.
- **Nervous stimulation** and **gastrin** stimulate the production of acidic gastric juices. Gastrin is produced in response to food in the stomach. **Secretin** and **cholecystokinin** are secreted in response to acidic chyme in the duodenum and stimulate the liver and pancreas to secrete digestive juices. **Enterogastrin** then inhibits acidic secretion in the stomach.
- Amino acids, disaccharides and monosaccharides are absorbed into the blood capillaries of the villi by diffusion, active transport and possibly facilitated diffusion. Fatty acids and glycerol can pass directly into the blood or may recombine in the lining of the villi to form **chylomicrons** – droplets of lipoprotein. These pass in the lymphatic system to the blood where they are again converted into fatty acids and glycerol for uptake by the cells.
- Blood comes to the small intestine from the **mesenteric artery** and blood carrying the products of digestion is then taken to the liver via the **hepatic portal vein**. Digested food is stored in the liver or converted to other molecules which leave the liver by the **hepatic vein** to be distributed to the body cells.
- The **large intestine** is thin walled and larger in lumen but shorter than the small intestine. Water is reabsorbed into the blood from the large intestine and the remaining **faeces** are stored in the **rectum** until **egestion** through the **anus**.
- A balanced human diet consists of carbohydrates, fats, proteins including essential amino acids, mineral salts, vitamins, water and fibre. Lack of essential amino acids, certain minerals or vitamins can result in **deficiency diseases**.

- **Malnutrition** can result from eating too little and deficiency diseases are a common result. **Overnutrition** causes obesity which can lead to coronary heart disease, high blood pressure and other disorders.

- The structure of the alimentary canal is adapted to suit different diets. Carnivores have shorter alimentary canals than omnivores since most of their food is protein, and easy to digest, and they egest small quantities. Herbivores have long guts to maximise the absorption area and some are specially adapted to digest cellulose. Symbiotic bacteria living in their guts, in the **rumen** or enlarged **appendix**, secrete the enzyme cellulase. Ruminants chew their food several times during the digestive process while rabbits eat their faeces and redigest the food.

QUESTIONS

1 a Explain the terms heterotroph, ingestion, digestion and egestion.
  b Produce an annotated sketch of the organs of the human alimentary canal and describe the role played by each in the process of digestion.
  c Describe some of the ways in which the functioning of the different areas of the gut are coordinated.

2 All heterotrophs obtain their food from other organisms. Describe some of the adaptations of heterotrophic organisms which enable them to obtain their food.

3 People (unlike other animals) are affected by a variety of eating-related disorders. These include constipation, ulcers, lactose intolerance, anorexia nervosa and bulimia, and vitamin deficiencies. What is the basis for these disorders? Do they indicate a fundamental problem with the human digestive system? Discuss.

All living things need energy, which they get from food. Autotrophic organisms make their own food, frequently using the energy of the Sun to do so. Heterotrophic organisms eat and digest other organisms to get their food. However it is obtained, food is used to provide energy for all the metabolic reactions which occur in a cell or organism. The energy source which is used by the cells is ATP. In this section we shall look at the ways in which the energy in the food molecules is transferred to the molecules of ATP needed by the cell.

ENERGY VALUES OF FOODS

Different types of foods contain different amounts of energy stored within their chemical bonds. The amount of energy contained in any individual food can be revealed using a process known as **calorimetry**, shown in figure 3.4.1.

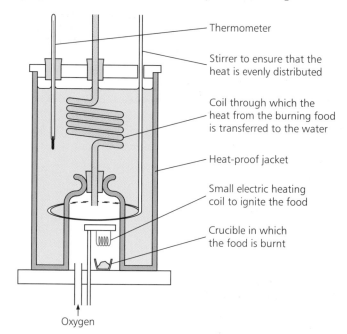

- Thermometer
- Stirrer to ensure that the heat is evenly distributed
- Coil through which the heat from the burning food is transferred to the water
- Heat-proof jacket
- Small electric heating coil to ignite the food
- Crucible in which the food is burnt

Oxygen

Figure 3.4.1 Food is burnt in pure oxygen, so that it is completely oxidised. The energy released as the chemical bonds are rearranged is transferred to the surrounding water as heat. The resulting rise in the temperature of the water can be measured and used to calculate the energy value of the food, based on the fact that 4.2 J of heat energy raise the temperature of 1 g of water by 1 °C.

Calorimetry measures the amount of energy released when a known quantity of food is completely oxidised by burning it in pure oxygen. Carbohydrates have an energy value of 17.2 kJ g^{-1}, fats 38.5 kJ g^{-1} and proteins 22.2 kJ g^{-1}. Food tables exist which show the energy values of an enormous range of foods.

How much energy do we need?

We need a certain amount of energy every day to keep our metabolism 'ticking over', to sustain life at a fairly basic level. The amount of energy required per day has been worked out by measuring the temperature changes which result from the heat production of a human body over a period of hours or days in a heat-proof room – another type of calorimeter. The minimum amount of energy on which the body can survive is known as the **basal metabolic rate** (BMR).

Calories

When measuring the energy in food using a calorimeter, the units of energy previously used were calories and kilocalories. Indeed, most confusingly, what many non-scientists referred to as a Calorie was in fact a kilocalorie! Old habits die hard, and the 'Calorie' is still a common term in everyday language, particularly when people are talking about weight-reducing diets. Biologists no longer refer to calories at all, but most food packaging indicates the energy value of food in both kilocalories and kilojoules. The relationship between the two is very simple:
4.2 joules = 1 calorie.

It has been worked out that an 'average' man needs to take in about 7500 kJ per day to maintain his BMR, and an 'average' woman needs about 5850 kJ per day. This assumes that the person concerned lies on a bed all day and night and expends no extra energy above that needed to breathe and excrete – not even to feed. Further measurements have been made which show that the amount of energy required by any individual depends on a variety of factors which include their age, sex and the type of life they lead. Some figures are shown in table 3.4.1.

Sex and age/yr	Activity level (where appropriate)	Energy requirement/kJ
Boy or girl 0–1		3300
Boy 9–12		10500
Girl 9–12		9600
Boy 13–18		12600
Girl 13–18		9600
Man 18–35	Moderately active	12600
Man 75 and over	Sedentary	8800
Woman 18–35	Most activity levels	9200
Woman 75 and over	Sedentary	8000
Woman 3–9 months pregnant		10000
Woman breastfeeding		11300

Table 3.4.1 The energy demands of the body vary quite dramatically at different times in our lives, regardless of whether we are male or female.

In the developed world, food is produced in abundance and the majority of people eat more than is necessary to supply the metabolic needs of the body. In the process of evolution many animals, ourselves included, have developed the ability to convert excess food energy into a store of fat ready for times of food shortage. Thus people in the affluent and developed areas of the world can become overweight. The tables of energy values of foods are largely used by overweight people trying to lower their total intake of energy and lose weight.

In stark contrast, around two-thirds of the world's population do not get enough food to provide them with the recommended minimum daily energy intake, and for many their food barely yields sufficient energy to cover the BMR.

CELLULAR RESPIRATION

The energy in food is of no value until it is transferred from the chemical bonds in the food to the phosphate bond in ATP. **Cellular respiration** is the process by which organisms use oxygen to release energy from their food and transfer it into molecules of ATP. The energy in the bonds of the ATP formed can then be used to drive all the other biochemical reactions of life. Carbon dioxide and water are formed as waste products. Cellular respiration, like the calorimeter, oxidises the food as completely as possible.

Respiratory quotients

The amounts of oxygen used and carbon dioxide produced during cellular respiration change depending on the level of activity of the organism, the type of food being respired and other factors. By measuring the amounts of carbon dioxide produced and oxygen used up by an organism in a given time period, we can produce what is known as the **respiratory quotient**:

$$\text{Respiratory quotient (RQ)} = \frac{\text{carbon dioxide produced}}{\text{oxygen used}}$$

The respiratory quotient helps us to develop a picture of the type of foods which are being oxidised in the body at a particular time. In theory at least, carbohydrates give an RQ of 1, fats of 0.7 and protein of 0.9. Under normal conditions protein is not much used to provide energy, so an RQ of around 1 suggests that a large proportion of carbohydrate is being used in cellular respiration, and an RQ of less indicates that a combination of carbohydrate and fat is being respired. If the RQ of an organism is greater than 1 then anaerobic respiration may well be taking place, with relatively little oxygen being used compared with the carbon dioxide produced. Very low RQs tend to be found in photosynthetic organisms, when much of the carbon dioxide produced is used up in making new sugars and so cannot be measured.

The process of cellular respiration

Cellular respiration takes place in both autotrophic and heterotrophic organisms. The process involves a complex series of reactions, many of them involving oxidation or reduction. Hydrogen is removed from the food molecules and, in the same way as we have seen earlier, split into protons and electrons which are passed along an electron transfer chain to result in the formation of ATP. As in all biochemical pathways, the reactions are controlled by enzymes. A sequence of reactions is controlled, often by various types of enzyme inhibition, so that one reaction does not occur unless a preceding one has taken place. The process of gaseous exchange, carried out at the respiratory systems considered in section 2.4, provides the oxygen needed and removes the carbon dioxide produced, enabling the chemical reactions to proceed.

The complex process of cellular respiration can be summed up in the following simple equations. It is interesting to note that the equation for respiration is basically the equation for photosynthesis in reverse:

$$\text{Glucose} + \text{oxygen} \rightarrow \text{ATP (energy)} + \text{water} + \text{carbon dioxide}$$
$$C_6H_{12}O_6 + 6O_2 \rightarrow \text{ATP} + 6H_2O + 6CO_2$$

Requirements for cellular respiration

The description used so far for cellular respiration is a simplification. The process takes place in two distinct phases. The first part of the process, called **glycolysis**, does *not* require oxygen. It produces a little ATP, but more importantly it primes the food molecules ready for entry into the second stage of the process which produces more ATP. This second stage is called the **Krebs cycle**, and for this set of reactions to proceed oxygen *is* needed. Thus the most important requirement for respiration is food, as some energy at least can be obtained from glycolysis without oxygen. This food is usually glucose, although as we shall see later, other substances can be used when there is a glucose shortage. In autotrophs the glucose is usually the product of photosynthesis, in heterotrophs it comes from digested food material. Oxygen is required for the Krebs cycle and so oxygen is needed for cellular respiration to proceed completely, so producing the maximum amount of ATP from the breakdown of the food molecules.

Aerobic and anaerobic respiration

Most organisms depend on **aerobic respiration**, which means that they use oxygen for the Krebs cycle in order to provide them with sufficient energy to survive. They may be able to cope with a temporary lack of oxygen, but only in the very short term.

Some organisms can survive without oxygen – they rely on **anaerobic respiration**. **Facultative anaerobes** only respire without oxygen when it is strictly necessary. However, there are a few groups which cannot use oxygen at all and may in fact be killed by it. They are known as **obligate anaerobes**. Figure 3.4.2 gives examples of an aerobic and an anaerobic organism.

Demonstrating cellular respiration

It is not easy to demonstrate the requirements for cellular respiration practically without sophisticated biochemical techniques. Depriving aerobic organisms or cells of oxygen simply kills them. Apparatus such as spirometers or the respirometer in figure 3.4.3 can give some valuable information but has limitations. Depriving an organism of its food is also not informative – most organisms will self-digest for some time before dying. It is easier to observe cellular respiration by considering the waste products of the reactions.

The products of cellular respiration

ATP is the desired end product of cellular respiration. But along with this, carbon dioxide and water are also produced. These waste products, particularly the potentially toxic carbon dioxide, are then removed from the cell or organism. The production of carbon dioxide by an actively respiring organism is the most readily observed feature of cellular respiration, by the use of limewater (see figure 3.4.4), although again making precise measurements can be difficult.

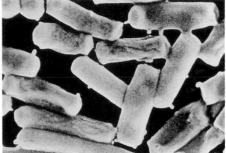

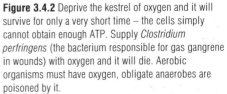

Figure 3.4.2 Deprive the kestrel of oxygen and it will survive for only a very short time – the cells simply cannot obtain enough ATP. Supply *Clostridium perfringens* (the bacterium responsible for gas gangrene in wounds) with oxygen and it will die. Aerobic organisms must have oxygen, obligate anaerobes are poisoned by it.

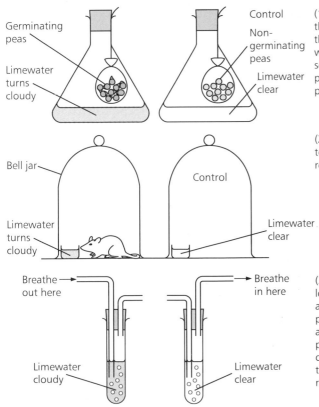

(1) Germinating seeds such as these peas can be used to show that carbon dioxide is produced when plants respire. By using seeds rather than fully grown plants the complications of photosynthesis are avoided.

(2) Small animals can be shown to produce carbon dioxide as a result of respiration.

(3) People take in air which is low in carbon dioxide and exhale air which contains a much higher proportion of the gas. The amount of carbon dioxide produced varies with the amount of exercise undertaken and therefore the amount of cellular respiration taking place.

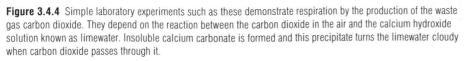

Figure 3.4.4 Simple laboratory experiments such as these demonstrate respiration by the production of the waste gas carbon dioxide. They depend on the reaction between the carbon dioxide in the air and the calcium hydroxide solution known as limewater. Insoluble calcium carbonate is formed and this precipitate turns the limewater cloudy when carbon dioxide passes through it.

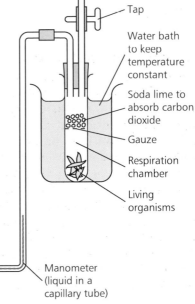

Figure 3.4.3 Any movement of the liquid in the tube indicates an uptake of oxygen by the organism. This type of respirometer can be useful for comparing the effect of, for example, temperature on the uptake of oxygen. However, a little thought will show that there are many limitations to apparatus of this type.

The site of cellular respiration

As we have seen, glycolysis is the first part of the respiratory pathway and does not require oxygen. Glycolysis provides comparatively little ATP but it occurs in all cells. It is not associated with any particular cellular organelle – the enzymes controlling glycolysis are found in the cytoplasm.

The second part of respiration, the Krebs cycle, involves oxygen and yields considerably more ATP than glycolysis. The biochemical events of the Krebs cycle and the electron transfer chain involved in producing ATP are found in the mitochondria. The structure of these organelles was discussed in section 1 and their role as the site of ATP production in section 3.1. The partially permeable outer membrane allows the products of glycolysis to enter the mitochondria. The matrix seems to contain the enzymes of the Krebs cycle. Those cells with very low energy requirements, for example, fat storage cells, tend to contain very few mitochondria. On the other hand, cells which are very active such as those of the muscles and the liver have very large numbers of mitochondria packed into the cytoplasm.

THE BIOCHEMISTRY OF RESPIRATION

As with photosynthesis, the equation summarising the process of respiration is a vast oversimplification. We have seen that it consists of two stages, glycolysis occurring in the cytoplasm and the Krebs cycle in the mitochondria. The two pathways of glycolysis and the Krebs cycle are part of a coordinated sequence of reactions bringing about the oxidation of glucose and the production of ATP, but for ease of understanding we shall consider them separately and then look at the overall process.

Glycolysis

Glycolysis literally means 'sugar-splitting' and in this initial part of the respiratory pathway glucose, a 6-carbon sugar, is split by a series of reactions into two molecules of the 3-carbon compound pyruvate. It is this pyruvate which is then taken into the mitochondria to supply the enzymes of the Krebs cycle. The main stages of glycolysis are shown in figure 3.4.5.

The first steps in glycolysis actually use up some ATP, phosphorylating the glucose molecule by the addition of two phosphate groups. This makes the glucose more reactive, and also ensures that it can no longer be transported readily across the cell membrane. The phosphorylated sugar is then split to give two 3-carbon (triose) sugar molecules. One is **glyceraldehyde 3-phosphate (GALP)** which continues on along the glycolytic pathway. The other triose sugar is dihydroxyacetone phosphate, which is readily converted to GALP.

In the next stage each molecule of GALP is converted by several steps into a molecule of **pyruvate**. Two hydrogen atoms are removed from the GALP and taken up by NAD. Because this is occurring in the cytoplasm, the other members of the electron transfer chain are not immediately available, but the reduced $NADH_2$ can be shunted through the outer mitochondrial membrane into the electron transfer chain of the inner mitochondrial membrane and so has the potential to produce ATP. Three molecules of ATP could result from the $NADH_2$ from each triose sugar in this way, as figure 3.1.6 (page 174) shows. We shall return again to the electron transfer chain when considering the Krebs cycle.

There is also some direct ATP formation from an energy transfer when the GALP is converted to pyruvate. For each GALP molecule converted (via several intermediates), 2 ATP molecules are made. So for each glucose molecule glycolysed, 2 ATP molecules are used and 4 produced (2 from *each* triose sugar). There is a net gain of 2 molecules of ATP per molecule of glucose and the potential to make 6 more via the electron transfer chain (3 from *each* triose sugar).

Pyruvate marks the end point of the reactions of glycolysis. In the presence of oxygen the pyruvate enters the mitochondria and is used in the aerobic reactions of the Krebs cycle. If there is insufficient oxygen for this, the pathway continues to form the end products of anaerobic respiration with no further ATP production.

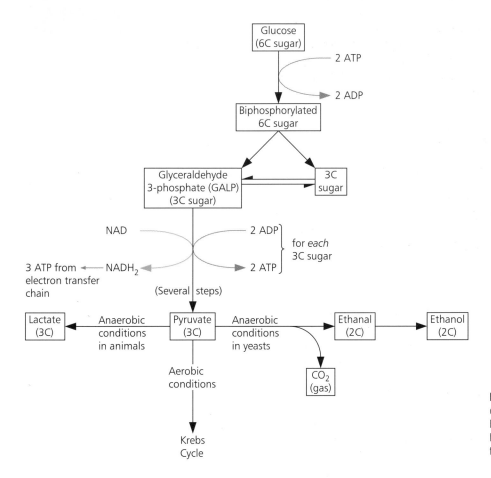

Figure 3.4.5 With oxygen in plentiful supply pyruvate continues on into the mitochondria and the Krebs cycle. But when oxygen is not available, yeasts and animals have alternative strategies which will allow them, on a temporary basis at least, to respire by anaerobic means.

Glycolytic products of anaerobic respiration in animals and fungi

Animals

We are all familiar with the pain which develops in our muscles if we overdo exercise. This is the result of insufficient oxygen reaching the muscles to supply the demand – we are using more ATP than we can produce by aerobic respiration, so the muscles are forced to respire anaerobically. The body is said to be in a state of **oxygen debt**. At the end of the glycolytic pathway pyruvate, instead of entering the Krebs cycle which depends on oxygen, is converted to **lactate**. Lactate is toxic and causes pain in the muscles as it builds up, until eventually we give in and stop exercising. As the body returns to normal this lactate has to be converted back to pyruvic acid and broken down to carbon dioxide with the use of oxygen, or otherwise metabolised, so paying off the oxygen debt.

Fungi

In fungi and plants too anaerobic respiration may occur. It is then known as **fermentation**. Here the result is not lactate but ethanol – an end product which people have been exploiting for many hundreds of years to make intoxicating drinks. The pyruvate is converted first to the intermediate product ethanal, by the removal of carbon dioxide which is given off as a gas. It is this carbon dioxide which is used to make bread rise and which gives beer its 'head', as figure 3.4.6 illustrates. The ethanal is then further reduced to ethanol.

Controlling the rate of glycolysis

Although ATP is often regarded as an energy store for the reactions of the body, it is perhaps better to regard it as a means of transferring energy from the food molecules to the molecules of the organism itself. ATP is not stored as such – it is made as and when it is needed. This means that it is important for the rate of glycolysis (and therefore the Krebs cycle which produces more ATP) to be tightly controlled. When energy demands are high, glycolysis needs to occur rapidly to supply plenty of pyruvic acid for the Krebs cycle. When the energy demands are low, less oxygen needs to be taken in and glycolysis must slow down so the Krebs cycle reactions proceed more slowly too.

Each individual step in the process of glycolysis is controlled by an enzyme, and these enzymes are sensitive to various substrates and products of the pathway, giving a greater or lesser degree of control. One enzyme is worth particular mention. **Phosphofructokinase** is an allosteric enzyme which catalyses one of the early reactions in the conversion of glucose to GALP:

$$\text{Fructose-6-phosphate} \xrightarrow{\text{PFK}} \text{fructose-1,6-diphosphate}$$

Phosphofructokinase is activated by high concentrations of ADP, and of its own substrate. It is also inhibited by high levels of ATP and citrate, an intermediate in the Krebs cycle. So when there is plenty of ATP or the components of the Krebs cycle begin to build up, the whole process of glycolysis is slowed down. Conversely, when the cell needs energy and the components of the Krebs cycle are low, glycolysis is speeded up to remedy the situation. In changing the rate of glycolysis the rate of the whole process of cellular respiration is controlled. Most biochemical pathways have particular enzymes which, like this one, play a vital role in controlling the rate of the entire pathway. They are called **regulatory enzymes**.

Figure 3.4.6 The anaerobic respiration of yeast, with the glycolytic pathway ending in the production of ethanol along with carbon dioxide, is utilised by the brewing industry to produce millions of gallons of beer and lager each year.

Evidence for glycolysis

It took many years for the pathways of glycolysis and the closely associated process of alcoholic fermentation to be worked out. There are several landmarks along the way.

(1) In 1897 Eduard Buchner discovered that an extract of yeast without any cells in it could still convert glucose to ethanol. This showed that the enzymes of glycolysis and fermentation are not associated closely with the cell structure.

(2) In the early 1900s Arthur Harden and W. J. Young showed that phosphate was needed for the pathway to proceed, and also that two particular elements of the yeast extracts were needed before fermentation could go ahead. One of these was inactivated by heat – it contained the enzymes. The other was not affected by heat in the same way and contained NAD, ADP and ATP.

(3) After work with inhibitors had allowed some of the pathway intermediates to be studied, the German biochemists Gustav Embden and Otto Meyerhof worked out much of the rest of the sequence.

(4) By the 1940s, Embden and Meyerhof, along with important contributions from others, had worked out the individual steps of the glycolysis pathway.

The Krebs cycle

The reactions of the Krebs cycle were worked out by Sir Hans Krebs. They have had an enormous impact, not just on our understanding of cellular respiration but also on our understanding of biochemical pathways in general. The cycle is also known as the **tricarboxylic acid cycle** (from the types of chemicals involved) or the **citric acid cycle** (from one of the main components of the pathway). The reactions occur in the matrix of the mitochondrion in the presence of oxygen, and are shown in figure 3.4.7.

The 3-carbon molecule pyruvate produced by the glycolysis pathway crosses the mitochondrial membrane and is immediately converted to a 2-carbon acetyl group which is linked to a coenzyme to form **acetylcoenzyme A** (**acetylco A**). A molecule of carbon dioxide is removed in this reaction, along with a molecule of hydrogen which reduces NAD to $NADH_2$ and is split and passed into the electron transfer system resulting in the formation of ATP. The enzymes in this step which remove carbon dioxide are known as **decarboxylases** and those which remove hydrogen are **dehydrogenases**.

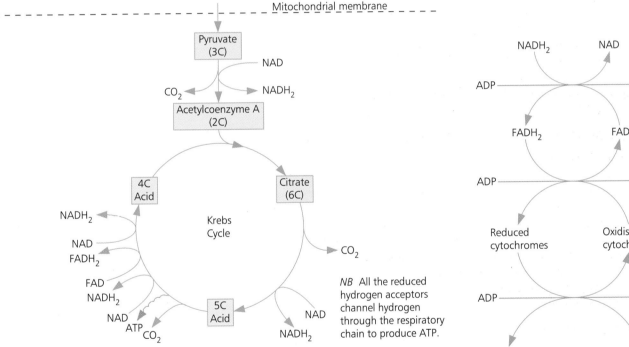

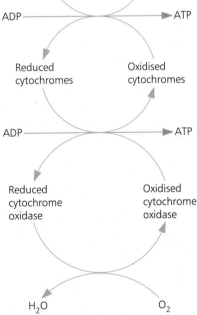

Figure 3.4.7 The Krebs cycle turns continuously to provide the cells with energy, and the rate of its turning is carefully controlled to ensure that the right amount of ATP is produced to meet the demands of the body.

Acetylco A enters the Krebs cycle by combining with a 4-carbon compound forming the 6-carbon compound **citrate**. The Krebs cycle is a cyclical series of reactions during which the 6-carbon citrate is broken down to give the original 4-carbon compound. This then combines with more acetylco A and the cycle turns again. As the cycle progresses two further molecules of carbon dioxide are removed, to be given off as a waste product. Also, four of the steps involve the removal of hydrogen atoms and thus the reduction of a carrier molecule. These can then be split and the electrons passed along the electron transfer chain and used to produce ATP, as in figure 3.4.8. The final hydrogen acceptor of the electron transfer chain is oxygen. This explains the need for oxygen by the Krebs cycle, and also the production of water as a waste product of respiration.

Figure 3.4.8 Another look at the electron transfer chain. The carrier molecules sited on the inner mitochondrial membranes are alternately reduced and oxidised and these reactions are used to drive the synthesis of ATP. Oxygen is required in aerobic respiration as the final hydrogen acceptor with the production of water.

Depending on which carrier accepts the hydrogen atoms (for four of the reactions it is NAD but for one it is FAD), either 3 or 2 molecules of ATP will result. The repeated oxidation and reduction sequence beginning with $NADH_2$ yields 3 molecules of ATP each time. It must be remembered that for each molecule of glucose which enters the glycolytic pathway, the Krebs cycle turns twice (6-carbon glucose giving 2 molecules of 3-carbon pyruvate). The reactions of the cycle itself take place in the matrix of the mitochondrion, but the ATP is produced from the electron transfer system in the stalked particles on the inner mitochondrial membranes.

How much ATP is gained from cellular respiration?

The whole process of cellular respiration, from the beginning of glycolysis to the release of carbon dioxide and water as the Krebs cycle turns, has evolved to produce energy in the form of ATP for use in the cells. The fact that the process is the same in almost all living organisms suggests that it is a very effective method of doing just that. But exactly how much ATP is gained during the oxidation of one molecule of glucose in its journey along the respiratory pathways? The easiest way to look at this is to consider where ATP results in the whole process, and figure 3.4.9 does this. The average amount is 38 molecules of ATP assuming that glucose enters the cycle and that oxidation is complete. If this is compared with the meagre 2 molecules of ATP which result when the breakdown of glucose is purely anaerobic, the importance of the oxygen-using process becomes clear.

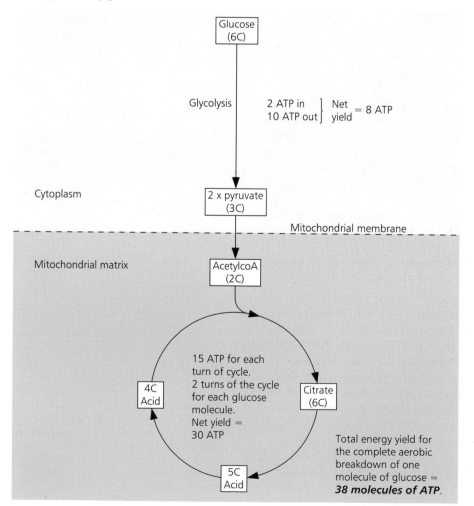

Figure 3.4.9 The ATP gained by the complete oxidation of a molecule of glucose in cellular respiration.

Evidence for the Krebs cycle

Hans Krebs first put forward his ideas for the now famous cycle in 1937. It was the result of brilliant reasoning and experimentation in the preceding years, both by Krebs and by others.

(1) In the period 1910–20 several biochemists showed that dehydrogenases are active in minced animal tissues, transferring hydrogen atoms from certain organic acid ions known to occur in cells.

(2) In 1935 A. Szent-Györgyi produced a sequence of enzymic reactions showing the oxidation of the organic ion succinate, and it was then shown that citrate is converted to succinate in cells.

(3) At this stage Krebs stepped in to show that only certain organic acid ions are oxidised by cells, and that certain inhibitors could bring the oxidations to a halt. After much work he came up with the sequence we now know as the Krebs cycle. He also showed that all his suggested reactions could take place at a fast enough rate to account for the known use of pyruvate and oxygen by the tissue, implying that his pathway was the main if not the only pathway for the oxidation of foodstuffs.

Respiration of other substrates

There is not always sufficient glucose to provide all the ATP needed in a cell. When this is the case, other substances can be respired, in particular other carbohydrates and fats. It is easy to see that, for example, disaccharide sugars could be broken down and enter the pathway, but where do the other substances fit in?

Fats are an excellent source of energy. They are split by lipases into their constituent fatty acids and glycerol. The glycerol is phosphorylated and feeds into the glycolytic pathway as GALP. The fatty acids are passed through a series of reactions which split off 2-carbon sections of acetylco A. Hydrogen atoms are removed in the process and are passed through the electron transfer system to form ATP even before the acetylco A enters the Krebs cycle. The fatty acid then goes through the reaction series again to remove the next 2-carbon fragment. A single fatty acid can yield a large amount of ATP – stearic acid gives about 180 molecules. As you can see, the complete oxidation of fat gives a great deal of energy and this is why fat is a very important energy source, particularly for active tissues such as heart muscle, liver and kidneys.

Protein is not usually used as a respiratory substrate. It is broken down only when supplies of both carbohydrate and fat are very low – basically when the body perceives itself to be starving. The amino acids must be deaminated before the residues can be used as a substrate for respiration.

Figure 3.4.10 shows that the respiratory pathway is not only the route by which carbohydrates are broken down to provide energy. It also provides a sort of biochemical crossroads for interconversions from storage compounds such as fats, starch and glycogen to glucose, and also from carbohydrates to fats.

Even a relatively brief look at the biochemistry of photosynthesis and respiration such as we have taken in this section shows us both the complexity of the processes and the importance of controls over the rate at which the various reactions occur. This need for control is present in all aspects of living organisms, and in the next section we shall consider in more detail the ways in which this control can be brought about.

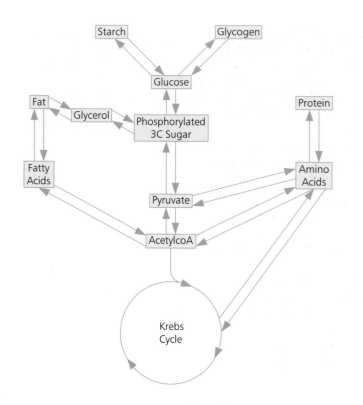

Figure 3.4.10 A biochemical 'Spaghetti Junction' – the respiratory pathways bring many strands of metabolism together.

SUMMARY

- The energy values of different foods can be determined by **calorimetry**.

- The minimum amount of energy taken in in food that will keep the body alive is the **basal metabolic rate** (**BMR**). The actual energy requirement of a person depends on age, sex and lifestyle.

- The energy in food molecules is transferred to energy in ATP for use by cells by the process of **cellular respiration**, which uses oxygen and produces carbon dioxide and water.

- The **respiratory quotient** (**RQ**) is the carbon dioxide produced divided by the oxygen used by a person in a given time period. RQ values can help show which food types are being respired predominantly at a given time.

- Cellular respiration is a complex series of reactions involving two stages – **glycolysis** which can take place without oxygen and produces a little ATP, and the **Krebs cycle** which uses oxygen and the product of glycolysis to produce much more ATP.

- **Aerobic organisms** respire by the reactions of the Krebs cycle so need oxygen, though they may be able to respire anaerobically for short periods. **Anaerobic organisms** rely on glycolysis only to produce their ATP. **Facultative anaerobes** respire anaerobically only when oxygen is not available, while **obligate anaerobes** can only respire anaerobically.

- Glycolysis takes place in the cytoplasm of the cell, while the reactions of the Krebs cycle take place in the matrix of the mitochondria. ATP production involves the electron transfer chain, which is sited on the stalked particles of the cristae in the mitochondria.

- In glycolysis, a molecule of glucose is biphosphorylated using two molecules of ATP and then splits to form two 3-carbon molecules which are **GALP** and another intermediate which can be converted into GALP. GALP is converted into **pyruvate**, a process which produces 2 molecules of ATP and one of $NADH_2$ for each GALP molecule. $NADH_2$ has the potential to generate 3 molecules of ATP via the electron transfer chain, if oxygen is available as the final hydrogen acceptor.

- Pyruvate may pass into the mitochondria for the Krebs cycle in aerobic conditions. In anaerobic conditions pyruvate is converted to **lactate** in animals. In plants it is fermented to **ethanol**.

- The rate of glycolysis is controlled by the **regulatory enzyme phosphofructokinase**. In the Krebs cycle the 3-carbon pyruvate is converted into 2-carbon **acetylcoenzyme A** with the loss of carbon dioxide and the production of $NADH_2$. Acetylco A combines with a 4-carbon acid to form 6-carbon **citrate**. This loses carbon dioxide to form a 5-carbon acid and another $NADH_2$. The 5-carbon acid loses another carbon dioxide to form the 4-carbon acid again. ATP is formed at this stage along with two molecules of $NADH_2$ and one of $FADH_2$. The 4-carbon acid can then combine with more acetylco A and the cycle repeats.

- The reduced hydrogen acceptors $NADH_2$ and $FADH_2$ channel electrons through the electron transfer chain to form ATP. The final acceptor is oxygen which is reduced to form water.

- Glycolysis yields 8 ATP molecules for each glucose molecule, and the Krebs cycle 30, giving a total of **38 ATPs per molecule of glucose**.

- In the absence of glucose, fats are converted to fatty acids and glycerol. Fatty acids are split into 2-carbon sections to form acetylco A and glycerol forms GALP, so both can feed into the Krebs cycle. A fatty acid molecule gives a great deal of energy in the form of ATP. In the absence of fats or carbohydrates, amino acids may be deaminated and respired.

- The respiratory pathway provides a means of interconversions from fats, starch and glycogen to glucose, and from carbohydrates to fats.

3

QUESTIONS

1 a What is cellular respiration?
b Where does cellular respiration take place?
c What is the function of each of the following in the metabolism of a cell?
  i NAD
  ii cytochromes
  iii oxygen

2 Describe the events in the breakdown of a molecule of glucose in the muscles of an athlete:
a in the absence of oxygen
b in the presence of oxygen.

3 a Where are the enzymes involved in glycolysis found?
b Where are the enzymes of the electron transport system found?
c What is the evidence for the stated positions of both sets of enzymes?

1 Outline the way in which each of the following obtain their carbohydrates:
a a green plant
b an herbivorous animal.

2 Read through and copy the following account of cellular respiration, and then write on the dotted lines the most appropriate word or words to complete the account.
The initial phase in the breakdown of glucose, a process known as .... , takes place in the .... of the cell and eventually results in the production of two molecules of .... from each molecule of glucose. In most organisms, this product then enters the second phase of cellular respiration, known as the .... cycle. This cycle occurs under .... conditions in specific organelles, called the .... During both phases, hydrogen atoms are removed from the substrate and passed to coenzymes such as .... These reactions are catalysed by enzymes called .... .In the respiratory process, energy is released and is used to synthesise energy rich molecules of .... from .... and .... , thereby storing energy for future use. **(Total 11 marks)**
(ULEAC June 1991)

3 The brewing of alcoholic drinks involves adding yeast cultures to solutions of sugars. At the end of the fermentation the excess yeast is removed from the reaction vessel and can be converted into animal feed.
a i Describe in outline how yeast ferments sugars into ethanol.
(3 marks)
ii State why the ethanol produced accumulates in the reaction vessel. **(1 mark)**
iii Suggest why fermentation of sugar by yeast does not normally produce ethanol concentrations higher than 12%.
(1 mark)
b An investigation was planned to find out if excess yeast from breweries could be used as a protein food source for pigs. Twenty young pigs were used for the trial, and fishmeal was used as an alternative protein source for purposes of comparison.
i Describe a suitable procedure for the investigation, and suggest precautions necessary to ensure valid conclusions.
(4 marks)
ii Suggest a suitable criterion by which the growth of the pigs could be measured. State *one* advantage and *one* disadvantage of this criterion. **(3 marks)**
c i Explain why different sources of dietary protein may lead to different patterns of growth in mammals such as pigs.
(2 marks)
ii Suggest *two* factors other than dietary protein which may affect the growth of mammals. **(2 marks)**

d The yeast used as pig food was found to be deficient in vitamin D (calciferol).
i Describe the physiological role of vitamin D in mammals.
(2 marks)
ii Suggest how the growth of pigs fed on yeast might be affected by a lack of vitamin D. **(2 marks)**
(Total 20 marks)
(ULEAC June 1991)

4 a Explain what is meant by the term 'heterotrophic nutrition'.
(4 marks)
b Describe how the following obtain their food materials:
i a herbivorous mammal **(7 marks)**
ii dodder. **(4 marks)**
c In what ways are saprophytes important in ecosystems?
(5 marks)
(ULEAC June 1991)

5 Below is a diagram representing some of the changes which occur in the light-dependent stage of photosynthesis in plants.

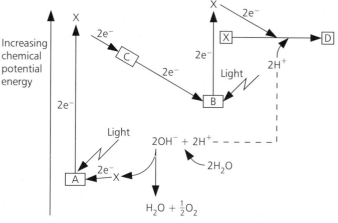

X represents a number of different electron acceptors.

a What terms are given to the light-absorbing complexes A and B?
(2 marks)
b What are the chemical substances represented by C and D?
(2 marks)
c i Besides oxygen, water and substance D, what other substance is produced in this process? **(1 mark)**
ii This substance and substance D are both used in the light-independent stage of photosynthesis together with carbon dioxide. Outline the steps in which these substances combine to form a sugar. **(4 marks)**
d i Excluding plants, which organisms can also photosynthesise?
(1 mark)
ii What process can these organisms perform that is of direct nutritional benefit to plants? **(1 mark)**

e What *two* methods of nutrition may bacteria use to exploit the biomass generated by plants? **(2 marks)**

f What other major group of organisms also uses plant material in *both* of these ways? **(1 mark)**

g Almost all living organisms use the products of photosynthesis for respiration.

   i What is the pathway called that oxidises pyruvate (α-keto-propanate) to carbon dioxide? **(1 mark)**

   ii Name *two* co-enzymes which are involved in hydrogen transport. **(2 marks)**

   iii Briefly describe how these two reduced co-enzymes are re-oxidised. **(4 marks)**

(Total 21 marks)

(O & C June 1992)

6 In 1965, uniform one-year-old oak seedlings were planted in two 1 m × 1 m plots (25 seedlings per plot) under light conditions which, in one plot, were approximately equivalent to 85% of full daylight and in the other to 30% of full daylight. Other conditions were similar for both groups. Annual measurements were made of the number and height of seedlings which survived. The results are given in the table below.

Year of measurement	Mean height/cm		Percentage survival	
	85% light	*30% light*	*85% light*	*30% light*
1965	6.3	6.8	100	100
1966	10.8	9.6	94	92
1967	14.4	11.5	92	84
1968	16.7	13.6	82	76
1969	20.4	18.1	74	32
1970	23.8	21.4	72	24
1971	32.9	22.5	72	18
1972	35.4	26.0	72	16

a Plot the data for the mean height on graph paper. **(4 marks)**

b **i** For each light intensity, calculate the yearly rate of increase in the mean height of the plants from 1969 to 1972. **(3 marks)**

   ii Calculate the mean rate of increase in height for each light intensity over this period. **(2 marks)**

c The results indicate that the yearly rate of increase varied over these years. Suggest *three* reasons for this variation. **(3 marks)**

d Suggest why:

   i the higher light intensity produced taller plants **(2 marks)**

   ii the mean height of the plants recorded in 1965 for the 85% light intensity was lower than that recorded for the lower light intensity. **(2 marks)**

e There was a large increase in growth during the year 1970–71 at 85% light intensity but a smaller increase, compared with that of other years, at 30% light intensity. Suggest reasons to explain this. **(3 marks)**

f Would the difference in survival between the two light intensities have an effect on the mean heights recorded for the plants? Give a reason for your answer. **(2 marks)**

(Total 21 marks)

(O & C June 1992)

7 Hundreds of different chemical changes are involved in digesting and absorbing food. To ensure the well-being of the gut, these changes must be carefully coordinated.

[Keith Burdett, *Biological Sciences Review*, Vol. 3 No. 4]

a Define *digestion*. **(2 marks)**

b List *three* ways in which the secretion and activation of enzymes is controlled. **(3 marks)**

c **i** Proteins are digested in the mammalian gut in two stages: by endopeptidases and then by exopeptidases. Explain, with *two* examples of each, the differences between these two groups of enzymes. **(6 marks)**

   ii What is the functional advantage of digesting proteins in two stages? **(1 mark)**

d Compare the immediate fate, after absorption, of the products of protein and carbohydrate digestion with that of the products of fat digestion. **(4 marks)**

e How does bile assist digestion? **(2 marks)**

f Indicate *two* ways in which the gut is protected from damage by its own digestive processes. **(2 marks)**

(Total 20 marks)

(O & C June 1992)

8 The efficiency of food absorption was investigated in two desert herbivores of similar size, one a reptile, the other a mammal. The table below shows some of the findings.

Aspect studied	Reptile	Mammal
Average daily food intake (g dry mass)	0.55	9.7
Retention time of food in gut (hours)	125	4
Length of small intestine (cm)	20	42
Gross internal area of small intestine (cm^2)	18	40
Microscopic area of small intestine (cm^2)	72	506
Extraction efficiency (%)	47	52

Note: Extraction efficiency is derived from:

$$\frac{\text{food dry mass} - \text{faeces dry mass}}{\text{food dry mass}} \times 100$$

a Explain briefly any *two* practical precautions you would take to ensure meaningful comparisons in such an investigation. **(2 marks)**

b Suggest briefly how retention time could have been determined. **(2 marks)**

c **i** Explain why the investigation was concerned especially with the small intestine.

   ii Suggest a reason for the difference between the gross and microscopic internal area of the small intestine. **(2 marks)**

d Account for the difference in average daily food intake between the two animals given that they both eat the same sort of food. **(2 marks)**

e **i** State *one* problem that you would expect to cause inaccuracy when estimating the extraction efficiency.

   ii Do you find it surprising that the extraction efficiencies of the two animals are comparable? Give reasons for your answer. **(3 marks)**

(Total 11 marks)

(NEAB June 1989)

9 Fifty discs were cut from the frond of a brown seaweed using a cork borer. These discs were divided into five sets of ten, each set being placed in a beaker containing 50 cm³ of seawater. All the discs sank to the bottoms of the beakers. The beakers were then placed at different distances from a lamp. The times taken for *three* discs in each beaker to rise to the surface were recorded. The results are shown in the table.

Distance of beaker from lamp/cm	Mean time taken for three discs to rise to surface/minutes
5	8
7	8
10	14
15	36
20	62

a i Explain why the discs rose when the beakers were illuminated.
  ii Suggest *two* different reasons why discs *in the same beaker* took different times to rise.
  iii Describe how the heating effect of the lamp might be prevented from interfering with this type of experiment.
  (5 marks)

The beakers were then left until at least five discs in each beaker had risen to the surface. The lamp was then switched off, and the times for three discs to sink to the bottom of each beaker were taken. These times were similar for all the beakers.

b i Suggest why the discs eventually sank when they were no longer illuminated.
  ii Suggest why the times were similar for all the beakers.
  (2 marks)

The data from the table were processed to produce the following graph.

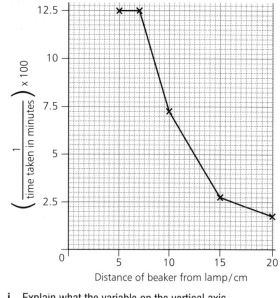

c i Explain what the variable on the vertical axis
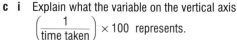
$\left(\dfrac{1}{\text{time taken}}\right) \times 100$ represents.

ii From the graph, determine the mean time you would predict for three discs to rise to the surface of a beaker placed 12.5 cm from the lamp. Show any calculations you make.
iii The light intensity on the beaker with the lamp 10 cm away was *X* units. What would be the value of the light intensity with the lamp 20 cm away from the beaker? **(5 marks)**
(Total 12 marks)

(NEAB June 1991)

10 Isolated muscle was used in a series of experiments to discover the sequence of reactions involved in a particular respiratory pathway. The tissue was maintained in a suitable bathing medium and provided with various substances thought to be intermediates in the pathway. The rate of respiration after each substance had been added was measured by the fall in oxygen concentration over a set time.

a i Suggest *one* advantage of using isolated muscle tissue instead of a whole organism.
  ii Could a suspension of mitochondria have been used instead? Explain your answer. **(3 marks)**
b i State *one* assumption made in using oxygen uptake as the measure of respiration rate.
  ii Name *two* factors, other than oxygen uptake, which could have been used instead to indicate rate of respiration.
  iii Describe *two* components of the bathing medium, other than water, which would be needed to keep the muscle tissue alive throughout the experiment. **(5 marks)**

It was found that the initial rapid rate of respiration in isolated muscle preparations could be restored by adding small amounts of succinate or fumarate. Table 1 shows the effects of these additions.

| Substance added | Change in level of substance in muscle | | |
	Succinate	Fumarate	Citrate
Succinate	—	increase	increase
Fumarate	increase	—	increase

Table 1

Two different reaction pathways were suggested to explain the observed results.
Hypothesis I: succinate → fumarate → citrate (Succinate is converted to fumarate, which is later converted to citrate.)
Hypothesis II: fumarate → succinate → citrate (Fumarate is converted to succinate, which is later converted to citrate.)
In an attempt to eliminate one of these hypotheses, a chemical was added which inhibits the enzyme catalysing the reversible reaction between succinate and fumarate.

c i Copy and complete Table 2 to show the effects you predict (fall, rise or no change) in *citrate level* when mixtures of succinate and the inhibitor, or fumarate and the inhibitor, are added to muscle that has reached a slow rate of respiration.

| Hypothesis assumed to be correct | Change in level of citrate in muscle | |
	when succinate plus inhibitor added	when fumarate plus inhibitor added
I		
II		

Table 2

ii Suggest *one* chemical effect (other than a reduction in the rate of respiration) that could be expected if the living muscle were supplied with the inhibitor on its own. **(5 marks)**
(Total 13 marks)
(NEAB June 1991)

11 Figure 1 illustrates the effect of light intensity on the rate of photosynthesis of a suspension of *Chlorella* cells, at combinations of two different temperatures, and two carbon dioxide concentrations.

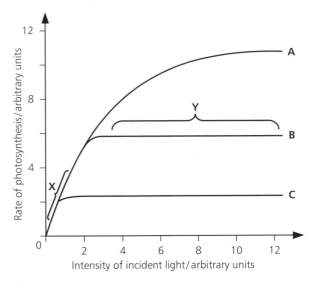

Figure 1

A, 25 °C and 0.4% CO_2
B, 15 °C and 0.4% CO_2
C, 25 °C and 0.01% CO_2
a i What factor is limiting the rate of photosynthesis in the zone marked **X** on the graphs? **(1 mark)**
ii Suggest an explanation for the plateau region marked **Y**. **(2 marks)**
iii The plateau in the rate of photosynthesis, in natural environments, is reached at a light intensity of about 100 W m^{-2}. The amount of sunlight falling on a clear summer day at midday in many parts of the world is about 1000 W m^{-2}. What does this suggest about light as a factor affecting the rate of photosynthesis? **(3 marks)**
b Explain the fact that at low light intensities the rate of photosynthesis is the same at 15 °C as at 25 °C, but at higher light intensities the rate of photosynthesis is much higher at 25 °C than at 15 °C. **(4 marks)**
c The average carbon dioxide content of the atmosphere is 0.035%. Using this fact, and the information given in Figure 1, what conclusions can be made about carbon dioxide, as a limiting factor in photosynthesis in natural environments? **(3 marks)**
d State *two* external factors, other than light intensity, carbon dioxide and temperature, which might limit the rate of photosynthesis. Indicate how each might limit the rate. **(4 marks)**
e Name the direct products of the light reactions of photosynthesis. **(1 mark)**
(Cambridge June 1991)

12 a With reference to photosynthesis, explain what is meant by
i *absorption spectrum* and ii *action spectrum*. **(5 marks)**
b Describe, giving practical details, how you would measure the rate of photosynthesis of an aquatic plant in light of different wavelengths. **(11 marks)**
c Suggest why brown algae can grow at greater depths than green algae. **(4 marks)**
(ULEAC June 1990)

13 A person eats a piece of cheese consisting mainly of fats and protein. Describe the processes which enable these fats and proteins to be digested and absorbed. **(20 marks)**
(NEAB June 1991)

14 a Define the term 'hydrolysis'. **(2 marks)**
b Describe the processes involved in the digestion of a piece of lean meat by a mammal. How are the products of this digestion absorbed? **(13 marks)**
c Plants contain starch and lipids as storage compounds. How can these be made available as respiratory substrates? **(5 marks)**
(ULEAC January 1991)

15 Read through and copy the following account of what happens during the early stages of digestion in humans, and then write on the dotted lines the most appropriate word or words to complete the account.
Saliva in the human mouth contains the enzyme .... which catalyses the hydrolysis of .... to reducing sugars. It also contains .... which lubricates the food and enables it to be made into the .... which is swallowed. After swallowing, the low pH of the stomach causes the enzyme in saliva to be .... by alteration of its .... structure. This low pH also activates the precursor .... of the enzyme which catalyses hydrolysis of .... bonds in protein. **(Total 8 marks)**
(ULEAC June 1990)

CONTROL SYSTEMS 4

Communication

For large organisms, communication between the cells of the body is extremely important. In the animal world it is not just the coordination of cellular activities that matters — communication between individual animals and between groups of animals becomes increasingly important too.

Social behaviour

The development of a variety of methods of communication is most clearly seen in those organisms which organise themselves into social groups. A large number of caterpillars may be found on a single leaf, but they do not interact in a cooperative, social way. They are simply in the same place because that is where the female laid her eggs. On the other hand, a fish swimming in a school is far less likely to be eaten than a fish swimming individually, and a wolf in a pack is more likely to capture prey than a solitary individual. These are **social animals** and they exhibit **social behaviour**, interacting with individuals in the group. The cornerstone on which all social behaviour is built is the ability to communicate.

The need to communicate

The great advantage of sociality can be summed up as **strength in numbers**. An individual antelope is less likely to be caught unawares by a predator because it is depending not only on its own ears, eyes and nose to detect evidence of the hunter, but on the sense organs of the whole herd. Similarly hyenas, hunting in groups, are far more likely to be successful in overcoming a sizable prey animal than would be an individual hunting alone. But in social situations such as these there must be effective communication – any antelope detecting a predator must relay that information to the rest of the herd, and the pack of hyenas must all chase the same animal.

What is communication?

Communication – the transmission of information from one organism to another – is a specialised kind of behaviour. It is behaviour produced by one individual which alters the behaviour of another individual or group of individuals. For communication to be successful the communicator must possess a signalling device and the recipient must have appropriate sensory organs. A loud roar,

Table 1 The pros and cons for an individual animal of living in a society

Advantages	Disadvantages
Better detection of, escape from or repulsion of predators	More competition with other members of the group for food, water, breeding sites and other resources
Better defence of limited resources	Higher risk that group members will kill an individual's progeny
More efficient foraging	Higher risk that group members will exploit parental care
Better care of offspring through communal feeding and protection	Higher risk that disease and parasites will be passed through the group

for example, would be useless unless other animals possessed ears capable of hearing the sound and brains capable of interpreting it.

The most basic types of communication consist of a simple **sign stimulus** which releases a particular behaviour in another individual. Many female moths, for example, produce a chemical known as a female sex pheromone. A male moth detecting only a few molecules of the pheromone will immediately begin to fly towards the source of the pheromone to mate with the female.

The basic patterns of species-specific behaviour, with the addition of some intricate learned behaviours, have enabled an enormous number of more complex, socially useful communication methods to evolve. These include visual, acoustic, tactile and chemical signals along with the sense organs necessary to receive the messages.

Insect communication

Certain groups of insects are well known for their social structures. Ants, termites, some wasps and bees live in complex communities with individuals carrying out different tasks for the benefit of the whole community. Sounds and chemical messages make up an important part of the insect communication system. Ants follow scent trails across the woodland floor, for example. But one of the most astonishing examples of complex communication in the insect world is that of the 'waggle dance' of the honeybee. Honeybees live in communities either in an artificial hive or in a natural situation such as a hollow tree. There is one fertile female, the **queen bee**, up to 1000 fertile male bees or **drones** whose main function is to mate with the queen, and the remainder of the community (up to 50 000 bees) is made up of infertile

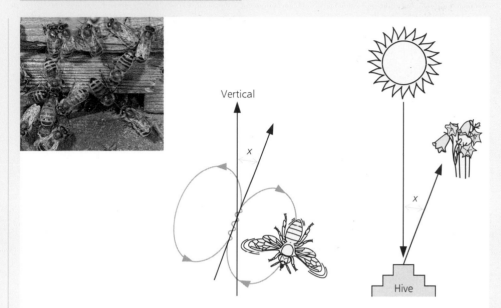

Figure 1 The waggle dance of the honeybee. Bees accumulate complex information about their food source and then convey it to other members of the hive by dancing. The length of the waggle run gives information about the distance from the hive, and the orientation to the vertical indicates the angle between the Sun and the food source. The other bees can even learn of the type and amount of the food – lots of waggling means lots of food!

females known as **workers**. They run the hive, caring for the developing larvae, cleaning the hive, building the comb and foraging for food.

Foraging bees have been shown to memorise details about a food source which include its precise location, geometry with respect to the hive and the odour of the food. For this sort of detailed information to be of use in the social context, the bee must be able to communicate it to the others. This communication is brought about by a complex dance (see figure 1) whereby the foraging bee informs others of both the distance of the food source from the hive and its direction. The message is received as a tactile stimulus – in the dark hive interior it could not be seen – and the other bees press against the dancing worker to pick up the message.

Placed away from the hive, honeybees can navigate both to food sources and back to the hive. Even though the brain of a bee is extremely small, bees are thus obviously capable of

more than simply innate patterns of behaviour – learning is taking place too. Highly developed communication systems such as those shown by bees are necessary for any large and complex social community to work.

Communication between birds

Communication in the bird world is based on sound and sight, as illustrated in figure 2. Birdsong is well known and used for a variety of purposes. The alarm call of many birds, sounded to warn of an approaching predator, is very similar. As a result, birds from a number of species benefit from the alarm of a single bird. Birds can also benefit from the alarm calls of other animals. Hummingbirds do not have an alarm call of their own but respond to the chattering of chipmunks and ground squirrels when danger is near. Visual displays are also very important in the communication of birds. The wide gaping bill of the newly hatched chick triggers the parent bird to feed its offspring.

Many birds have elaborate courtship rituals with movements and coloration both to attract females and to warn off possible rivals. Visual displays in birds with small breeding territories have the advantage of avoiding constant physical battles for space.

Many species of birds have very specific songs. It was originally thought that these were instinctive,

Figure 2 A combination of sounds and visual cues enables birds to communicate effectively.

but it has since been shown that birds sing with local dialects which have to be learned. There is an innate tendency to sing, but to learn their specific song baby birds must hear not only the parent birds singing during the early stages of their development but also their own efforts as they learn to sing, as shown in figure 3.

Communication in humans

For thousands of years human beings have lived in groups and have accomplished an enormously wide variety of developments, many of which could never have taken place without our advanced ability to communicate. At first thought 'human communication' means language, and this is of course of vital importance. But there are many strands to human communication, some of which have clear origins in the communications of other animal groups.

Pheromones (the type of chemicals used for communication in insects as already mentioned) play a largely unrecognised part in our everyday lives. Chemical signals give out information about our sex and our mood, and we respond to these messages from others without knowing it. Experimenters sprayed certain chairs in a doctor's waiting room with female pheromones. The chairs were otherwise no different from the other chairs in the room. Male patients always selected to sit in these seats. When male pheromones were applied, female patients sat in the sprayed seats! Thus our interactions with other people are affected by communication systems of which we are unaware.

Tactile messages are also important in basic human communication. Gestures of comfort to small children almost always involve the close physical contact which can also be observed in the interactions between many primates and their offspring, whilst modified forms of these gestures are used between adults to give similar messages. Visual cues too, such as facial expressions, gestures and body

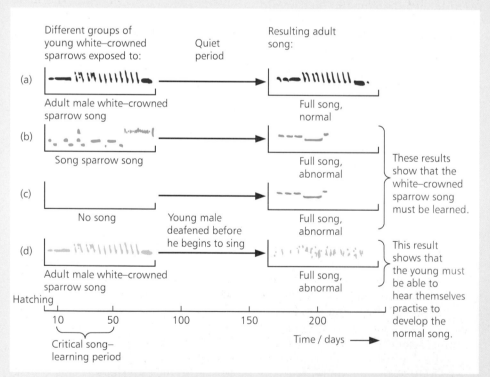

Figure 3 White-crowned sparrows learn their songs between 10 and 15 days of age. The results of these experiments showed that the birds need to hear their parents singing to be able to learn the song of their species, and that they also need to be able to hear their own efforts and then match them to the original. It is interesting to see that after the exposure of the birds to the adult song there is a quiet period where they assimilate the information before attempting the song themselves. If the birds do not learn their particular song during this early period of their lives, they cannot learn it later on, however many times they are exposed to it.

postures convey a great deal of information. Received by the sense organs, these messages are carried to the brain where they are interpreted in the light of past experience. We do not need to be told verbally when someone is bringing bad – or good – news. The posture and way of moving as well as the facial expression communicate this clearly. Experience affects our interpretation of non-verbal communication, however. The child who is used to affection and love will see an outstretched hand as welcoming and take hold of it happily. The child used to physical cruelty will see the same gesture and shrink away.

Human babies show patterns of behaviour which seem to be species specific – the patterns of early development are strikingly similar in all human infants. For example, crying is a universal message of distress. A crying infant first and foremost demands the reassurance of physical contact, and then food, or changing, or warmth. All babies, even if they are profoundly deaf, cry. It is a form of communication which seems to have evolved to ensure that adults notice the needs of babies. Similarly smiling, which develops in the first weeks of life, occurs regardless of whether a baby can see. The first smiles are not a mimicking of adult expressions, they are a form of innate communication which seems to act as a 'reward' to guarantee the infant the continued interest and care of the adults around.

Much of this non-verbal communication can also be seen in the communications of other animal groups. It is in the development of language that humans seem to differ most from other species. The human ability to express ideas and discuss thoughts, to write music and to learn immensely complex sequences of behaviour does seem to be unique to our particular species. But the mechanisms by which this language is acquired are far from unique. Biologists are virtually certain that the human ability to acquire language is innate (species specific). For example, all the thousands of human languages draw on the same 40 consonant sounds (standard English uses about two dozen of them) and every normal human baby can distinguish them all. As the child matures and gains experience of one particular language the ability to distinguish the other consonants fades and seems to be completely gone by the age of two.

Just like songbirds, in humans there seems to be a sensitive period during which each individual assimilates the basic elements of the native language. Sounds not heard during this sensitive period are always difficult to learn in future. The babbling of babies is also thought to demonstrate the innate roots of language. All babies babble, even if they are totally deaf. Babies who can hear learn during this period to control their voices and make first the sounds and then the words of their language. The deaf baby, unable to hear its own babbling, makes no progress towards talking and gradually ceases to babble. But once a child can talk, subsequent deafness will not remove the ability to speak.

A combination of sophisticated signalling mechanisms, effective sensory organs and a highly developed brain enables people all over the world to communicate with each other. The ability of humans to speak, to communicate complex ideas and, what is more, to write them down and so communicate them to future generations, has made it possible for the human race to reach its current state of world domination. Whether it will also enable that domination to continue, or will lead to the destruction of much of the planet which supports us, only time and future history books will tell.

Figure 4 Human language provides us with a very sophisticated form of communication, but even without words we can make our basic feelings known.

All living organisms have a basic need for control systems. The complex biochemistry of all cells needs to be controlled to ensure that the right products are available at the right time. Processes such as digestion in animals need to be controlled to make sure that food is digested and body tissues are not. Growth must occur at the appropriate time and place, food and mates must be found and danger avoided. One of the most important requirements for controlling all of these factors is a system of rapid internal communication.

CHEMICAL AND NERVOUS COMMUNICATION

Messages may be transmitted in two main ways within a living organism to provide a communication system. Specific chemical messengers may be released by cells. Within single-celled organisms these can carry information by diffusion very rapidly over the small distances involved. Chemical messages are used for communication between the different parts within a plant, and some aspects of control in animals are carried out in this way too. We shall be looking at chemical control in more detail in sections 4.4 and 4.5. But for many animals a relatively large size, combined with the need to move around and react rapidly to the external as well as the internal environment, means that chemical communications alone are not adequate. Such organisms have evolved a **nervous system** which uses electrical rather than chemical signals. Even simple, sedentary animals such as sea anemones have systems of specialised cells known as **nerves**, whose function is to carry electrical messages around the organism in a faster and more targeted way than would be possible with a chemical message.

WHAT IS A NERVOUS SYSTEM?

A nervous system is made up of interconnected **nerve cells** specialised for the rapid transmission of messages throughout the organism. The nerve cells carry messages from special **receptor cells**, giving information to the organism about both the internal and the external environment. Nerve cells also carry messages to specialised **effector cells** – often muscles – which then bring about the appropriate response.

The organisation of nervous systems

At its most simple, a nervous system consists of receptor cells and effector cells connected by nerve cells, as shown in figure 4.1.1. However, in many organisms the nervous system is much more complex than this. Groups of receptors have evolved to work together in **sensory organs** such as the eye and the ear. Simple diffuse nerve nets are replaced by complex nerve pathways. Some nerves carry information in one direction only, from the internal or external environment into the central processing areas of the nervous system. These are known as **sensory nerves**. **Motor nerves** carry messages only to the effector organs. More specialised concentrations of nerve cells develop in some animals to give rise to a **central nervous system (CNS)**, an area where incoming information is processed and coordinated and from where messages

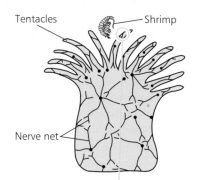

The tentacle of a sea anemone detects the presence of prey. Messages are sent through the nerve net to control the other tentacles, causing them to bend over and help to capture the shrimp.

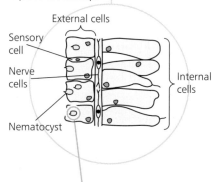

Messages travelling through the nerve net also stimulate the firing of the 'attack or defence' system of the sea anemone. Cells called nematoblasts may release a poisoned dart from their nematocyst to paralyse the prey or sticky threads to help hold it tight.

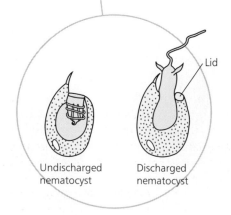

Figure 4.1.1 The nerve net of the sea anemone allows coordinated action by the whole animal.

are sent out into the motor nerves. In vertebrates the central nervous system consists of the **brain** and **spinal cord**.

Nerve cells

Nerve cells or **neurones** are the basic unit of a nervous system – millions of neurones work together as an integrated whole in mammals such as ourselves. Neurones are cells specialised for the transmission of electrical signals (**impulses**). They have a cell body which contains the cell nucleus, mitochondria and other organelles along with **Nissl's granules**, prominent groups of ribosomes for protein synthesis. The cell body has slender finger-like processes called **dendrites** which connect with neighbouring nerve cells. The most distinctive feature of all nerve cells is the **nerve fibre**. This is called the **axon** if it transmits impulses away from the cell body, and the **dendron** if it transmits impulses towards the cell body. The nerve fibre is extremely long and thin, as shown in figure 4.1.2, and carries the nerve impulse.

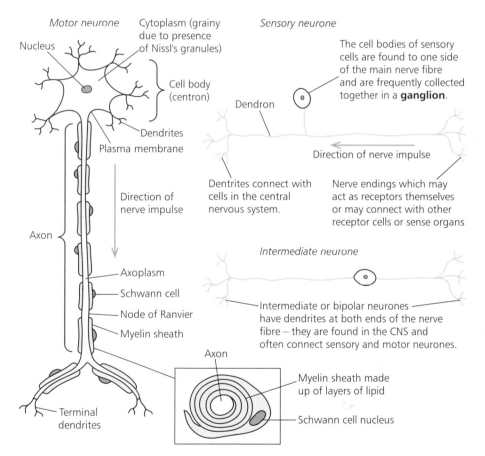

Figure 4.1.2 All nerve cells have the same basic structure of a cell body, dendrites and a nerve fibre. The detailed arrangements vary in motor, sensory and intermediate neurones.

Some vertebrate neurones are associated with another specialised type of cell, the **Schwann cell**. The Schwann cell membrane is wrapped repeatedly around the neurone, forming a fatty layer known as the **myelin sheath**. This sheath has gaps in it known as the **nodes of Ranvier**. The myelin sheath is important for two reasons – it protects the nerves from damage, and speeds up the transmission of the nerve impulse, as we shall see later.

Nerve fibres are bundled together to form **nerves**. Some carry only motor fibres and are known as **motor nerves**, some carry only sensory fibres and are known as **sensory nerves**, whilst others carry a mixture of motor and sensory fibres and are called **mixed nerves**.

Speed of transmission in neurones

The role of neurones is the passage of electrical messages from one area of an organism to another as fast as possible. The speed at which the messages can be carried depends largely on two things. The first is the diameter of the nerve fibre. In general, the larger the fibre, the more rapidly impulses travel along it. The second is the presence or absence of a myelin sheath. Myelinated nerve fibres can carry impulses much faster than unmyelinated ones.

Invertebrates do not have myelin sheaths on any of their nerve fibres, and many of their fibres are less than 0.1 mm in diameter, so in general invertebrate nerve impulses travel quite slowly, at around 0.5 m s^{-1}. But there are times when even a relatively slow-moving invertebrate needs to react quickly to avoid danger, and to allow for a more rapid passage of impulses many groups have evolved **giant axons**. These are nerve fibres with diameters of around 1 mm which allow impulses to travel at around 100 m s^{-1}, fast enough for most escape strategies to have a chance of success.

Vertebrates have both myelinated and unmyelinated nerves. The voluntary motor nerves that transmit impulses to voluntary muscles, for example to control movement, are myelinated while the autonomic nerves that control involuntary muscles such as those in the digestive system have some unmyelinated fibres. There is more about the voluntary and autonomic nervous systems in section 4.2. The effect of the myelin sheath is to speed up the transmission of a nerve impulse without the need for giant axons. A more versatile network of relatively small nerve fibres can carry messages extremely rapidly, at speeds of up to 120 m s^{-1}.

NERVE IMPULSES

What is a nerve impulse?

The nervous system carries nerve impulses very rapidly from one part of the body to another. But what is a nerve impulse? The attempt to discover the answer to this question began many years ago. Long before Georg Ohm and Michael Faraday made their contributions to the understanding and measurement of electricity at the beginning of the nineteenth century, two other famous scientists had a dispute over the electrical activity of the body. In 1791 Luigi Galvani discovered that the muscles in severed frogs' legs twitched when touched by brass and iron simultaneously. He thought this was the result of what he called 'animal electricity'. A few years later Alessandro Volta showed that the effect was in fact due to the difference in electrical potential between the two metals, and nothing to do with animal electricity in the muscles at all. This dispute at the end of the eighteenth century was the starting point for a huge range of experiments in physiology, physics and physical chemistry which has led to our present day understanding of the nature of the nerve impulse.

The nerve impulse is a minute electrical event which is the result of charge differences across the membrane of the nerve fibre. It is based on ion movements through specialised protein pores and by an active transport mechanism. To look at the events of a nerve impulse we shall consider a 'typical' axon – ignoring for the moment size, myelination or type.

The resting neurone

The membrane of an axon, like any other cell surface membrane, is partially permeable. It is the difference in permeability of this membrane to sodium and potassium ions which sets neurones apart from other cells and gives them their special conducting properties. The axon membrane is relatively impermeable to sodium ions, but quite freely permeable to potassium ions. It also contains a very active sodium/potassium pump which uses ATP to move sodium ions out of the axon and potassium ions in. The effect of this is to reduce the concentration of sodium ions inside the axon – they are pumped out and cannot diffuse back in. At the same time, potassium ions are moved in – but then diffuse out again along a concentration gradient. As a result, the inside of the cell is left slightly negatively charged relative to the outside – it is **polarised**, as shown in figure 4.1.3. There is a potential difference across the membrane of –70 mV which is known as the **resting potential**.

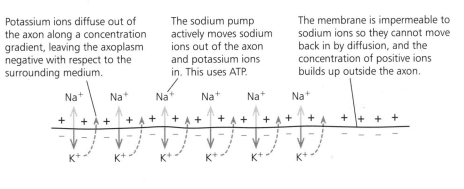

Potassium ions diffuse out of the axon along a concentration gradient, leaving the axoplasm negative with respect to the surrounding medium.

The sodium pump actively moves sodium ions out of the axon and potassium ions in. This uses ATP.

The membrane is impermeable to sodium ions so they cannot move back in by diffusion, and the concentration of positive ions builds up outside the axon.

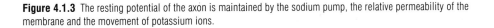

Figure 4.1.3 The resting potential of the axon is maintained by the sodium pump, the relative permeability of the membrane and the movement of potassium ions.

The active neurone

The resting potential represents the normal situation in the nervous system. What happens when an impulse travels along an axon? The key event in an active nerve fibre is a change in the permeability of the cell surface membrane to sodium ions. This change occurs in response to a **stimulus** which in a living organism could be one of a variety of things – light, sound, touch, taste or smell, for example. In the experimental laboratory situation the stimulus is usually a minute and precisely controlled electrical impulse.

When a neurone is stimulated the axon membrane shows a sudden and dramatic increase in its permeability to sodium ions. Specific **sodium channels** or **sodium gates** open up, allowing sodium ions to rush in along both concentration and electrochemical gradients. As a result the potential difference across the membrane is briefly reversed, the cell becoming positive on the inside with respect to the outside. This **depolarisation** lasts about 1 millisecond. The potential difference across the membrane at this point is about +40 mV. This is known as the **action potential**.

At the end of this brief depolarisation, the sodium channels close again and the excess sodium ions are rapidly pumped out by the sodium pump. Also, the permeability of the membrane to potassium ions is temporarily increased so that potassium ions diffuse out along an electrochemical gradient. It takes a few milliseconds before the resting potential is restored and the nerve fibre is ready to carry another impulse, as figure 4.1.4 shows. It is this **refractory period** which ensures that the nerve impulse only travels in one direction. Until the resting potential is restored, the part of the nerve fibre that the impulse has just left cannot conduct another impulse, so the impulse can only continue travelling in the same direction.

An action potential passing along an axon

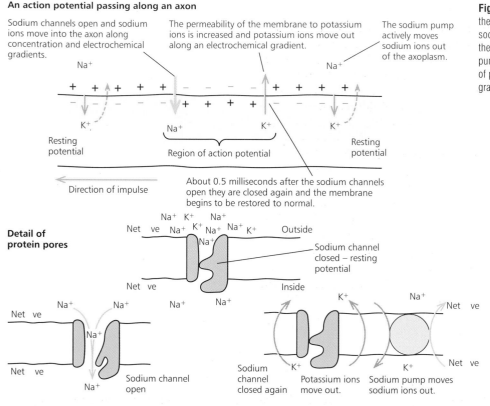

Sodium channels open and sodium ions move into the axon along concentration and electrochemical gradients.

The permeability of the membrane to potassium ions is increased and potassium ions move out along an electrochemical gradient.

The sodium pump actively moves sodium ions out of the axoplasm.

Resting potential

Region of action potential

Resting potential

Direction of impulse

About 0.5 milliseconds after the sodium channels open they are closed again and the membrane begins to be restored to normal.

Detail of protein pores

Sodium channel closed – resting potential

Sodium channel open

Sodium channel closed again

Potassium ions move out.

Sodium pump moves sodium ions out.

Figure 4.1.4 The action potential is brought about by the movement of sodium ions through the opened sodium channels. The resting situation is restored by the closing of the channels, the action of the sodium pump removing excess sodium ions and the movement of potassium ions out along an electrochemical gradient.

Evidence for the nerve impulse mechanism

Some of the most convincing evidence for this model of the nerve impulse comes from work done using poisons. A metabolic poison such as dinitrophenol prevents the production of ATP. It also prevents the nerve fibre from functioning properly. But how does this confirm the active pumping out of sodium ions and the movement of potassium ions due to differential permeability?

(1) When an axon is treated with a metabolic poison the sodium pump runs down and the resting potential is lost at the same rate as ATP is used up by the poison. This suggests that ATP is being used to power the pump – when it runs out, the pump no longer works.

(2) If the poison is washed away, the metabolism returns to normal and ATP production begins again. The resting potential is restored, suggesting that the sodium pump started up again with the return of ATP (see figure 4.1.5).

(3) If a poisoned axon is supplied with ATP by experimenters, the resting potential will be at least partly restored. This again confirms our model, suggesting that the poison is acting by

depriving the sodium pump of energy rather than by interfering with the membrane structure and its permeability. If the latter were the case, then supplying ATP to the pump would have no effect because ions would move freely across the membrane and a potential difference could not be maintained.

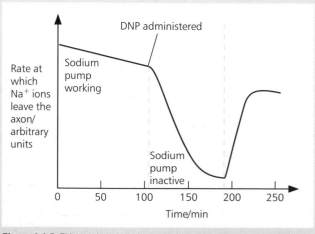

Figure 4.1.5 This graph, based on work by Hodgkin and Keynes in 1955, illustrates clearly the effect of dinitrophenol on the removal of sodium ions from the axon of a cuttlefish.

Investigating nerve impulses

Because the nerve impulse is an electrical event, albeit a very small one, an effective way of investigating it is to record and measure the electrical changes in a nerve. This is done using apparatus sensitive to small electrical changes, usually a **cathode ray oscilloscope**. A pair of recording electrodes is placed on a nerve which is then given a controlled stimulus. The impulses which result are detected by the electrodes and passed into an amplifier which magnifies them. They are then passed to the oscilloscope and displayed on a screen.

Much of the earliest work on nerves and nerve impulses was carried out using this method. The recordings of nerve impulses were taken from the outsides of entire nerves, made up of large numbers of different nerve fibres. These fibres are of varying diameter and sensitivity, and so the results of the recordings can be difficult to interpret. As most nerve fibres are around 20 μm in diameter, making a recording from inside is not an easy procedure. The breakthrough came around 40 years ago when Alan Hodgkin and Andrew Huxley began work on the giant axons of the squid. These unmyelinated nerve fibres are around 1 mm in diameter. They supply the mantle muscles of the squid and allow for very rapid nerve transmission in situations when the squid needs to move quickly, either when hunting or when escaping. The development of electrodes which could be inserted inside these giant axons allowed a far greater understanding of the events of the action potential.

Action potentials – the inside story

For work on the squid giant axons, very fine glass microelectrodes were made which could be inserted into the giant axon. Another electrode recorded the electrical potential from the outside. This combination of an internal and an external electrode allowed the electrical changes during the passage of an individual nerve impulse to be accurately recorded for the first time, as figure 4.1.6 shows. Since the early revolutionary work by Hodgkin and Huxley, they and others have refined the techniques so that now internal electrodes can be used with almost any nerve fibre.

Figure 4.1.6 Comparing the traces for the internal electrode with that made with external electrodes from a whole nerve shows the impact of internal electrodes on our understanding of nerve impulses.

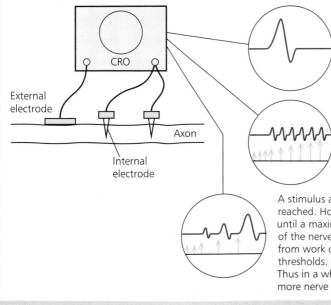

The events of a single action potential (also known as the 'spike' due to the shape of the oscilloscope trace). The trace shows clearly the change in the potential difference with the inrush of sodium ions followed by a return to the resting potential as the permeability of the membrane changes again.

A stimulus applied to a single axon gets no response until it reaches a certain threshold level. Beyond that threshold, the size of the response is always the same. However much the stimulus increases in size, the impulse in the nerve fibre is identical. This is the 'all-or-nothing law' – a nerve fibre either carries an impulse or it does not.

A stimulus applied to a whole nerve also gets no response until a threshold is reached. However, as the strength of the stimulus increases, so does the response until a maximum is reached. Without further knowledge, it looks as if the strength of the nerve impulse increases in response to the strength of the stimulus. But from work on individual axons we know that different nerve fibres have different thresholds. Once the threshold is reached they obey the 'all-or-nothing law'. Thus in a whole nerve, as the stimulus is increased, the threshold of more and more nerve fibres is reached until all the fibres in the nerve are responding.

A closer look at the nerve impulse

Work using techniques such as those just described has allowed us to build up the detailed picture of the events of the action potential shown in figure 4.1.7, including the timing and the cause of the refractory period. It has also led to an understanding of many other observed aspects of nervous transmission.

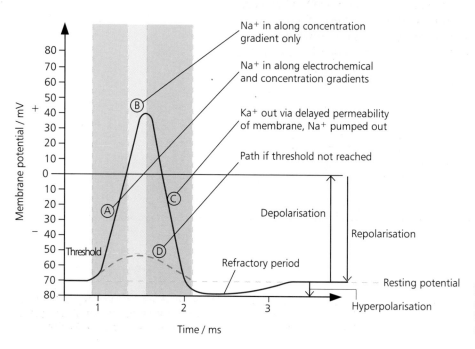

Figure 4.1.7 A representation of an action potential combining information from recordings made using an internal electrode and the movements of ions which cause the observed electrical changes

The threshold

The **threshold** for any nerve fibre is the point at which sufficient sodium channels open such that the rush of sodium ions into the axon is greater than the outflow of potassium ions. Once the threshold has been reached, the action potential occurs. The size of this action potential is always the same – it is an **all-or-nothing** response.

The refractory period

The **refractory period** is the time it takes for an area of the axon membrane to recover after an action potential, that is, the time it takes for ionic movements to repolarise the membrane and restore the resting potential. As we have seen, this is brought about by the sodium pump and the membrane permeability to potassium ions. For the first millisecond or so after the action potential it is impossible to restimulate the nerve fibre – the sodium channels are completely blocked and the resting potential has not yet been restored. This is known as the **absolute refractory period**. After this there is a period of several milliseconds during which the nerve fibre may be restimulated, but it will only respond to a much stronger stimulus than before – the threshold has effectively been raised. This is known as the **relative refractory period**.

The refractory period is important in the functioning of the nervous system as a whole. It limits the frequency with which impulses may flow along a nerve fibre to 500–1000 each second. It also ensures that impulses flow in only one direction along nerves, making it possible to have motor and sensory systems with no internal confusion.

The propagation of the nerve impulse

So far we have considered the action potential as an isolated event in one area of a nerve fibre. In fact, once an action potential has been set up in response to a stimulus, it will travel the entire length of that fibre, which may be many centimetres or even metres long. How is the impulse **propagated** (spread)?

The movement of the nerve impulse along the axon is the result of local currents set up by the ion movements at the action potential itself. They occur both in front of and behind the action potential, and depolarise the membrane sufficiently to cause the sodium channels to open in front of the action potential, as shown in figure 4.1.8. (The refractory period prevents the sodium channels opening behind the spike.) In this way the impulse is continually propagated in the required direction.

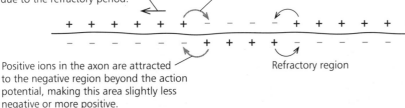

The result of these ion movements is to depolarise the membrane sufficiently to set up a new action potential in one direction only. It cannot go backwards due to the refractory period.

Positive Na$^+$ ions are attracted to the negative region of the action potential, leaving the membrane in this region slightly negative.

Positive ions in the axon are attracted to the negative region beyond the action potential, making this area slightly less negative or more positive.

Refractory region

Direction of impulse

Figure 4.1.8 Tiny local currents propagate the action potential along a nerve fibre. The refractory period ensures that it moves in only one direction.

Saltatory conduction

In myelinated vertebrate nerves, the mechanism of propagation is slightly more complex. Ions can only pass freely into and out of the axon at the nodes of Ranvier, which are about 1 mm apart. This means that action potentials can only occur at the nodes, and so they appear to jump from one node to the next, as figure 4.1.9 shows. The effect of this is to speed up transmission, as the ionic movements associated with the action potential occur much less frequently, taking less time. This conduction is known as **saltatory conduction** from the Latin verb *saltare*, which means 'to jump'.

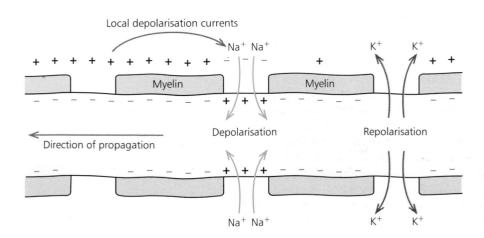

Local depolarisation currents

Na$^+$ Na$^+$ K$^+$ K$^+$

Myelin   Myelin

Depolarisation   Repolarisation

Direction of propagation

Na$^+$ Na$^+$ K$^+$ K$^+$

Figure 4.1.9 By 'jumping' from node to node along a myelinated nerve fibre, the nerve impulses can travel very rapidly along very narrow axons – allowing the development of complex but compact nervous systems.

LINKING THE SYSTEM

The nerves are the basic units of the nervous system, adapted for the rapid passage of electrical impulses from one region to another. But nerves must be able to intercommunicate. Receptors must pass their information into the sensory nerves, which in turn must relay the information to the central nervous system. Information needs to pass freely around the central nervous system and the messages sent along the motor nerves must be communicated to the effector organs so that action can be taken. How is this intercommunication brought about?

Synapses

Wherever two nerve cells meet they are linked by a **synapse**, as shown in figure 4.1.10. Every cell in the central nervous system is covered with synaptic knobs from other cells – several hundred in some cases. Neurones never actually touch their target cell, so a synapse is a gap between two nerve cells which the nerve messages must somehow cross.

The electrical nature of the nerve impulse was deduced long before it could be accurately recorded and measured. Similarly, it was suspected that transmission at the synapses was not electrical but chemical long before the electron microscope and other techniques could demonstrate this clearly. Once the structure of the synapse had been seen using the electron microscope the synaptic gap could be measured. This settled the argument – the gap is simply too wide for an impulse the size of an action potential to jump across. Synaptic transmission had to be chemical – and all the available evidence confirms this.

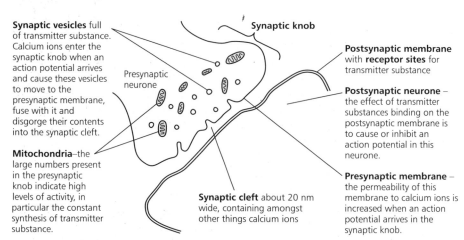

Synaptic vesicles full of transmitter substance. Calcium ions enter the synaptic knob when an action potential arrives and cause these vesicles to move to the presynaptic membrane, fuse with it and disgorge their contents into the synaptic cleft.

Mitochondria–the large numbers present in the presynaptic knob indicate high levels of activity, in particular the constant synthesis of transmitter substance.

Presynaptic neurone

Synaptic knob

Postsynaptic membrane with **receptor sites** for transmitter substance

Postsynaptic neurone – the effect of transmitter substances binding on the postsynaptic membrane is to cause or inhibit an action potential in this neurone.

Presynaptic membrane – the permeability of this membrane to calcium ions is increased when an action potential arrives in the synaptic knob.

Synaptic cleft about 20 nm wide, containing amongst other things calcium ions

Figure 4.1.10 The structure of the synapse as revealed by the electron microscope. Once these details were revealed, the way in which the system functions could be worked out in detail.

The synapse at work

The arrival of an impulse at the synaptic knob increases the permeability of the presynaptic membrane to calcium ions. Calcium ions therefore move into the synaptic knob along a concentration gradient. The effect of these calcium ions is to cause the synaptic vesicles containing transmitter substance to move to the presynaptic membrane. Each vesicle contains about 3000 molecules of transmitter. Some of the vesicles fuse with the membrane and release the transmitter substance into the synaptic cleft. The transmitter diffuses across the gap and becomes attached to specific protein receptor sites on the postsynaptic membrane. As a result, ion channels are opened and there is usually a local depolarisation and influx of sodium ions, causing an **excitatory postsynaptic potential (EPSP)** to be set up. If there are sufficient of these potentials the positive charge in the postsynaptic cell builds up to the threshold level and an action potential is set up which then travels on along the postsynaptic neurone.

In some cases the transmitter has the opposite effect. Channels allowing the inward movement of negative ions are opened in the postsynaptic membrane, which makes the inside more negative than the normal resting potential. An **inhibitory postsynaptic potential** results, which makes it less likely that an action potential will occur in the postsynaptic fibre.

Once the transmitter has had its effect it is destroyed by enzymes. This is very important because unless the transmitter is removed from the synaptic cleft, subsequent impulses would have no effect, as the receptors on the postsynaptic membrane would all be bound.

The transmitter substances

The most common transmitter substance, found at the majority of synapses, is **acetylcholine (ACh)**. It is synthesised in the synaptic knob using ATP produced in the many mitochondria present. Nerves using acetylcholine as their transmitter are known as **cholinergic nerves**. Once the acetylcholine has done its job it is very rapidly hydrolysed by the enzyme **cholinesterase**. This ensures that it no longer affects the postsynaptic membrane, and it also releases the components to be recycled – they pass back into the synaptic knob and are resynthesised into more acetylcholine.

Some vertebrate nerves, particularly those of the sympathetic nervous system (see section 4.2), produce **noradrenaline** in their synaptic vesicles and are known as **adrenergic nerves**.

Neuromuscular junctions

Nerves have to communicate not only with each other, but with receptors and effectors as well. Motor nerves need to communicate with muscles. Where a motor nerve and muscle fibre meet, a special kind of synapse is found known as a **neuromuscular junction**. The membrane of the muscle fibre (the sarcolemma) is very folded in this region and forms a structure known as an **end plate**, to which the end of the motor nerve joins. Electron microscopy shows us that the structure of the neuromuscular junction is remarkably similar to that of any other synapse, as figure 4.1.11 shows. The end of the motor neurone is full of mitochondria and synaptic vesicles which contain acetylcholine. It appears that when an impulse arrives at the end of the motor neurone acetylcholine is discharged into the synaptic cleft. As a result of its effect on the postsynaptic membrane an **end plate potential** is set up which can be recorded. If sufficient end plate potentials are set up an action potential is fired off in the muscle fibre, spreading through the **T tubules** and leading to a contraction of the muscle as described in section 2.6.

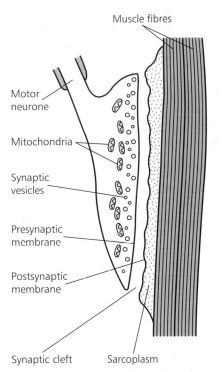

Figure 4.1.11 The neuromuscular junction is very similar to any other synapse. The end result is not an action potential propagated along the postsynaptic fibre, but the contraction of the muscle.

Evidence for the functioning of synapses

There are several convincing strands of evidence supporting the current model of the working of the synapse.

(1) Apart from the basic structural details revealed by the electron microscope, micrographs taken after a nerve has been strongly stimulated for some time show a lack of synaptic vesicles. This reflects the observed fact that after a period of stimulation a

nerve can no longer respond (it becomes accommodated, see page 243) and suggests that the reason for this is that all the transmitter substance has been used up. Thus it is reasonable to deduce that normal transmission across the synapse results from the release of transmitter from a small number of vesicles.

(2) A variety of drugs and poisons interfere with the working of synapses or neuromuscular junctions. A look at the effect of some of these substances throws light onto the normal working of the system (see table 4.1.1). Any substance which interferes with the formation of acetylcholine, stops it being released, prevents it interacting with the postsynaptic receptors or reduces the rate at which it is broken down will in turn have a major effect on the nervous communication system of the body.

Substance	Where does it act?	What does it do?
Botulinus toxin	Affects the presynaptic membrane and stops the release of acetylcholine	Prevents transmission of impulses across synapses and so prevents the nervous system working
Nicotine	Mimics the action of acetylcholine on post-synaptic membranes	Stimulates the nervous system
Strychnine, eserine, organophosphorus compounds used as weedkillers and insecticides, some nerve gases	Inactivate cholinesterase at the postsynaptic membrane and so prevent the breakdown of acetylcholine	Enhance and prolong the effects of acetylcholine as it is no longer destroyed. This means that nerves fire continuously, or at the slightest stimulus, and muscles are sent into tetanus
Curare (used on arrow tips by South American Indians)	Interferes with the action of acetylcholine and stops the depolarisation of the postsynaptic membrane	Causes paralysis as the muscles can no longer be stimulated by the nerves

Table 4.1.1

Coordination and control of neurones

Summation and facilitation

Neurones interact in a variety of complex ways. Sometimes a single nerve fibre will carry an action potential to a synapse with another cell, and transmission across that synapse will set up the next action potential. But in many cases the situation is much more complex than this. Often a single synaptic knob does not release enough transmitter substance to set up an action potential in the postsynaptic fibre. However, if two or more synaptic knobs are stimulated and release transmitter at the same time onto the same postsynaptic membrane the effects add together and a postsynaptic action potential results. This is known as **spatial summation**, illustrated in figure 4.1.12.

In other cases, a single knob does not release enough transmitter substance to stimulate the postsynaptic nerve fibre, but if a second impulse is received from the same knob in quick succession an action potential results. This effect is known as **temporal summation** (adding over time). It involves **facilitation** – in other words, the first impulse does not trigger off a response but it has an effect which makes easier (facilitates) the passage of the next impulse.

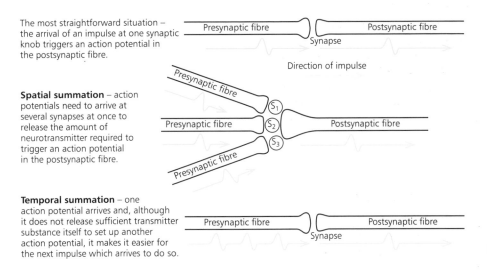

The most straightforward situation – the arrival of an impulse at one synaptic knob triggers an action potential in the postsynaptic fibre.

Presynaptic fibre — Synapse — Postsynaptic fibre

Direction of impulse

Spatial summation – action potentials need to arrive at several synapses at once to release the amount of neurotransmitter required to trigger an action potential in the postsynaptic fibre.

Presynaptic fibre — S_1
Presynaptic fibre — S_2 — Postsynaptic fibre
— S_3
Presynaptic fibre

Temporal summation – one action potential arrives and, although it does not release sufficient transmitter substance itself to set up another action potential, it makes it easier for the next impulse which arrives to do so.

Presynaptic fibre — Synapse — Postsynaptic fibre

Figure 4.1.12 Different types of synaptic transmission

Accommodation

On first applying perfume or aftershave we tend to be very aware of the smell ourselves. After a short time we lose that awareness, and it is other people who notice how pleasant we smell! If we reapply our scent another day, we can smell it again. This reaction is the same as that of a sea anemone which, when poked with a pointer, will withdraw its tentacles. If the sea anemone is poked repeatedly, the response is lost. If left alone for a while, the sea anemone reacts to the pointer again. Both of these examples are the result of a process known as **accommodation**.

If a nerve is repeatedly stimulated it eventually loses the ability to respond. Each time an impulse arrives at a synapse, vesicles full of transmitter discharge their contents into the synaptic cleft. The transmitter can only be synthesised at a certain rate. If the synapse is used too often, all of the vesicles are discharged into the synaptic cleft and the rate of synthesis simply cannot keep up. At this point the nerves can no longer respond to the stimulus – they are said to have **accommodated** or **fatigued**. A short rest restores the response as new vesicles and transmitter molecules are made. Some synapses never fatigue – they have an extremely rapid resynthesis rate – whilst others accommodate very quickly.

The implications of the organisation of the nervous system

The nerve fibres and synapses which we have been considering in isolation make up enormously complicated systems. Bundles of nerve fibres form nerves capable of carrying vast numbers of messages in different directions to gather all the available information and control all the actions of the body. Neurones and synapses in the central nervous system collate information and send out instructions. Synapses, whilst susceptible to both fatigue and drugs, allow for great flexibility, intercommunication between cells, facilitation and inhibition. They also play a vital and incompletely understood role in the brain, closely linked with both learning and memory.

Nerves give rapid communication. They also give the ability, in people at least, for long and involved nervous activity to take place in the brain before a particular action is undertaken. But for most simpler organisms most nervous activity and behaviour involves reflex actions which have a minimum of input from the central nervous system. Even human beings are ruled by reflexes to a remarkable extent. In section 4.2 we shall look in more detail at what happens in a reflex, and how other specialised parts of the nervous system such as the sense organs work.

SUMMARY

- Control and coordination may be achieved in an organism by **chemical messages** or by **electrical impulses**. Electrical or **nerve** impulses are carried within a **nervous system** and allow quicker and more specific communication.

- **Neurones**, cells specialised to transmit electrical impulses, interconnect to form a nervous system. Messages from the internal or external environment are received at **receptor cells** and carried in **sensory neurones**. **Motor neurones** carry impulses to **effector cells** which bring about a **response**. In larger animals a **central nervous system** coordinates and processes information.

- Neurone cell bodies contain **Nissl's granules** for protein synthesis and have projections called **dendrites**. There is one long **nerve fibre**, the **axon** in motor neurones and the **dendron** in sensory neurones. This carries the impulses. **Myelinated** nerve fibres have a fatty **myelin sheath** formed by a series of **Schwann cells** wrapped around the neurone. The sheath is interrupted at the **nodes of Ranvier**.

- Invertebrates have only unmyelinated fibres which transmit impulses more slowly than myelinated fibres of the same diameter. **Giant axons** are unmyelinated nerve fibres about 1 mm in diameter found in some invertebrates. They transmit vital impulses relatively quickly.

- The membrane of a nerve fibre is impermeable to sodium ions but more permeable to potassium ions. Sodium ions are actively pumped out of the fibre and potassium ions pumped in. Potassium ions then diffuse out again along a concentration gradient. The inside of the fibre is negatively charged relative to the outside – it has a **resting potential** of –70 mV.

- When an impulse or **action potential** is set up, sodium channels open allowing sodium ions to move in. The fibre becomes **depolarised** – it is more positive on the inside than the outside by about +40 mV. The sodium channels then close and sodium ions are pumped out as usual. The permeability to potassium ions is temporarily increased so that they diffuse out more quickly. The resting potential is thus restored after a **refractory period**.

- An action potential is an **all-or-nothing** response – it either happens or not, and is always the same size. It is only activated once the **threshold** has been reached – the point at which sufficient sodium channels are open for there to be more sodium ions entering the cell than potassium ions leaving by diffusion.

- During the **absolute refractory period** the sodium channels are blocked and the resting potential has not yet been restored. The neurone cannot be restimulated. During the **relative refractory period** restimulation is possible but the threshold is raised. The refractory period limits the frequency at which impulses can travel along a neurone, and also ensures that impulses travel in one direction only. This speeds up conduction.

- An action potential is **propagated** by local currents caused by ion movements around the action potential. These cause the sodium channels to open in front of the action potential (but not behind it because of the refractory period).

- In myelinated fibres **saltatory conduction** occurs – the impulse jumps from node to node, where ions can pass freely in and out of the fibre.

- Nerve fibres are linked by **synapses**, gaps between neurones across which chemical messages called **neurotransmitters** pass. When an impulse arrives at the **synaptic knob**, the **presynaptic membrane** becomes permeable to calcium ions which move into the synaptic knob. **Synaptic vesicles** then fuse with the presynaptic membrane and release neurotransmitter into the **synaptic cleft**. The transmitter diffuses across the cleft and binds to receptor sites on the **postsynaptic membrane**. This causes channels to open, local depolarisation and an influx of sodium ions causing an **excitatory postsynaptic potential**. These EPSPs build up to a threshold level and an action potential is set up in the postsynaptic neurone. The effect of the neurotransmitter binding to the postsynaptic membrane is sometimes to allow negative ions to enter resulting in an **inhibitory postsynaptic potential**.

- Neurotransmitters include **acetylcholine** (ACh) in **cholinergic nerves** and **noradrenaline** in **adrenergic nerves**. Acetylcholine is hydrolysed by **cholinesterase**.

- Neurones meet muscle fibres at a **neuromuscular junction** which contains **motor end plates** similar to synapses. Acetylcholine is discharged into the cleft and causes an **end plate potential** in the muscle fibre. These potentials summate to fire an action potential in the muscle fibre.

- In **spatial summation**, several synaptic knobs release neurotransmitter on the same postsynaptic membrane, and their effects add to produce a postsynaptic action potential.

- **Temporal summation** involves the same synaptic knob producing bursts of transmitter in succession. The first burst **facilitates** (makes easier) the passage of the next impulse across the cleft.

- **Accommodation** happens when a nerve is repeatedly stimulated and loses its ability to respond. This is due to depletion of the neurotransmitter which takes time to be resynthesised.

QUESTIONS

1 a Why are communication systems necessary?
  b Describe the structure of a typical motor nerve cell and explain how its structure is related to its function.
  c Explain what is meant by the terms:
     i central nervous system
     ii peripheral nerves.

2 a What is the resting potential of a nerve cell and how is it maintained?
  b Describe an action potential and explain how it is propagated along a nerve cell.
  c Explain briefly how nerve impulses are investigated.

3 a What is a synapse?
  b How is the structure of a synapse related to its function?
  c What is the role of:
     i myelin sheath
     ii ATP
     iii refractory period
     in the transmission of nerve impulses?

At the most basic level, the nervous systems of animals are made up of receptor cells and interconnected nerve cells which eventually stimulate effector cells. In mammals nervous systems are complex, involving highly specialised sensory organs containing many receptors, nerves made up of large numbers of nerve fibres bringing in information or carrying out instructions, and highly developed central nervous systems containing millions of interconnecting nerve cells capable of extremely sophisticated thought processes.

In simpler organisms, almost all actions take place without conscious thought – they are simply **reflex responses** to particular stimuli. Surprisingly, a large number of the actions of more complex animals are also the results of unconscious reflex actions – and this is true for ourselves as well as for any other mammal. Thus before considering the details of the nervous system such as the workings of sensory organs or the major nerve types, it is important to understand this most basic type of nervous response.

REFLEXES

Unconditioned reflexes

Unconditioned reflexes are responses that are not learned, but inborn. The contracting of the tentacles of a sea anemone or the writhing of a worm when touched are examples of these reflexes in invertebrates. Well known examples in humans include moving a hand or foot away rapidly from a hot or sharp object, swallowing as food moves down towards the back of the throat, blinking if an object approaches the eyes and the contracting and dilating of the pupil of the eye in response to light levels.

Figure 4.2.1 Reflex responses to very specific stimuli shown by newborn infants are quickly lost as conscious control grows, but they illustrate how the nervous systems of even highly developed mammals are based around the unconditioned reflex. Left – the **Moro reflex** is a response to a startling stimulus such as a sudden loud noise or movement, or the loss of a feeling of support. The infant throws up his arms and legs, and brings them together in a grasping movement. It is thought that this reflex is useful to the infants of monkeys and apes, allowing them to grasp the fur of the mother. In humans it has obviously lost its usefulness and fades over the first couple of months. Right – the **stepping reflex** is present only for the first few days of life. If a newborn infant is held upright with his feet touching a firm surface, he will take quite deliberate 'steps', placing one foot in front of the other as long as the body weight is supported.

An unconditioned reflex involves a fixed response (an **unconditioned response**) to a particular stimulus (an **unconditioned stimulus**). Such reflexes are controlled by the simplest type of nerve pathway in the body, known as a **reflex arc**. In vertebrates, including the mammals, this involves at its simplest a receptor, a sensory neurone and a motor neurone connected to an effector cell. Part of the pathway takes place in the central nervous system, often the spinal cord. The reflex arc may involve simply a motor neurone and a sensory neurone with a synapse between them, or there may be a third small **relay** or **intermediate neurone**, situated in the central nervous system, as shown in figure 4.2.2.

However the reflex arc is organised physically, its function is to bring about an appropriate response to a particular stimulus as rapidly as possible without the time delay which occurs when the conscious centres become involved.

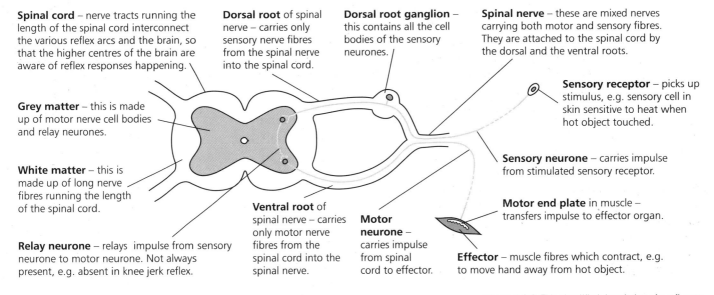

Spinal cord – nerve tracts running the length of the spinal cord interconnect the various reflex arcs and the brain, so that the higher centres of the brain are aware of reflex responses happening.

Grey matter – this is made up of motor nerve cell bodies and relay neurones.

White matter – this is made up of long nerve fibres running the length of the spinal cord.

Relay neurone – relays impulse from sensory neurone to motor neurone. Not always present, e.g. absent in knee jerk reflex.

Dorsal root of spinal nerve – carries only sensory nerve fibres from the spinal nerve into the spinal cord.

Ventral root of spinal nerve – carries only motor nerve fibres from the spinal cord into the spinal nerve.

Motor neurone – carries impulse from spinal cord to effector.

Dorsal root ganglion – this contains all the cell bodies of the sensory neurones.

Spinal nerve – these are mixed nerves carrying both motor and sensory fibres. They are attached to the spinal cord by the dorsal and the ventral roots.

Sensory receptor – picks up stimulus, e.g. sensory cell in skin sensitive to heat when hot object touched.

Sensory neurone – carries impulse from stimulated sensory receptor.

Motor end plate in muscle – transfers impulse to effector organ.

Effector – muscle fibres which contract, e.g. to move hand away from hot object.

Figure 4.2.2 shows clearly the structure of a vertebrate reflex arc. When a stimulus is detected by the receptor cell an impulse is set up in the sensory nerve fibre and an action potential travels along the spinal nerve to the dorsal root of the spinal nerve. The impulse enters the grey matter of the spinal cord. Synaptic transmission then takes place which results in another action potential being set up. This may be directly in a motor neurone, or first in a relay neurone and then, via another synapse, in the motor neurone. The impulse then travels out through the ventral root along the spinal nerve to the effector organ, causing the unconditioned response to occur.

Figure 4.2.2 This simplified description of a reflex arc shows how the information can be carried rapidly from the sensory system into the central nervous system and back out through the motor system, resulting in an almost immediate response by the appropriate effector organ.

Conditioned reflexes

Unconditioned reflexes are not learned – they are present from birth and do not at any point involve the conscious areas of the brain, other than the knowledge that a reflex action has taken place. **Conditioned reflexes** on the other hand are learned, although often without us realising it. If an unconditioned stimulus, such as the sight of food which produces the unconditioned salivation reflex, is associated with a second unrelated stimulus such as the ringing of a bell, an animal can unconsciously learn to react to the second stimulus alone – it salivates at the ringing of a bell when no food is present. Some of the best-known experimental work on this was done by Pavlov on the salivation reflex in dogs (see section 3.3, page 196).

Conditioned reflexes are important as they allow us to develop control of our bladder and bowel sphincters amongst other things, and at a more basic level they are an important part of the learning process throughout the animal kingdom. (Learning is dealt with in more detail in section 4.3.)

The importance of reflexes

Reflexes are the single most common form of nervous control throughout the animal kingdom. However, there are two other reasons for their importance in animals such as mammals. One has already been mentioned. By avoiding the areas of the nervous system involved in conscious thought, reflex actions can occur very rapidly. This means that they are of great importance when it comes to potentially dangerous situations. Evasive or avoiding action can be

The role of the lens

As the light enters the eye the process of focusing it onto the retina begins. The rays of light have to be bent (**refracted**) sufficiently first to pass through the pupil and also to arrive focused on the retina. Most of this focusing is in fact carried out by the cornea and the fluid through which the light passes, but the degree of refraction is the same for light from every object. When considering the working of the eye most attention is given to the working of the lens. Although the lens is not responsible for much of the bending of the light, it plays a very important role in giving fine, accurate focusing of light seen from objects both distant and close at hand.

The lens is a transparent elastic disc. Its shape can be changed by the action of the **ciliary muscles**, as described in the box below.

Changing the shape of the lens

The **ciliary muscles** are arranged circularly around the ciliary body. The effects of their contractions and relaxations are relayed to the lens by the **suspensory ligaments**. The lens itself is elastic and its unstretched shape is relatively short and fat.

Figure 4.2.8 A newborn baby can only see clearly at a certain distance from his eyes – roughly where his mother's face will be when he is feeding. With age, the ability to focus on objects at a great variety of distances develops. The shape of the lens is changed by the action of the ciliary muscles. To allow light from objects at different distances to be brought into focus, the thickness of the lens can be varied between the two extremes shown here. As the ability to do this develops, so does our visual understanding of the world around us.

The usual section through the eye makes it difficult to visualise what is happening to the lens. By looking at the lens and ciliary body from the front it is easier to see how the changes are brought about.

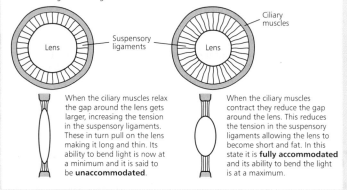

When the ciliary muscles relax the gap around the lens gets larger, increasing the tension in the suspensory ligaments. These in turn pull on the lens making it long and thin. Its ability to bend light is now at a minimum and it is said to be **unaccommodated**.

When the ciliary muscles contract they reduce the gap around the lens. This reduces the tension in the suspensory ligaments allowing the lens to become short and fat. In this state it is **fully accommodated** and its ability to bend the light is at a maximum.

Light rays from an object spread out in all directions – they are said to **diverge**. When we look at an object which is close to us a cone of diverging light enters our eyes. When we look at objects from further away the light rays are spreading less – in fact they appear almost parallel, as shown in figure 4.2.9. The light entering the eye is refracted by its passage through the conjunctiva, cornea, aqueous humour and vitreous humour in exactly the same way regardless of whether it is from a near or a distant object. But by changing the shape of the lens the degree of bending of the light can be altered. Light from distant objects needs relatively little bending to bring it into focus on the retina and so the lens has to be thin. To bring light from near objects into focus on the retina more refraction is needed and so the lens must be short and fat. This ability to focus light from objects at various distances is known as **accommodation** and the way it is brought about is shown in figure 4.2.9.

The role of the retina

As we have seen, light from an object is focused onto the retina. The retina must then perceive that light and inform the brain of its presence. In order to do this the retina contains about a hundred million light-sensitive cells (**photoreceptors**), along with the neurones with which they synapse. There are two main types of photoreceptors in the retina, known as the **rods** and the **cones**, shown in figure 4.2.10. Both types are secondary exteroceptors.

Rods are spread evenly across the retina except at the fovea where there are none. They provide black and white vision only and are used mainly for seeing in low light intensities or at night. The total number of rods is estimated as about 1.2×10^8. Rods are very sensitive to light, even of relatively low intensity.

Light rays from close objects are diverging or spreading out as they reach the eye.

Light from close objects is focused by the contraction of the ciliary muscles, allowing the lens to shorten and fatten, bending the light onto the retina.

Light rays from a distant object are effectively almost parallel by the time they enter the eye.

Light from distant objects is focused by the relaxation of the ciliary muscles, pulling the lens longer and thinner and so reducing its refractive powers.

As a result of the way the light rays travel, the image which forms on the retina is upside down. The brain effectively inverts it again so that we perceive things the right way up.

Figure 4.2.9 The ability to accommodate – focus on objects which are different distances away – develops gradually after birth and is important for giving us a clear and accurate view of the world.

They contain a single **visual pigment** called **rhodopsin** (visual purple). Rods are not very tightly packed together and several of them synapse with the same sensory neurone. This means that they do not give a particularly clear picture, but makes them extremely sensitive both to low light levels and to movements in the visual field, because several small generator potentials can trigger an action potential to the central nervous system.

Cones, on the other hand, are found tightly packed together in the region known as the fovea. There are only around 6×10^6 of them. They are used principally for vision in bright daylight and may have one of three visual pigments, so provide colour vision. As a result of their tight packing in the fovea, and the fact that each cone usually has its own sensory neurone, cones provide a picture of great visual acuity. In fact, it is only when light falls directly on the fovea that it can be clearly in focus.

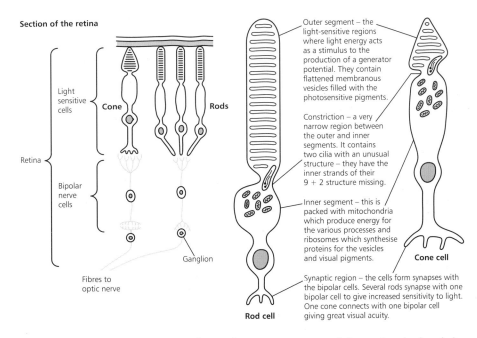

Figure 4.2.10 The two types of receptor cells along with their different arrangements of synapses in the retina give a visual system which combines great sensitivity to low levels of light with high visual acuity and clarity of vision.

An interesting point to note about the arrangement of the retina is that it is 'back to front'. The outer segments are actually next to the choroid, and the neurones are at the interior edge of the eyeball. This is why there is a blind spot where all the neurones pass through the layers of the eye to go into the brain. The light has to pass through the synapses and the inner segments before reaching the outer segments containing the visual pigments. The reason for this somewhat unexpected arrangement is the origin of the retinal cells in the embryo and the way in which the eye is formed during embryonic development. To add to the confusion, the optic nerves carrying the visual information cross over on their way to the visual cortex in the brain so that the information seen with the right eye is taken to the left side of the brain for processing!

How the rods work

Both rods and cones work in a similar way, based on the reactions of the visual pigments with light. In the rods this visual pigment is rhodopsin, which is formed from two components. These are **opsin** and **retinene**. Opsin is a lipoprotein, and retinene is a light-absorbing derivative of vitamin A. Retinene exists in two isomeric forms, *cis* retinene and *trans* retinene. In the dark, it is all in the *cis* form. When a photon of light hits a molecule of rhodopsin, it converts the *cis* retinene into *trans* retinene and the rhodopsin then breaks up into opsin and retinene. This breaking up of the molecule is referred to as **bleaching**. The bleaching of the rhodopsin sets up a generator potential in the rod, and if this is large enough or if several rods are stimulated at once an action potential is set up in the receptor neurone.

Once bleaching of the visual pigment has occurred, the rod cannot be stimulated again until the rhodopsin is resynthesised. It takes energy from ATP produced by the many mitochondria in the inner segment to convert the retinene back to the *cis* isomer and rejoin it to the opsin. In normal daylight the rods are almost entirely bleached and can no longer respond to dim light – the eye is said to be **light adapted**. After about 30 minutes in complete darkness the rhodopsin will be almost fully reformed. The eye is now sensitive to dim light and is said to be **dark adapted**.

Cones and colour vision

Cones work in a very similar way to rods, except that their visual pigment is known as **iodopsin**. There appear to be three types of iodopsin, each sensitive to one of the primary colours of light. Iodopsin needs to receive more light energy than rhodopsin in order to break down, and so it is not sensitive to low light intensities. But the cones provide colour vision, because the brain interprets the numbers of different types of cones stimulated as different colours, as shown in table 4.2.1. This model of how cones sensitive to the three primary colours of light can provide a wide range of colour vision is known as the **trichromatic theory** of colour vision.

Light stimulates			Colour perceived
red cones	green cones	blue cones	
✓	✗	✗	Red ●
✗	✓	✗	Green ●
✗	✗	✓	Blue ●
✓	✓	✗	Orange/yellow ●
✗	✓	✓	Cyan ●
✓	✗	✓	Magenta ●
✓	✓	✓	White ○

Table 4.2.1 The perception of different colours by the brain, as explained by the trichromatic theory of colour vision

Vision in other animals

Mammals as a group generally have eyes based on the same pattern as humans, although many of them do not possess the cones necessary for colour vision. For some mammals, the ability to judge distance is of great importance. The apes and monkeys need to do this very accurately in order to climb trees and swing from branch to branch. Also, along with their close relatives the humans, these animals use tools and need delicate manipulative skills which are greatly enhanced by good judgement of distances. Many carnivores also need to judge distance in order to leap on their prey without missing. In all of these animals the eyes are sited at the front of the head, both looking forward. This gives an overlapping area of vision where sufficient information is fed into the brain to result in a three-dimensional picture, giving excellent distance judgement.

Many other animals have the opposite problem – they are vulnerable to attack by predators. For these animals the important factor is not distance judgement but all-round vision and sensitivity to movement, increasing the chances of spotting a predator before it is close enough to attack. Because of this most herbivores have eyes on the sides of their heads, often giving a 360° field of vision.

Most arthropods including the insects have **compound eyes**. These are made up of hundreds or thousands of tiny units called **ommatidia**, each providing part of the whole picture. Thus the field of vision of an insect is made up of thousands of overlapping images. The visual acuity this gives is poor, but there is great sensitivity to movement – hence the difficulties in swatting a fly!

THE HUMAN EAR

Sensitivity to vibrations is another basic sense. For many higher animals sensitivity to vibrations in the air is interpreted in the brain as hearing sounds. Hearing is often of vital importance for survival, and many species have far more acute hearing than we do ourselves. But while our ears may not be the most sensitive in the animal kingdom, we rely heavily on speech in our everyday communications so that to be without hearing is perceived as a disadvantage in society.

The structure of the human ear

The ear is the human organ of hearing. Because it detects a mechanical wave in the air, the ear contains mechanoreceptors. However, the ear is not simply sensitive to sound. It also detects both gravity and movement, again using specially adapted mechanoreceptors. As in the case of the eye, the structure of the ear is closely related to its functions, as shown by figure 4.2.11.

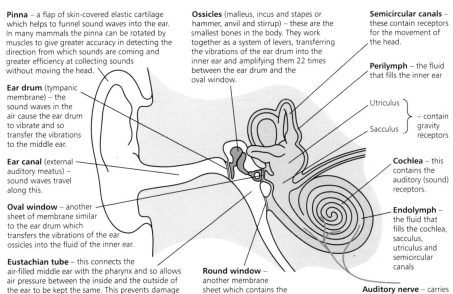

Pinna – a flap of skin-covered elastic cartilage which helps to funnel sound waves into the ear. In many mammals the pinna can be rotated by muscles to give greater accuracy in detecting the direction from which sounds are coming and greater efficiency at collecting sounds without moving the head.

Ear drum (tympanic membrane) – the sound waves in the air cause the ear drum to vibrate and so transfer the vibrations to the middle ear.

Ear canal (external auditory meatus) – sound waves travel along this.

Oval window – another sheet of membrane similar to the ear drum which transfers the vibrations of the ear ossicles into the fluid of the inner ear.

Eustachian tube – this connects the air-filled middle ear with the pharynx and so allows air pressure between the inside and the outside of the ear to be kept the same. This prevents damage to the ear drum due to pressure differences.

Ossicles (malleus, incus and stapes or hammer, anvil and stirrup) – these are the smallest bones in the body. They work together as a system of levers, transferring the vibrations of the ear drum into the inner ear and amplifying them 22 times between the ear drum and the oval window.

Round window – another membrane sheet which contains the fluid of the inner ear.

Semicircular canals – these contain receptors for the movement of the head.

Perilymph – the fluid that fills the inner ear

Utriculus

Sacculus
} – contain gravity receptors

Cochlea – this contains the auditory (sound) receptors.

Endolymph – the fluid that fills the cochlea, sacculus, utriculus and semicircular canals

Auditory nerve – carries impulses to the brain.

Figure 4.2.11 The human ear – an organ sensitive not only to sound but also to movement and gravity

How do we hear?

For us to hear, the ear has to collect sound waves from the air and funnel them into the region containing the sound receptors – that is, the inner ear. The **pinna** helps to do this and the sound waves are funnelled along the **ear canal** and set up vibrations in the **ear drum**. These vibrations in turn set the ear **ossicles** rocking against each other, with the smallest bone, the stapes, rocking against the **oval window** and setting up vibrations there which are transferred to the **perilymph**, the fluid of the inner ear. The area of the ear drum is relatively large compared with that of the oval window, and so vibrations

What is sound?

What we perceive as sound is usually the result of vibrations (mechanical waves) travelling through air. These sound waves can also travel through liquids such as water or solids such as the laboratory bench or the wall of a house. Sound is difficult to define because sounds do not exist until a mechanical wave in the air is picked up by the ear and interpreted by the brain – the term 'sound' describes the sensation that we hear.

The frequency of vibration of an object determines the frequency of the vibrations travelling through the air. The human ear is sensitive to frequencies of between 40 and 20 000 Hz (cycles per second). Low-frequency sound waves are perceived as deep sounds and high-frequency sound waves as high-pitched sounds.

transferred across the middle ear are amplified by a factor of 22. The movements set up in the perilymph are picked up by sensory receptors in the **cochlea** and the information sent to the brain as impulses in the **auditory nerve**. The brain interprets these messages as sounds.

The role of the cochlea

The cochlea is the organ which is sensitive to sound within the ear. It is coiled rather like the shell of a snail, but to understand how it works it is easier to consider the structure uncoiled. Figure 4.2.12 shows a simplified diagram of the cochlea. The oval window and the round window are the two ends of a narrow perilymph-filled canal around the cochlea. The upper part is called the **vestibular canal** and the lower part is the **tympanic canal**. Vibrations run through the perilymph from the oval to the round window. The cavity within the cochlea, called the **median canal**, is filled with endolymph.

The **Reissner's membrane** separates the endolymph from the vestibular canal and the **basilar membrane** separates it from the tympanic canal. Running through the centre of the whole cochlea is a third, rather rigid membrane known as the **tectorial membrane**. The basilar membrane has a bulge running along its entire length, and this bulge contains receptor cells. The bulge is called the **organ of Corti**. The receptor cells of the organ of Corti have sensory projections which are embedded in the tectorial membrane (see figure 4.2.12).

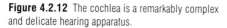

Figure 4.2.12 The cochlea is a remarkably complex and delicate hearing apparatus.

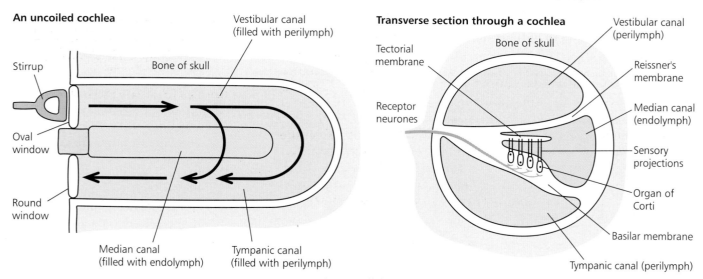

When a sound wave causes the oval window to vibrate, the vibrations of the membrane result in pressure waves passing along the perilymph in the vestibular canal. At some point along the cochlea, depending on the frequency of the vibration, these waves cause the Reissner's membrane to vibrate. This causes vibrations in the endolymph which in turn causes vibration in the basilar membrane. The tectorial membrane is too rigid to vibrate. This means that as the basilar membrane vibrates it moves relative to the tectorial membrane. This stretches the sensory projections between them which causes a generator potential to be set up in the receptor cells in the organ of Corti. The generator potential is transmitted across synapses to receptor neurones where action potentials are set up which carry information to the brain.

The basilar membrane is thinnest and so vibrates most easily near the oval window. This is where high-frequency, high-pitched sounds are picked up. Deeper sounds are detected further along the cochlea towards the round window, where the basilar membrane is thicker. Information about the pitch of a sound is given to the brain by the position along the cochlea of the receptor cells stimulated. Information about the loudness comes from how many receptor cells are stimulated at a particular site. Quiet sounds stimulate only a few receptor cells whereas loud sounds stimulate many. Spatial summation occurs to give the brain detailed information about the sounds heard. The information is passed to the brain along the auditory nerve.

The other senses of the ear

The ear contains many receptors sensitive to sound in the cochlea, but it also contains receptors sensitive to movement and gravity in other areas of the inner ear. These senses are vital to both our sense of balance and our coordination. Anyone who has experienced disorientation after a rapidly rotating fairground ride will recognise the problems of staying upright and walking straight which can result from only a minor disturbance of these systems. So how are our senses of gravity and movement achieved? We shall look at the two systems separately.

The perception of gravitational fields

The **sacculus** and **utriculus** (shown on figure 4.2.11) are the organs involved in the perception of gravitational fields. We do not have a direct awareness of this sense – we might comment on the brightness of a light or the loudness of a sound but we do not usually discuss the state of gravity! However, we do know which way up we are, or whether our heads are on one side or upright, and this is the result of our gravitational sense.

Figure 4.2.13 shows the internal structure of the utriculus, and the sacculus works on the same plan. Both organs are filled with endolymph, like the cochlea. There is a bulge or mound called the **macula** containing receptor cells. These receptor cells have sensory projections which are embedded in the **otoliths**, chalky crystals in a mass of jelly-like material. When the head is upright there is no pull on the sensory projections, but as the head tilts in one direction the otoliths on that side move away from the macula under the influence of gravity. This causes a strain on the sensory projections and generator potentials are set up in the receptor cells of the macula. Varying numbers of generator potentials and differing strengths give information which the brain can interpret to produce a picture of the amount and direction of the tilting of the head.

The awareness of movement

Movements of the head in any direction can be detected by the **semicircular canals**. These sense organs are arranged so that each one is at right angles to the other two, making comprehensive detection of movement possible. Each semicircular canal is filled with endolymph and has a swelling called the **ampulla**. Inside the ampulla is a mound of receptor cells known as the **crista**, and these have sensory projections embedded in a jelly-like structure, the **cupula**, as shown in figure 4.2.14. When we turn our heads, the semicircular canals move with us. The movement of the cupula is opposed by the inertia of the endolymph, which tends to stay still. This results in a distortion of the cupula and consequently a strain on the sensory projections. This pull sets up generator potentials in the receptor cells on one side of the ampulla, followed by action potentials in the sensory nerves, giving the brain information as to in which direction the head is turning.

Processing sensory information

The eyes and ears, although major and very important sense organs, supply only a part of the huge amount of sensory information constantly gathered by

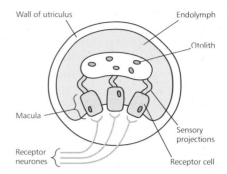

When the head is upright there is no pull on the sensory projections in the utriculus.

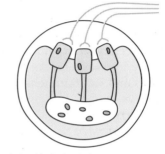

When the head is tilted or, in the extreme example shown here, upside down, the otoliths fall away from the macula. As a result some or, as seen here, all of the sensory projections are stretched. This causes generator potentials to be set up in the receptor cells and messages to be sent to the brain giving information about the orientation of the head.

Figure 4.2.13 A simplified section of a utriculus shows how the organ works.

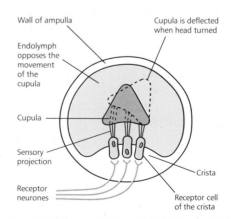

Figure 4.2.14 A simplified section through a cupula in one of the semicircular canals

the body. This is carried by sensory nerve fibres to the central nervous system. In many cases these fibres are part of reflex arcs and the body responds in an immediate, preprogrammed way. But in many other cases the sensory information is sent to the higher centres of the brain and the information from many sources is put together and processed. As a result of all this information, messages need to be sent to effector cells and organs. These messages are sent out along motor nerves in the peripheral nervous system, which we shall now go on to consider in more detail.

THE PERIPHERAL NERVOUS SYSTEM

The mammalian nervous system is made up of the **central nervous system** (brain and spinal cord, discussed in section 4.3) and the **peripheral nervous system** extending from and coordinated by the central nervous system. The peripheral nervous system is made up of the sensory or afferent nerves carrying messages from the receptors into the central nervous system, and the motor or efferent nerves carrying messages out to the effectors. All the sensory nerves function in much the same way, but the motor nerves can be divided into two main types.

The motor system

The voluntary motor system

The **voluntary** motor nerves, as the name suggests, are under voluntary or conscious control. The higher areas of the brain are involved and they function as a result of conscious thought. When we consider an action, such as picking up a cup of tea or switching on a CD player, the instructions which need to be issued to the muscles will be carried along voluntary nerve fibres.

The autonomic motor system

The other major division of the motor nervous system is the **autonomic** or **involuntary** nerves. The nerves of the autonomic nervous system are involved with the control of involuntary activities – bodily functions such as the movements and secretions of the gut, sweating, breathing and dilating or constricting blood vessels, which are normally not dealt with by the conscious area of the brain.

The autonomic nervous system can itself be subdivided into the **sympathetic nervous system** and the **parasympathetic nervous system**. The differences between these two groups are both anatomical and functional. The structural differences can be seen in figure 4.2.15. The main functional differences relate

Dizziness

The endolymph in the movement-detecting organs resists movement in normal circumstances. But if we spin round rapidly for a prolonged time – as on a fairground ride – the endolymph eventually begins to move. Just as it takes time to start moving, it also takes time to stop moving again. For some moments after the spinning has stopped, the cupula is distorted by the moving endolymph, although now the pressure is on the other side. Our brains receive false information, implying that we are still spinning, although in the opposite direction. The information from our eyes denies this and the conflict of inputs results in a sensation of disorientation and failure of our usual balance control.

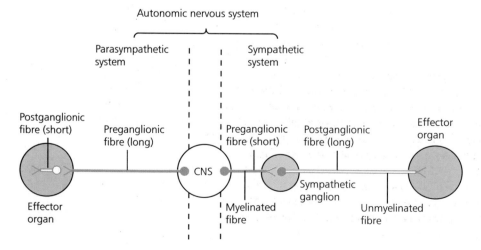

Figure 4.2.15 In both the parasympathetic and the sympathetic systems, myelinated preganglionic fibres leave the central nervous system and synapse within a ganglion with unmyelinated postganglionic fibres. In the sympathetic system the ganglia are very close to the CNS, so the preganglionic fibres are short and the postganglionic fibres are long. In the parasympathetic system the situation is reversed. The ganglia are near to or in the effector organ, so the preganglionic fibres are very long and the postganglionic ones very short.

to the neurotransmitters at the synapses. The sympathetic system produces noradrenaline, whilst the parasympathetic system produces acetylcholine.

Most of the body organ systems are supplied by both the sympathetic and the parasympathetic systems. So what are the functional effects of the differences between them? Put simply, the sympathetic nervous system usually has an **excitatory effect**, whilst the parasympathetic system has an **inhibitory effect**. What this means in terms of individual organs can be seen in figure 4.2.16.

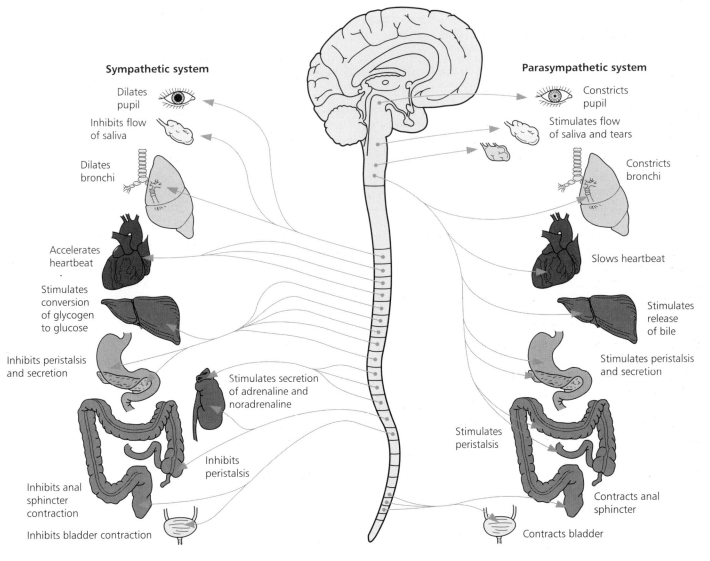

Sympathetic system

Dilates pupil

Inhibits flow of saliva

Dilates bronchi

Accelerates heartbeat

Stimulates conversion of glycogen to glucose

Inhibits peristalsis and secretion

Stimulates secretion of adrenaline and noradrenaline

Inhibits peristalsis

Inhibits anal sphincter contraction

Inhibits bladder contraction

Parasympathetic system

Constricts pupil

Stimulates flow of saliva and tears

Constricts bronchi

Slows heartbeat

Stimulates release of bile

Stimulates peristalsis and secretion

Stimulates peristalsis

Contracts anal sphincter

Contracts bladder

Figure 4.2.16 The opposing effects of the sympathetic and parasympathetic nerves on body systems

Several bodily functions which we might consider to be under voluntary control – opening the bowel and bladder sphincters, for example – are shown on the figure as under the control of the autonomic system. The nervous system is of enormous sophistication and complexity and many areas of the body are supplied with voluntary as well as involuntary nerves. Most of us have control over our bladder and bowels, and can control our breathing rate if we wish to. It is relatively easy to control the heart rate to some degree, and mystics the world over have shown control over many other normally involuntary activities. The nervous system still has many elements of mystery about it, and the central nervous system which we go on to consider next is the least understood of all.

SUMMARY

- An **unconditioned reflex** is a fixed or **unconditioned response** to a particular or **unconditioned stimulus**. Such responses are controlled by a nerve pathway called a **reflex arc**.

- In a vertebrate reflex arc a stimulus in the receptor sets up an impulse in the sensory neurone which travels along the spinal nerve to the grey matter of the spinal cord. Synaptic transmission sets up another impulse in a motor neurone, either directly or via a **relay neurone**. The impulse leaves the spinal cord via the ventral root and passes along the spinal nerve to the effector organ.

- Reflex actions happen quickly and so are useful to evade danger. They also control bodily functions and free the central nervous system to deal with non-routine activities.

- A **primary receptor** is a neurone with a specialised dendrite which detects a stimulus and sets up an action potential in its fibre. A **secondary receptor** is a modified epithelial cell which synapses with a sensory neurone.

- Receptors have a negative resting potential maintained by sodium pumps. A stimulus interferes with the sodium pump and ions move into the neurone setting up a generator potential of varying size – generator potentials do not obey the all-or-nothing law. If the generator potential is large enough it sets up an action potential in the neurone.

- Several secondary receptor cells may synapse with a single receptor neurone. In **convergence**, generator potentials summate to trigger an action potential. A graded response is achieved by varying the rate at which action potentials pass along the sensory neurone.

- A receptor exposed to an unchanging stimulus **adapts** – the generator potentials produced gradually die away until the stimulus changes, when they build up again.

- **Sense organs** such as the eye and the ear are specialised regions where receptors sensitive to a particular stimulus are collected together.

- The human eye is surrounded and protected by the **sclera**, inside which is the **choroid** which prevents internal reflection of light. The **conjunctiva** is an epithelium covering the **cornea**, a transparent area at the front of the eyeball enclosing the **aqueous humour**. The coloured **iris** surrounds the **pupil**, an aperture through which light passes to the **lens**. Light then passes through the **vitreous humour** to the light-sensitive **retina** at the back of the eyeball. The **fovea** is the most sensitive part of the retina while the **blind spot** is the point where the **optic nerve** leaves the eye.

- The **radial** and **circular** muscles of the iris control the size of the pupil and thus the amount of light entering the eye.

- Light is refracted by the cornea, the aqueous humour, the lens and the vitreous humour. The lens carries out the fine focusing necessary to ensure that the image is formed on the retina. The shape of the lens is adjusted by the action of the **ciliary muscles** in the **ciliary body** surrounding the lens, which is supported in the ciliary body by the **suspensory ligaments**. This adjustment or **accommodation** allows light from objects at different distances to be focused.

- The retina contains **photoreceptors** called **rods** and **cones**. Rods provide black and white vision. They contain a visual pigment called **rhodopsin**.

Several rods synapse with the same sensory neurone giving sensitivity to low levels of light. Rhodopsin is formed from **opsin** and **retinene**. Retinene exists in the *cis* form in the dark and the *trans* form in the light. When retinene is converted to the *trans* form in the light, the rhodopsin breaks up or **bleaches**, setting up a generator potential. In the light the rods are bleached and the eye is **light adapted**. In the dark rhodopsin is reformed and the eye becomes **dark adapted**.

- Cones are concentrated at the fovea and produce colour vision in bright light. Each cone has its own sensory neurone and contains one of three visual pigments.

- The human ear consists of the external **pinna** which funnels the sound waves through the **ear canal** to the **ear drum**. This transmits the vibration to the **ossicles**, small bones which amplify the vibrations and transmit them to the **oval window** of the **cochlea**. Vibrations run through the **perilymph** in the cochlea to the **round window**. Three membranes run the length of the cochlea and between two of them are the sensory projections of the **organ of Corti**. These vibrate when the sound wave matches their own particular frequency and set up a generator potential in the receptor cell.

- The ear perceives gravitational fields by the action of the **sacculus** and **utriculus**. The **macula** of both these organs contains receptor cells and sensory projections embedded in the **otoliths**, crystals which move under the influence of gravity.

- Movement of the head is perceived by three **semicircular canals** arranged at right angles. The **ampulla** contains a mound of receptor cells called the **crista** with sensory projections embedded in the **cupula** which exerts a strain on the sensory projections when the head is moved.

- The **peripheral nervous system** consists of sensory and motor nerves carrying impulses between the central nervous system and the rest of the body. The motor nerves are divided into the **voluntary system**, under conscious control, and the **autonomic** (involuntary) **system**.

- The autonomic system is further subdivided into the **sympathetic** and **parasympathetic** systems. The sympathetic system has adrenergic nerves while the parasympathetic has cholinergic nerves. In both systems myelinated preganglionic fibres leave the central nervous system and synapse in a ganglion with unmyelinated postganglionic fibres. The ganglia are close to the central nervous system in the sympathetic system and close to the effector in the parasympathetic system. The two systems coordinate to control body systems, the sympathetic system generally having an excitatory effect and the parasympathetic an inhibitory effect.

QUESTIONS

1 Describe a typical mammalian reflex response. Discuss the role of reflexes in the survival of different groups of animals.

2 Sketch a diagram of the structure of the human eye. Annotate the diagram to indicate the main features and how their structure is related to their function.

3 **a** Glue ear is a condition found in young children. It is the result of sticky secretions produced in the middle ear being unable to drain away through the Eustachian tube, due to blockage or to the angle of the tube in the immature skull. Explain why this may result in temporary deafness.

   b Why may ear infections be associated with sensations of giddiness and loss of balance?

4.3 Brains and behaviour

A nervous system made up of receptors, nerves and effectors provides sufficient sensitivity to an animal's surroundings for it to capture food and avoid at least some predators. But complex animals need a 'sorting station' where information can be processed and from where instructions can be issued to give fully coordinated responses to a wide range of situations. Evolution has provided the **central nervous system** (**CNS**), and in particular the **brain**.

THE BRAIN

The development of complex nervous systems

As nervous systems became increasingly complex, one of the first developments was **ganglia** – collections of cell bodies and synapses in one enclosed area. These gave increased coordination and efficiency, enabling larger numbers of nerve fibres to exchange information. The development of a central nerve cord (the **spinal cord** in vertebrates) was another major step in the evolution of nervous systems. A central major nerve pathway facilitates whole-body communications and gives a specific route for incoming and outgoing messages. Ganglia and a central nerve cord made it possible for animals to respond in a much greater variety of ways to their environment. As an example, the brains of insects may seem relatively rudimentary, yet in combination with a central nerve cord and ganglia in every body segment they enable some extremely sophisticated patterns of behaviour to be developed in the insect world.

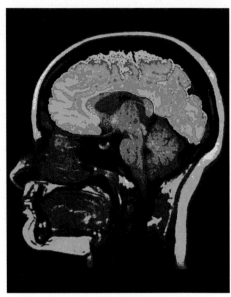

A CAT scanner allows us to see sections of the living brain.

Figure 4.3.1 The brain is a remarkable organ of enormous complexity. So far our knowledge of it is somewhat primitive and rudimentary.

The **forebrain**. The **endbrain** at the end contains the olfactory lobes, and in higher vertebrates forms the cerebral hemispheres.

The **'tweenbrain** forms the pineal gland and the pituitary gland.

The **midbrain** contains the optic lobes.

The **hindbrain** forms the cerebellum and the medulla.

Cerebral hemispheres (cerebrum) – the seat of intelligence, this area controls the voluntary behaviour of the body along with learning, memory, personality and the ability to reason.

Thalamus – this processes all the sensory impulses before directing them to the appropriate area of the brain. It is also involved in the perception of pain and pleasure.

Hypothalamus – this coordinates the autonomic nervous system. It monitors the chemistry of the blood and controls the hormone secretions of the pituitary gland. It also controls thirst, hunger, aggression and reproductive behaviour.

Pituitary gland

Pons – literally 'bridge' – relays impulses to the cerebellum.

Medulla – the most primitive part of the brain, containing reflex centres which control functions such as the heart rate, blood pressure, breathing rate, coughing, sneezing, swallowing, saliva production and peristalsis.

Meninges – protective membranes which cover the brain. Between them is the **cerebrospinal fluid** which surrounds the central nervous system and fills any available spaces.

Corpus callosum – this area connects the left and right cerebral hemispheres, giving communication between the two.

Corpora quadrigemina – this controls the reflexes associated with sight and hearing. Its origin is the midbrain.

Cerebellum – the 'tree of life', so called because of the patterns of white and grey matter within it. It coordinates smooth movements and uses information from the muscles and ears to maintain posture and balance.

However, it is the development of increasingly large and complex brains that has led to the success of some groups of vertebrates. In the vertebrates the central nervous system develops as a hollow tube of nervous tissue which forms the **spinal cord**, containing the **grey matter** made up of the neurone cell bodies and the **white matter** consisting of the tracts of nerve fibres. At the front or **anterior** end of a vertebrate embryo, this tube swells and to some extent folds back on itself to form a **brain**. In some vertebrates this brain remains fairly simple, with areas very specific to particular functions such as sight or smell. The brain has three distinct areas – the **forebrain**, **midbrain** and **hindbrain**.

The human brain

In other vertebrates such as humans the brain becomes a remarkably complex structure. The original simple arrangement into three areas is very difficult to see because a part called the **cerebral cortex** is folded back over the entire brain. There are areas of the human brain with very specific functions concerned with the major senses and control of basic bodily functions. Equally there are many regions of the brain where the precise functions and interrelationships with other areas are not clearly understood. The basic pattern of a vertebrate brain along with a simplified representation of the human brain are given in figure 4.3.1.

Investigating the brain

Our understanding of the human brain has been developed using evidence from two different sources – the brains of other animals, and the human brain itself.

The brains of other animals

Over the years many investigations have been carried out on the brains of other animals from insects to monkeys. Some results have been related directly to the human brain, although this is not always possible. Examples include the effect of removing the cerebral hemispheres from animals such as dogs and monkeys, and also implanting electrodes to see the effect on behaviour of artificially stimulating a region of the brain. Many people find the idea of this type of research distasteful, and there are moves to minimise the experiments that are permitted. On the other hand, much information has been gained which has been put to good use in human medicine.

The brains of humans

Experiments such as those described above are obviously not carried out on humans. However, volunteers have allowed areas of their brains to be artificially stimulated during brain surgery under local anaesthetic. The volunteers have described the resulting sensations to the experimenters. As a result we have a fairly clear picture of how certain areas of the brain are associated with very particular functions (**localisation of function**).

Most of our information about the functions of the human brain has come from situations where parts of the brain are damaged or missing either at birth or as the result of illness or injury.

The brain at work

The way in which the brain works is still not fully understood. We know that it contains several hundred million cells working together. We also know that the great nerve tracts from the spinal cord cross over as they enter and leave the brain, so that the left-hand side of the brain receives information from and controls the right-hand side of the body, and vice versa. We are aware that the cerebral cortex is only about 3 mm thick, and yet controls most of those functions which make us what we are. Damage to the cortex affects our memory, intelligence, learning ability and decision-making skills. Sensory awareness and speech may be affected too. **Centres** or **nuclei** in the brain are made up of cell bodies which may have hundreds of synapses, making intercommunication between thousands and indeed millions of cells possible.

The role of the brain

Brains perform a vital role in coordinating the activities of many multicellular organisms. They give unconscious control over many vital bodily functions. They receive information from a vast array of sensory inputs, correlate it all and send out instructions for an appropriate response. But perhaps one of the most important functions of a brain is to allow the development of increasingly complex forms of behaviour and communication.

ANIMAL BEHAVIOUR

Intelligence and behaviour

Some invertebrates have very simple behaviour patterns which are almost entirely predictable. Take the sea anemone which we saw in figure 4.1.1, page 232. If you poke the sea anemone its tentacles withdraw. It will repeat this behaviour until it adapts to the stimulus and stops responding. If it is then left for a while its neurotransmitters will be replenished and the same pattern of behaviour will be repeated. In animals with bigger and more complicated brains, this type of simple, repetitive behaviour is seen less frequently. The elaboration of original behaviour patterns enables animals to become more efficient hunters, to avoid attack, to develop successful foraging strategies and to attract mates – they become generally more successful. Communication skills are also developed, making cooperative behaviour possible and improving group survival chances.

What do we mean by behaviour?

We continually observe animal behaviour in the lives of those species which coexist with us such as cats and dogs, hamsters and goldfish, birds and insects, but most particularly in the members of our own species. Not only do we observe the behaviour of others, we also behave ourselves. **Behaviour** can be defined as an action in response to a stimulus which modifies the relationship between the organism and the environment.

This sounds rather daunting – it means that animals respond to any change in their environment which might affect them, and attempt to make sure that their situation either remains the same or improves. A simple example is to imagine a fly buzzing around your head. You might brush it away, or swat it, or fetch the fly spray – you would act in some way to get rid of the irritating stimulus and return the situation to normal. Animal behaviour is not always as clear cut as this, but the principle behind it is the same.

Studying behaviour

Our knowledge of animal behaviour is based largely on observations of the way animals behave, either in their natural environment or in the laboratory. Both these approaches have advantages and disadvantages. Whilst animals are most likely to behave normally in their natural environment, they might travel great distances in a day, or live somewhere inaccessible to people, or be almost impossible to see. It is also impossible to control all the variables in a natural environment. In the laboratory all the surrounding conditions can be carefully controlled and observation is easy, but the animals are in an artificial setting and may not behave in a normal way. In vertebrates, more of these factors have to be considered when evaluating any observations made. Flatworms behave no differently whether the water they are in is part of a pond or in a Petri dish. Monkeys that are used to a natural environment behave quite differently in a cage or in a forest.

Figure 4.3.2 A problem in studying animal behaviour is our tendency to put ourselves in the place of others, and to interpret animal actions in the light of human responses (anthropomorphism). Cats used to a caged existence, or rats which frequently explore mazes in the interests of science, may well be healthy, well-fed individuals displaying normal behaviour for their species. But our natural response tends to be to feel sorry for them, perhaps even deciding that they look 'unhappy' – because we are mentally putting ourselves in their place and endowing them with our reactions to the situation. To draw any worthwhile conclusions about animal behaviour, we must be objective and dispassionate in our observations.

Much animal behaviour is either very rapid or very slow, and it is also frequently repetitive. Thus as well as direct human observations, time-lapse photography and video recordings are often used to analyse the sequence of events in a piece of behaviour. Because animal behaviour is more open to subjective interpretation and anthropomorphism than most other areas of biology, it is important to realise that observations and conclusions in this area of biology are still very much open to discussion.

In observing how animals progress from the reception of a stimulus to a behavioural response, two major categories of behaviour have emerged. The first is **innate** or **species-characteristic behaviour**. The second is **learned** or **individual-characteristic behaviour**. We shall consider the two types separately, but this distinction is largely for our convenience, as there is considerable overlap between the categories.

INNATE BEHAVIOUR

What is innate behaviour?

Innate behaviour is a large collection of responses which are usually seen in every member of a particular species, hence the term species-characteristic behaviour. This type of behaviour is not learned but is a genetically determined response to a particular stimulus. It occurs as a result of very specific nerve pathways laid down in the embryo from the instructions in the DNA of the organism. Innate behaviour covers an enormous range of types of response, from the simplest avoidance reflexes to highly complex courtship and territorial displays. The stimulus for a piece of innate behaviour will always elicit the same response, which has been selected over generations for its survival value. Some examples are shown in figure 4.3.3. In animals with relatively highly developed brains, innate behaviour frees the conscious areas of the brain for dealing with situations which present new problems for solution.

The courtship display of the bowerbird is a complex sequence of behaviour. Yet it is completely unlearned, an example of a piece of behaviour where it is vital that all members of the species respond in the same way to the stimulus. If only some birds knew the ritual, or not all completed the sequence, successful pair-bonding and mating would not occur.

Figure 4.3.3 Innate behaviour is not learned, and is usually important to the survival of the individual. Inborn responses vary from simple reflexes to complex sequences of behaviour.

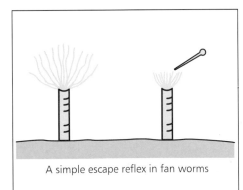

A simple escape reflex in fan worms

Woodlice demonstrate kinesis, a very simple type of innate response, when the protective cover of their rotten log is removed. They all move and rotate relatively rapidly in random directions until they arrive once more in a moist environment, when the movements slow down and stop.

Types of innate behaviour

Innate behaviour can be divided into categories which again are not rigid divisions, but consider the levels of complexity involved.

1   **Taxes** involve the whole organism moving in response to an external, directional stimulus – in other words, it moves towards or away from the stimulus. Taxic responses can be seen to a wide variety of stimuli in the animal kingdom. Examples of **phototaxis** (responses to light), **geotaxis** (to gravity), **chemotaxis** (to chemicals) and **rheotaxis** (resistance to movement) amongst others can easily be demonstrated. When the organism moves towards the stimulus it is a positive (+) taxis, and when it moves away it is a negative (–) taxis. Examples include earthworms and woodlice moving away from light (negative phototaxis) and moths flying into the wind (positive rheotaxis).

2   **Kineses** involve the whole organism moving in response to a stimulus, but not in a directional way. The rate of movement is related to the *intensity* of the stimulus, not to the direction from which it comes. For example, when woodlice are placed in a dry environment their rate of random movement and turning increases until they find themselves back in a more humid, damp area when the rate of movement decreases again.

3   **Simple reflexes** have been discussed in section 4.2. A very rapid response is made to a stimulus such as a potentially damaging situation or the presence of food.

4   **Instinct** is the most complex form of innate behaviour. Konrad Lorenz, one of the great researchers in animal behaviour, defined instinct as 'unlearned, species-specific motor patterns'. Instinctive actions may be inborn, but they can be very sophisticated. However, they show a high degree of stereotyping – the same behaviour is seen in all members of a species with little or no individual differences, and the same pattern of behaviour is always produced in response to a particular stimulus. Instinctive patterns of behaviour are usually of very direct survival value to an animal, involved with courtship, mating or defending territory for feeding or reproduction.

Invertebrates are frequently very short lived. They do not have time to rely on trial-and-error learning for their basic responses if they are to complete their life cycles successfully. Instinctive behaviour patterns equip them to cope with most of the situations they will meet. Vertebrates tend to live longer, and certainly some of them have plenty of time for learning. But innate behaviour is still important. Infants and young animals need instincts to enable them to survive until learning and experience take over. And even in adults, innate behaviour is an economy measure, a ready-made set of responses to a given situation leaving the higher areas of the brain free for other less basic functions.

LEARNED BEHAVIOUR

What is learned behaviour?

Learned behaviour is an adaptive change in the behaviour of an individual which occurs as the result of experience. This is why it is also known as individual-characteristic behaviour. Individuals learn as a result of their own experience, and modify their behaviour accordingly. No two members of a group or species will have identical experiences, and so learned behaviour is specific to each individual. An example of learned behaviour is the reaction of a child to touching a hot oven door. The hand will be withdrawn rapidly as a result of a piece of reflex, innate behaviour. But the child will learn from the

Figure 4.3.4 The permanence of any learned behaviour depends on the memory, and not everything that is learned remains in the memory very long. Facts swotted up for an examination may well be forgotten soon afterwards – but the knowledge of how to ride a bicycle, once learned, is usually with you for life.

experience not to touch the door again deliberately. This learning will vary from child to child. Some will need only one experience of the heat to modify their behaviour. Others will try the experiment several times before the change in behaviour is made. Some individuals may not even need to touch the hot door themselves – seeing another child's reaction or being warned by an adult may be enough.

Memory

For learning to occur, there must be an ability to **memorise** or store information. There appear to be two types of memory – **short-term memory**, which lasts only a few minutes, and **long-term memory** which is much more stable and can last for many years. The mechanism by which memory is laid down is not yet fully understood. It appears to involve synaptic changes and the synthesis of proteins. RNA is involved, as demonstrated by the experiments shown in figure 4.3.5. By injecting different fractions from previously trained flatworms, it was shown that only RNA-containing fractions had an effect on the learning ability of the recipients. But for an experience to bring about a long-term change in the behaviour of an organism, a memory of either the experience, the modified behaviour pattern or both is necessary.

Types of learned behaviour

We tend to assume that learning and learned behaviour happen only in vertebrates, and particularly in the mammals. In fact learned behaviour occurs in the vast majority of animal groups, as we shall see in considering some of the types of learned behaviour.

1 **Habituation** occurs when a stimulus is repeated many times and nothing happens – there is neither 'punishment' nor 'reward'. The stimulus is then ignored. This is not a simple adaptation of the sensory system like accommodation because once a response is habituated or lost it does not return. Examples include birds learning to ignore a scarecrow and babies learning not to 'startle' at every sudden noise. Habituation is particularly important in the development of young animals, as they have to learn not to react to the neutral elements in the world around them. For instance, the movement and noise of the wind must be ignored by a large number of animals or their nervous systems would be constantly firing off 'false alarms'.

2 **Conditioned reflexes** are the result of animals learning to associate new stimuli with an existing unconditioned reflex, as in the case of Pavlov's dogs mentioned in section 4.2.

3 **Trial-and-error (operant) learning** occurs when a piece of trial behaviour on the part of the animal is either rewarded (for example, food is found) or punished (for example, the animal is hurt). If the animal associates the outcome of a piece of behaviour with a reward, that behaviour is likely to be repeated. If the behaviour is associated with punishment, it is less likely to be tried again. The American psychologist B. F. Skinner did extensive studies on this type of learning using pigeons.

4 **Imprinting** is a simple and specialised sort of learning which only occurs in very young animals. At one receptive stage the young animal identifies with another organism, which is usually the parent, or if no parent is available on another large object. It will then follow this object and relate to other similar objects throughout its life.

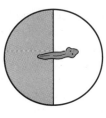

Flatworms move from a light area to a dark one. By giving a mild electric shock each time the dark area is entered, this tendency can be reversed so that the 'trained' worms move from dark to light.

Fragments of a 'trained' flatworm are injected into an untrained worm.

The injected flatworm learns dark avoidance more rapidly than the original worm did. Some of the 'memory' may have been passed on in chemicals from the injected tissues of the 'trained' worm.

Figure 4.3.5

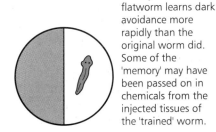

The importance of rewards and punishments in operant learning was demonstrated using pigeons and Skinner boxes. The use of these boxes has been extended to studies of other birds.

When young animals imprint on an adult of the wrong species, all kinds of problems can result!

Figure 4.3.6 Learning takes place in a great variety of ways throughout the animal kingdom. Humans use most of the available methods to increase our knowledge and understanding of the world around us.

5   **Exploratory (latent) learning** takes place when an animal explores new surroundings and learns them, without any immediate reward or punishment. The information may then be useful at another time.

6   **Insight learning** is based on thought and reasoning. It is mainly seen in the mammals, particularly the primates. It is regarded as the highest sort of learning. Once a problem has been solved, the solution is then remembered.

Some examples of these types of learned behaviour are illustrated in figure 4.3.6.

How behaviour is modified

Think about your own behaviour – do you always react in the same way to the same situation? The answer is almost certainly no. The behaviour of any animal is modified by the circumstances of the moment. The **motivational state** of an animal depends on a range of factors. The length of time since the last meal, the reproductive state, the presence of other individuals who may be dominant or lower in status and whether or not the animal is on its own territory will all affect both the innate and learned behaviour of an individual. An animal will not indulge in exploratory learning if it is very hungry. Sexual desire, aggression and fear may interfere with foraging or hunting. It is almost impossible to predict how an animal will behave unless all the factors of its mental and physiological state are known.

If an animal is torn between fighting and running away, or feeding and displaying to a potential mate, or any situation where two strong drives are involved, it will often show **displacement behaviour**. This involves an activity such as grooming or exploring, to take the pressure off until the situation resolves itself one way or the other and a course of action becomes clear.

The study of behaviour is fascinating and in a book such as this we can only scratch the surface. But the behaviour of animals is the end result of all that we have looked at in terms of nervous coordination and control – the importance of the sensory inputs, the computations of the brain and the messages to the effector organs are seen in the way that living organisms behave.

SUMMARY

- The **central nervous system** processes information and coordinates responses. Vertebrate central nervous systems consist of a well-developed **brain** and the **spinal cord**. The insect central nervous system comprises a **central nerve cord** and **ganglia**, collections of cell bodies and synapses.

- The vertebrate spinal cord has **grey matter** made up of neurone cell bodies and **white matter** containing nerve fibres.

- The vertebrate brain has specialised areas – the **forebrain** contains olfactory lobes and forms the cerebral hemispheres, the **'tweenbrain** forms the pineal gland and pituitary gland, the **midbrain** contains the optic lobes and the **hindbrain** forms the cerebellum and medulla.

- In the human brain the **cerebrum** controls voluntary behaviour and conscious thought. The **thalamus** coordinates the autonomic nervous system. The **cerebellum** coordinates smooth movements, posture and balance. The **medulla** contains reflex centres that control bodily functions.

- **Behaviour** is an action in response to a stimulus which modifies the relationship between the organism and the environment – it is the action of an organism to try to maintain or improve its situation. The study of behaviour is called **ethology**.

- In **innate (species-characteristic) behaviour**, an organism has a genetically determined (unlearned) response to a particular stimulus. **Taxes** are movements of the whole organism in response to an external directional stimulus. **Kineses** are non-directional movements of the whole organism to a stimulus. **Simple reflexes** involve a rapid response for example to a potentially damaging stimulus, and **instinctive behaviour** consists of sophisticated inborn actions specific to a species.

- **Learned** or **individual-characteristic behaviour** is an adaptive change in the behaviour of an individual as a result of experience, and depends on **memory**. **Habituation** is the ignoring of a stimulus which has been repeated with neither punishment nor reward. A **conditioned reflex** is the association of a new stimulus with an unconditioned reflex response. **Operant learning** is the result of trial and error, when an organism is either punished or rewarded. **Imprinting** occurs in young animals when they identify with a parent. **Latent learning** takes place when the organism explores and learns its environment. **Insight learning** is based on thought and reasoning.

- Behaviour is modified by circumstances (the **motivational state**). **Displacement behaviour** occurs when the most appropriate behaviour is unclear due to conflicting circumstances.

QUESTIONS

1 Identify and briefly describe the main areas of a mammalian brain.

2 Compare and contrast species-characteristic behaviour with individual-characteristic behaviour.

3 Discuss the following types of learning, giving examples of each:
   a trial-and-error learning
   b imprinting
   c insight learning.

4

Chemical control systems

Chemical control was mentioned in section 4.1 as an alternative to the nervous system. The nervous system is extremely effective at carrying messages rapidly from one specific place to another. However, in order to carry these messages relatively large amounts of energy need to be expended in the production of transmitter substances. The messages are carried along distinct pathways, and to give a maintained stimulus over a long period of time a constant stream of nerve impulses has to be sent.

Some of the functions of the body require long-term stimulation of tissues, for example, growth and sexual development, and in other cases it is necessary to send messages which have an effect on many different areas of the body simultaneously. Plants rarely need rapid responses but they have to coordinate and control their cells just as animals do. In these situations chemical messages are economical for the system to produce as they can have an effect over a long period of time. They can also reach the entire body as they are carried to their target organs in the transport system of the animal or plant. In plants, chemical control is the main system of coordination. In animals it interacts with and complements the nervous system.

CHEMICAL CONTROL IN ANIMALS

Hormones

Chemical control is brought about in animals by the action of **hormones**. These are organic chemicals produced by the body which are released into the blood or body fluid and bring about widespread changes. The changes your body undergoes during puberty and the sensations you experience before an interview or an examination are the result of hormone action.

Hormones are usually either proteins, parts of proteins such as polypeptides, or steroids. They are secreted by glands. The glands which produce the secretions of the gut release their juices along small tubes or **ducts**, and are known as **exocrine glands**. The glands which produce hormones do not have these ducts – they release the hormones directly into the bloodstream and are known as **endocrine** or **ductless glands**. Once a hormone enters the bloodstream it is carried around the system and will reach the target organ or organs, as figure 4.4.1 shows. The cells of the target organs have specific receptor molecules on the surface of their membranes which bind to the hormone molecules. This brings about a change in the membrane and elicits a response.

Most of the hormones described in this section will be mammalian hormones, although we shall also consider the role of hormones in the moulting of insects.

The positions of the endocrine glands

The endocrine glands are found around the body, often in association with other organ systems. Several of the glands have more than one function – for example, the ovaries produce ova as well as hormones, and the pancreas is both an exocrine gland producing digestive enzymes and an endocrine gland

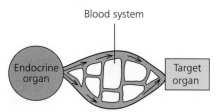

Figure 4.4.1 The pathway followed by a hormone from the endocrine gland where it is produced to the cells of the target organ. There are similarities to the pathway taken by a nerve impulse from the sensory neurone to the effector cell, but the chemical pathway is much less specific.

producing the hormone insulin. The glands all have rich blood supplies, with plenty of capillaries within the glandular tissue itself so that the hormones can pass directly into the blood when needed. The sites of the main glands in humans and the hormones they secrete are shown in figure 4.4.2.

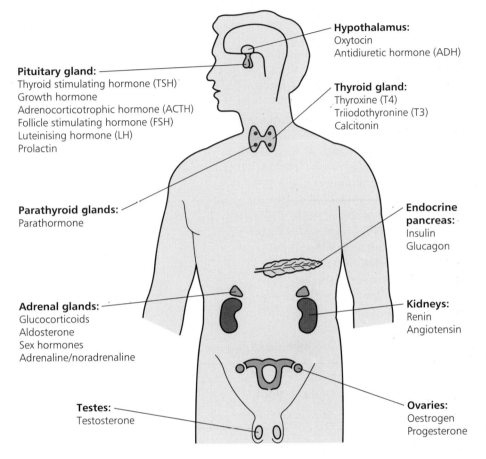

Hypothalamus:
Oxytocin
Antidiuretic hormone (ADH)

Pituitary gland:
Thyroid stimulating hormone (TSH)
Growth hormone
Adrenocorticotrophic hormone (ACTH)
Follicle stimulating hormone (FSH)
Luteinising hormone (LH)
Prolactin

Thyroid gland:
Thyroxine (T4)
Triiodothyronine (T3)
Calcitonin

Parathyroid glands:
Parathormone

Endocrine pancreas:
Insulin
Glucagon

Adrenal glands:
Glucocorticoids
Aldosterone
Sex hormones
Adrenaline/noradrenaline

Kidneys:
Renin
Angiotensin

Testes:
Testosterone

Ovaries:
Oestrogen
Progesterone

Figure 4.4.2 The endocrine organs of the body. Although small, these organs have a profound effect on the processes of life.

Differences between nervous and endocrine control

Table 4.4.1 summarises how nervous and endocrine control differ.

Nervous system	Endocrine system
Messages travel fast – generally have a rapid effect	Messages transported less rapidly – generally take longer to have an effect
Usually a short-lived response	Often a long-lasting response
Very localised effects as the impulse is transmitted to individual effector cells	Effects often widespread as the hormone is carried throughout the body in the bloodstream. A very specific response can be achieved by the siting of receptors.
Relatively few neurotransmitters – acetylcholine and noradrenaline most commonly used	Variety of hormones produced by the different organs, each hormone producing very specific effect

Table 4.4.1

Release of hormones

Hormones are released from the endocrine glands into the bloodstream in response to specific stimuli. The endocrine system interacts very closely with the nervous system. Some glands release their secretions as a result of direct stimulation by nerves. For example, the adrenal medulla of the adrenal glands releases adrenaline when it is stimulated by the sympathetic nervous system. The tissue of the adrenal medulla is so similar to the cells of the nervous system that it seems likely that they both form from the same origins in the embryo.

Many hormones are released from the endocrine glands in response to another hormone in the blood. As we shall see in more detail later, the pituitary gland in the brain secretes several hormones which directly stimulate other endocrine glands. Raised levels of certain chemicals such as glucose and salt in the blood can also stimulate the release of hormones, which in turn act to regulate the levels of the chemicals.

When hormones are released in response to nervous stimulation, the control of the release is simple. If the gland is stimulated, hormone is released. If it is not stimulated, no hormone is released. The level of stimulation determines the level of response. The situation is slightly more complex when hormones are released in response to a chemical stimulus such as another hormone or glucose. In this instance secretion is controlled by a **negative feedback loop**. The presence of the appropriate chemical in the blood stimulates the release of the hormone. As the hormone levels rise, the amount of stimulating chemical in the blood drops, as shown in figure 4.4.3. As a result of this, the endocrine gland receives less stimulation and so the hormone levels drop. This kind of feedback loop is very common, and it gives a sensitive level of control which can be constantly adjusted to the needs of the body. Negative feedback loops are also a common control feature in mechanical systems – for example, the thermostat of a central heating system works in this way.

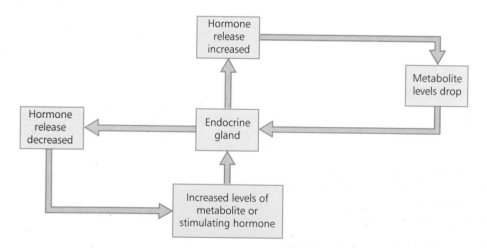

Figure 4.4.3 The basic principle of negative feedback. More specific examples will be seen when we consider the hormones insulin and thyroxine, and how their levels are controlled.

HUMAN HORMONES IN ACTION

The hormones of the pancreas

The level of glucose in the blood, often referred to as the blood sugar level, is of great importance to the cells of the body because they use glucose for respiration. If the level of glucose falls too low, the body cells are starved of energy. The cells of the brain are particularly vulnerable and a coma quickly results. Too much sugar in the blood affects the osmotic balance and water is lost from the body cells. Ideally a level of around 80–100 mg of glucose per 100 cm^3 of blood needs to be maintained for the optimum working of the body systems. This is achieved by the interactions of several hormones with the glycogen stores in the liver. Two of the most important of these hormones are produced by the pancreas.

The pancreas plays an important role in digestion by producing a mixture of enzymes which are released into the small intestine, as we saw in section 3.3. This is the function of the **exocrine pancreas**. But the pancreas is also a vital endocrine organ. Scattered within the enzyme-producing cells are groups of endocrine cells known as the **islets of Langerhans**. The islets contain two different types of cells and produce two different hormones. The large α (alpha) cells produce the hormone **glucagon**. The smaller β (beta) cells produce **insulin**. Both these hormones are relatively short-chain polypeptides. Insulin is well known for its role in the control of blood sugar levels, and the less well-known hormone glucagon is also involved in the regulation of the blood sugar level. Glucagon has the opposite effect to insulin.

Insulin and the β cells

The β cells of the islets of Langerhans are sensitive to a rise in blood sugar levels such as occurs after a meal has been digested. Raised blood sugar levels are known as **hyperglycaemia**. The effect of this is to stimulate the secretion and release of insulin by the β cells. When the blood sugar levels fall (**hypoglycaemia**) the secretion of insulin also falls. This interaction between glucose levels and insulin secretion is an example of a negative feedback loop – see figure 4.4.4.

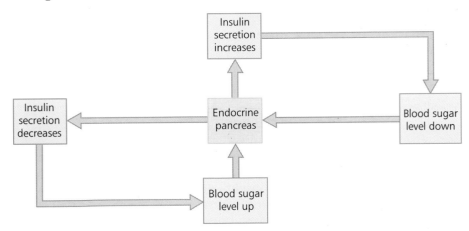

Figure 4.4.4 Negative feedback systems such as this allow the body to make constant small adjustments to the blood sugar level and so maintain it at the optimum level.

The majority of body cells have insulin receptors on their membranes, although there are a few exceptions such as the red blood cells. Once insulin is bound to the receptor sites on the membrane it lowers the blood sugar in one or more of four different ways, depending on the type of cell:

1 The rate of cellular respiration goes up, increasing the use of glucose.

2 The rate of conversion of glucose into the carbohydrate store glycogen is increased in cells such as the liver and the muscles.

3 The rate of conversion of glucose to fat goes up in adipose (fat storage) tissue.

4 The rate of glucose absorption goes up in muscle cells in particular.

The overall effect of all these strategies is to remove glucose from the blood and so reduce the level of hyperglycaemia. Insulin is the only hormone in the body which lowers the blood glucose level.

Glucagon and the α cells

The α cells are also sensitive to blood glucose levels, but they respond to a drop in blood glucose by secreting increased amounts of glucagon. Only the cells of the liver have receptors for this hormone. When the molecules of glucagon bind to the membranes of the liver cells the level of glucose in the blood is increased in two ways:

1 Glucagon increases the rate of conversion of glycogen stores in the liver into glucose.

2 Glucagon also increases the rate at which new glucose is formed from amino acids.

Figure 4.4.5 shows how glucagon and insulin interact to maintain a steady blood glucose level. Other hormones, particularly adrenaline, also increase the levels of glucose in the blood.

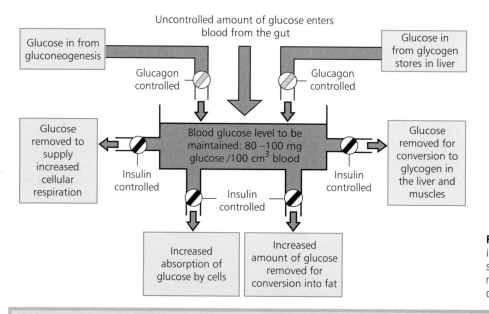

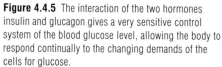

Figure 4.4.5 The interaction of the two hormones insulin and glucagon gives a very sensitive control system of the blood glucose level, allowing the body to respond continually to the changing demands of the cells for glucose.

Diabetes and its effects

In some people the pancreas does not function properly and insulin is not produced in sufficient quantity to control the level of glucose in the blood. In others, the cells of the body do not appear to have insulin receptors on their membranes. Whatever the cause, the effect is the same – the body cannot control the blood glucose level. After a meal high in carbohydrates the glucose level rises to a point at which the kidneys cannot cope (section 4.6 gives more details on the functioning of the kidney) and glucose appears in the urine. This, along with copious amounts of urine, is one of the classic symptoms of **diabetes mellitus** (literally, 'sweet fountain'). Obviously, diabetics need treatment of some sort as their cells cannot function properly without glucose as a fuel for cellular respiration.

When diabetes is caused by a failure of the β cells to produce sufficient hormone, the treatment is regular injections of insulin, originally extracted from pigs. More recently human insulin has been produced by bacteria as a result of genetic engineering. Diabetics need to monitor their diets carefully to avoid an overload of carbohydrates. They also need to inject insulin at an appropriate time to avoid going into a coma due to lack of glucose (hypoglycaemic coma) when the hormone is given without an intake of food. Insulin cannot usually be taken orally as, being a polypeptide chain, it is digested by the body, hence the need for injections.

Once diabetics become used to managing their own insulin and blood glucose levels they can lead perfectly normal and active lives, though they may suffer from long-term effects which are as yet unavoidable.

The hormones of the thyroid gland

The rate of cellular metabolism affects many aspects of life. Growth and development are closely linked to the metabolic rate and if this is too low, severe abnormalities can result. In adult life body weight and both physical activity and mental attitudes can be affected by the rate of the metabolism. The metabolic rate is controlled by the hormones of the **thyroid gland**.

The thyroid gland is shaped rather like a bow-tie and found in the neck in roughly the position where a bow-tie would be worn (see figure 4.4.2). It produces three hormones – **thyroxine** (T4), **triiodothyronine** (T3) and **calcitonin**. The first two of these are closely involved in the control of metabolism. Their secretion is controlled by a special releasing factor produced by the hypothalamus of the brain, and thyroid stimulating hormone produced by the anterior pituitary gland. This is a complex negative feedback system which is shown in more detail in figure 4.4.6.

Calcitonin plays an important role in calcium metabolism, lowering the blood levels of calcium ions by speeding up the absorption of calcium ions by the bones. Yet again a negative feedback loop is in operation – raised blood calcium levels stimulate the release of calcitonin, which in turn lowers the blood calcium levels and so reduces the amount of hormone secreted.

Iodine and the thyroid gland

Thyroxine and triiodothyronine are both synthesised by the follicle cells of the thyroid gland using iodine. All the iodide ions taken in by the body in the diet end up in the thyroid gland. This is useful in the treatment of overactive thyroid glands. Radioactive iodine-containing compounds can be administered in the sure knowledge that they will be concentrated in and destroy part of the thyroid gland, and other parts of the body will not be affected.

If the diet is lacking in iodine, the thyroid gland cannot make sufficient thyroxine. The feedback system goes into overdrive, continually stimulating the thyroid gland which gets larger and larger in a vain attempt to produce enough thyroxine. A swollen neck resulting from the enlarged gland is typical of this condition, known as **simple goitre**. Nowadays iodide is added to table salt in the developed world to prevent this condition, but it was so common during European history that it was regarded in some countries as normal. There are even paintings of the Madonna and child where Mary, presumably modelled on a local beauty of the time, shows a distinctly goitrous neck.

The hormones of the hypothalamus and pituitary

The **pituitary gland** in the brain has been described as 'the conductor of the endocrine orchestra' because of the role it plays in controlling the secretions of the other endocrine glands. The pituitary gland has an **anterior lobe** and a **posterior lobe**. It produces and releases secretions which affect the activity of most of the other endocrine glands in the body. However, whilst the pituitary is most frequently referred to as the 'master gland', control of the pituitary itself falls largely to the **hypothalamus**.

Control of the pituitary by the hypothalamus

The hypothalamus is a small area of brain directly above the pituitary gland. It carries out a variety of functions, one of which is to monitor the blood levels of a number of metabolites and hormones. In response to the levels of these chemicals the hypothalamus controls the activity of the pituitary gland.

Figure 4.4.7 shows the anatomical relationship between the hypothalamus and the pituitary. As the embryo forms, the posterior lobe develops as an

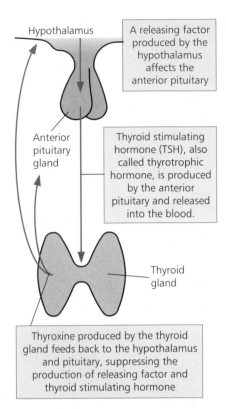

Figure 4.4.6 The hypothalamus is sensitive to the requirements of many areas of the body. Its secretions are involved in the control of the production of both thyroxine and triiodothyronine. Levels of thyroxine in the blood affect the levels of secretion of both the releasing factor and thyroid stimulating hormone, giving a very sensitive control system.

outgrowth of the hypothalamus itself, whilst the anterior lobe grows out from the roof of the mouth. Then the two parts fuse and the connection with the roof of the mouth is lost. But the two different origins of the parts of the gland are reflected in their different functions.

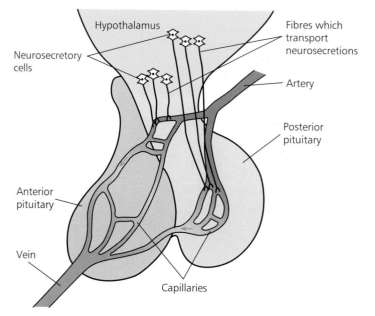

Figure 4.4.7 The close structural relationship of the hypothalamus and pituitary is reflected in their functions, the hypothalamus producing neurosecretions which control both lobes of the pituitary.

The hypothalamus contains some unusual cells known as **neurosecretory cells**. These are nerve cells which produce secretions from the ends of their fibres. One group of these cells (neurosecretory cells 1) produce substances which stimulate or inhibit the release of hormones from the anterior pituitary. They are known as **releasing factors** or **release-inhibiting factors**, depending on their function (see figure 4.4.6 for the role of a releasing factor on the control of thyroxine secretion). The other group of neurosecretory cells (neurosecretory cells 2) produce secretions which are stored in the posterior pituitary and then later released as hormones.

Secretions of the pituitary

The pituitary gland, under the control of the hypothalamus, produces six hormones from the anterior lobe and two from the posterior lobe. These range in function from controlling the secretions of the thyroid gland to the control of growth, from sexual development to the control of urine volume. The hormones produced and the roles they play are described in figure 4.4.8.

There are many other hormones which play important roles in communication and control in the body. We shall meet some of these later when we consider in more detail how the internal environment of the body is kept as constant as possible.

How do hormones have their effects?

Hormones act by binding to specific receptor sites on the membranes of their target cells. The hormone then affects the target cell in some way to bring about the desired change in activity. There appear to be three main ways in which hormones may have their effect:

1 The binding of the hormone molecule to a receptor site may result in the formation of a second chemical messenger inside the cell. This second messenger then takes effect by activating enzymes within the cell and so altering its metabolism. The most common second messenger is a

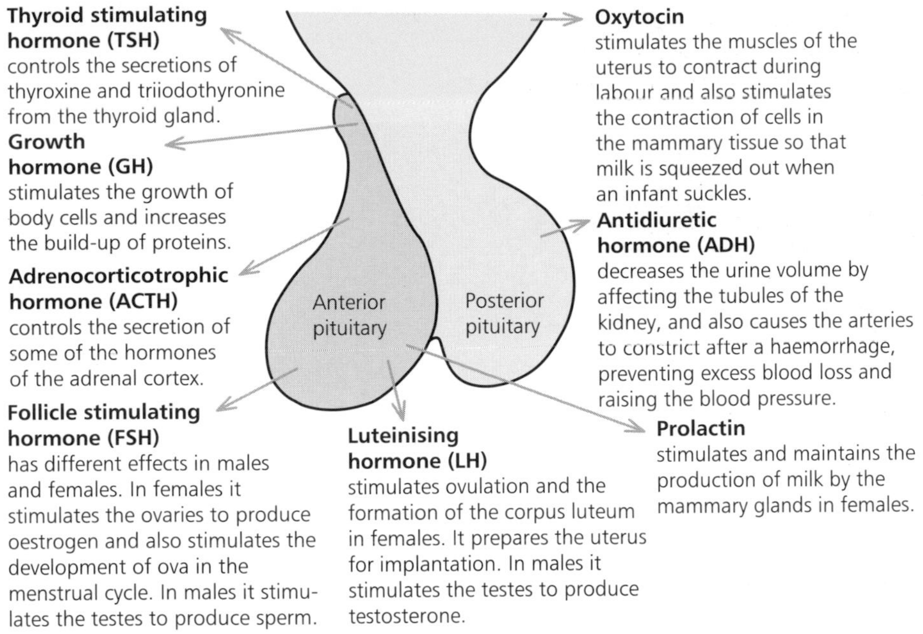

Thyroid stimulating hormone (TSH) controls the secretions of thyroxine and triiodothyronine from the thyroid gland.

Growth hormone (GH) stimulates the growth of body cells and increases the build-up of proteins.

Adrenocorticotrophic hormone (ACTH) controls the secretion of some of the hormones of the adrenal cortex.

Follicle stimulating hormone (FSH) has different effects in males and females. In females it stimulates the ovaries to produce oestrogen and also stimulates the development of ova in the menstrual cycle. In males it stimulates the testes to produce sperm.

Anterior pituitary

Posterior pituitary

Luteinising hormone (LH) stimulates ovulation and the formation of the corpus luteum in females. It prepares the uterus for implantation. In males it stimulates the testes to produce testosterone.

Oxytocin stimulates the muscles of the uterus to contract during labour and also stimulates the contraction of cells in the mammary tissue so that milk is squeezed out when an infant suckles.

Antidiuretic hormone (ADH) decreases the urine volume by affecting the tubules of the kidney, and also causes the arteries to constrict after a haemorrhage, preventing excess blood loss and raising the blood pressure.

Prolactin stimulates and maintains the production of milk by the mammary glands in females.

Figure 4.4.8 The hormones of the pituitary gland, particularly those from the anterior lobe, have their effect by stimulating another endocrine organ elsewhere in the body.

substance called **cyclic AMP**, which is formed from ATP. Adrenaline is thought to have an effect in this way.

2   The hormone may have a more direct effect, for example, it may change the permeability of the cell membrane to particular substances. Insulin works in this way, increasing the activity of glucose carriers across the membrane.

3   The hormone linked to its receptor may pass through the membrane and act as the internal messenger itself. In this case the hormone usually reaches the nucleus of the cell and turns on or off sections of the DNA. The lipid-soluble steroid hormones such as oestrogen and testosterone can pass through the membrane and act in this way.

HORMONES IN INSECTS

The control of moulting

Hormones and their mechanisms of action in most mammals are relatively similar to those in humans. But some of the most interesting light thrown on the action of hormones has come from considering the control of moulting (**ecdysis**) in insects. The hard exoskeleton of insects imposes limits on growth, so the development of many insects such as butterflies, fruit flies and bees occurs as a series of transformations from an egg through a variety of larval stages to the adult insect (**imago**). Each time an insect sheds its exoskeleton (moults) and 'grows' it becomes more mature. The moulting is controlled by two hormones:

1   **Ecdysone**, the 'moulting and metamorphosis' hormone, controls the events of the moult itself. It is a steroid hormone which was first extracted from the pupae of silkworms – it took 3 tons of silkworms to extract 100 mg of hormone!

2   **Juvenile hormone** determines the kind of moult that occurs. When juvenile hormone is present, another larval form results, and as juvenile hormone levels get lower more adult characteristics occur. When there is no juvenile hormone the pupa becomes an adult, as shown in figure 4.4.9.

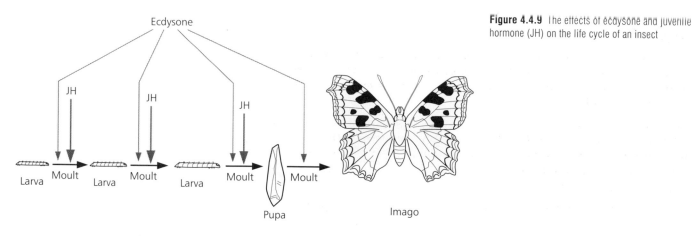

Figure 4.4.9 The effects of ecdysone and juvenile hormone (JH) on the life cycle of an insect

The way in which ecdysone has its effect has been studied using the larvae of *Drosophila* (fruit flies) and *Chironomus* (midge). In the cells of the salivary glands of these insects there are **giant chromosomes**, 100 times thicker and 10 times longer than normal chromosomes and easily visible with the light microscope. Bands visible on these chromosomes are thought to represent genes or small groups of genes. When insects are undergoing a moult, or when ecdysone is injected artificially into an insect, 'puffs' result on the chromosomes, as shown in figure 4.4.10. These chromosome puffs appear to be areas of genetic material made available for transcription, and they are very rich in RNA. Presumably they carry information about new proteins to be formed in a more adult stage of the life cycle. This supports the theory that some hormones such as steroids can have a direct effect on the DNA of a cell.

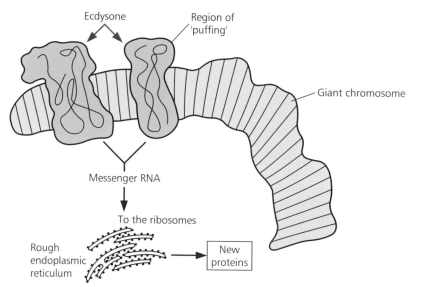

Figure 4.4.10 Chromosome puffs appear as a result of the hormone ecdysone. The puffing of the chromosome is followed by much RNA synthesis, and this in turn moves to the ribosomes and gives rise to the formation of new proteins. It is thought that many steroid hormones have their effect in this way. Unfortunately, not many organisms possess giant chromosomes for us to observe.

The role of hormones in animals

We have seen the basic roles played by hormones in animals, the feedback loops which control the secretion of the hormones and the ways in which they bring about their effects. There are many more hormones than those which we have studied here. Many others will be highlighted in later sections, particularly those on homeostasis and reproduction, where they will be considered more as part of an integrated whole. As we have seen, for animals hormones are but a part of a complex system of intercellular communication and control. For the plants which we go on to consider next, hormones give the only means of control.

SUMMARY

- Chemical control often has a long-term general effect on an organism, and in animals it is brought about by **hormones**. These are secreted by **endocrine** (ductless) glands in response to specific stimuli, which may be nervous or chemical.

- The release of hormones is often controlled by a **negative feedback loop**. The presence of a chemical in the blood stimulates the release of a hormone which reduces the level of the stimulating chemical. This reduced level then results in reduced stimulation of the gland and so less hormone is produced.

- The hormones of the **pancreas** play a major role in the control of the blood glucose level. The β cells of the **islets of Langerhans** secrete **insulin** when the blood glucose levels rise (**hyperglycaemia**). Insulin brings about a reduction in blood glucose. The α cells respond to a drop in blood glucose levels (**hypoglycaemia**) by secreting **glucagon**. This brings about a rise in the blood glucose levels. Insulin and glucagon interact to maintain a steady blood glucose level.

- The hormones of the **thyroid gland** control the rate of metabolism. The secretion of **thyroxine** (T4) and **triiodothyronine** (T3) are controlled by the **hypothalamus** in a complex negative feedback loop. **Calcitonin** lowers the calcium concentration in the blood and is controlled by a negative feedback loop. T4 and T3 are both synthesised using **iodine** and iodine is concentrated in the thyroid gland. Goitre is a swelling of the thyroid gland which may result from a lack of iodine in the diet.

- The **pituitary gland** controls the secretions of many of the endocrine glands. The pituitary is itself controlled by the **neurosecretions** (releasing factors and release-inhibiting factors) of the **hypothalamus**. The pituitary gland has an **anterior** and a **posterior lobe** and many of its secretions stimulate other endocrine glands in the body.

- Hormones bind to receptor sites on the membranes of their receptor cells. This may result in the formation of another messenger chemical inside the cell such as **cyclic AMP**. The hormone may have an effect on the cell such as changing its membrane permeability to a particular substance. The hormone–receptor complex may pass through the cell membrane and act as an internal messenger by acting on a part of the DNA.

- In insects hormones control **ecdysis** (moulting). **Ecdysone** controls the moulting, and **juvenile hormone** results in another larval form. Absence of juvenile hormone results in adult characteristics.

QUESTIONS

1 a What is a hormone?

b What are the functional and chemical similarities and differences between the nervous system and the endocrine system?

2 Describe the role played by the pancreatic hormones in glucose metabolism. What other glands and hormones affect the metabolism of glucose?

3 The pituitary gland is often referred to as the 'middleman' in endocrine functions. Explain why it is described in this way.

4.5 Control systems in plants

In animals the nervous and endocrine systems interact to ensure a high degree of coordination and control. The body of an animal can respond rapidly to the smallest change in either the internal or the external environment whilst long-term growth and development patterns are also maintained. For plants the situation is somewhat different. They do not, in general, need fast responses to small changes in their immediate environment. But plants do need to respond to factors such as light levels and direction, gravity and seasonal changes in conditions. They also need to coordinate growth of the cells in different areas. Plants do not appear to have nervous systems – their sensitivity and coordination is the result of chemical control alone.

PLANT RESPONSES

Chemical messages affecting growth

Plants respond to a variety of stimuli by producing or moving chemical messages. Many of these messages are similar to animal hormones – they are produced in one area of the plant, transported around the body of the plant and have their effect on cells elsewhere. Animals can respond to nervous and chemical messages in a variety of ways which include the release of further chemicals, the contraction of muscle cells and growth. The main way in which plants respond to their chemical messages is by growth. In some cases growth is stimulated while in others it is inhibited to bring about an appropriate response to the original stimulus. Sometimes one side of a plant grows more than the other, resulting in the bending of shoots or roots in response to a particular stimulus.

How plants grow

Growth is a permanent increase in the size of an organism or of some part of it. It is brought about by **cell division**, the **assimilation** of new material into the cells which result from the division and the **cell expansion** which follows. This is shown in figure 4.5.2, and there is more about growth in section 5.1. Cell expansion is particularly noticeable in plants, where rapid enlargement can occur as a result of water taken up by osmosis before the cell wall becomes rigid.

The main areas of cell division in plants are known as the **meristems**. These are areas which occur just behind the tip of a root or shoot. Not only are the meristems the main areas of growth, they are also particularly sensitive to the chemical messages produced. These chemical messages seem to make it easier for the cellulose walls to be stretched.

Stimuli affecting plants

Plants are a major part of our environment. Because their movements are usually invisible to the naked eye, we tend to look upon plants as being living but inert.

In fact plants respond to a variety of stimuli. They are sensitive to light, and not simply its presence or absence. Plants respond to the direction from which light comes, the intensity of the light and the length of daily exposures to it.

Figure 4.5.1 Chemical control allows plants to maximise their opportunities for photosynthesis and coordinate their reproductive cycles with the most opportune times of the year.

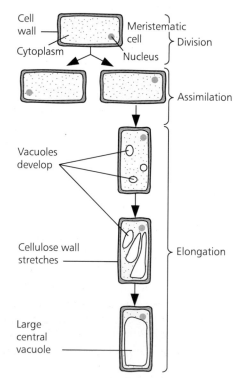

Figure 4.5.2 Plants respond to a variety of stimuli by differential growth. The growth regulating chemicals generally seem to affect cell elongation, making it easier for the cellulose cell wall to be stretched, although they also affect the cell division stage, increasing the number of divisions that occur.

They are also sensitive to gravity, to water, to temperature and in some cases to chemicals. Different parts of the same plant may react differently to the same stimulus (for example, shoots grow towards light but roots grow away from it). Not all plant responses involve growth – stomatal opening and flowering are both affected by light stimuli, for example. As well as these responses to external stimuli plants also respond to internal chemical signals. Most of the responses of plants are concerned either directly or indirectly with maximising the opportunities for photosynthesis and reproduction.

CHEMICAL CONTROL IN PLANTS

Tropic responses in plants

Once a seed begins to germinate in the soil, the shoot and root must keep growing if the developing plant is to survive. But growth must take place in the right direction. The shoot must grow up towards the light source which will provide the energy for its cells via photosynthesis. The roots must grow downwards into the soil which will provide support, minerals and water for the plant. The movements of these parts of the plant take place in direct response to external stimuli. The direction of the response is related to the direction from which the stimulus comes. Responses such as these are known as **tropisms**. Simple observations tell us that shoots bend towards the light, and even when developing seedlings are deprived of light they still grow upwards, away from the pull of gravity.

Evidence for tropisms

Much of the evidence for and work on tropisms has been carried out using germinating seeds and very young seedlings. This is because they are easy to work with and manipulate. As they are growing rapidly any changes in their growth show up quickly and tend to affect the whole organism rather than a small part as might be the case with a mature plant. The most widely used seedlings are those of monocotyledonous plants, usually cereals such as oats and wheat. This is because the shoot, when it emerges, is a single spike with no leaves apparent. This makes manipulation and observations easier than in dicotyledonous shoots. The newly emerged oat shoot is known as a **coleoptile**, although the more general term 'shoot' will be used. It must be remembered that these early shoots are relatively simple plant systems and that the control of the responses to light in an intact adult plant may well be more complex than our basic model allows. Figure 4.5.3 shows some simple experiments that demonstrate **geotropism** – the response to gravity.

Regardless of the orientation of a seed, the shoot will grow upwards and the root will grow downwards. This could be a response to light, or gravity, or both.

If the stimulus of light is removed, the shoot still grows upwards and the root downwards. This is a response to gravity and is called **geotropism**. Roots are said to be **positively geotropic** – they grow towards the force of gravity. Shoots are said to be **negatively geotropic** – they grow away from the force of gravity.

The stimulus of gravity is removed by placing a developing seedling on a klinostat. By rotating the drum at a constant speed the effects of gravity are applied evenly to the whole seedling – and the response of the root and shoot is lost.

Figure 4.5.3 As seeds generally germinate underground away from the stimulus of light, it seems likely that they should orientate as a result of gravity. This is in fact the stimulus involved, as these simple experiments show.

Phototropism

If plants are grown in bright, all-round light they thrive and grow more or less straight upwards. If plants are grown in even but low light, they also grow straight upwards, and in fact grow faster and taller than those in bright light.

But if the light is brighter on one side of the plant than another or only shines from one side (**unilateral light**) then the shoots of the plant will bend towards that light and the roots, if they are at all exposed, will grow away from it. Shoots are said to be **positively phototropic** and roots are **negatively phototropic**. This response has an obvious survival value for a plant. It helps to ensure that the shoots receive as much all-round light as possible, allowing the maximum amount of photosynthesis to take place. Also, if the roots should emerge from the soil – as they might do after particularly heavy rain, for example – they will rapidly return to the soil. But how are phototropisms brought about?

To answer this satisfactorily we need a picture of how growth is controlled in a shoot under conditions of all-round light. The control of growth was shown to be by chemical messages early in this century, by the Dutch plant scientist Went. He had the simple idea of attempting to block or collect any message that passed from the tip to the growing region behind it, and thus to show the nature of the message. He hypothesised that a chemical carrier could be collected in small blocks of agar and the effects then demonstrated on other shoots. Figure 4.5.4 gives a brief résumé of some of the work done on the mechanisms of phototropisms.

Figure 4.5.4 Experiments such as these gave rise to our understanding of the control of growth in plants. Substances referred to as growth hormones are produced at the tip of a shoot and move back to the dividing and elongating regions of the apical meristems, where they have their effect.

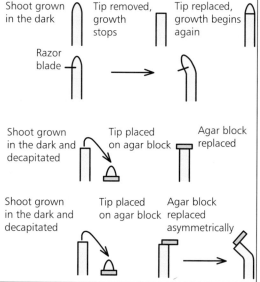

A shoot germinated in the dark grows straight upwards. When the tip is removed, the upward growth stops. If the tip is then replaced, growth begins again. This simple experiment shows that the tip of the shoot exerts an influence on the region of growing cells behind it.

A razor blade inserted into one side of the shoot just behind the tip stops the growth of the shoot on that side. As a result the shoot bends over. This shows that whatever normally stimulates growth is blocked by the razor blade, and therefore the message must be a chemical one, as an electrical signal would pass through the metal blade.

The tip of a shoot growing in the dark is removed, placed on an agar block and left for several hours. The decapitated shoot does not grow. The agar block is then placed on the cut end of the shoot, and normal growth is resumed. This again shows that a chemical message is produced in the shoot tip. It has diffused into the agar block and then diffuses from there into the rest of the shoot, stimulating normal growth. Subsequent experiments with blocks of cocoa butter showed no response in the decapitated shoot, demonstrating that the chemical message is water soluble and not fat soluble.

The tip of a shoot grown in the dark is removed and placed on a agar block. After several hours the block is placed asymmetrically on the decapitated shoot. The side with the agar block on it grows more than the other side, so that the shoot bends away from the stimulated side. Went showed that the amount of bending of the shoot is directly related to the amount of chemical messenger in the block. The biological activity of the messenger chemical can be measured or **assayed** in terms of the amount of bending observed in the shoot. This is known as the Went bioassay.

The people behind the theories

Work on the tropic movements of plants began a long time ago, and a succession of people developed experiments, each of which made the picture clearer. The following list mentions a few of these scientists and the contributions they have made to our understanding of how tropisms work.

- Nineteenth century: Charles Darwin and his son Francis carried out some experiments on oat coleoptiles and showed that the phototropic response of plants to light was due to some sort of message being passed from the root tip to the growing region.

- 1913: Boysen-Jensen, a Danish plant biologist, inserted a thin, impermeable mica plate into coleoptiles which appeared to act as a barrier to the message from the tip. This suggested that the message was chemical, but as mica does not conduct electricity these experiments could not rule out the possibility of electrical, nerve-like impulses.

- 1928: the Dutch plant physiologist Went proved the presence of a chemical transmitter substance in experiments such as those described in figure 4.5.4.

Unilateral illumination

So far we have considered shoots kept entirely in the dark or in full illumination. However, plants are usually in a situation where the light from one side is stronger than from the other. Experiments done on shoots illuminated from one side only (unilateral light) confirm the results from earlier experiments and add more detailed information of their own, as figure 4.5.5 shows.

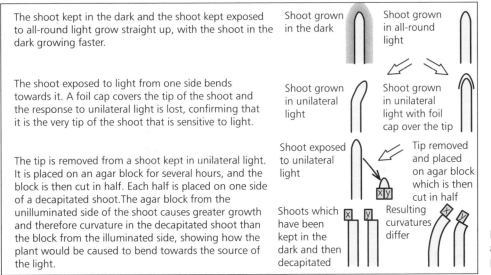

The shoot kept in the dark and the shoot kept exposed to all-round light grow straight up, with the shoot in the dark growing faster.

The shoot exposed to light from one side bends towards it. A foil cap covers the tip of the shoot and the response to unilateral light is lost, confirming that it is the very tip of the shoot that is sensitive to light.

The tip is removed from a shoot kept in unilateral light. It is placed on an agar block for several hours, and the block is then cut in half. Each half is placed on one side of a decapitated shoot. The agar block from the unilluminated side of the shoot causes greater growth and therefore curvature in the decapitated shoot than the block from the illuminated side, showing how the plant would be caused to bend towards the source of the light.

Figure 4.5.5 As a result of experiments such as these a more detailed hypothesis for the mechanism of phototropisms can be built up.

Auxins

The responses which we know as phototropisms are the results of chemical messages made in the tip of the shoot and transported to the growing region where they have an effect. The messages are known as **plant hormones** or **growth substances**. The growth substances involved in phototropisms are called **auxins**. Auxins are powerful growth stimulants and are effective in extremely low concentrations. The first auxin discovered was IAA (indoleacetic acid). The term 'auxins' covers a group of substances similar to IAA which have the same effect.

Auxins are now produced commercially. They can be bought in garden centres to help cuttings to root and are proving important in agriculture for improving the yields of crops.

How does light bring about its effects?

In any garden or woodland, plants can be seen responding to unilateral light. Where plants are partially shaded the shoots bend towards the light and then grow on straight towards it. This response seems to be the result of the way auxin moves within the plant under the influence of light.

Figure 4.5.6 shows a model in which the side of a shoot exposed to light contains less auxin than the side which is not illuminated. It appears that light causes the auxin to move laterally across the shoot, so there is a greater concentration on the unilluminated side. This in turn stimulates cell elongation and so growth on the dark side, resulting in the observed bending towards the light. Once the shoot is growing directly towards the light, the unilateral stimulus is removed. The transport of auxin stops and the shoot then grows straight towards the light. The original theory was that light destroyed the auxin, but this has been disproved by experiments along the lines of those in figure 4.5.6, which show that the levels of auxin in shoots are much the same regardless of whether they have been kept in the dark or under unilateral illumination.

How do plants grow more rapidly in the dark than in the light?

The mechanism of this response is not well understood. The old theory that auxin was destroyed by light gave a good model of how plants in the dark might grow more rapidly, as their auxin was not being destroyed. In contrast, plants in all-round illumination would be losing a lot of auxin and so growing less rapidly. However, the current theory based on evidence such as that shown in figure 4.5.6 does not provide such a clear explanation of the phenomenon and work on this continues.

1 The current model for the effect of unilateral illumination assumes that the auxin moves away from the light source. This gives a greater concentration of auxin on the shaded side of the shoot, which in turn causes increased cell elongation and thus bending of the shoot.

2 This experiment involved maize coleoptiles, some of which were kept intact and some divided. They were either exposed to unilateral light or kept in the dark. The tips were then removed and placed on agar blocks for several hours, before the agar blocks were placed on decapitated coleoptiles to assay the auxin levels.

When the shoots were kept in the dark it made no difference whether the shoot was split or not – the amount of auxin produced was virtually the same.

When the shoot was illuminated unilaterally the total amount of auxin was much the same whether or not the shoot was divided, and was also very similar to the level when the shoot was kept in the dark.

When the shoot was illuminated unilaterally but *not* divided, auxin accumulated on the dark side. In the divided shoot the auxin levels on both sides were the same. This evidence suggests that the normal situation in a shoot exposed to unilateral light is for auxin to be transported laterally (sideways) across from the lit side to the shaded side.

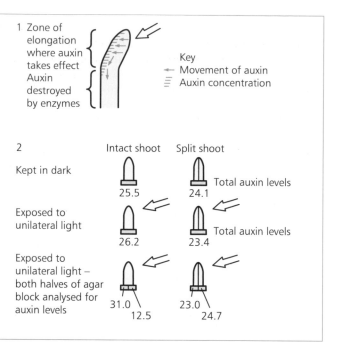

Figure 4.5.6 It appears that, whilst light may alter the distribution of auxin within the plant, it does not destroy it.

Etiolation

The fact that plants grow more rapidly in the dark than when they are illuminated can at first seem to be illogical. However, careful consideration shows that this is in fact useful to the plant. If a plant is in the dark it needs to grow upwards as rapidly as possible to reach the light and so be able to photosynthesise. Once it is in the light, a slowing of upward growth is valuable as it allows resources to be used for synthesising leaves, strengthening stems and generally consolidating the position.

This aspect of the response of plants to light is utilised by gardeners to 'force' plants. In order to give plants a good start or to develop a crop such as rhubarb particularly early, gardeners cover the plants so that they demonstrate this rapid upward growth known as etiolation, as shown in figure 4.5.7.

Geotropisms

The response of plants to gravity can be seen in the laboratory when seedlings placed on their sides are grown either in all-round light or in the dark (to eliminate the response to directional light). Figure 4.5.3 showed how geotropic responses can be demonstrated. Under normal conditions shoots are **negatively geotropic** and roots are **positively geotropic**. This makes good sense, as roots need to grow down into the soil and shoots need to grow up.

Figure 4.5.7 Rhubarb grown under a cover such as an upturned dustbin undergoes rapid growth (etiolation). These young rhubarb leaves are pale because they have been deprived of light. The end result of forcing in this way is advantageous to the gardener as it allows the crop to be brought forward and so better prices obtained.

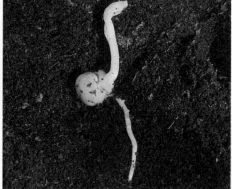

Figure 4.5.8 Both in 'real life' and in the laboratory plants can be seen to respond to gravity – shoots grow away from it and roots down towards it.

As we saw in the box 'Evidence for tropisms' on page 281, the response can be removed using a rotating drum known as a klinostat. Instead of receiving an effectively unilateral gravitational stimulus, by growing the plants on a klinostat rotating at about 4 revolutions per hour, the stimulus is applied evenly to all sides of the plant, and the root and shoot grow straight.

How does gravity bring about its effects?

The classic theory for the mechanism of geotropisms follows closely the pattern for phototropisms. It is thought that gravity causes auxin to build up on the lower side of the root and shoot, and that the cells of the two areas respond differently to the hormone. The cells on the lower side of the shoot are stimulated by the auxin to elongate and grow faster than the cells on the upper surface. This causes the shoot to bend upwards. In the root, on the other hand, the cells are inhibited by the raised auxin levels and so the growth and bending is downwards, as shown in figure 4.5.9.

This description of geotropism is very neat and fits in with what we know about phototropisms, but it begins to appear that auxin may not be the whole story. In fact, auxin may play only a minor part in the geotropic response, particularly in shoots. Many experimenters have been unable to find auxins in the root tips of oat coleoptiles, one of the most commonly used plants for work on tropisms. Two other plant growth regulators have been found and appear to be involved. **Abscisic acid** is a growth inhibitor. It has been shown to occur in the areas where differential growth takes place to give a response to gravity. Also in these regions, and particularly in the rapidly growing sides of roots and shoots, **gibberellins** (growth promoters) are found. (These growth inhibitors and promoters are discussed in more detail on pages 287–9.) Thus the response of plants to gravity may well be more complex than was at first thought and research into this continues.

Plant sensitivity to gravity

The way in which plants appear to sense gravity is not dissimilar from the method used by mammals (see section 4.2, page 257). It depends on particles being affected by gravity and that effect in turn causing a response in the organism. In some plants the particles which move in response to gravity appear to be very large starch grains which occur in certain root and shoot cells. They are known as **starch statoliths** and are contained within organelles called **amyloplasts**. The cells containing these starch grains are the **statocytes** – the gravity receptors. The starch grains fall to the lower sides of the cells under the influence of gravity. This aggregation of amyloplasts shown in figure 4.5.10 affects the distribution of growth substances in the region and results in a geotropism. The way in which this is brought about is not yet clearly understood. Also, there are plants without observable statoliths which nevertheless respond to gravity.

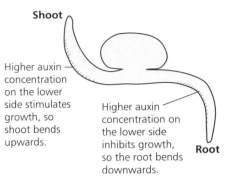

Figure 4.5.9 Geotropism has been assumed until recently to be due to the auxin response shown here – but recent research suggests there may be more to it than this.

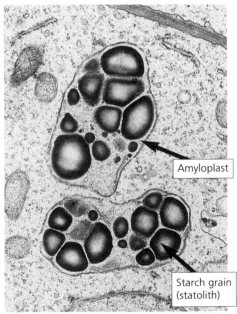

Figure 4.5.10 Raising the temperature of a seedling increases the rate at which the starch grains fall to the lower surface of the cells, and also increases the speed at which the seedling responds to gravity. This and other evidence confirms that these statoliths are the sensory mechanism for gravity in plants, but exactly how their movement is converted into a geotropic response is still not fully understood.

Does auxin have the same effect on all cells?

The effect that auxin has on the cells of the shoot and the root depends on its concentration:

- Relatively high levels of auxins stimulate growth in the shoot. Levels of auxins that would stimulate root growth are too low to have an effect on the cells of the stem.
- Relatively low levels of auxins stimulate growth in the roots. The higher levels of auxins that stimulate growth in the shoot actually inhibit growth in the root.

How are auxins used?

So far we have looked at auxins involved in the tropic responses of plants to both light and gravity. Their involvement in plant responses to their environment is far greater than this, however. They are needed for a wide variety of processes to occur successfully. Biologists have studied the natural role of these remarkable chemicals and then produced both naturally occurring and artificial versions in order to manipulate the 'auxin effect' in a variety of areas of plant life. The following list shows just some of these.

1   Auxin from the main apical bud inhibits the growth of side branches from lateral buds on stems. This is known as **apical dominance**. If the apical bud is removed the inhibition on the growth of side buds is lifted and lateral buds develop. If auxin is placed experimentally on the cut apical stem, the inhibition is continued. Gardeners prune plants regularly to remove the apical buds, which results in lateral growth (bushiness). They may not all understand the underlying plant physiology, but they know that it works.

2   Auxin stimulates the growth of adventitious roots from a cut stem. This is used commercially and by gardeners when taking plant cuttings. Dipping the end of a stem cutting in auxin-containing powder dramatically increases the chances of the cutting developing roots and 'taking'. However, excess rooting hormone may inhibit lateral root growth.

3   Auxin helps fruit to set – spraying with auxin greatly increases the natural success rate for pollination and fertilisation. It can also bring about fruit formation even in the absence of fertilisation, which has the added bonus of producing seedless fruits.

4   Auxin helps to prevent the fruit of a plant falling before it is ripe, which is important for successful reproduction.

5   Some synthetic auxins have a greatly exaggerated and abnormal effect on the growth of plant cells. They also affect the metabolism of the cells, making them respire excessively. These effects cause the death of the plant. As a result, synthetic auxins can be used as very effective weedkillers. They are absorbed much more effectively by broad-leaved (dicotyledonous) plants than by monocotyledons and are therefore particularly useful for removing broad-leaved weeds from monocotyledonous cultures such as lawns and wheat fields.

OTHER PLANT RESPONSES

Nasties

Although many plant responses are tropisms to light, gravity, chemicals or touch, plants also exhibit another type of response known as a **nasty** or **nastic response**. Nastic responses are to rather general stimuli, and involve differential growth in one part of a plant which results in a localised response. Nastic movements are often observed using techniques such as time-lapse photography. Although thermonasties and photonasties are well described, the mechanism by which they are brought about is not yet known.

Thigmotropisms

Some of the most interesting and least understood plant responses are in response to touch. Plants such as vines and sweet peas have tendrils which coil round other plants or sticks in response to one-sided contact or touch. These **thigmotropisms** seem to involve similar mechanisms of auxins and differential growth as the other tropic responses we have seen. But some plants react to touch in a much more unexpected way. Some of the carnivorous plants such as the Venus flytrap in figure 4.5.11 respond very rapidly, snapping shut to capture their prey insect. *Mimosa pudica*, better known as the 'sensitive plant', also reacts very rapidly to touch. The leaflets fold up immediately and, if the touch is really rough, the stems collapse too. The mechanisms of these rapid responses are not understood, although some workers have recorded electrical activity in the cells. Recently, proteins with contractile properties closely related to the muscle proteins actin and myosin have been found. Some plants may have a response system which is closer to animal nerves and muscles than had previously been considered possible.

PLANT GROWTH REGULATORS

As well as giving relatively immediate responses to stimuli such as unilateral light, gravity and touch, plants respond to changes on a much longer time scale. The growth of plants follows a seasonal pattern, with growth occurring in the spring, and fruit developing and then ripening at the appropriate time. Deciduous trees lose their leaves before the adverse conditions of winter, when the light levels are so low that photosynthesis would not produce sufficient food to support a large plant structure. Seeds exhibit dormancy, a period of inactivity before they germinate and begin to grow. These are all responses to the seasonal changes in light levels, length of daylight and temperature. They are brought about by a variety of plant growth regulators and chemical messages. Substances which regulate growth in plants may be naturally occurring plant hormones such as IAA, gibberellic acid and abscisic acid, or they may be synthetic compounds which, although not naturally occurring within the plant, nevertheless affect its growth.

Gibberellins

Gibberellins are involved in the growth of stems in particular. The effect of these growth promoters is best demonstrated by their absence. Dwarf plants are bred so that they do not produce gibberellins, as figure 4.5.12 shows. Auxins have no effect on the growth of dwarf plants but if gibberellins are applied to them, they will grow to normal height.

Gibberellins were discovered in the 1920s as a result of work done in Japan on a fungus which attacked the rice crop. Affected rice seedlings grew very tall and spindly, then either died or produced a very low yield. Chemicals extracted from culture solutions on which the fungus *Gibberella* had grown were found to give this effect and were named 'gibberellins'. They have subsequently been found to be naturally occurring in many plants.

Gibberellins have their effect on plant stems in the same way as auxins, by stimulating cell elongation in particular. They also seem to have an effect on the dormancy of seeds, with gibberellin levels rising towards the end of dormancy. Gibberellins are also produced in the germinating seeds of some cereal seeds, where they in turn stimulate the production of enzymes which break down the food stores in the seed so that the embryonic plant can develop.

Cytokinins

Cytokinins are a group of plant growth regulators which are found particularly in regions of very active cell division. They occur in very small quantities and are most readily extracted from fruits and seeds where they seem to be involved in the growth of the embryo. Cytokinins are involved in

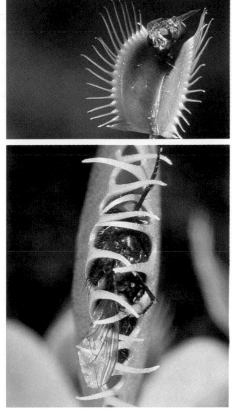

Figure 4.5.11 The Venus flytrap needs a series of three touches to the sensitive hairs inside its jaws before it snaps shut, trapping the fly or other insect attracted by its sweet sticky secretions. Anyone who has tried to swat a fly will know how rapid this movement must be to be successful. The requirement for three touches is a defence against unnecessary responses to touches by a leaf or by an insect which immediately flies away.

Normal bean plants

Dwarf bean plant

Figure 4.5.12 The difference between the plants is the result of an absence of gibberellins in the dwarf plant.

many plant responses because of their effect on plant growth. They stimulate cell division, but only if auxin is present too. As auxin equally cannot stimulate cell division without cytokinins, it appears that the two substances interact. They work together to affect the dividing cells. This is an important difference between plant and animal hormones. Animal hormones work quite independently of each other. They may be complementary or antagonistic to each other, but they are produced and have an effect without other hormones necessarily being present. In plants, the hormones bring about their effects as a result of interactions with each other. A single hormone is usually ineffective.

Abscisic acid

The growth regulating substances we have considered so far are all growth promoters – they stimulate the growth of the plants in one way or another. **Abscisic acid** (**ABA**) is a growth inhibitor. It was discovered in the 1960s during investigations into the loss of fruits from the cotton plant. It has an inhibitory effect on auxins, gibberellins and cytokinins, and seems to be involved in the production of a weakened area of cells at the base of a fruit or leaf which finally breaks as the fruit or leaf falls. Abscisic acid also seems to be involved in dormancy in seeds and the prevention of germination. Recent research indicates that abscisic acid may play a role in geotropisms, and also that its role in leaf fall may be considerably less than was originally thought.

Ethene

The chemicals which we have so far considered as plant growth regulators are all fairly complex (see box 'The chemicals that control plants') and are carried around the body of the plant in solution. But there is another substance produced by plants which does not fit into this general pattern – the gas **ethene**. This is produced by plants in small amounts and seems to be involved in several responses. The effect of ethene on the plant is to raise the respiration rate, and this causes the ripening of fruit. Imported fruits are often transported unripe, when they are less likely to become damaged or over-ripe and so be unsaleable. They are then exposed to low levels of ethene to ripen them for the supermarket shelves. Ethene can also cause leaf fall, and the release of buds and seeds from dormancy.

Traces of ethene are produced by oil-fired central heating boilers, which can have unfortunate effects on both house plants and bowls of fruit. The plants lose their leaves and the fruit gets over-ripe very quickly.

Synergism and antagonism

Many of the growth substances just described do not work on their own but by interaction with other substances. By this means, very fine control over the responses of the plant can be achieved. The growth regulators interact in one of two ways. If they work together, complementing each other and giving a greater response than each regulator alone, the interaction is known as **synergism**, for example, the effect of auxin on growth is much more dramatic if gibberellin is present as well. If the substances have opposite effects, for example one promoting growth and one inhibiting it, the balance between them will determine the response of the plant. This is known as **antagonism**.

Our knowledge of plant growth regulators is still far from complete. The mechanisms by which they have an effect are far from understood, and there may yet be more regulators to be discovered.

The chemicals that control plants

Table 4.5.1 summarises the main plant growth regulators and their structures.

Regulator	Chemical structure	Main functions
Auxins	Indole / Ethanoic acid, CH_2COOH	Promote stem growth by cell elongation. Stimulate root growth at very low concentrations. Involved in apical dominance and tropisms.
Gibberellins		Promote stem growth of the internodes by cell elongation. Promote fruit growth. Break seed dormancy and involved in germination.
Cytokinins		Promote cell division in the apical meristems and the cambium. Interact with auxins in these areas, but promote lateral bud growth, helping overcome apical dominance.
Abscisic acid		Inhibits cell division and growth in stem and root. Promotes dormancy in both buds and seeds.
Ethene		Inhibits growth of stem and root. Promotes fruit ripening and fruit and leaf fall.

Table 4.5.1

THE EFFECTS OF LIGHT ON PLANTS

The importance of light

As we have already seen, light is fundamental to the existence of plants in a wide variety of ways. Light-dependent photosynthesis is the ultimate source of food not only for plants but for animals as well. Chlorophyll formation depends on light and the phototropic responses already discussed are made in response to light. Day length is the cue which determines changes such as bud development, flowering, fruit ripening and leaf fall in many plants. Without light the metabolism of a plant is severely disrupted and prolonged light deprivation causes death.

'Sensory' systems in plants – phytochrome

There must be a mechanism for sensing the presence of light in order for a plant to respond to it. Plants do not have a nervous system. But just as the pigments in the rods and cones in the human eye undergo chemical changes in the presence of light, so plants seem to have evolved a chemical photoreceptor system. The details of the way in which plants sense and respond to light have not yet been fully worked out, but it appears to involve a photoreceptor known as **phytochrome**.

For many years it has been known that the seeds of many plants will only germinate if they are exposed, even very briefly, to light. Much of the research on this was done using some varieties of lettuce seeds which need only a very short flash of light to trigger germination. Researchers in the US Department of Agriculture did some classic work in the 1950s which showed that red light (wavelength 580–660 nm) is most effective at stimulating germination whilst far red light (wavelength 700–730 nm) actually inhibits germination.

If the lettuce seeds are exposed to a flash of red light, they will germinate. If they are exposed to a flash of red light followed by a flash of far red light, they will not germinate. If experimenters expose the seeds to a series of flashes of light, it is the colour of the final flash which determines whether or not the seeds will germinate (see figure 4.5.13). As a result of this and other work it was hypothesised that plants contain a pigment which reacts with the different types of light, and then in turn affects the responses of the plant. In 1960, the American group carrying out the research isolated this theoretical pigment from plants and called it **phytochrome**.

The phytochrome light receptor system

Phytochrome is a blue-green pigment which exists in two interconvertible forms. P_R or P_{660} absorbs red light and P_{FR} or P_{730} absorbs far red light as shown in figure 4.5.13. When one form of the pigment absorbs light it is converted reversibly into the other form. The length of time it takes for one form of the pigment to be converted into the other depends on the light intensity. In low light intensity it takes minutes, in high light intensity it takes seconds.

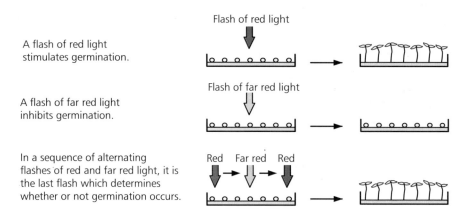

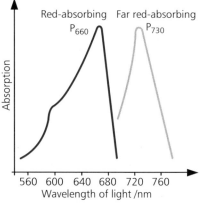

The absorption spectra of the two forms of phytochrome – the peaks of absorption give the two forms their symbols, P_{660} and P_{730} (after Hendricks).

Figure 4.5.13 The idea of a phytochrome light receptor system in plants developed from a theory to a fact once the theoretical light-sensitive chemical was extracted and analysed in 1960.

When P_R absorbs red light it is converted rapidly into P_{FR}. When P_{FR} absorbs far red light it is converted rapidly into P_R, and this conversion also takes place very slowly in the dark. P_R is the more stable form of the pigment, but it is P_{FR} which is biologically active.

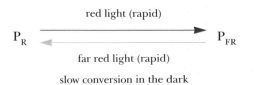

As normal sunlight contains more red light than far red light, the usual situation in a plant during daylight hours is for most of the phytochrome to be in the far red form. During the night it is all converted back into the red form. The control of the germination of lettuce seedlings by flashes of red and far red light seen in figure 4.5.13 is due to a flash of red light producing P_{FR}, the biologically active form of the phytochrome which then initiates germination, but a flash of far red light converting P_{FR} back to the inactive form.

In some cases phytochromes have a stimulating effect on growth in plants while in others they have an inhibitory effect, as table 4.5.2 shows. How the phytochromes influence the responses of plants is not fully understood. It appears that the presence of phytochromes may stimulate the production of other growth regulators and plant hormones, and thus the response to light is brought about.

Part of plant affected	Effect of red and far red light
Seed	Red light stimulates germination, far red light inhibits germination.
Stem	Stem elongation is stimulated by far red light and inhibited by red light. Exposure to far red light gives the same effect as etiolation.
Leaf	Leaf expansion is stimulated by red light and inhibited by far red light.
Lateral roots	Growth of lateral roots is stimulated by far red light and inhibited by red light.
Flowering	The stimulus to flower depends on alternating periods of light and dark and involves the phytochrome system.

Table 4.5.2 The main effects of red and far red light on various areas of a plant give an indication of the importance of the phytochromes.

Photoperiodism

One of the best known plant responses involving phytochromes is the control of flowering in some plants, known as **photoperiodism**. At the equator the days and nights are always of approximately the same length, around 12 hours each. In temperate areas of the world, at some distance from the equator, the lengths of the days and nights change throughout the year so that the period of daylight can vary between around 9 and 15 hours out of 24. In these temperate areas the lengths of the days and nights give important physiological cues to both animals and plants, directing their growth, development and behaviour. In plants one of the most clearly affected activities is flowering.

The day length affects the flowering of many plants. Some plants only flower when the days are short and the nights are long. They are known as **short-day plants** (**SDPs**) and examples include strawberries, chrysanthemums, cockleburs and the tobacco plant. Others flower when the days are relatively long and the nights short. These are the **long-day plants** (**LDPs**) which include snapdragons, cabbages and henbane. Short-day plants will not flower with more than a critical amount of daylight, and long-day plants will not flower with less than a critical amount of daylight. It can be very difficult to decide whether a plant is a short- or long-day plant, as the two groups merge. Yet other plants are unaffected by the length of the day. Plants such as cucumbers, tomatoes and pea plants which flower regardless of the photoperiod are known as **day-neutral plants**.

The value of these different flowering patterns to plants seems to be to take advantage of different circumstances. In temperate regions woodland short-day plants tend to flower in spring and autumn, taking advantage of the Sun at times when the full canopy of leaves either has not developed or has fallen off. Short-day plants are also found near the equator, where the days are never longer than about 12 hours. On the other hand, long-day plants flower in the summer in temperate regions, and are found further from the equator where there are very long days for part of the year.

In spite of the naming of these different types of flowering plants as 'long-day' or 'short-day', it has subsequently been discovered that it is the length of the period of darkness rather than the length of daylight which affects flowering. For example, if short-day plants have their long night interrupted by flashes of light, they do not flower. Some experiments demonstrating this are summarised in figure 4.5.14.

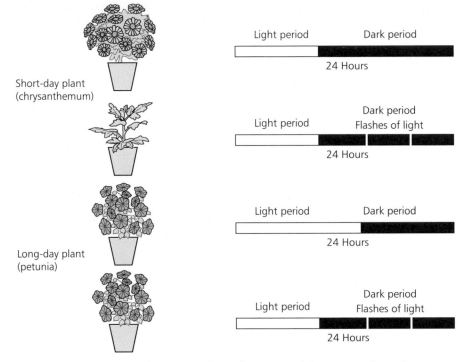

Short-day plant (chrysanthemum)

Long-day plant (petunia)

Figure 4.5.14 Experiments such as these show that it is the length of the period of darkness rather than of light which determines when plants will flower.

This discovery has led to a number of commercial ventures based on changing the time at which a plant flowers by manipulating the periods of darkness to which it is exposed. As a result, breeders have crossed varieties of plants which would otherwise not have flowered simultaneously, and at Christmas time we are awash with poinsettias and chrysanthemums which would not naturally be flowering.

How is the day length signal received?

All the research points to the involvement of the phytochromes in the sensitivity of plants to the photoperiod. The changes in flowering patterns which can be brought about by disturbing the dark periods can also be affected by red or far red light alone. It has been found that red light inhibits the flowering of short-day plants. If the red light is followed by far red light, the inhibition is lifted. What is the implication of this?

Red light leads to the formation of P_{FR}. Far red light converts this back to P_R. The current hypothesis is that P_{FR} inhibits flowering in short-day plants, and the lack of P_{FR} when it is exposed to far red light allows flowering to occur. It is thought that it is the lack of P_{FR} rather than the build-up of P_R which allows the flowering to go ahead. As the two forms of phytochrome are almost always present to some degree in a plant, it is the balance between them which is affected by varying periods of light and dark, and which in turn affects flowering. In long-day plants the situation is reversed. It appears that a build-up of P_{FR} during the daylight hours stimulates flowering.

The detection of the photoperiod seems to take place in the leaves of the plant. Experiments have been carried out where the whole plant has been kept in the dark apart from one leaf which is exposed to the appropriate periods of light and dark. Flowering has occurred as normal, whereas a plant kept in total darkness does not flower. How the message received in the leaves by the phytochromes is carried to the flower buds is not yet understood. The presence of a plant hormone known as **florigen** has been hypothesised, as illustrated in figure 4.5.15. This would be made in response to the levels of phytochromes and be carried in the plant transport system to the flower buds. However, no one so far has been able to demonstrate or isolate this hormone successfully, and so the mechanism for the control of flowering remains to be fully clarified.

Thus we can see that plants which may seem unresponsive in fact have complex systems of coordination and control. These systems appear to be almost entirely chemical in nature, and some are very similar in their mechanisms to the hormones used in animal responses. By interactions between their various growth regulators, plants are well adapted to respond to small changes in their environment, maximising their opportunities for survival and successful reproduction.

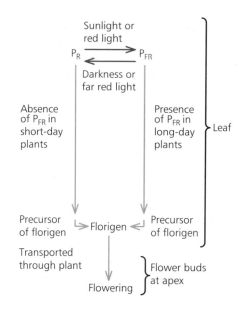

Figure 4.5.15 Until further research either confirms or disproves the theory, this is the best picture we have for the effect of the photoperiod on the flowering of long- and short-day plants.

SUMMARY

- Plants are controlled by chemical messages which bring about their actions by controlling growth. Growth is brought about by **cell divisions**, the **assimilation** of new material into the cells and then **cell expansion**. The **meristems** are the main areas of growth and are most sensitive to chemical control.

- Directional responses in plants to directional stimuli are called **tropisms**.

- Shoots are **positively phototropic** (they bend in unidirectional light towards the light source) and roots are **negatively phototropic** (if exposed to light they grow away from it).

- Experiments have shown that tropisms are controlled by **plant hormones** produced at the tip of a shoot which are transported to the meristems where they have an effect.

- **Auxins** are growth substances that bring about phototropism. It is thought that light causes auxin to move across a shoot to the dark side and promote growth there, causing the shoot to bend towards the light.

- Plants grow more rapidly in the dark (**etiolation**) than in the light so that they can have more chance of reaching the light.

- Shoots are **negatively geotropic** (they grow away from gravity) and roots are **positively geotropic.** It is thought that this is brought about by the action of auxin along with **abscisic acid**, a growth inhibitor, and **gibberellins**, growth promoters. Plants appear to sense gravity by large starch grains called **statoliths** which fall under the influence of gravity inside organelles called **amyloplasts** in cells called **statocytes**.

- Pruning of apical buds promotes side branches to grow, normally inhibited by auxin produced at the apical bud. Auxin stimulates growth of adventitious roots from stems so is used in plant rooting preparations. Auxin helps fruits to set, and prevents them falling prematurely. Some auxins are used as weedkillers, as they cause abnormal exaggerated growth.

- **Nastic responses** are localised responses to general stimuli, such as flowers opening in response to rising temperature.

- **Thigmotropisms** are responses to touch, such as tendrils coiling around supports or the Venus flytrap snapping shut.

- **Gibberellins** are plant growth regulators which affect stems particularly. **Cytokinins** stimulate cell division in the presence of auxin. **Abscisic acid** has an inhibitory effect on auxin, gibberellins and cytokinins and is thought to be involved in the weakening of fruit and leaf stems before they fall. **Ethene** raises the respiratory rate and causes ripening of fruit, leaf fall and the release of buds and seeds from dormancy.

- The complementary action of growth regulators is called **synergism**, while the inhibition of one regulator by another is **antagonism**.

- Plants respond to light by the action of a photoreceptor called **phytochrome**. P_R absorbs red light and is converted to P_{FR}, which absorbs far red light and is reconverted to P_R. P_{FR} also converts slowly in the dark to P_R. P_R and P_{FR} have opposing biological activities.

- Phytochromes control **photoperiodism**, the flowering of plants in response to the day length. **Short-day plants** flower when the nights are long and **long-day plants** flower when the nights are short. **Day-neutral plants** flower regardless of the photoperiod. It is thought that P_{FR} inhibits flowering in short-day plants, and lack of P_{FR} allows flowering to happen. In long-day plants a build-up of P_{FR} stimulates flowering. The detection of the photoperiod takes place in the leaves and may be transmitted to the flower buds by a hypothetical hormone called **florigen**.

QUESTIONS

1 a How does growth occur in plants?
  b To what stimuli do plants respond?
  c Explain how plants respond to unilateral light.

2 a List *five* plant growth regulators and briefly describe their role in plants.
  b Plant growth substances are frequently synergistic or antagonistic to each other. What does this mean?
  c The response of plants to gravity is known as geotropism. How are plants sensitive to gravity?

3 a How are plants sensitive to light?
  b Produce a leaflet for sale in a garden centre explaining how people might bring on or delay the flowering of particular plants. You will need to explain how flowering is initiated in plants and what is meant by photoperiodism.

Sections 4.1–4.5 have considered the communication systems available to animals and plants which allow them to respond to changes in both their internal and their external environments. Nervous systems and hormones give animals the control they need, whilst the responses of plants to their environment are brought about mainly as the result of chemical coordination. But why are these systems of coordination and control so important?

HOMEOSTASIS

Cells are extremely sensitive to changes in their environment. They only function properly within relatively narrow limits of temperature and pH, they require particular concentrations of nutrients such as glucose and they can tolerate very little build-up in the levels of waste products. To minimise any changes in the cellular environment, there are several systems which work to keep the internal conditions of the body as stable as possible.

The maintenance of a steady internal state is known as **homeostasis** (from the Greek *homoios* meaning 'like' and *stasis* meaning 'state'), and it is vital to the successful functioning of any organism. Homeostasis in animals involves a high level of coordination and control. Any changes in the composition of the blood are detected by a sensor or receptor. This causes an effector to work to reverse the change. Feedback systems provide a very sensitive way of maintaining the concentration of a substance within a very narrow range. The nervous and hormonal systems interact to maintain a steady state in the body. There are four main homeostatic systems in mammals:

1   the control of the water balance of the body (**osmoregulation**), brought about largely by the kidney

2   the control of the levels of nitrogenous waste products and other solutes, brought about by the kidney and the liver

3   the control of the blood sugar level by the pancreas and the liver (discussed in section 4.4)

4   the control of temperature, brought about largely by the skin (**thermoregulation**).

In this section we shall look at osmoregulation in animals, and the role of the mammalian kidney in osmoregulation and in the excretion of nitrogenous waste products. We shall also consider osmoregulation in plants. Section 4.7 looks at the control of other waste products by the liver, and thermoregulation.

OSMOREGULATION

The need for osmoregulation in animals

As we saw in section 1.4, water moves into and out of cells by the process of osmosis. If the balance of water and solutes inside and outside a cell is not correct, water may enter, causing the cell to swell and burst. On the other hand, water may leave so that the cytoplasm becomes shrunken and unable to function.

The problem of osmoregulation affects animals living in the sea rather differently from those in fresh water or on the land. The water potential of the sea is similar to that of cells, although the sodium chloride concentration is higher. Many marine invertebrates cannot osmoregulate, and have no need to do so. For many marine vertebrates the problem is getting rid of excess salt taken in when the salt water is drunk. In fresh water, there is a great tendency for water to move into cells by osmosis, whilst on land the conservation of water becomes a priority. Animals have evolved to overcome the challenges of their particular environment, but species which can move freely between the different environments are relatively few and need sophisticated osmoregulatory systems.

Figure 4.6.1 The contractile vacuole is an effective way of controlling the composition of the cytoplasm for a protoctist such as *Amoeba*. It removes the excess water which enters by osmosis in an energy-dependent process, and so prevents osmotic damage. If the protoctist is given a metabolic poison which interferes with the utilisation of ATP the contractile vacuole no longer fills, the water entering by osmosis is not removed and the cell bursts.

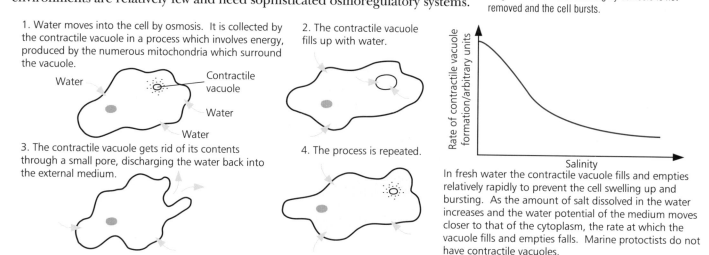

1. Water moves into the cell by osmosis. It is collected by the contractile vacuole in a process which involves energy, produced by the numerous mitochondria which surround the vacuole.

2. The contractile vacuole fills up with water.

3. The contractile vacuole gets rid of its contents through a small pore, discharging the water back into the external medium.

4. The process is repeated.

In fresh water the contractile vacuole fills and empties relatively rapidly to prevent the cell swelling up and bursting. As the amount of salt dissolved in the water increases and the water potential of the medium moves closer to that of the cytoplasm, the rate at which the vacuole fills and empties falls. Marine protoctists do not have contractile vacuoles.

Unicellular organisms living in fresh water have a specialised **contractile vacuole** for osmoregulation, as shown in figure 4.6.1. Because the whole cell is immersed in water and the surface area:volume ratio is high, a great deal of water enters the cell. In contrast, many larger animals have relatively few cells in contact with water and a reduced surface area:volume ratio. Water still enters the cells but this can be overcome by **kidneys** producing large quantities of very dilute urine.

Osmoregulation in fish

Fish demonstrate clearly the different problems which face organisms living in fresh and salt water, as figure 4.6.2 describes.

Figure 4.6.2 The osmotic problems facing fish in fresh and salt water, and the ways in which the difficulties are overcome.

Water balance in a salt-water teleost fish

The body fluids of salt-water teleost fish are hypotonic to their surroundings. As a result water tends to move out of the fish by osmosis, a process which would quickly lead to dehydration of the tissues.

Sea water is swallowed to make up fluid and the filtration rate of the kidney is slow, so relatively little urine is produced. The urine is isotonic with the sea water.

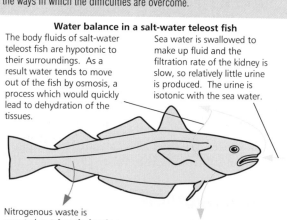

Water balance in a freshwater teleost (bony fish)

Salts are reabsorbed in the kidney tubules to prevent excess loss in the copious urine. The urine is very hypotonic to the blood.

The body fluids of freshwater teleost fish are hypertonic to their environment. As a result water moves into the fish by osmosis across the gills and through the lining of the mouth and pharynx.

Chloride secretory cells in the gills actively take up salts from the external medium to make up for those lost in the copious urine. Chloride ions are 800 times more concentrated in the blood of the fish than in their watery environment as a result.

Large amounts of very dilute urine are produced by the kidneys to remove excess water.

Nitrogenous waste is excreted as **ammonia**, highly toxic but very dilute in the large volumes of water excreted.

Nitrogenous waste is excreted as **trimethylamine oxide** which is very soluble but non-toxic. This means it can be excreted in a concentrated solution without harming the fish or upsetting the water balance.

Salt is actively removed by the **chloride secretory cells** in the gills, moving chloride ions against a concentration gradient in the opposite direction to that in freshwater fish.

Osmoregulation in mammals

Land-dwellers need to conserve water as they must drink all that they need. In these organisms osmoregulation involves the internal environment rather than the interface between the cell and the outside world. The cells of a land-living mammal are surrounded by tissue fluid which comes from the blood capillaries running close to them. By controlling the water potential of the blood (its water content and solute concentration), the body can control the water potential of the tissue fluid and so protect the cells from osmotic damage. Osmoregulation in mammals is largely brought about by the kidneys.

STRUCTURE OF THE MAMMALIAN KIDNEY

The kidneys of mammals are capable of producing waste fluid which is more concentrated than (hypertonic to) the body fluids, and this allows mammals to conserve water and inhabit relatively inhospitable environments.

In humans, as in other mammals, the kidneys are a dark reddish brown in colour and are attached to the back of the abdominal cavity. They are surrounded by a thick layer of fat which helps to protect them from mechanical damage. They control the water potential of the blood that passes through them, removing substances which would affect the water balance as well as getting rid of **urea**, the nitrogenous waste product of protein breakdown. (The production of urea by the liver is discussed in section 4.7, page 310.) The kidneys produce a fluid called **urine**, which is collected and stored in the **urinary bladder**. This is emptied at intervals when full. The position of the kidneys in the body and the overall structure of the organs is shown in figure 4.6.3.

THE KIDNEY AT WORK

Functions of the kidney

The kidney carries out three main functions in its osmoregulatory role:

1 **ultrafiltration**
2 **selective reabsorption**
3 **tubular secretion.**

In humans, blood passes through the kidneys at a rate of 1200 cm^3 per minute, which means that all the blood in the body travels through the kidneys and is filtered and balanced approximately once every 5 minutes. The functions of the kidney are carried out by the individual nephrons. Each nephron is about 12–14 mm long, and there are about 1.5 million of them in each kidney. This means that there are many kilometres of tubules in the kidneys, all engaged in filtering and balancing the blood. The three major functions occur in different areas of the kidney. The first stage is ultrafiltration, and this takes place in the Malpighian bodies.

Ultrafiltration

Ultrafiltration occurs due to a combination of very high blood pressure in the glomerular capillaries, and the structure of the Bowman's capsule and glomerulus. The high blood pressure develops in the capillaries because the diameter of the afferent vessel is greater than that of the efferent vessel. The high pressure squeezes the blood plasma out through the pores in the capillary wall. All the contents of the plasma can pass out of the capillary – its wall retains only the blood cells. The basement membrane between the

Figure 4.6.3 The kidney is a complex organ carrying out a vital role in homeostasis. It is involved in osmoregulation, in the removal of nitrogenous waste products, in the pH balance of the blood and in the removal of certain drugs from the system.

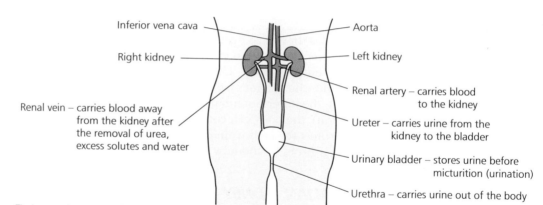

Inferior vena cava

Right kidney

Renal vein – carries blood away from the kidney after the removal of urea, excess solutes and water

Aorta

Left kidney

Renal artery – carries blood to the kidney

Ureter – carries urine from the kidney to the bladder

Urinary bladder – stores urine before micturition (urination)

Urethra – carries urine out of the body

The human urinary system. Blood from the body is passed through the kidneys where urea, excess salts and water are removed and form urine. The urine is stored in the bladder and released from the body at intervals.

The internal structure of the kidney. The relationships between structure and function become clearer when the details of the nephron are revealed.

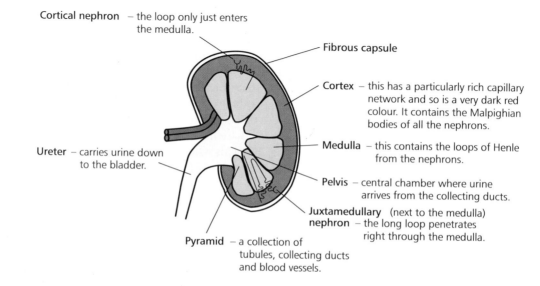

Cortical nephron – the loop only just enters the medulla.

Fibrous capsule

Cortex – this has a particularly rich capillary network and so is a very dark red colour. It contains the Malpighian bodies of all the nephrons.

Medulla – this contains the loops of Henle from the nephrons.

Pelvis – central chamber where urine arrives from the collecting ducts.

Juxtamedullary (next to the medulla) **nephron** – the long loop penetrates right through the medulla.

Ureter – carries urine down to the bladder.

Pyramid – a collection of tubules, collecting ducts and blood vessels.

The nephron of the kidney – a highly complex structure which removes waste products from the blood and conserves water for the body.

Bowman's capsule – the site of ultrafiltration of the blood. The ultrafiltrate from the glomerulus passes into the Bowman's capsule through specialised cells known as **podocytes**.

Malpighian body – found in the cortex, made up of the Bowman's capsule and the glomerulus. About 200 μm in diameter, it can just be seen with the naked eye.

Afferent vessel bringing blood into the glomerulus

Efferent vessel taking blood away from the glomerulus – the blood has been reduced in volume by 15–20% by ultrafiltration.

Glomerulus – a tangled knot of capillaries in the Bowman's capsule. The pressure of blood in the glomerulus is high as the efferent vessel carrying blood away from the Malpighian body is narrower than the afferent vessel which brings the blood into it. Fluid is forced out of the capillaries under this pressure, passed through the basement membrane which retains the plasma proteins and then the remaining ultrafiltrate passes into the Bowman's capsule and so into the rest of the nephron.

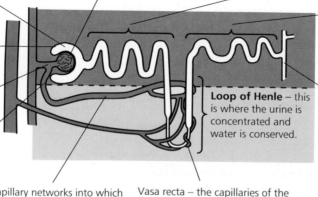

Capillary networks into which substances are reabsorbed

Vasa recta – the capillaries of the medulla which run closely with the loops of Henle.

Loop of Henle – this is where the urine is concentrated and water is conserved.

Proximal or **first convoluted tubule** – active reabsorption of the glucose and of about 67% of the sodium ions which have left the blood occurs here, which also causes passive reabsorption of water and chloride ions. Microvilli on the lining epithelial cells give a greatly increased surface area for reabsorption to occur. The filtrate is considerably reduced in volume before it reaches the loop of Henle.

Distal or **second convoluted tubule** – water is lost from the urine in this part of the tubule under the influence of the hormone ADH, and certain ions and drugs are actively secreted into it from the blood. It is involved in the concentration of the urine, the removal of unwanted substances from the blood and also in the control of the blood pH.

Collecting duct – water is reabsorbed from this region by osmosis. The amount of water lost depends in part on the permeability of the tubule to water, which in turn is affected by the hormone ADH.

In the body the kidney is surrounded by a thick, protective cushion of fat.

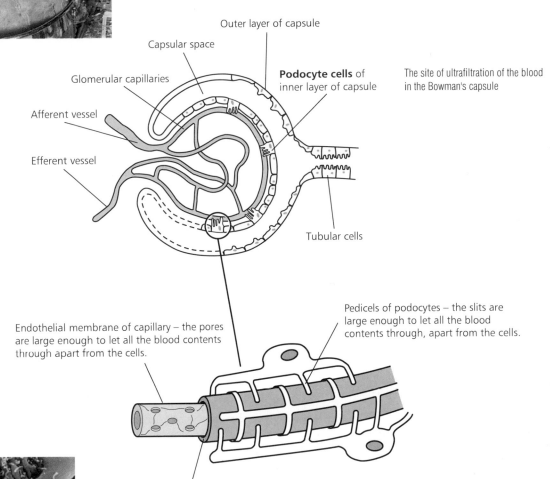

Capsular space

Outer layer of capsule

Glomerular capillaries

Afferent vessel

Efferent vessel

Podocyte cells of inner layer of capsule

The site of ultrafiltration of the blood in the Bowman's capsule

Tubular cells

Endothelial membrane of capillary – the pores are large enough to let all the blood contents through apart from the cells.

Pedicels of podocytes – the slits are large enough to let all the blood contents through, apart from the cells.

Basement membrane of capillary – the site of ultrafiltration as plasma proteins cannot pass through this membrane.

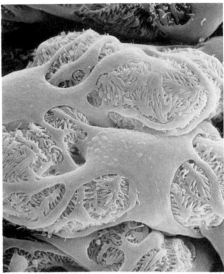

The complexity of the cellular structures in the Malpighian bodies can be seen on this electron micrograph. About 100 litres of blood are filtered through the pedicels every hour.

capillary and the Bowman's capsule acts as the filter, preventing platelets and the large plasma proteins from passing through. The structure of the wall of the Bowman's capsule ensures that any cells which may have left the capillary do not get through into the tubule itself. The filtrate which enters the capsule contains glucose, salts, urea and many other substances in the same concentrations as in the blood plasma, as table 4.6.1 shows.

Substance	Approximate concentration/g dm^{-3}	
	In plasma	*In filtrate*
Water	900.0	900.0
Protein	80.0	0.0
Inorganic ions	7.2	7.2
Glucose	1.0	1.0
Amino acids	0.5	0.5
Urea	0.3	0.3

Table 4.6.1 The compositions of human plasma and glomerular filtrate

The process of ultrafiltration is very efficient, with up to 20% of the water and solutes from the plasma entering the nephron. If all of this filtrate was then passed out of the body, we would produce around 200 dm^3 of urine a day – the equivalent of about 350 milk bottles! Obviously then, ultrafiltration is only the first step in the process, and the kidney has more work to do before the urine finally leaves the organ.

Selective reabsorption

Ultrafiltration removes urea, the waste product of protein breakdown, from the blood. However, it also removes a lot of water and all the glucose, salt and other substances which are present in the plasma. Much of the water and salt may be needed by the body. Glucose is a vital energy supply for the body and is never, under normal circumstances, excreted.

The ultrafiltrate is hypotonic to (less concentrated than) the blood plasma. Therefore, after the ultrafiltrate has entered the nephron, the main function of the kidney tubule is to return to the blood most of what has been removed.

The first convoluted tubule

The cells of the first or proximal convoluted tubule are covered with microvilli which greatly increase the surface area through which substances can be absorbed. Figure 4.6.4 shows that the cells also have large numbers of mitochondria, indicating that they are involved in active processes.

In this first part of the nephron over 80% of the glomerular filtrate is reabsorbed into the blood. All of the glucose, amino acids, vitamins and hormones are taken back into the blood by active transport. About 85% of the sodium chloride and water are reabsorbed as well. The sodium ions are actively transported, and the chloride ions and water follow passively along diffusion gradients. Once the substances are removed from the tubule they pass by diffusion into the extensive capillary network which surrounds the tubules. The blood is then removed so there is a constant concentration gradient along which this diffusion occurs. By the time the filtrate reaches the loop of Henle it is isotonic with the tissue fluid which surrounds the tubule. The amount of reabsorption that occurs in the first convoluted tubule is always the same – the fine tuning of the water balance takes place further along the nephron.

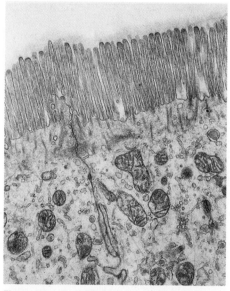

Figure 4.6.4 The cells of the first convoluted tubule show a high level of adaptation to their function. The brush border of microvilli provides a large surface area and the mitochondria a source of energy (ATP).

The loop of Henle

The loop of Henle is found in the medulla of the kidney, in close contact with the network of capillaries known as the vasa recta. The loop of Henle is the region of the kidney tubule which enables mammals to produce hypertonic urine, an important adaptation to the mammals as a group. The way in which this is brought about is complicated and not fully understood, but the current understanding is as follows.

The loop of Henle is shown in figure 4.6.5. It has a **descending limb** which leads from the first convoluted tubule. This limb is permeable to water and runs down into the medulla. A hairpin bend at the bottom of the loop leads into the **ascending limb**. This passes back up through the medulla until it joins the second convoluted tubule. The ascending limb of the loop is effectively impermeable to water. The second convoluted tubule leads into the collecting duct which then travels through the medulla to the pelvis of the kidney.

The concentrated urine is formed because the loop of Henle creates a diffusion gradient in the tissue of the medulla, which means that when water is in short supply it leaves the collecting duct by diffusion as it passes through the medulla and enters the blood vessels of the vasa recta. How does it work?

The simplest way to consider what happens in the loop of Henle is to begin with the filtrate entering the descending limb. This is isotonic with the blood. As the filtrate travels down the limb, the external concentration of sodium and chloride ions in the tissue fluid of the medulla increases. As a result, water diffuses out of the loop into the tissue fluid and then back into the blood of the vasa recta. Sodium and chloride ions also move into the limb by diffusion along concentration gradients. By the time the filtrate reaches the hairpin bend at the bottom of the loop it is very concentrated – hypertonic to the arterial blood.

As the filtrate moves up the ascending limb of the loop, chloride ions are actively secreted out of the tubule into the tissue fluid of the medulla, and sodium ions follow passively along an electrochemical gradient. It is this which causes the tissues of the medulla to have such high salt concentrations. However, because the ascending limb is impermeable to water, water cannot follow the ions and so the filtrate left in the ascending limb becomes increasingly dilute again, although the volume is not substantially changed. These processes are summarised in figure 4.6.6.

Once the dilute filtrate reaches the top of the ascending limb it is hypotonic to the blood again, and it then enters the second convoluted tubule and collecting duct, discussed below.

The lengths of the loops of Henle present in the kidney of a mammal are a good indication of the water stress under which it lives. Animals such as jerboas which live in deserts and have very little to drink have extremely long loops of Henle which enable them to retain the maximum amount of fluid and produce very small quantities of extremely concentrated urine.

The second convoluted tubule

The second convoluted tubule is permeable to water, but the permeability varies with the levels of the hormone ADH. It is here and in the collecting duct that the balancing of the water needs of the body takes place. If there is too little salt in the body, sodium ions may be pumped out of the tubule with chloride ions following. Water also leaves by diffusion if the walls of the tubule are permeable.

The collecting duct

The permeability of the collecting duct is strongly affected by the hormone ADH. Water moves out of the collecting duct as it passes through the medulla, with the urine becoming steadily more concentrated. Because the level of

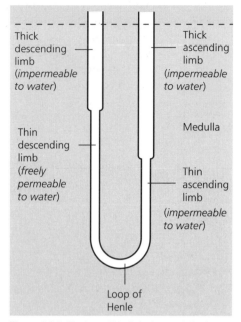

Figure 4.6.5 The loop of Henle is the specialised part of the nephron that enables mammals to produce urine which is more concentrated than their own blood.

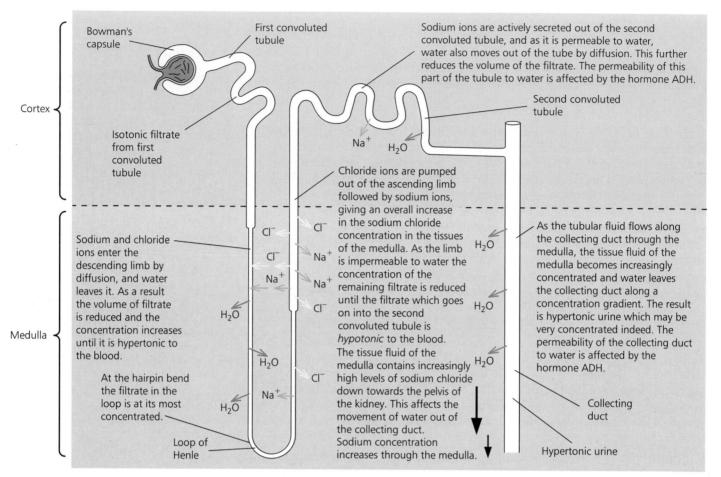

Bowman's capsule

First convoluted tubule

Sodium ions are actively secreted out of the second convoluted tubule, and as it is permeable to water, water also moves out of the tube by diffusion. This further reduces the volume of the filtrate. The permeability of this part of the tubule to water is affected by the hormone ADH.

Second convoluted tubule

Cortex

Isotonic filtrate from first convoluted tubule

Na$^+$ H$_2$O

Chloride ions are pumped out of the ascending limb followed by sodium ions, giving an overall increase in the sodium chloride concentration in the tissues of the medulla. As the limb is impermeable to water the concentration of the remaining filtrate is reduced until the filtrate which goes on into the second convoluted tubule is *hypotonic* to the blood. The tissue fluid of the medulla contains increasingly high levels of sodium chloride down towards the pelvis of the kidney. This affects the movement of water out of the collecting duct. Sodium concentration increases through the medulla.

Sodium and chloride ions enter the descending limb by diffusion, and water leaves it. As a result the volume of filtrate is reduced and the concentration increases until it is hypertonic to the blood.

Cl$^-$ Cl$^-$
Cl$^-$ Na$^+$
Na$^+$ Na$^+$
H$_2$O   Cl$^-$

H$_2$O

As the tubular fluid flows along the collecting duct through the medulla, the tissue fluid of the medulla becomes increasingly concentrated and water leaves the collecting duct along a concentration gradient. The result is hypertonic urine which may be very concentrated indeed. The permeability of the collecting duct to water is affected by the hormone ADH.

Medulla

H$_2$O

H$_2$O   Cl$^-$

At the hairpin bend the filtrate in the loop is at its most concentrated.

Na$^+$

H$_2$O

Collecting duct

Loop of Henle

Hypertonic urine

Figure 4.6.6 The role of the loop of Henle in the reabsorption of water is still not fully understood. However, this complex system known as a **countercurrent multiplier** is very effective at producing concentrated urine.

sodium ions increases through the medulla, so water may be removed from the collecting duct all the way along, leading to very hypertonic urine when it is necessary to conserve water.

Tubular secretion

The composition of the urine is also affected by substances actively secreted into it. This secretion occurs mainly in the second convoluted tubule. Potassium ions and creatinine (a nitrogenous excretory product) may be secreted from the blood into the tubule, and certain drugs such as penicillin are also removed from the body in this way. But perhaps the most important aspect of active secretion in the second convoluted tubule is its role in balancing the pH of the blood. If the pH of the blood starts to fall (if it becomes more acidic) hydrogen ions are secreted from the second convoluted tubule into the filtrate. Also, the kidney cells make and secrete ammonium ions which combine with the anions brought to the kidney and are excreted as ammonium salts. Equally, if the pH rises the tubule secretes hydrogencarbonate ions into the filtrate. As a result of this tubular secretion the pH of the urine may vary from 4.5 to 8.5, whilst the pH of the blood is maintained within a very narrow band.

The urine

The fluid produced by the kidney tubules is collected first in the pelvis of the organ. It then passes along the ureters to the bladder, where it is stored until the bladder is sufficiently stretched to stimulate micturition.

The urine contains variable amounts of water and salts, and large quantities of urea. Substances such as glucose or protein appearing in the urine indicate either that there are problems elsewhere in the body – in the pancreas, for example – or that the kidneys themselves are not working properly.

Control of the kidney and homeostasis

The kidney is of vital importance in the balance of both water and solutes in the body. The removal of the waste product urea is a continuous and important part of this role as urea is produced all the time by metabolic processes. However, levels of other important substances vary according to the situation of the individual. The functioning of the kidney must be adjusted to cope with changes in circumstances. A night out at the pub, a long walk on a hot summer's day or a very salty meal all threaten the steady state of the body. How is the functioning of the kidney controlled to bring about homeostasis?

Osmoregulation

The water potential of the blood is maintained at a more or less steady level by balancing the fluid taken in by eating and drinking with the fluid lost by sweating, defecation and in the urine. It is the concentration of the urine which is most important in this balancing act, and this is controlled by **antidiuretic hormone** (**ADH**). As we saw in section 4.4, ADH is produced by the hypothalamus and secreted into the posterior lobe of the pituitary where it is stored. ADH increases the permeability of the second convoluted tubule and the collecting duct to water.

When water is in short supply, or a lot of sweating takes place, or very salty food is eaten, the blood concentration rises – its water potential becomes more negative. If this was allowed to continue the osmotic balance of the tissue fluids would be disturbed and cell damage would occur. However, the change in blood concentration is detected in the hypothalamus by receptors called **osmoreceptors**. These send impulses to the posterior pituitary which in turn releases stored ADH. The ADH increases the permeability of the second convoluted tubule and the collecting duct to water. As a result water leaves the tubules by diffusion into the concentrated tissues of the medulla and then moves into the blood vessels of the vasa recta. This means that more water is returned from the filtrate to the blood, and only small amounts of concentrated urine are produced.

When large amounts of liquid are taken in the blood becomes more dilute – its water potential becomes less negative. Again the change is detected by the osmoreceptors of the hypothalamus, and in this case the release of ADH by the pituitary is inhibited. The walls of the second convoluted tubule and the collecting duct remain impermeable to water and so little or no reabsorption takes place. Thus the concentration of the blood is maintained and large amounts of very dilute urine are produced. This control of water excretion is shown in figure 4.6.7.

If ADH is not produced at all then a clinical condition known as **diabetes insipidus** occurs. The diabetes mellitus ('sweet fountain') described in section 4.4 is caused by insufficient insulin being produced, resulting in large amounts of urine containing sugar. In diabetes insipidus ('dilute fountain') large amounts of very dilute urine are formed continuously as the tubules are permanently impermeable to water. There is no sugar in the urine, however. The patient feels extremely thirsty and has to drink large quantities of liquid to avoid severe dehydration.

Evidence for the way the kidney functions

How do we know what happens in the kidney tubules, and that the concentration of the filtrate changes as just described? Electron microscopy has allowed a detailed picture of the cells in the different parts of the nephron to be built up. Our knowledge of the changes in the filtrate as it passes along the tubule comes from work done by a number of investigators using a micropipette to withdraw fluid from various points along the nephron in a living organism.

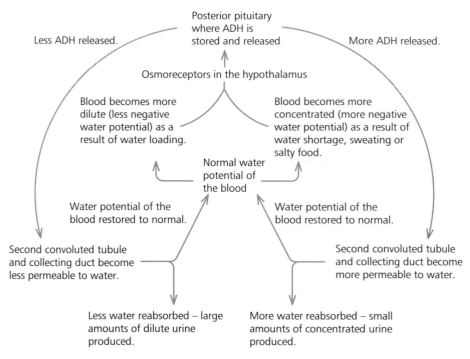

Figure 4.6.7 A negative feedback system maintains the osmotic potential of the blood within very narrow limits. It is easy to test the effectiveness of this system – simply drink several pints of water or squash over a short period of time and wait for the results!

Regulation of salt concentrations

The other major homeostatic system involving the kidney is in the control of the salt concentrations of the body. This is brought about by the hormone **aldosterone** produced by the cortex of the adrenal glands. It causes the active uptake of sodium ions in the second convoluted tubule from the filtrate into the plasma, and water follows by osmosis, increasing the blood volume and consequently the blood pressure.

If sodium ions are lost from the blood, for example after copious sweating, then the blood water potential becomes less negative and water tends to be lost from the blood into the tissue fluid and cells. This causes a slight drop in blood pressure, which is detected by a group of cells in the kidney found between the second convoluted tubule and the afferent arteriole. These cells are called the **juxtaglomerular complex**, and when the blood pressure drops they produce an enzyme called **renin**. This acts on a protein in the blood to produce **angiotensin**, a hormone which in turn stimulates the release of aldosterone from the adrenal glands. Once aldosterone is being produced the adrenal glands are stimulated by adrenocorticotrophic hormone (ACTH) from the pituitary gland as well. This is a complex system which helps to fine tune the concentration of sodium ions and the volume of the blood, as illustrated by figure 4.6.8.

OSMOREGULATION IN PLANTS

The maintenance of a steady internal environment is important to plants as well as to animals. For example, the enzymes controlling the biochemistry of plant cells are just as sensitive to temperature as are those in animal cells, though they have much lower optimum temperatures. Water balance is of great importance for the maintenance of turgor and the transport of substances within the plant. However plant cells are, by their nature, less vulnerable than animal cells to osmotic damage because of the cellulose cell wall. Osmoregulation in plants tends to involve strategies for preventing water loss rather than a specific controlling organ such as the mammalian kidney.

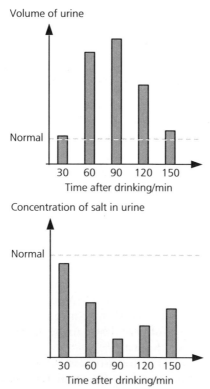

Figure 4.6.8 These graphs show the effect of drinking a given volume of distilled water on both the volume and salt concentration of the urine. Urine was collected at 30-minute intervals after drinking the water. This shows the sensitivity of the response to an imposed change – the water load is removed but without an unwanted loss of salt.

Plant osmotic habitats

Plants live in a wide range of habitats, and these affect the water balance of the organism. Plants which live in fresh water are known as **hydrophytes** and they obviously have no shortage of water. Plants living in a normal terrestrial environment are known as **mesophytes**. They have to balance water loss by transpiration with water uptake from the soil, but conditions are usually such that this can be achieved without major specialised adaptations. **Xerophytes** live in conditions of great water shortage such as desert regions, and so they require drastic measures to avoid water loss. And finally there are plants living where there is plenty of water, but water which is unsuitable for them to use – the **halophytes** live on seashores and salt marshes, surrounded by salty water with a more negative water potential than their tissue fluids. As a result they exist in a state of physiological drought.

Methods of conserving water

There are a variety of ways in which plants conserve water, some common to most species, others very specific. Some of the main strategies for maintaining water balance in plants are given below.

1 The **waxy cuticle** reduces water loss from the surfaces of leaves and is very important in the conservation of water by plants.

2 **Stomata** are common to all plants, except those submerged in water. As we have seen in sections 2.2 and 2.3, these pores are opened and closed to allow gaseous exchange to take place for photosynthesis. A side effect of stomatal opening is that water is lost by evaporation. Plants which are under particular water stress have evolved ways of reducing water loss by evaporation from stomata. These include:

   a most stomata being sited on the underside of the leaves, minimising their exposure to the Sun

   b hairy leaves – the hairs trap a layer of still air which reduces water loss (see figure 2.2.4, page 104)

   c stomata sunk in pits, creating a microenvironment of still air

   d reduced leaves such as the very narrow 'needles' of conifers. Plants which live in extremely dry conditions may have leaves reduced to almost nothing and use the stem as the main photosynthetic organ, for example cacti with leaves reduced to spines (see figure 2.2.4, page 104)

   e curled leaves, such as those of the marram grass found colonising sand dunes, creating a microenvironment and so reducing water loss.

   Some of these strategies are shown in figure 4.6.9.

3 **Succulents** store water in specialised parenchyma tissues in their stems and roots. Water is stored when it is in plentiful supply and then used in times of drought.

4 **Root adaptations** for coping with water stress include very deep roots which can absorb water from deep down in the soil, and superficial roots close to the surface, spread out a long way from the plant. These provide a large surface area to absorb water when it penetrates the top layers of the soil.

5 Avoiding the problem entirely is an alternative solution to the problem of water balance. In winter time frosts may make it impossible to obtain water from the soil. Deciduous trees lose their leaves in the winter and drop their metabolism to a very low level. Many other plants survive dry periods by forming seeds, spores or underground perennating organs capable of withstanding long periods of desiccation.

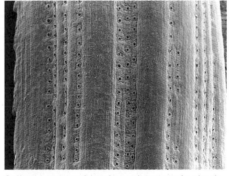

Sunken stomata are in a microenvironment of moist air.

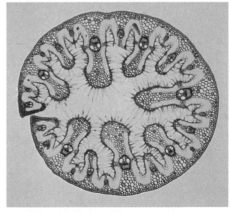

A curled or rolled leaf provides a protected microenvironment.

Figure 4.6.9 Some of the strategies plants have evolved to reduce water loss by evaporation

Thus we can see that osmoregulation is a major homeostatic function in both plants and animals. Section 4.7 looks at other homeostatic mechanisms and how they are controlled.

SUMMARY

- **Homeostasis** is the maintenance of a steady internal state, and is necessary for cells to function properly. Homeostasis involves **osmoregulation** (control of water balance), control of nitrogenous and other waste products, control of blood glucose levels and **thermoregulation** (control of body temperature).

- Unicells such as *Amoeba* have a **contractile vacuole** for osmoregulation which collects and expels excess water.

- In mammals the **kidneys** are the main osmoregulatory organs. They can produce urine hypertonic to the body fluids in order to conserve water. The kidney has three main areas – the **cortex**, **medulla** and **pelvis**. The cortex contains the **Malpighian bodies** of the **nephrons**, and the **loops of Henle** extend down into the medulla.

- Blood from the **renal vein** enters the Malpighian body of the nephron. It passes at high pressure into a knot of capillaries called the **glomerulus** inside the **Bowman's capsule**. **Ultrafiltration** takes place here – the fluid is forced out of the capillaries and into the Bowman's capsule through cells called **podocytes**. Blood cells are retained by the capillary wall and plasma proteins and platelets by the **basement membrane**.

- The ultrafiltrate (hypotonic to blood) is passed into the first convoluted tubule where **selective reabsorption** occurs. Glucose, amino acids, hormones, vitamins and sodium ions are reabsorbed into the blood and chloride ions follow by passive diffusion.

- The filtrate is concentrated in the loop of Henle. The **descending limb** is permeable to water. As it passes down into the medulla there is a high concentration of sodium chloride in the surrounding tissue and water passes out of the tubule and into the blood vessels of the **vasa recta**, and sodium and chloride ions pass into the tubule – the filtrate becomes hypertonic. As the filtrate moves up the **ascending limb**, which is impermeable to water, chloride ions are secreted out of it but water cannot leave it – it becomes hypotonic again.

- In the **second convoluted tubule** and **collecting duct**, sodium ions may be pumped out or water may leave the filtrate – the permeability of the tubule and duct is controlled by the hormone ADH. Active secretion of potassium ions, creatinine and drugs takes place in the second convoluted tubule. The pH balance of the blood is also controlled here – hydrogen ions and ammonium ions are secreted into the tubule to increase blood pH, and hydrogencarbonate ions secreted to reduce blood pH.

- As the filtrate passes into the collecting duct and again through the medulla, water again passes out under the control of ADH and the urine may become very concentrated (hypertonic).

- Changes in blood concentration are detected by **osmoreceptors** in the hypothalamus which send impulses to the posterior pituitary so that it releases ADH. This increases the permeability of the second convoluted tubule and collecting duct to water, which therefore passes out into the vasa recta, and the urine is concentrated. If the blood becomes too dilute, the osmoreceptors inhibit the release of ADH from the pituitary and the tubule walls are impermeable to water, which remains in the urine and is excreted.

- **Aldosterone** produced by the cortex of the adrenal glands causes the active uptake of sodium ions (and also water by osmosis) from the filtrate. If sodium ions are lost from the blood there is a drop in blood pressure which is detected by the **juxtaglomerular apparatus**. This in turn produces the enzyme **renin** which acts on a protein in the blood to produce **angiotensin**. This stimulates the adrenals to produce aldosterone, and ACTH then also stimulates the adrenals.

- Plants may be classified according to their osmotic habitat. **Hydrophytes** live in water and **mesophytes** on land. **Xerophytes** live in arid conditions and **halophytes** in salty water.

- Plants have various measures to prevent excessive water loss. A waxy cuticle prevents water loss from the leaf surface. Plants under water stress may have stomata on the undersides of leaves or sunken into pits. Hairy leaves trap a layer of still air and reduced or curled leaves also cut down transpiration. Succulents store water in specialised tissues. Roots may be very deep, or shallow and spreading. Dropping leaves and producing spores or seeds allow plants to survive prolonged dry periods.

QUESTIONS

1 a What is meant by the term *homeostasis*?
  b What is meant by the term *osmoregulation*?
  c Describe how the following osmoregulate:
     i freshwater amoebae
     ii salt-water fish.

2 Teenagers newly diagnosed as suffering from severe kidney disease face diet restrictions, dialysis and possible transplants. Produce a leaflet explaining clearly and concisely the role of the kidney in the body and how it works. Make clear the reasons for dietary and drinking restrictions, and for the artificial filtering of the blood they may have to undergo.

3 a Explain how the kidneys of desert animals are adapted to cope with a lack of water available for drinking.
  b Explain how plants living in deserts are adapted to dry conditions.

4.7 Homeostasis: Control of waste products and temperature

In section 4.6 we saw the importance of maintaining a steady internal state for the successful functioning of a living organism. The balance of water and solutes is particularly important because of the dramatic effect osmotic changes can have on cells. However, the concentrations of various waste products and the temperature of the organism also have a major effect on the functioning of cells. In mammals the main organ involved in water balance is the kidney. The liver plays a major role in other aspects of homeostasis.

STRUCTURE OF THE MAMMALIAN LIVER

The liver is one of the major body systems concerned with homeostasis. It is the largest individual organ contained within the body, a reddish-brown organ which makes up about 5% of the total body mass. It lies just below the diaphragm and is made up of several lobes. The cells of the liver are surprisingly simple and uniform in appearance, considering the variety and complexity of the functions of the organ. Liver cells are called **hepatocytes**, shown in figure 4.7.1. The appearance of the hepatocytes shows that they are metabolically active cells – they have large nuclei, prominent Golgi apparatus and many mitochondria.

The liver has a unique blood supply within the body. Oxygenated blood is supplied to the liver by the **hepatic artery** and removed from the liver and returned to the heart in the **hepatic vein**. However, the liver is also supplied with blood by a second vessel, the **hepatic portal vein** which carries blood from the intestines straight to the liver (see figure 4.7.1). Blood from the hepatic artery and the hepatic portal vein is mixed in spaces called **sinusoids** which are surrounded by hepatocytes. This mixing increases the oxygen content of the blood from the hepatic portal vein, supplying the hepatocytes with sufficient oxygen for their needs. The hepatocytes secrete bile from the breakdown of the blood into spaces called **canaliculi**, and from these the bile drains into the bile ductiles which take it to the gall bladder.

THE LIVER AT WORK

The liver carries out about 500 different tasks, which can be considered under a manageable number of major roles. Some of these we have met before and so will only discuss briefly here. Others are of major homeostatic importance and so will be considered more fully.

Carbohydrate control

All the hexose sugars which are absorbed from the gut are converted into glucose by the hepatocytes. Also, the hepatocytes are closely involved in the homeostatic control of glucose levels in the blood under the control of insulin and glucagon. When blood glucose levels rise, hepatocytes convert glucose to the storage carbohydrate glycogen under the influence of insulin. Similarly, when blood sugar levels start to fall, the hepatocytes convert the glycogen back to glucose under the influence of the hormone glucagon. The action of these hormones was discussed in section 4.4.

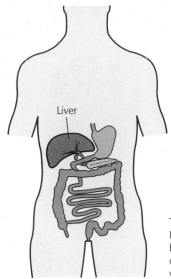

The position of the liver in the body. It is the largest of the organs in the body cavity and is very fast growing – damaged areas generally regenerate very quickly.

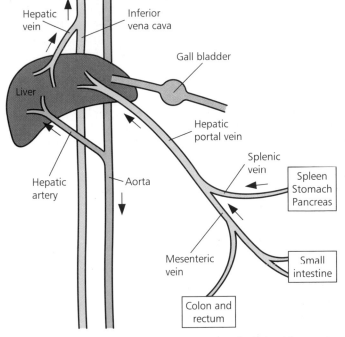

The liver has a very rich blood supply – about 1 dm³ of blood flows through it every minute. Up to three-quarters of this blood comes from the gut via the hepatic portal vein and is loaded with the products of digestion and absorption. These are the raw materials for the many metabolic activities of the liver.

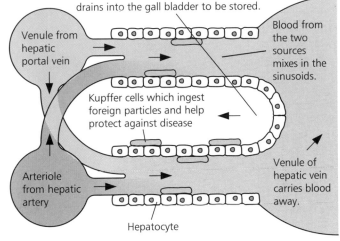

Bile canaliculus – the hepatocytes secrete bile into these canaliculi and from there it drains into the gall bladder to be stored.

Blood from the two sources mixes in the sinusoids.

Venule from hepatic portal vein

Kupffer cells which ingest foreign particles and help protect against disease

Arteriole from hepatic artery

Venule of hepatic vein carries blood away.

Hepatocyte

The arrangement of the tissues in the liver, giving all the hepatocytes close contact with the blood for both removing and adding materials

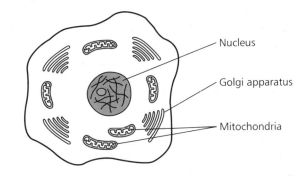

One of the unspecialised hepatocytes which carries out the metabolic and homeostatic functions of the liver

Figure 4.7.1 The microscopic structure of the liver is relatively simple and surprisingly uniform considering its roles in the metabolism of carbohydrates, proteins and fats and its major part in the maintenance of a steady state within the body.

Lipid control

Some carbohydrate (about 100 g) can be stored as glycogen by the liver, and some glycogen is stored in the muscles. Any excess carbohydrate eaten over and above this amount is converted to lipids by the hepatocytes for storage elsewhere in the body. The hepatocytes are also involved in removing **cholesterol** and various other lipids from the blood and breaking them down or modifying them for use elsewhere. Cholesterol is needed by the body for the formation of cell membranes, particularly in the nerve cells, and for the production of hormones. Excess cholesterol is excreted in the bile and may precipitate to form gall stones in the gall bladder or bile duct. A raised blood cholesterol level can cause problems as cholesterol may be deposited in the blood vessels and cause angina or heart attacks. The homeostatic function of the liver in removing excess cholesterol from the blood is therefore most important to the health of the organism as a whole, as illustrated by figure 4.7.2.

Protein control

The liver plays a vital role in protein metabolism. It produces certain plasma proteins. The hepatocytes also carry out **transamination**, the conversion of one amino acid into another. The diet does not always contain the required balance of amino acids, but transamination can overcome many potential problems this might cause.

The most important role of the liver in protein metabolism is in **deamination**. The body cannot store either protein or amino acids. Any excess protein eaten would be excreted and therefore wasted were it not for the action of the hepatocytes. They deaminate the amino acids, removing the amino group and converting it first into the very toxic **ammonia** and then to **urea**. Urea is less poisonous and can be excreted by the kidneys, as we have seen in section 4.6. The remainder of the amino acid can then be fed into cellular respiration or converted into lipids for storage.

The ammonia produced in the deamination of proteins is converted into urea in a set of enzyme-controlled reactions known as the **ornithine cycle** (see box below).

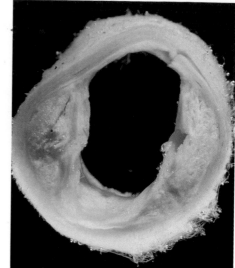

The biochemistry of deamination

Removing the amino group from amino acids and converting the resulting highly poisonous ammonia into the less toxic and more manageable compound urea involves some complex biochemistry. This can be simplified and summarised as in figure 4.7.3.

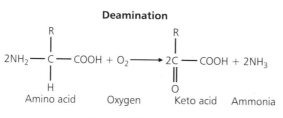

Deamination

$$2NH_2 - \underset{\underset{H}{|}}{\overset{\overset{R}{|}}{C}} - COOH + O_2 \longrightarrow 2\underset{\underset{O}{\|}}{\overset{\overset{R}{|}}{C}} - COOH + 2NH_3$$

Amino acid   Oxygen   Keto acid   Ammonia

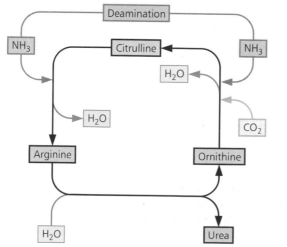

The ornithine cycle

Deamination

NH₃ — Citrulline — NH₃

H₂O

H₂O — CO₂

Arginine — Ornithine

H₂O — Urea

Figure 4.7.3 Excess amino acids in the body cannot be stored for later use. They have to be broken down, so that as little of the molecule as possible is wasted. This is carried out by the cells of the liver in the process known as deamination, with the waste products being converted into urea by the reactions of the ornithine cycle.

Figure 4.7.2 An excess of cholesterol in the blood can lead to circulatory problems as the lipid is deposited on the walls of blood vessels. The liver works to prevent such problems by regulating the levels of cholesterol in the blood and excreting excess into the bile. However, this can cause its own problems, as these gall stones show.

Red blood cell control

The red blood cells of a developing fetus are made by the fetal liver, although by the time the baby is born this function has been taken over by the bone marrow. The Kupffer cells in the sinusoids (see figure 4.7.1) break down red blood cells that are at the end of their 120-day life. The haemoglobin is broken down to form the bile pigments **biliverdin** (green) and **bilirubin** (brown) which are excreted in the bile and are responsible for its colour, and so indirectly for the colour of the faeces.

Bile production

The production of bile by the hepatocytes is important in homeostasis for several reasons. Bile contains:
- bile salts which have a valuable role in the digestion of fats
- bile pigments which excrete the breakdown products of haemoglobin
- cholesterol, the removal of which helps to control the blood level within narrow boundaries.

Storage

The liver stores a variety of substances, releasing them as necessary to maintain a steady concentration in the blood. Substances stored include glycogen, with its important role in sugar metabolism, the lipid-soluble vitamins A, D, E and K and the water-soluble vitamins of group B and vitamin C. Along with these the liver is also a store for iron, copper, zinc and cobalt ions. Polar bear liver contains so much vitamin A that it is actually toxic for humans to eat!

Hormone control

It is important that hormones are broken down once they have carried their message to the appropriate areas of the body. The liver is the site where hormones, particularly steroids such as the sex hormones, are broken down. Some are sent for excretion by the kidney whilst others are excreted in the bile.

Control of toxins

Toxins are constantly produced in the body. Apart from urea which we have already considered, many other metabolic pathways produce potentially poisonous substances. We also take in a wide variety of poisons on a voluntary basis, such as alcohol and drugs like paracetamol. The liver is the site where all these substances are absorbed and detoxified.

A classic example is the way the liver deals with **hydrogen peroxide**, a by-product of various metabolic pathways. Hepatocytes contain the enzyme catalase, one of the most active of all known enzymes, which splits the hydrogen peroxide into oxygen and water.

Temperature control

The liver has a very high metabolic rate. For many years it has been assumed that as a result of the number of biochemical reactions which occur in the liver and its large size, much heat was generated there and distributed round the body by the extensive blood supply. More recently this view has been questioned and it is now thought that under normal circumstances endothermic and exothermic reactions cancel each other out and the liver produces little excess heat.

The liver has many important homeostatic functions – it is involved in the control of the blood levels of a wide range of substances and in preventing the build-up of a wide range of toxic substances. But even the actions of the liver and the kidney combined do not complete the homeostatic picture, even at this relatively simple level. The third major element of homeostasis, particularly in mammals, is the control of temperature.

THERMOREGULATION

The need for thermoregulation

The chemical reactions which occur in cells only take place within a relatively narrow range of temperatures. As we have seen in section 1.5, this is largely due to the sensitivity of the enzymes which control the reactions. For example, most enzymes have an optimum temperature and once temperatures rise above 40 °C most enzymes are denatured as their tertiary and quarternary structure is destroyed. Because of this sensitivity to temperature, many organisms control their internal body temperature – **thermoregulation** is important to their survival and is a major aspect of their homeostasis.

Losing and gaining heat

When we discuss the 'temperature' of an organism, we are considering the core temperature of the body. The surface temperature of an animal or plant fluctuates rapidly, but it is the internal temperature which is relevant to enzyme activity. Living organisms are continually losing heat. They are also continually gaining heat. The balance between the gains and losses determines whether the core temperature of the organism rises, falls or stays the same. This is illustrated in figure 4.7.4.

Most of the ways in which animals exchange heat with their surroundings are affected by the additional factor of the size of the animal. Small animals

Trapping a layer of air in the feathers helps small birds to avoid heat loss by convection.

The thick layer of blubber on a whale helps to prevent heat loss to the water by conduction.

Basking in the Sun on a warm rock allows the lizard to gain heat by both radiation and conduction.

The elephant cools down by losing heat via the evaporation of water it applies to its skin and through radiation from its large ears.

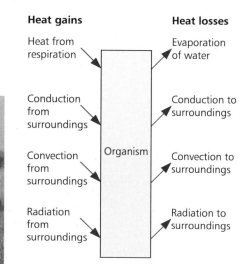

Figure 4.7.4 The heat gains and losses experienced by an organism. It is the balance of these which determines the core temperature, and organisms use a variety of means to shift the balance and allow them to gain or lose heat as needed.

have a large surface area:volume ratio and this means that they lose and gain heat more rapidly than a larger organism.

There are several ways in which organisms gain and lose heat.

- Heat is gained as a by-product of *metabolism*. Heat produced by exothermic reactions warms the core of the organism.
- Heat is lost by the *evaporation* of water from the body surfaces. A certain amount of heat is always lost in this way from the mouth and respiratory surfaces of land-dwelling animals, and this cannot generally be controlled. Sweating and behavioural patterns such as wallowing can increase this heat loss.
- Heat may be lost to or gained from the environment by *radiation*. This is the transfer of energy in the form of electromagnetic waves. Infra-red radiation is the most important form of heat energy absorbed and radiated by animals.
- Heat may be lost to or gained from the environment by *convection*. This is the transfer of heat by currents in air or water. Convection currents are set up around objects that are warmer than their surroundings (warm air or water rises, cold air or water sinks), and so adaptations to prevent heat loss by these currents are common in animals.
- Heat may be lost to or gained from the environment by *conduction*. This is the transfer of heat by the collisions of molecules. Conduction is particularly important between organisms and the ground or water, as air does not conduct heat well. Neither does fat, so fat deposits in adipose tissue are therefore a valuable insulator preventing heat exchange with the environment.

How do animals control their body temperatures?

Not all animals need to control their body temperatures. Protoctists and small animals living in water have no means of temperature regulation. For those organisms living in large water masses such as the sea this raises no difficulties, because the temperature of their environment is very stable. Similar organisms living in small ponds either develop tolerance of temperature fluctuations, or have survival strategies to cope with adverse conditions, such as forming cysts and emerging when the temperature is suitable again.

However, larger animals in a wide variety of habitats need to regulate their temperatures, either to avoid damage to cells or to enable them to have an active way of life. Animals may be classified by their thermoregulation mechanisms in two different ways.

One approach, which has been used for many years, is to consider the stability of the body temperature of an animal.

- **Homoiothermic animals** – mainly the mammals and birds – regulate their body temperatures within a very narrow range (mammals 37–8 °C, birds 40 °C).
- **Poikilothermic animals** – such as the fish and reptiles – allow their body temperature to fluctuate much more widely, approximately following the external environment.

However, more recently there has been dissatisfaction with this way of looking at thermoregulation in animals. Many so-called homoiothermic animals in fact have body temperatures which fluctuate quite widely. Equally, many poikilothermic animals regulate their temperatures within a fairly narrow band. The number of species which have large fluctuations in their core temperature following the environment is relatively small. As a result, animals are often now classified according to the major source of their body heat.

- **Endothermic animals** rely on their own metabolic reactions to provide at least some of their body heat.
- **Ectothermic animals** depend largely on the environment for their body heat.

Endotherms

Endotherms produce at least some of their own body heat and usually have a body temperature higher than the ambient temperature. They are adapted to conserve their body heat when necessary, and also to take advantage of warmth from the environment when possible. This means that there are few environments where they cannot remain active, and the main groups of endotherms, the mammals and birds, are found in an extremely wide variety of environmental niches. They can cope with high external temperatures and can also live in areas with very low ambient temperatures where ectotherms would not be able to remain active. In order to maintain the body temperature against adverse environmental conditions, the metabolic rate has to be high – endotherms on average have a metabolic rate five times higher than an ectotherm of comparable size – and this means they have to eat considerably more food to supply their metabolic needs.

Ectotherms

Ectotherms may produce a significant part of their own body heat, but rely on behavioural and structural modifications that take advantage of the environment in order to maintain a reasonably steady temperature. When cold they may bask in the Sun, press themselves to warm surfaces or have special areas of skin which they erect in order to maximise their absorption of radiation from the Sun. To cool down they may move into the shade, or into water or mud. (Endotherms also modify their behaviour to aid thermoregulation. There is more about this on pages 318–20.) By a variety of strategies such as these, many ectotherms maintain a body temperature which is almost as stable as that of an endotherm, although they are always more vulnerable to fluctuations. Because their metabolic rate is much lower, ectotherms need less food, an advantage in many environments where food may be in short supply.

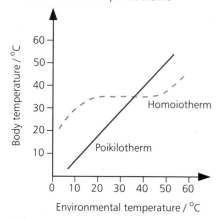

Relationship between body temperature and environmental temperature for homoiotherms and poikilotherms

Figure 4.7.5 Endothermic control of the body temperature allows large animals such as the polar bear to lead an active life where ectotherms of similar size would have difficulty, while both groups can exploit the warmer regions of the world. The graph shows why the majority of organisms have evolved to show at least some degree of homoiothermy – the fluctuations in temperature shown by a true poikilotherm make survival in anything but a very stable environment difficult in the extreme.

THERMOREGULATION IN MAMMALS

The mechanisms of temperature control

Mammals and birds are endothermic animals. Temperature control in humans is a good example of homeostasis in mammals as we are not only endotherms but also homoiotherms – we regulate our body temperature within a very narrow range. How is this control brought about? The main source of heat is from our metabolism, but humans survive in almost every area of the world and have effective ways of both losing and conserving heat.

The skin

The major homeostatic organ involved in thermoregulation is the skin. Heat loss through the gut and the respiratory surfaces occurs and cannot be prevented, but the skin has evolved to provide an enormous surface area which can be modified either to conserve or to lose heat, as figure 4.7.6 shows.

The skin is the largest single organ. It covers the entire surface of the body in a waterproof layer, providing protection both against mechanical damage and from the ultraviolet radiation of the Sun. The skin can be regarded as a sense organ, containing thousands of sensory nerve endings providing the brain with information about many aspects of the outside world, including temperature, touch and pain. It is also an excretory organ, with urea, salt and water being lost in the sweat. But the major homeostatic function of the skin is in thermoregulation and this is summarised in table 4.7.1.

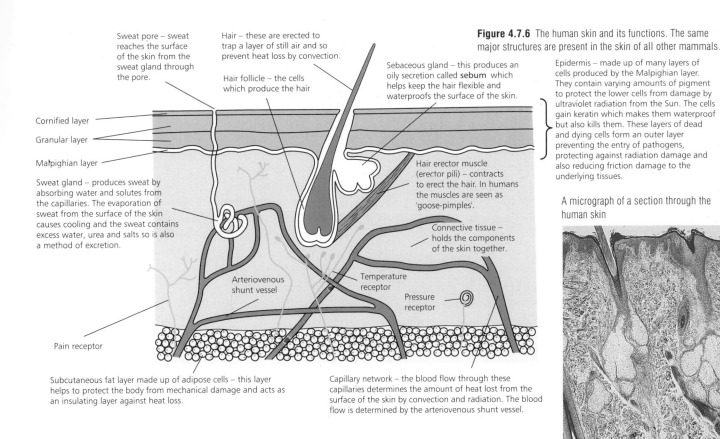

Sweat pore – sweat reaches the surface of the skin from the sweat gland through the pore.

Hair – these are erected to trap a layer of still air and so prevent heat loss by convection.

Hair follicle – the cells which produce the hair

Sebaceous gland – this produces an oily secretion called **sebum** which helps keep the hair flexible and waterproofs the surface of the skin.

Figure 4.7.6 The human skin and its functions. The same major structures are present in the skin of all other mammals.

Epidermis – made up of many layers of cells produced by the Malpighian layer. They contain varying amounts of pigment to protect the lower cells from damage by ultraviolet radiation from the Sun. The cells gain keratin which makes them waterproof but also kills them. These layers of dead and dying cells form an outer layer preventing the entry of pathogens, protecting against radiation damage and also reducing friction damage to the underlying tissues.

Cornified layer

Granular layer

Malpighian layer

Sweat gland – produces sweat by absorbing water and solutes from the capillaries. The evaporation of sweat from the surface of the skin causes cooling and the sweat contains excess water, urea and salts so is also a method of excretion.

Hair erector muscle (erector pili) – contracts to erect the hair. In humans the muscles are seen as 'goose-pimples'.

Connective tissue – holds the components of the skin together.

Arteriovenous shunt vessel

Temperature receptor

Pressure receptor

Pain receptor

A micrograph of a section through the human skin

Subcutaneous fat layer made up of adipose cells – this layer helps to protect the body from mechanical damage and acts as an insulating layer against heat loss.

Capillary network – the blood flow through these capillaries determines the amount of heat lost from the surface of the skin by convection and radiation. The blood flow is determined by the arteriovenous shunt vessel.

Measures to encourage heat loss when the external temperature rises

The arteriovenous shunt closes to increase the blood flow through the capillaries and so increase heat loss by conduction and radiation.

Skin capillaries

Arteriovenous shunt closed

The rate of sweat production by the sweat glands increases. As more sweat is released onto the skin surface, heat is lost as the water evaporates. Almost 1 dm^3 of sweat is usually produced and lost in a day, but this can rise to 12 dm^3 in very hot conditions.

The erector pili muscles are relaxed and the hairs lie flat against the body. In humans this has little effect, but in hairy mammals the insulating layer of air trapped by the hair is reduced and so more heat can be lost by convection.

The metabolic rate drops so that less heat is produced by the body – animals tend to be less active in hot weather.

Measures to conserve heat when the external temperature falls

The arteriovenous shunt opens, reducing the blood flow through the capillaries and so reducing the heat lost from the surface of the skin by conduction and radiation.

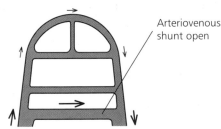

Arteriovenous shunt open

Sweat production is reduced and so heat loss by evaporation is also reduced.

The erector pili muscles are contracted, pulling the hairs upright. In humans this is visible as 'goose-pimples' and has little effect on temperature regulation, but in hairy mammals it traps an insulating layer of air which reduces heat loss by convection.

The metabolic rate of the body rises, producing extra heat. This takes place particularly in the liver and the muscles. Shivering – involuntary contractions of the skeletal muscles – also helps to generate metabolic heat. Special **brown fat** may also be metabolised. Raising the body temperature by increased metabolism in this way is particularly important in emergence of animals from hibernation and for temperature control in very young human babies.

Table 4.7.1 Some major homeostatic activities of the skin

Temperature limits

Experiments carried out on human subjects in calorimeters, along with observations on seriously ill patients and people who have been exposed to extreme cold, have allowed us to build up a picture of the temperature tolerance of the human body. The subject must be naked so that any responses to changes in temperature are purely physiological and do not involve the physical properties of clothes or coverings. The environmental temperature is changed, and the internal core temperature of the subject is recorded.

As the external temperature drops a range of thermoregulatory mechanisms are used to conserve heat. Below a certain point simple measures such as controlling the blood supply to the skin surface and raising the body hairs are no longer sufficient, and the metabolic rate starts to go up. This point is known as the **low critical temperature**. If the

temperature keeps on falling the metabolic rate will rise to produce sufficient heat to maintain the core temperature. Eventually the chemical reactions can no longer take place and the subject dies – this is the **low lethal temperature**.

As the external temperature rises, thermoregulating mechanisms such as sweating can keep the core temperature stable until the **high critical temperature** is reached. This is very variable as it also depends on the level of humidity – sweating is more efficient in dry air. If the external temperature continues to increase, the metabolic rate starts to go up as the body's reactions double their rate for every 10 °C rise in temperature. Once this happens, positive feedback occurs, because the faster the metabolic rate, the more heat is produced. Death usually occurs when the core temperature reaches about 42 °C – the **high critical temperature** – when the enzymes controlling cell reactions fail.

Control of blood temperature

In any homeostatic system there are receptors sensitive to changes in the system. In the case of temperature regulation there are two types of receptors. Receptors in the brain directly monitor the temperature of the blood, whilst receptors in the skin detect changes in the external temperature. This allows for great sensitivity not only to actual changes in the core temperature but to potential changes too as a result of changes in the environmental temperature.

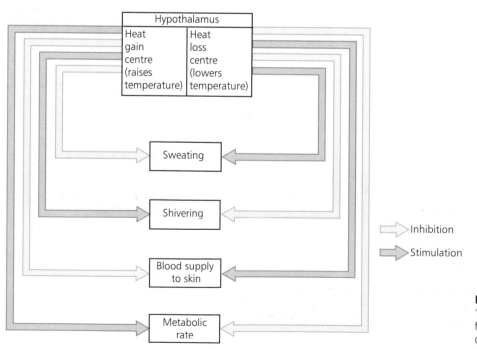

Figure 4.7.7 The hypothalamus acts as the 'thermostat' of the body. As a result of sensitive feedback mechanisms, the temperature of the body is controlled to within 1 °C of its normal temperature.

The temperature receptors in the brain are sited in the hypothalamus. If the temperature of the blood flowing through the hypothalamus drops, the **heat gain centre** reacts by sending nerve impulses to the skin. These cause a reduction in the blood flow through the capillaries in the skin, along with a reduction in the production of sweat and contraction of the erector pili muscles to erect the hairs. Messages from the heat gain centre also stimulate involuntary muscle contractions (shivering) and raise the rate of metabolism.

When the temperature of the blood flowing through the hypothalamus rises, the **heat loss centre** is activated and sends out impulses which increase the blood flow through the skin and increase sweating. The erector pili muscles are relaxed so that the hairs lie flat and shivering stops. The metabolic rate is reduced to generate less heat in the body. Thus the core temperature is usually maintained within very narrow limits, as figure 4.7.7 shows.

Responses to the skin temperature receptors are also important in the homeostatic control of temperature. Messages from skin receptors inform the brain if the environment is hot or cold, and the behaviour of the animal is modified accordingly. For example, as the skin heats up an animal will seek shade, whilst cooling of the skin surface might give rise to increased activity to raise the amount of metabolic heat produced.

Temperature control in premature infants

The mammalian fetus does not have to control its own body temperature – a stable temperature is maintained for it in the uterus by the maternal control systems. Therefore when a baby is born very prematurely it is poorly prepared for the regulation of its body temperature. There is a minimal amount of insulating subcutaneous fat. The shivering response is poorly developed and brown fat has not been laid down in any quantity. The skin is very thin and so the blood flows close to the surface. The skin is also extremely porous so water evaporates freely from it, causing cooling of the blood. Because of the small size of such infants, the surface area:volume ratio means that all types of heat loss occur very readily.

How are these problems overcome? One of the main priorities in the care of premature infants is to allow as much of the energy intake as possible to be used for the growth and maturation of the body systems. Thus it is vital to minimise heat loss and prevent the expenditure of energy on attempting to keep warm. Incubators are designed to maintain a constant temperature for the air surrounding the baby. The air may also be humidified to reduce cooling by the evaporation of water from the skin, and thin plastic and aluminium sheeting wrapped around the baby for the first few weeks will reduce this effect still further.

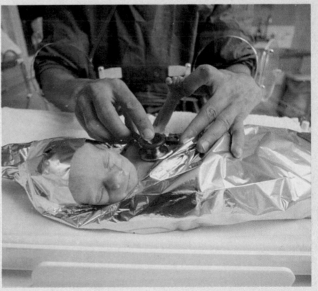

Figure 4.7.8 The temperature control mechanisms which adults take for granted are simply not ready to function in a premature baby.

ADAPTATIONS OF ENDOTHERMS

Heat loss and its control are important elements in homeostatic temperature regulation. The rate of heat loss is related to the surface area:volume ratio of an animal, and this affects the type of endotherms found in particular environments.

Cold environments

In a cold environment mammals and birds are usually larger than their equivalent in a warmer climate. This reduces their surface area:volume ratio and so reduces their heat loss. Animals develop thick layers of subcutaneous fat which insulates against heat loss. They also tend to have small extremities such as ears. Parts that stick out are particularly vulnerable to heat loss, and so the smaller they are the better, in terms of heat conservation. Finally, some mammals and birds living in cold external conditions have developed **countercurrent heat exchange systems** in their limbs which allow body heat to be conserved, as shown in figure 4.7.9.

Hibernation

Some endotherms cannot generate enough heat for them to survive in cold conditions, generally small animals living in Arctic or temperate regions, where the winter temperatures are considerably lower than the summer ones. This problem can be overcome by **hibernating**. In hibernation the animal goes into a very deep sleep. Its metabolic rate slows down greatly and the core body temperature is substantially lowered. Many hibernating animals, for example dormice and hamsters, allow their body temperature to fall just so far and then maintain it at this lower level, making substantial energy savings. A few such as bats allow their body temperature to follow that of the external environment. This saves even more energy, but if the temperature drops too low the tissues might freeze, killing the animal.

Animals usually go into hibernation as a result of both low ambient temperatures and a shortening of the day length. Warm temperatures and lengthening days bring them out of hibernation again. The restoration of the metabolic rate needs to occur quickly to allow the animal to feed and to make sure that it does not become easy prey for predators. Stores of **brown fat** (fat stored in tissues with a particularly rich blood supply) are conserved during hibernation and used up rapidly to produce metabolic heat at the end of hibernation.

Hot environments

A hot environment poses different problems. As losing heat is not a problem, animals do not need to be particularly large in order to conserve heat. Indeed, one of the problems for animals in very hot areas is to lose sufficient heat to prevent themselves from overheating. This is why the extremities of many such animals are large and thin with a rich blood supply, for example an elephant's ears. Heat can be lost to the environment relatively easily through adaptations such as these. There is very little insulating fat, so increasing the amount of heat that can be lost.

Some endotherms simply cannot lose enough heat to maintain a steady internal temperature, and they have evolved the ability to tolerate much larger fluctuations in body temperature than most organisms. The camel is a classic example.

Living in the desert, their major priority is to conserve water. Camels do this most effectively, largely because they do not sweat. This saves water but also removes a major cooling mechanism. As a result the body temperature of a camel climbs to around 40 °C during the day. This is high but not lethal, and the cells can continue to function at this temperature. Desert nights are very cold and the camel then loses heat so that the body temperature falls to a low of 34 °C – cooler than most mammals but still within a range where the cells can function. The low night temperature ensures that the temperature does not climb too high during the day as the temperature rise starts from a low

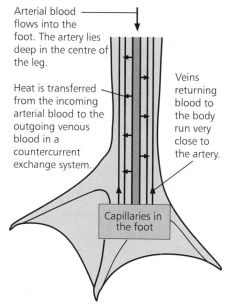

Arterial blood flows into the foot. The artery lies deep in the centre of the leg.

Heat is transferred from the incoming arterial blood to the outgoing venous blood in a countercurrent exchange system.

Veins returning blood to the body run very close to the artery.

Capillaries in the foot

Figure 4.7.9 In a countercurrent heat exchange system such as those found in ducks' legs and dolphins' flippers, the arterial blood cools as it flows to the extremity and the venous blood is warmed as it returns to the body. This minimises the heat loss to the environment and so conserves energy.

base. By tolerating these fluctuations in temperature the camel conserves water, achieving osmoregulation. At the same time, by maintaining the temperature fluctuations within the limits of normal functioning, temperature damage to its cells is avoided.

Control of body temperature by behaviour

As well as physiological controls, modification of behaviour can help avoid major changes in core temperature. Endotherms use behaviour to control their body temperature to some extent (wallowing in mud or water, panting to increase heat loss by evaporation, etc.). In ectotherms the major thermoregulatory mechanism is the modification of behaviour to control body temperature.

Behavioural modifications which help maintain animal body temperatures within manageable limits include:

- **Basking** – desert lizards are among many groups of animals which bask in the Sun when their body temperature is tending to fall. By orientating the body differently to the Sun, or by erecting special areas of tissue which have evolved to absorb heat from the Sun, the body temperature may be raised.
- **Sheltering** – when the core temperature is rising too high many animals shelter from the direct heat of the Sun in burrows, holes or crevices in rocks. They may also attempt to lose extra heat by conduction, pressing the body against the cool earth.
- **Evaporation** – by panting and so exposing the moist tissues of the mouth, by licking the body surface or by wallowing in mud or water an animal can increase evaporation and so heat loss from the body.

By absorbing as much heat from the Sun as possible, an ectotherm such as this locust can warm up its body sufficiently to allow for early morning flight. Depending on the temperature of the body and the heat from the Sun the insect will climb higher or lower, and turn its body in different directions, to gain just the amount of heat needed.

Increasing the level of evaporation from the body allows for increased cooling.

Figure 4.7.10 Behavioural modifications such as these allow both ectotherms and endotherms to thermoregulate.

Cooling the brain

The brain is the part of most mammals which is particularly susceptible to damage as a result of overheating. A modification of a countercurrent exchange system helps to prevent this heat damage, as shown in figure 4.7.11. This system is valuable in many mammals, but is of particular importance and is extremely well developed in animals such as the camel which live in conditions of heat stress.

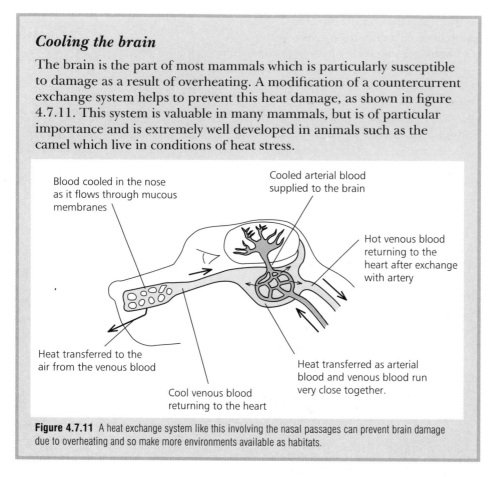

Blood cooled in the nose as it flows through mucous membranes

Cooled arterial blood supplied to the brain

Hot venous blood returning to the heart after exchange with artery

Heat transferred to the air from the venous blood

Cool venous blood returning to the heart

Heat transferred as arterial blood and venous blood run very close together.

Figure 4.7.11 A heat exchange system like this involving the nasal passages can prevent brain damage due to overheating and so make more environments available as habitats.

Moving into or out of the Sun is one of the most common behavioural means of regulating body temperature. It is used by both endotherms and ectotherms as part of their homeostatic control system. However, such simple methods of modifying the internal temperature are only available to animals. Plants cannot move around and so rely on physical and physiological adaptations to help them avoid heat damage.

TEMPERATURE CONTROL IN PLANTS

As we saw in section 4.6, it is just as important for plants to maintain a reasonably stable internal environment as it is for animals. Plants lose and gain heat by the same physical processes as animals – radiation, evaporation, conduction and convection. They have adapted to live in a variety of habitats and can cope with a wide range of temperatures. In general plant tissues are rather like those of the camel – they can tolerate relatively large fluctuations in temperature. However, plant enzymes, like any others, denature if the temperature gets too high, and most plants have optimum temperatures. How do plants thermoregulate?

Prevention of overheating

Transpiration is an inevitable result of stomatal opening for photosynthesis. But as plants lose water by evaporation from their leaves there is a valuable side-effect – the plant is cooled down. This is very effective in helping to maintain the internal temperature of the plant within working limits until water becomes limiting. If there has been insufficient rain and drought conditions exist, the plant cannot take up enough water through its roots to maintain transpiration at the high level needed to cool the plant. In this situation the plant will **wilt**, with the unspecialised parenchyma cells losing turgidity and the leaves and stems drooping. This has the effect of reducing the surface area of the leaves exposed to the Sun, and so the plant avoids overheating to damaging levels. In very hot regions many plants have evolved shiny cuticles which reflect up to 50% of the Sun's radiation and so help to keep the plant cool. Small needle-like leaves are also useful for preventing too much heat gain.

Prevention of freezing

Heat gain is less frequently a problem to plants. The leaves are orientated to take maximum advantage of the light at any one time. Most plants living where the temperature is sometimes low have evolved ways of coping with the problem of excess heat loss – usually by avoiding it, as shown in figure 4.7.12. By losing the easily damaged leaves or by producing seeds or spores which are very temperature resistant, plants survive long periods of very low temperatures without suffering tissue damage.

Thus in this section we have seen something of the range and complexity of the control systems that are found in both animals and plants. The sensory systems and nerves of the animals combined with hormonal systems give fine control over the internal environment of the organism. This interaction is clearly demonstrated when we look at homeostasis. And in plants, chemical messages allow coordination, control and the maintenance of a relatively steady state in the organism.

Figure 4.7.12 By taking avoiding action plants can prevent the tissue damage which can be caused by very low temperatures. Only if we plant them in unsuitable areas, where the temperature fluctuations are greater than they can cope with, do plants suffer the type of damage shown here on the right.

SUMMARY

- The liver is a large organ made of several lobes. **Hepatocytes** (liver cells) have large nuclei, prominent Golgi apparatus and many mitochondria.

- The liver is supplied with blood by the hepatic artery and the hepatic portal vein, which carries the products of digestion and absorption. The blood leaves the liver in the hepatic vein. Blood from the hepatic artery and the hepatic portal vein is mixed in **sinusoids** surrounded by hepatocytes which detoxify it and regulate the levels of various substances. **Kupffer cells** ingest foreign particles. Hepatocytes secrete bile into **canaliculi** and it drains into the gall bladder via the **bile ductiles**.

- The hepatocytes convert all hexose sugars to glucose. They convert glucose to glycogen under the control of insulin, and glycogen to glucose under the control of glucagon. Excess carbohydrate is converted to lipid. The hepatocytes remove cholesterol and other lipids from the blood.

4 Questions

1 a What is meant by the **central nervous system**? **(4 marks)**

b Describe the location and functions of each of the following:
  i cerebral hemispheres **(6 marks)**
  ii medulla oblongata. **(6 marks)**

c Comment on the survival value of reflex actions. **(4 marks)**
(ULEAC June 1991)

2 Compare hormonal control in plants and animals. **(20 marks)**
(ULEAC June 1991)

3 a State the exact position of the pancreas in the body of a named mammal. **(2 marks)**

b Describe the functions of the pancreatic secretions. (Reference to glucagon is *not* required.) **(12 marks)**

c Explain how the release of pancreatic secretions is controlled. **(4 marks)**
(Cambridge June 1990)

4 a State what is meant by a **short day plant**. **(2 marks)**

b Outline a possible sequence of events by which flowering might be initiated in a short day plant. **(7 marks)**
(Cambridge June 1991)

5 Describe the role played by hormones in the life cycle of a flowering plant under the following headings:
  a dormancy
  b growth
  c response to stimuli
  d fruit formation. **(20 marks)**
(NEAB June 1989)

6 a Describe the formation, chemical nature and biological significance of glycogen. **(7 marks)**

b How are the glycogen levels of the human body altered by the action of hormones? **(9 marks)**

c What are the implications of producing hormones, such as insulin, by genetic engineering? **(4 marks)**
(ULEAC June 1990)

7 a Describe the structure of a neurone as revealed by the light microscope. **(5 marks)**

b Explain the following processes involved in the transmission of impulses along a neurone:
  i the formation of the resting membrane potential
  ii the formation and transmission of an action potential. **(10 marks)**

c How do retinal cells transduce light energy into nerve impulses? **(5 marks)**
(ULEAC June 1990)

8 Explain how a series of impulses are initiated and transmitted along a neurone and across a synapse. **(12 marks)**
(NEAB June 1991)

9 a Describe the structure of the mammalian eye. **(8 marks)**

b Explain the changes occurring in the eye in the following situations:
  i entering a dark tunnel from sunlight
  ii trying to thread a needle having just looked out of the window. **(10 marks)**

c What is the role of the visual cortex in vision? **(2 marks)**
(ULEAC January 1991)

10 a Give a brief account of the methods by which heat can be lost by a mammal to its environment. **(5 marks)**

b Describe how a mammal:
  i detects changes in the temperature of its surroundings and changes within its body
  ii reduces heat loss and maintains a steady body temperature in a cold environment. **(15 marks)**
(NEAB June 1990)

11 Explain how the sensory receptor mechanisms of the eye or ear, as appropriate, enable discrimination between:
  a green and red colours **(4 marks)**
  b letters in small print **(4 marks)**
  c sounds of different pitch (frequency) **(4 marks)**
  d rotational movements of the head **(4 marks)**
  e varying positions of the head with respect to gravity. **(4 marks)**
(NEAB June 1990)

12 a With reference to *either* stems *or* roots in flowering plants, explain the difference between growth and differentiation. **(8 marks)**

b Describe how the height of flowering plants may be influenced by each of the following:
  i auxins **(4 marks)**
  ii gibberellins **(4 marks)**
  iii genes. **(4 marks)**
(ULEAC January 1992)

13 a Explain how each of the following is involved in the transmission of information across a synapse in the nervous system of a mammal:
  i synaptic vesicles **(2 marks)**
  ii calcium ions **(2 marks)**
  iii receptor sites on the post-synaptic membrane **(2 marks)**
  iv hydrolytic enzymes at the synapse. **(2 marks)**

b Indicate how the transmission of information in the nervous system may be modified by the presence of synapses. **(3 marks)**
(Total 11 marks)
(ULEAC January 1992)

14 Give reasoned explanations of each of the following observations.

  a The rate of discharge of the contractile vacuole of a freshwater protozoan falls when the protozoan is transferred to a dilute salt solution. **(3 marks)**

  b The gills of freshwater teleosts contain cells which actively absorb chloride ions from the surrounding water. **(3 marks)**

  c The length of the loop of Henle is relatively much longer in mammals inhabiting dry deserts than in mammals inhabiting places where more water is available. **(3 marks)**
  (Total 9 marks)
  (ULEAC January 1991)

15 The diagram below shows a vertical section of the brain of a mammal.

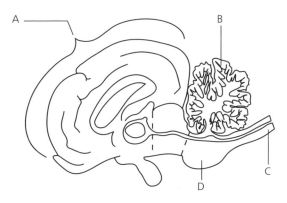

  a Name parts A, B, C and D and give *one* function for each. **(8 marks)**

  b What is the significance of the folding of part A? **(3 marks)**
  (Total 11 marks)
  (ULEAC January 1991)

16 a Make a large, labelled diagram of a motor neurone. **(5 marks)**

  b State *three* ways in which a motor neurone is adapted for its functions. **(3 marks)**
  (Total 8 marks)
  (ULEAC January 1991)

17 a Mammals are endothermic. Explain the meaning of this statement. **(2 marks)**

  b When mammals are exposed to hot conditions changes occur in their superficial blood vessels, their sweat glands and their fur. For each of these, describe *one* change which occurs and explain how this brings about cooling. Write your answers in a table. **(6 marks)**
  (Total 8 marks)
  (ULEAC January 1991)

18 This question is about a veterinary investigation of a sick cat, thought to be suffering from a malfunction of the thyroid gland.

  a Give *one* symptom that might have led the vet to suspect a thyroid hormone deficiency. **(1 mark)**

A sample of tissue was taken from the thyroid gland of the sick cat and a stained section prepared. This was compared with a similar section from a healthy cat. Drawings of parts of these sections are shown.

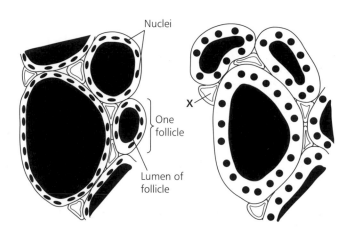

Sick cat   Healthy cat

  b Suggest a function for the lumen of the follicle. **(1 mark)**

  c i Name the tubular structure labelled **X**.

  ii Give *one* reason why there should be many such tubular structures (**X**) in thyroid tissue. **(2 marks)**

  d Describe *two* ways, apparent from the drawings, in which the thyroid of the sick cat differed from that of the healthy cat. **(2 marks)**

  e The investigator thought that the cat might be suffering as a result of a lack of iodine in its diet. Why should this element be needed for healthy thyroid function? **(1 mark)**

The investigator arranged for both cats to receive radioactive iodine compounds and then measured the levels of radioactivity in their blood. These levels were never high enough to cause harmful effects. The results are shown in the graph.

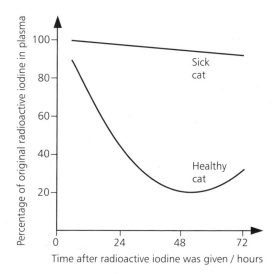

f i Do these results support the hypothesis that a dietary deficiency of iodine was the cause of the illness? Explain your answer.
  ii Suggest why the curve for the healthy cat rose after about 48 hours.
  iii Suggest a treatment which might enable the sick cat to recover.   **(4 marks)**
    (Total 11 marks)
    (NEAB June 1990)

19 The table below refers to pupil dilation and accommodation for near vision in the human eye.
Copy the table. If the statement is correct, place a tick (✔) in the appropriate box and if the statement is incorrect place a cross (✗) in the appropriate box.

Statement	Pupil dilation	Near accommodation
Is an example of a reflex action		
Is brought about by contraction of circular muscles		
Alters amount of light entering eye		
Is capable of voluntary control		
Alters refracting power of lens		

(Total 5 marks)
(ULEAC June 1991)

20 a Various activities in the body are controlled by hormones. Copy out and complete the following table by writing the name of a hormone involved in the activity described and the name of the gland which produces it.

Function	Hormone	Gland
1. decreases blood sugar levels		
2. controls the basal metabolic rate		
3. controls secretion of gastric juice		
4. increases rate of heart beat		
5. controls retention of sodium ions in kidney		
6. alters permeability of distal convoluted tubule		
7. controls growth of the ovarian follicle		
8. controls uterine growth and development		
9. involved in ovulation and maintenance of the corpus luteum		
10. controls male secondary characteristics		

(20 marks)

b State *two* advantages in the body of controlling physiological processes by the action of hormones compared with nervous control.   **(2 marks)**
c A small patch of tissue in a mammal is known to be secretory in function. Outline an investigation to discover whether the secretion acted as a hormone.   **(3 marks)**
    (Total 25 marks)
    (O & C June 1991)

21 a In a copy of the table below name the structures labelled 1–6 on the diagram of the skin and state the function each plays in the temperature control mechanism of a mammal.

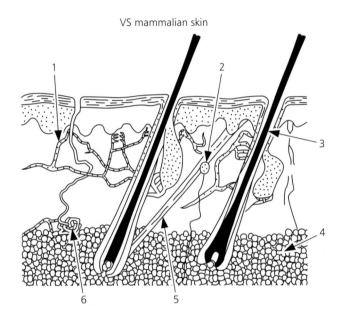

VS mammalian skin

	Name	Function in temperature control
1		
2		
3		
4		
5		
6		

(½ mark each)   **(1 mark each)**   **(9 marks)**

b In 1961 Benzinger, experimenting on human volunteers, obtained the results shown overleaf, after his subjects had ingested ice. Use your knowledge of the mechanism of temperature control to give an explanation of these results.   **(5 marks)**
c Endothermic animals maintain a constant body temperature.
  i State *four* advantages of endothermy.   **(4 marks)**
  ii State *two* disadvantages of endothermy.   **(2 marks)**
    (Total 20 marks)
    (O & C June 1991)

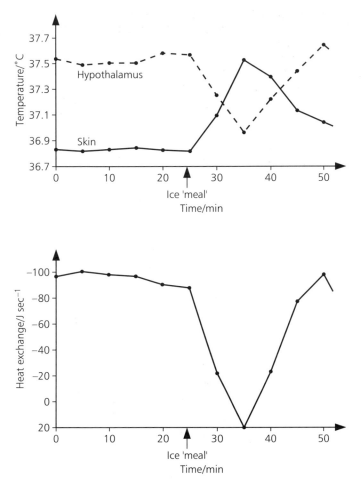

[Based on Simpkins & Williams,
Biology of the Cell: Mammal and Flowering Plant (Mills & Boon, 1980)]

22 An investigation into the factors affecting development of the common frog was made using four identical aquaria, labelled **A–D**. The aquaria contained different types of water and were kept at one of two temperatures. On April 10th, a single sample of frog spawn was collected, and a quarter placed in each aquarium. The subsequent development was observed and the results are shown below.

a Make a table to show the length of time taken from the start of the investigation for the tadpoles and adults to appear in each aquarium. **(3 marks)**

b Suggest *two* factors in this investigation which would need to be standardised to make it possible to compare the results. **(2 marks)**

c Suggest why the tadpoles in Aquarium **A** failed to develop into adults, whereas those in Aquaria **B** and **C** did. **(1 mark)**

d In terms of the results of this investigation, suggest explanations for the effect on the development of eggs (spawn) and tadpoles of:
i temperature
ii thyroxine. **(4 marks)**

e Suggest *two* possible reasons for the differences in size of the adults produced as a result of the different treatments. **(2 marks)**

f Describe briefly how you might determine whether the small size of adults in Aquarium **D** was the result of a permanent genetic change. **(2 marks)**
(Total 14 marks)
(NEAB June 1991)

Aquarium	Type of water	Temperature/°C	Date when first tadpole hatched	Date when first adult appeared	Mean final length of adult/mm
A	distilled	10	April 22nd	none appeared	—
B	tap	10	April 21st	May 25th	75
C	tap	15	April 19th	May 11th	80
D	tap with thyroxine	10	April 22nd	May 3rd	38

REPRODUCTION & GENETICS

5

Decisions, decisions – the future of human reproduction

The conception of a baby inevitably involves decisions. A couple may make a conscious decision to try and conceive a baby, or they may simple decide to have sexual intercourse and conception is the unplanned result! But as a consequence of rapidly developing reproductive technology, the decisions involved in human reproduction are moving far beyond the simple 'Shall we have a baby?'

Why is reproductive technology necessary?

In the developed world the mechanism of reproduction is taught in schools and colleges and discussed by parents and their children. The assumption is always made that producing babies is something which occurs very naturally and very easily – indeed, at times almost too easily. And for the majority of couples this proves to be the case – they achieve a pregnancy within a few months of deciding to start a family.

But at least one couple in 10 do not conceive after a year or more of trying. Alleviating the distress of those who want children but cannot have them has been the driving force behind much research into infertility treatments. Coupled with investigations into the human genome and manipulation of the genetic informa-

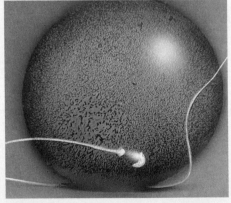

Figure 1 The moment of conception, when a sperm enters an ovum and a new life begins. But for some couples the journey to this moment is long and tortuous, complicated by the problems of reproductive systems which for one reason or another do not function in the normal way.

tion, a brave new world of reproductive technology has been opened up, enabling many people to have children who in the past would have

remained childless. Many of the developments have already brought much good – but equally there is the potential for great damage to be caused. Society as a whole, as well as the individuals concerned, needs to make many difficult decisions in the future as a result of these new technologies.

What causes infertility?

There are many reasons why a couple cannot produce children. In about 30% of cases there is a problem in the reproductive system of the female partner, and in another 30% the malfunctioning is in the male reproductive system. For the remaining couples both partners may be less fertile than average (**sub-fertile**), or both may be infertile, or there may be no apparent reason why pregnancy does not occur.

Infertility in women...

Difficulties in conception in a woman can be the result of a great variety of problems. If the ovaries do not contain ova then pregnancy is obviously impossible. This is very rare, however. Far more commonly the ovaries contain ova but they are not released in the normal monthly cycle – **ovulation** does not occur. The Fallopian tubes carrying the ova from the ovary to the uterus may be blocked, or the uterus may not accept a pregnancy due to an imbalance in the normal reproductive hormones. Some of the main causes are summarised in figure 2.

...and infertility in men

Infertility in the male has been recognised only relatively recently. Before that it was assumed that any inability to have children was the woman's problem. We now know that malfunctions of the male reproductive system are just as likely as those of the female, and the main problems occur with the process of sperm production known as **spermatogenesis**. Obviously if no sperm are produced, then fertilisation is impossible. More common than the complete absence of sperm is the presence of a high proportion of abnormal sperm. In the semen of a normal individual, about 15% of the sperm are abnormal, as illustrated in figure 3. If this proportion is substantially increased then the chances of successful conception drop.

Spermatogenesis takes about 70 days, and can be affected by a variety of factors such as the general health of the man, his alcohol intake and smoking. Spermatogenesis is under the control of hormones, particularly FSH, and so hormonal imbalance can affect sperm production. The testes need to be about 1 °C below body temperature for normal sperm production to occur, so if the testes are

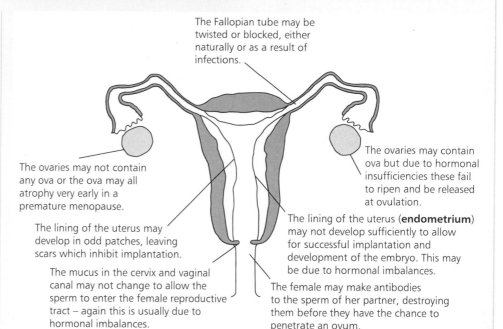

The Fallopian tube may be twisted or blocked, either naturally or as a result of infections.

The ovaries may not contain any ova or the ova may all atrophy very early in a premature menopause.

The ovaries may contain ova but due to hormonal insufficiencies these fail to ripen and be released at ovulation.

The lining of the uterus may develop in odd patches, leaving scars which inhibit implantation.

The lining of the uterus (**endometrium**) may not develop sufficiently to allow for successful implantation and development of the embryo. This may be due to hormonal imbalances.

The mucus in the cervix and vaginal canal may not change to allow the sperm to enter the female reproductive tract – again this is usually due to hormonal imbalances.

The female may make antibodies to the sperm of her partner, destroying them before they have the chance to penetrate an ovum.

Figure 2 The production of a fertile ovum every 28 days is a complex sequence of events coordinated by the pituitary hormones LH and FSH, and the female hormones oestrogen and progesterone. It is not surprising that sometimes the system fails.

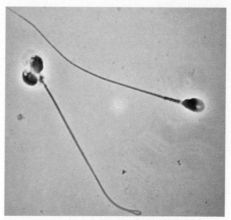

Figure 3 All men produce some abnormal sperm. But if the numbers of sperm with two heads, broken necks or several tails rises too high, then their chances of remaining suspended in the semen, reaching the ovum and achieving fertilisation become substantially reduced.

undescended (held in the body cavity) or if the temperature is increased in any other way, sperm production drops. Even the wearing of tight underpants and trousers has been indicated as a possible cause of subfertility in men because of the effect on testicular temperature. In a few cases the sperm cannot be released

due to a blockage in the vas deferens, the tube carrying them away from the testes towards the urethra.

Overcoming infertility

For many years the techniques available to help infertile couples were very limited. Artificial insemination, where sperm donated by an anonymous donor was inserted into the cervix, could help those couples where the malfunction was in the male sperm production. But AID (artificial insemination by a donor) could not help couples whose problem lay in the female reproductive system. In recent years many new methods of treatment have become available. Some of these are outlined below.

Using drugs to stimulate ovulation

Modern research means it has become possible to overcome some of the causes of infertility, particularly in women. As our knowledge of the

5

hormonal control of fertility has increased, so has our ability to manipulate it artificially. The inability to ovulate has been treated for some time now with what are commonly known as **fertility drugs**. These may be synthetic hormones such as Clomiphene, which stimulate the pituitary gland ofthe patient to produce more gonadotrophins to bring about ovulation, or they may be the gonadotrophins themselves, extracted from the urine of post-menopausal women. In the early days of this treatment the risk of multiple pregnancies was high, and women conceived as many as eight embryos at once. Although a few women did give birth to quads, quins and sextuplets, with most of the babies surviving, many more had their hopes dashed as they miscarried the unnaturally high number of embryos. And however badly a couple want children, four or five babies at the same time is an enormous physical, emotional and financial strain. Nowadays the treatment has been greatly refined and the drug preparations are far more sophisticated. As a result, although there is still a higher risk of twins and triplets resulting from treatment with fertility drugs than from an unassisted pregnancy, many couples are treated successfully and large multiple births are much less common.

In vitro fertilisation – a major breakthrough

The biggest breakthrough in the treatment of human infertility problems, and the technology which has opened a Pandora's box of possibilities for the future, is known in the popular press as the 'test tube baby' technique. More properly described as **in vitro fertilisation** (IVF), it involves the fertilisation of the ovum outside the mother's body and then the replacement of the embryo into the uterus to implant and develop as normal.

The first step in the process is to induce **superovulation**, giving fertility drugs to ensure that not just one but several ova ripen in the ovarian follicles. Just prior to natural ovulation these ova are harvested surgically and they are then mixed with sperm from the male partner in a petri dish. Examination under a high powered microscope shows which ova have been fertilised, and all of those that appear undamaged are incubated until several cell divisions have occurred successfully. The woman is meanwhile given further hormones to ensure that her uterus is ready to accept a pregnancy, and then several embryos are placed into the mother's body. Different medical centres replace varying numbers of embryos. Some replace only one, but should that one not implant the procedure needs to be repeated. Most centres replace two or three embryos, so that the failure of one or two to implant can still have a successful outcome. The risk of twins or triplets is thus increased when the technique works, as shown in figure 4.

The technique of *in vitro* fertilisation was first developed by Patrick Steptoe and Robert Edwards in the 1980s. It is still very expensive, with a relatively low success rate, and is not widely available. It has spawned other techniques which are cheaper, such as GIFT (**gamete intra-fallopian**

Figure 4 Test tube babies! Although far from emerging from a laboratory test tube, none of these children would have been born without the infertility treatment known as *in vitro* fertilisation.

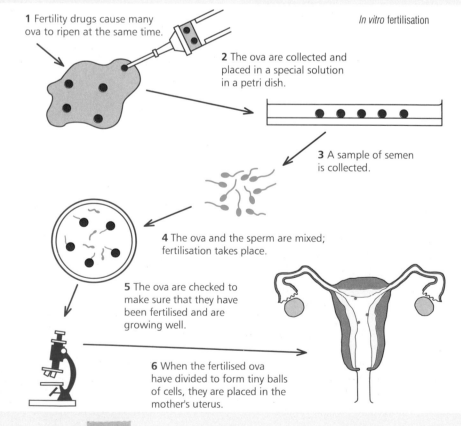

In vitro fertilisation

1 Fertility drugs cause many ova to ripen at the same time.

2 The ova are collected and placed in a special solution in a petri dish.

3 A sample of semen is collected.

4 The ova and the sperm are mixed; fertilisation takes place.

5 The ova are checked to make sure that they have been fertilised and are growing well.

6 When the fertilised ova have divided to form tiny balls of cells, they are placed in the mother's uterus.

transfer). Here the ova and sperm are collected and replaced immediately in the body again, but on the uterus side of the tubal blockage. Easier and cheaper than IVF, several eggs are placed in the Fallopian tube in an attempt to ensure successful fertilisation, but of course no control can then be exercised over the number of embryos which eventually implant. Also, to enable couples to donate ova only once (it is a time-consuming, expensive and uncomfortable process), work has been done on the freezing of both ova and embryos. (Sperm have been successfully stored in liquid nitrogen for many years.) It is now possible for a couple to donate ova and sperm, have some of the fertilised ova implanted immediately and have the remainder frozen to be implanted later when they want another child. It is the development of techniques such as IVF, GIFT and the freezing of embryos which leads to the possibility of conflict between those developments which are of benefit to the individuals concerned and those which benefit society as a whole.

Embryo experimentation – a moral dilemma

IVF has resulted in a great many 'spare' embryos. What is to be done with them? In most cases the parents give permission for them to be used for experimentation, to further our knowledge of reproduction and development. But in Britain and many other countries a limit has been set on the age of embryos used in this way. At 14 days the developing embryo is no longer likely to split and form identical twins, and the very beginnings of the nervous system are forming. Beyond this point no further experiments on *in vitro* human embryos are allowed. The problem is that now the technology is there, it is possible that someone, somewhere might continue these experiments to a later stage.

Genetic engineering

The most serious doubts and worries raised by the ability to fertilise ova outside the body and replace them inside the mother come when considering the combination of IVF and genetic engineering together. Our knowledge of the human genome is increasing all the time. Certain diseases are known to be carried in the genetic material and inherited – passed on from parent to child. Some of these genetic diseases such as haemophilia and Duchenne muscular dystrophy are known to be sex linked. The defective gene is carried on the X chromosome and so boys are far more likely to be affected than girls.

Until recently, the only measure that could be taken in families affected by these defective genes was to perform a test known as **amniocentesis** at around 16 weeks of pregnancy. This involves extracting a little of the amniotic fluid from around the fetus and then culturing some of the cells found in the fluid. A karyograph is produced to look at the chromosomes in order to determine whether the fetus is a boy or a girl. Male fetuses which are very likely to be affected can then be aborted if the parents wish. Not only is this procedure very stressful and distressing for the parents, the amniocentesis also runs the risk of causing a spontaneous abortion of a healthy fetus.

With IVF, cells can be removed from the developing early embryo without harmful effects and then sexed. Thus IVF means that families affected by sex-linked genetic diseases can have only female fetuses implanted, avoiding traumatic decisions later in the pregnancy.

Genetic engineering goes several steps further than this. The technique enables us to alter the genetic material of the cell. So far its use has been mainly in agriculture and drug manufacture. Recently, genes have been engineered into the cells of children with genetic diseases with the aim of affecting their immune systems, and before long this technique will be used to treat cystic fibrosis. At the moment no-one is allowed to engineer either germ cells – the ova and sperm – or the early cells of an embryo which is to be implanted back into its mother.

However, when we have the knowledge which would allow us to manipulate the genes of an early embryo and prevent the onset of genetic disease, should it not be used? In the wrong hands of course, it would also be possible to isolate the genes which cause bad temper or aggression, beautiful hair or intelligence, and begin to manipulate them too ... . There are many issues which need to be weighed in the balance. To the childless couple, or the family affected by genetic disease, all advances are welcomed. But for society as a whole the good needs to be weighed against the potential harm and controls put in place to prevent, as far as humanly possible, some of our latest great discoveries becoming the source of our undoing.

Figure 5 Points to ponder

At one level, IVF has enabled many couples to overcome the heartbreak of being unable to have much-desired children. However, there are many implications and possible misuses of the technique.
For example:
• If embryos are frozen for future use, what happens if the parents divorce? To whom do the embryos belong?
• If parents and their first child are killed there could be great difficulties over the estate. Do the frozen embryos inherit, and if so should they be implanted in a surrogate mother and allowed to develop in order to collect their inheritance?
• In some countries (e.g. the USA and Italy) IVF is now being used to allow women who are past the menopause, in their 50s and 60s, to have babies. This is being carried out with substantially higher success rates than IVF in younger, infertile women. IVF specialists in Britain feel that this is not an appropriate use of the technology because of the social issues it raises. What are these social issues? Men can father children into their 70s and even 80s – does it matter if women too prolong their reproductive lives?

5.1 Reproductive strategies and asexual reproduction

The need to reproduce is fundamental to all living things. Animals and plants grow, feed, respire and excrete for a certain length of time, and then become increasingly less efficient until eventually they die. For life to continue an organism must reproduce itself before it dies. Parent organisms age and die so they do not compete with their offspring for resources. The ability of individuals to reproduce successfully is also the main factor which decides whether a whole group of organisms survives.

Each group of animals and plants has very distinct and specific genetic material. Reproduction passes this genetic material to a new generation. Reproduction uses up many resources and may often be the cause, either directly or indirectly, of the death of the parent organism. In spite of this, it can be considered that all other life processes (respiration, feeding, etc.) exist simply to provide the resources for reproduction. Indeed, the success of an organism (its **biological fitness**) is measured in terms of the production of fertile offspring which survive to reproductive maturity, thus replacing their parents.

ONE PARENT OR TWO?

There are two main methods of reproduction. **Asexual reproduction** involves only one parent individual, be it plant or animal. Asexual reproduction has many advantages. It is safe, certain (there are no problems of finding a receptive mate) and can give rise to large numbers of offspring very rapidly. The offspring produced are almost all genetically identical to the parent organism, and so a successful genetic combination can be passed on without change. This is an important advantage of asexual reproduction until living conditions change in some way. However, the introduction of a new disease to an environment, a change in the temperature or human intervention can cause the total destruction of a group of genetically identical organisms, because if one cannot cope with the new environment, neither can all the others.

Sexual reproduction carries considerably more risk. It involves the combining of genetic information from two individuals. The meeting of the special sex cells (**gametes**) which carry this information is by no means certain and arranging for them to meet often carries an element of risk, for animals at least. But the great advantage of sexual reproduction is that it introduces **variety**, which is of great value when conditions are not stable. Variety increases the chance of survival because it provides the opportunity to evolve and take advantage of new conditions and surroundings.

REPRODUCTIVE STRATEGIES

Whether organisms employ asexual reproduction, sexual reproduction or a combination of the two, the most important factor is their reproductive strategy. In the plant world many species undergo both sexual and asexual reproduction as a matter of course. They reproduce sexually by flowering, but they also produce bulbs, corms, tubers, rhizomes, runners or suckers as asexual methods.

Most plants are hampered in terms of sexual reproduction by being rooted to the spot, but they have evolved a wide variety of methods to ensure that the male gametes (carried in the pollen) reach the female parts of the flower. A range of **vectors** is used to transfer the pollen to the female parts of the plant. These include the use of insects (attracted by colours, patterns, scents and food rewards such as nectar), the wind, small mammals and water.

Animals too have evolved a wide variety of strategies to bring about successful reproduction. Many simple organisms, such as *Amoeba*, reproduce in the simplest way possible and split in two. *Hydra* and other pond-water species may be seen with small, identical individuals forming as buds, but they also form gonads producing eggs and sperm for sexual reproduction.

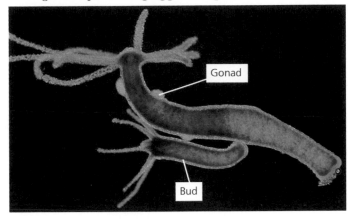

Figure 5.1.1 These organisms take full advantage of all the reproductive opportunities, both sexual and asexual, presented to them. The daffodil flower is a means of sexual reproduction, while the bulb reproduces by asexual means. The hydra too has both sexual organs and asexual reproductive buds.

Strategies for asexual reproduction

There are a variety of strategies for asexual reproduction which are outlined below.

Fission

Fission involves the splitting of an individual. It is a method of reproduction found in many invertebrate organisms. As two new individuals are usually formed it is also known as **binary fission**. Bacteria and protoctists such as amoebae undergo this form of asexual reproduction. Bacteria are capable of enormous and rapid increases in numbers, since under ideal conditions they may divide every 20 minutes. Bacteria also undergo a form of sexual reproduction or **conjugation** whereby some or all of the genetic material is exchanged (there is more about this in section 5.2). This introduces variety.

Amoebae do not rely solely on straightforward fission either. Under normal conditions the organisms divide when they have reached a certain size, but in adverse conditions an amoeba will form a cyst and then undergo a series of fissions so that a number of very small, identical organisms are released when conditions improve.

Although fission has its limits as a reproductive strategy, a similar process is used in cell reproduction for growth and repair in all living things.

Sporulation

Sporulation involves the production of asexual **spores** which are capable of growing into new individuals. Spores can usually survive adverse conditions, and are dispersed very readily. They are produced most commonly in very large numbers by fungi and plants, in organs known as **sporangia**, as shown in figure 5.1.3. The spores produced by bracken are **carcinogenic** (cancer causing) and the levels of spores in the air of bracken-covered hillsides can be so high that hill farmers and walkers are advised to wear masks to filter the air, thus removing the spores and so the risk of disease. Some protoctists also produce spores and these are called **zoospores**.

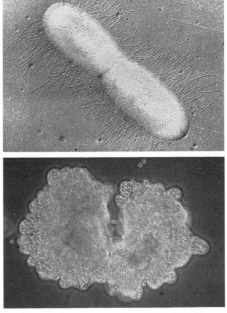

Figure 5.1.2 Fission is an effective reproductive strategy.

Figure 5.1.3 Spores are frequently produced in very large quantities. If each of the millions of spores produced by a puffball actually developed into another puffball, the world would rapidly disappear under the fungus.

Fragmentation and regeneration

Some organisms can replace parts of the body which have been lost. Many lizards shed their tails when attacked and then grow another. This is known as **regeneration**. Some organisms manage an even more spectacular form of regeneration – they can reproduce themselves asexually from fragments of their original body. Starfish can not only survive being chopped into pieces, but if conditions are suitable each piece will regenerate to form an entire new starfish (see figure 5.1.4). Some members of groups as diverse as fungi, flatworms, filamentous algae and sponges fragment and then regenerate as a preferred method of reproducing.

Budding

Budding in a reproductive sense does not in general include the production of buds by flowers. In reproductive budding, there is an outgrowth from the parent organism which produces a smaller but identical individual. This 'bud' eventually becomes detached from the parent and has an independent existence. Yeast cells reproduce by budding. In single-celled organisms like these the only recognisable difference between budding and binary fission is that in budding the parent cell is larger than the bud. Budding is relatively rare in the animal kingdom, occurring, for example, in *Hydra*, seen clearly in figure 5.1.5. As we have seen, budding is only part of the reproductive strategy of *Hydra* – they reproduce sexually as well.

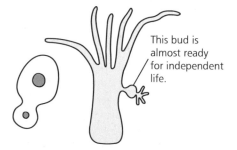

Figure 5.1.4 To protect oyster beds from destruction by starfish, oyster fishermen have often chopped up starfish and thrown them back into the sea, a measure which proves to be entirely counterproductive as each fragment regenerates to form another starfish!

Vegetative propagation

In some ways, **vegetative propagation** is a more sophisticated version of budding which occurs in flowering plants. A structure forms which develops into a fully differentiated new plant, identical with the parent. The new plant may be propagated from the stem, leaf, bud or root of the parent, depending on the type of plant. Vegetative propagation often involves **perennating organs**. These contain stored food made by photosynthesis and can remain dormant in the soil to survive adverse conditions. They are often not only a means of asexual reproduction, but also a way of surviving from one growing season to the next. Examples are shown in figure 5.1.6.

Vegetative propagation is exploited by gardeners to produce new plants. Splitting off new daffodil bulbs, removing strawberry plants from their runners and cutting up rhizomes all increase plant numbers cheaply, and the new plants have exactly the same characteristics as their parents.

This bud is almost ready for independent life.

Figure 5.1.5 Budding in yeast cells and in *Hydra*. Notice the asymmetry of the yeast cell and the bud which distinguish this process from binary fission in single-celled organisms.

Cloning

A **clone** is a group of cells or organisms which are genetically identical and have all been produced from the same original cell. In the 1950s a carrot was produced by Frederick Steward from a single carrot phloem cell which was grown in a rich nutritive medium. For this to happen the original cell needs to have retained the capacity, present in the cells of an embryo, to go through all the stages of development. Since it produced a normal adult, none of the genes in the phloem cell must have been permanently switched off. A cell which is capable of being used in this way is known as **totipotent**.

Cloning has two major uses. In horticulture, plants such as orchids are now almost always produced commercially by cloning. In research, it is of great value to have plants and animals to work on which are known to be genetically identical. Any differences in their development or behaviour can then reasonably be put down to the experimental variable.

Cloning of plants is now commonplace and relatively easy. The cloning of animals is less straightforward, and the day of the cloned human being is still a very long way off indeed – most adult human cells are not totipotent.

Figure 5.1.6 Examples of perennating organs – organs that allow plants to be perennial, or to flower year after year. They are also frequently organs of asexual reproduction, but only if more than one bud develops into a new shoot.

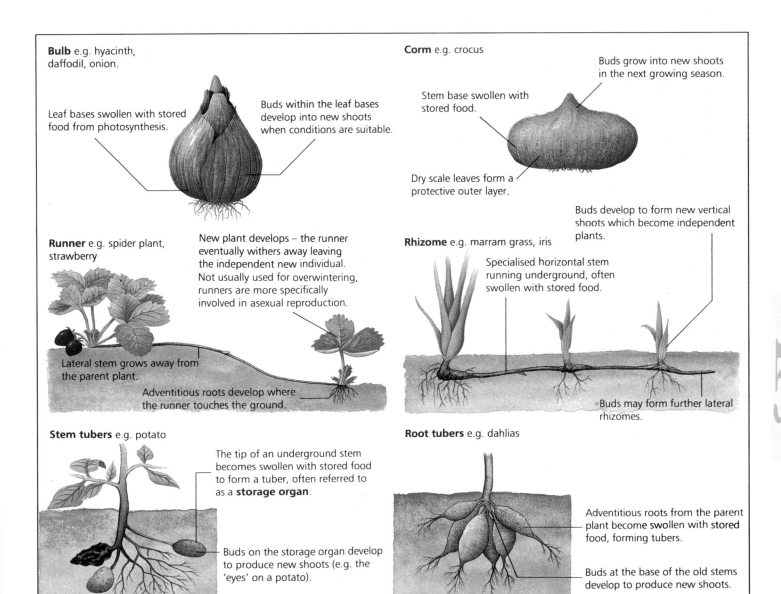

Bulb e.g. hyacinth, daffodil, onion.

Leaf bases swollen with stored food from photosynthesis.

Buds within the leaf bases develop into new shoots when conditions are suitable.

Corm e.g. crocus

Buds grow into new shoots in the next growing season.

Stem base swollen with stored food.

Dry scale leaves form a protective outer layer.

Runner e.g. spider plant, strawberry

New plant develops – the runner eventually withers away leaving the independent new individual. Not usually used for overwintering, runners are more specifically involved in asexual reproduction.

Lateral stem grows away from the parent plant.

Adventitious roots develop where the runner touches the ground.

Rhizome e.g. marram grass, iris

Buds develop to form new vertical shoots which become independent plants.

Specialised horizontal stem running underground, often swollen with stored food.

Buds may form further lateral rhizomes.

Stem tubers e.g. potato

The tip of an underground stem becomes swollen with stored food to form a tuber, often referred to as a **storage organ**.

Buds on the storage organ develop to produce new shoots (e.g. the 'eyes' on a potato).

Root tubers e.g. dahlias

Adventitious roots from the parent plant become **swollen** with **stored** food, forming tubers.

Buds at the base of the old stems develop to produce new shoots.

5

Strategies for sexual reproduction

Relatively few organisms rely solely on asexual reproduction. In almost every case there is at the very least a back-up system of sexual reproduction, often used in adverse conditions. In many organisms – most flowering plants, most arthropods and the vertebrates in particular – sexual reproduction is the main or only method. Some flowering plants and arthropods reproduce asexually as well.

Plants

Reproductive strategies developed to ensure the success of sexual reproduction are many and varied. Some plants have evolved complex combinations of colours, scents and rewards to attract appropriate pollinators. Others have a floral anatomy particularly suited to releasing their pollen into the wind. There are also a vast number of strategies for making sure that the seeds which result are dispersed as widely as possible from the parent plant, as figure 5.1.7 shows.

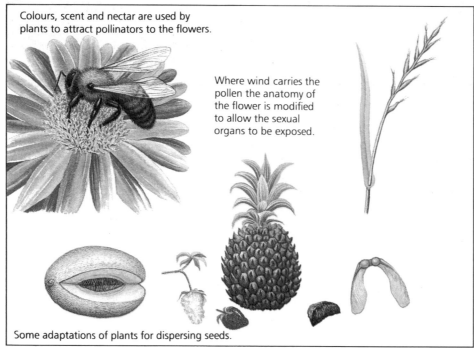

Colours, scent and nectar are used by plants to attract pollinators to the flowers.

Where wind carries the pollen the anatomy of the flower is modified to allow the sexual organs to be exposed.

Some adaptations of plants for dispersing seeds.

Figure 5.1.7 Some of the sexual reproductive strategies of plants

Animals

In animals reproductive strategies include linking sexual receptiveness to times of high fertility, regular sexual cycles in the female animal and lifestyles which range from constant coexistence of the male and female to the sexes living completely separately except when mating needs to occur. Fertilisation may occur outside the body (**external fertilisation**) or inside the body (**internal fertilisation**). Once fertilisation has taken place there are varying levels of parental care of the offspring. Many organisms just release eggs and sperm into the water and provide no care to the offspring at all. Amphibians, reptiles and birds lay eggs. On the whole, amphibians provide less parental care than reptiles, whilst birds are well known for the effort they expend raising their young. In mammals, the young develop within the body of the mother. Once they are born the female provides all the nutrition required by her offspring in the early stages of independent life in the form of milk from her own body. Examples of reproductive strategies in both animals and plants will be considered in more detail in sections 5.2–5.5.

Many amphibians ensure that eggs and sperm are deposited together by close association between the male and female. The young, however, generally receive little care. The midwife toad shown here is an exception – the fertilised eggs are placed in special 'pockets' in the skin of the back and the tadpoles develop there until the young toads emerge.

Some organisms such as this starfish release their sex cells into the water and provide no care of the resulting offspring.

Anatomical specialisation and complex mating rituals and manoeuvres have enabled land-living animals to ensure that eggs and sperm meet. Eggs provide the developing offspring with both food and protection.

Mammals mate to ensure egg and sperm meet. The offspring develop in the uterus of the mother and are provided with a food supply after birth.

Figure 5.1.8 Some reproductive strategies of animals

THE MECHANISM OF ASEXUAL REPRODUCTION – MITOSIS

How cells divide

During reproductive processes, both at the level of the whole organism and at a cellular level, like begets like. Buttercups produce new buttercups, amoebae produce more amoebae and liver cells generate more liver cells. Most of this new biological material comes about as a result of the process of nuclear division known as **mitosis**. This is usually followed by an equal division of the cytoplasm, resulting in two identical cells. The entire process is known as **cell division**. Asexual reproduction and growth are the result of mitotic cell division. The production of offspring by sexual reproduction is also largely dependent on mitosis to produce the new cells which form after the gametes have fused. It is only the formation of the sex cells themselves which involves a different type of nuclear division, **meiosis**, which we shall look at in section 5.2. In meiosis the chromosome numbers are halved and four, non-identical daughter cells result. Cell replication involving mitosis can be regarded as normal cell division.

Chromosomes

As we have seen in section 1.3, the chromosomes are found in the nucleus of the cell. They are made up of DNA and carry the blueprint for the proteins which determine the make-up of the cell and indeed the entire organism. The information is carried in the genes and is translated into living cellular material by the process of protein synthesis. A chromosome is made up of a mass of coiled threads of DNA and proteins. In a cell which is not actively dividing the chromosomes have relatively little structure and cannot easily be identified as individual entities. When the cell enters an active dividing stage the chromosomes 'condense' – they become much shorter and denser. They then take up stains very readily (this is the basis of the name 'chromosome' or 'coloured body') and individual chromosome pairs can be identified.

The cells of each species possess a characteristic number of chromosomes – in our own case there are 46. These chromosomes occur in matching pairs, one of which originates from each parent. Humans have 23 pairs of chromosomes in their cells, whilst turkeys have 41 pairs. In mitosis the cells resulting from the division will both receive the same number of chromosomes. Thus before a cell divides it must duplicate its original set of chromosomes. Then during mitosis these chromosomes are divided equally between the two new cells so that each has a complete and identical set of genetic information. A normal cell which contains the full set of chromosome pairs is known as **diploid**. Cells exist which have only one of each pair of chromosomes, as we shall see in section 5.2. Such cells are **haploid**.

During the active phases of cell division the chromosomes become very coiled and condensed. In this state they are relatively easy to photograph and a special display or **karyotype** can be made, as shown in figure 5.1.9.

Not all cells are haploid or diploid. In a surprising number of organisms the cells have three (triploid), four or more copies of each chromosome. They are known as **polyploid**. For organisms such as the cultivated strawberry where polyploidy is the normal situation, it causes no problems. But if polyploidy of one or all of the chromosomes occurs unexpectedly then the consequences for the organism are usually devastating, causing major abnormalities or death.

The cell cycle

Cells divide on a regular basis to bring about growth and asexual reproduction. They divide in a sequence of events known as the **cell cycle** which involves several different phases. There is a period of active division by mitosis, and a period of non-division called **interphase** which is when growth, replication of the DNA and normal cellular activities occur. The cell cycle can be very rapid, taking 24 hours or less to be completed, though it may only occur once every few years. In multicellular organisms it is repeated very frequently in almost all cells during development. However, once the organism is mature the cell cycle may be slowed down or stopped completely in some tissues. Figure 5.1.10 shows the main stages of the cell cycle.

The chromosomes occur in pairs and in most cases both members of the pairs are the same shape. These **homologous** chromosomes are referred to as the **autosomes**.

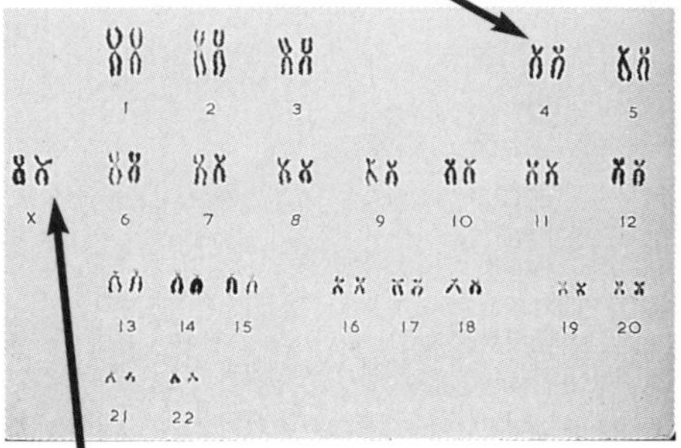

The sex chromosomes may be the same shape (XX), in which case the individual is female, or they may be different (XY) as in the human male.

Figure 5.1.9 A human karyotype shows the 22 pairs of autosomes and one pair of sex chromosomes which go to make up the 23 pairs of chromosomes found in every normal diploid human cell.

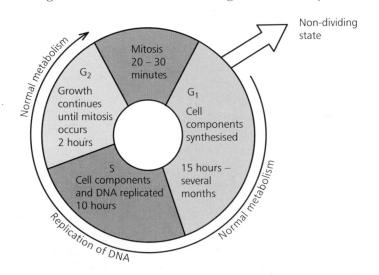

Figure 5.1.10 Phases of the cell cycle. In very actively dividing tissue the cycle is repeated often, whilst in other regions the gaps between successive divisions may be years.

The stages of mitosis

During the process of cell division the chromosomes are duplicated and then they and the remaining contents of the cell are divided up in such a way that two identical daughter cells are formed. Walther Fleming, a German cytologist, was the first to describe the 'dance of the chromosomes'. This is the description sometimes given to the complex series of movements which occur as the chromosomes jostle for space in the middle of the nucleus and then pull apart to opposite ends of the cell. The events of mitosis are continuous, but as in the case of so many biological processes it is easier to describe what is happening by breaking events down into phases. These are known as **prophase**, **metaphase**, **anaphase** and **telophase**, shown in figures 5.1.11–12 and described below.

Evidence for mitosis

Mitosis can be seen relatively easily in the cells of rapidly dividing tissues such as a growing root tip. Using an appropriate dye (such as acetic orcein) which stains the chromosomes, tissue squashes can be produced which show the stages of mitosis quite clearly, as shown in figure 5.1.11.

Stained tissue squashes are obviously dead, but living tissue can also be observed, and time-lapse photography has enabled dramatic recordings of the activity of chromosomes to be made. Such a view of the movements of the cell contents during mitosis shows why it is called the 'dance of the chromosomes'.

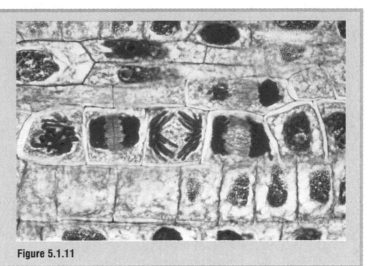

Figure 5.1.11

Interphase

A cell will be in interphase for much of its life. This used to be called the **resting phase**, but nothing could be further from the truth. During interphase not only do the normal metabolic processes of the cell continue, but also new DNA is produced as the chromosomes replicate, as we have seen in figure 5.1.10. Sufficient new proteins, cytoplasm and cell organelles are also synthesised so that the cell is prepared for the production of two new cells. ATP production is also raised at times to provide the extra energy needed by the cells for division by mitosis. Once all that is needed is present and the parent cell is large enough, interphase ends and mitosis begins.

Figure 5.1.12 These diagrams show the main phases of mitosis in a simplified animal cell which has only two pairs of chromosomes.

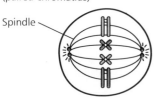

Interphase – Before mitosis, the tangled, uncoiled mass of chromosomes fills the nucleus. DNA is replicated during this stage.

Prophase – The chromosomes begin to condense. They replicate to form two **chromatids**. The nuclear membrane and the nucleolus disappear, and the centrioles begin to separate to form the **spindle**.

Metaphase – Spindles made of microtubules are formed by the centrioles. The chromatids line up on the metaphase plate.

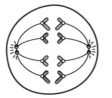

⬭ Chromosome inherited from mother
⬭ Chromosome inherited from father

Anaphase – The centromeres separate and each pulls its chromatid along a spindle tubule towards one of the poles.

Early telophase – The chromatids reach the poles of the cell and become chromosomes. The nuclear membrane begins to reform and the cytoplasm begins to divide.

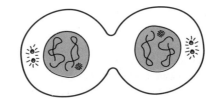

Late telophase – The chromosomes begin to unravel or 'decondense'. The nuclear membranes and nucleoli are fully reformed and centrioles are present again. The division of the cytoplasm continues until two new identical cells are formed which once more enter interphase.

Prophase

Before mitosis begins the genetic material has been replicated to produce exact copies of the original chromosomes. By the beginning of prophase both the originals and the copies are referred to as chromatids. In prophase the chromosomes coil up, take up stain and become visible. Each chromosome at this point consists of two daughter chromatids which are attached to each other in a region known as the **centromere**. The nuclear membrane breaks down and the nucleolus disappears. In animal cells the centrioles begin to pull apart to form the **spindle** of microtubules. In plant cells there are no centrioles, but spindles form at this stage in a very similar way.

Metaphase

The formation of a spindle made of microtubules is completed. The chromosomes appear to jostle about for position on the **metaphase plate** or **equator** of the spindle during metaphase. They eventually line up along this plate, with each centromere associated with a microtubule of the spindle.

Anaphase

The centromeres linking the two identical chromatids divide, and from then on the chromatids act as separate entities and effectively become new chromosomes. One chromatid from each pair is drawn towards opposite poles of the cell, the centromeres moving first. This separation occurs quickly, taking only a matter of minutes. At the end of anaphase the two sets of chromatids are at opposite ends of the cell.

Telophase

During telophase the spindle fibres disappear and nuclear membranes form around the two sets of chromosomes. Nucleoli and centrioles also reform. The chromosomes begin to unravel and become less dense and harder to see. The final step is the division of the cytoplasm, sometimes referred to as **cytokinesis**.

In animal cells a ring of contractile fibres tightens around the centre of the cell, rather like a belt tightening around a sack of flour. These fibres appear to be actin and myosin, the proteins found in animal muscles (see section 2.6). They continue to contract until the two cells have been separated. In plant cells the division of the cell occurs rather differently, with a cellulose cell wall building up from the inside of the cell outwards, as shown in figure 5.1.13. In both cases the end result is the same – two identical daughter cells which will then enter interphase and begin to grow and prepare for the next cycle of division.

Mitosis is the method by which organisms undergo asexual reproduction. It is also the source of all the new cells needed for organisms to grow and to replace worn-out cells. We shall take time here to consider what growth entails and how it is measured before moving on to look at sexual reproduction in more detail, in sections 5.2–5.5.

GROWTH

What is growth?

Growth is a permanent increase in the mass or size of an organism. There are three distinct aspects of growth, **cell division**, **assimilation** and **cell expansion**, already considered in section 4.5. Cell division or mitosis is the basis of growth. New plant cells are supplied with fresh material which is the product of photosynthesis in other, mature cells. Similarly new animal cells are provided with new material from the products of digestion. This is what is meant by assimilation, and the result is cell expansion. However, cells can

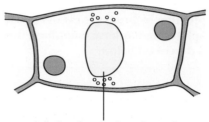

Cell plate forming in a plant cell

Animal cells dividing

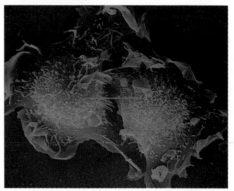

Figure 5.1.13 To complete the process of mitosis the two new sets of chromosomes need two new cells to house them. The cytoplasm of animal and plant cells is divided in different ways to bring this about.

How the spindle moves the chromosomes

Chromosomes cannot move on their own. They rely on the microtubules of the spindle to allow them to move. The spindle was for many years envisaged as a structure running from one end of the cell to the other. It is now known to be made up of overlapping microtubules containing contractile fibres not dissimilar to the actin and myosin of animal muscles. It is by the contraction of these overlapping fibres that movement of the chromosomes is brought about. This is an energy-using process, and is one of the main reasons why energy is made and stored in interphase.

expand in other ways, for example by taking in water. The increase in size which results from this may only be temporary and so we define growth as involving a *permanent* increase in size or mass.

How is growth measured?

The measurement of growth is important in many ways, both scientifically and medically. Growth may be affected by factors such as the availability of food, temperature and light intensity as well as the genetic make-up of the organism. The measurement of growth is not at all easy. A linear dimension such as height or circumference can be very deceptive – cake mixture will increase in both height and circumference as it cooks, but it has not grown! Measuring mass also has its problems – the water content of the cells may vary greatly, particularly in plants, and animals may have varying quantities of faecal material and urine held in their bodies.

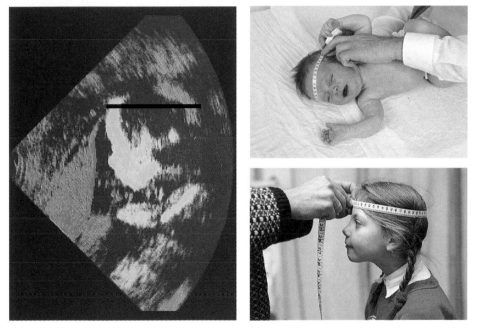

Figure 5.1.14 Starting long before birth, head circumference is measured at intervals as a child grows, to make sure all is well. Scientific as this may seem, an increase in circumference does not always indicate biological growth, as the example of the balloon shows.

As growth involves an increase in the cytoplasmic content of an organism, mass is the best and most commonly used measure. To avoid the difficulties mentioned above of varying water content, the **dry mass** can be measured. This gives an accurate picture of the amount of biological material present but has a major drawback in that removing all the water from an organism kills it, so that further growth cannot be measured. For dry mass to give useful results, large samples of genetically identical organisms need to be grown under similar conditions. Random samples are then taken for drying to a constant mass. This method is very useful for plants but has obvious limitations for animals. It is neither easily possible nor ethical to maintain large colonies of genetically identical animals and then kill and dry them, and so less reliable indicators of growth such as height and wet mass are used.

Growth patterns

In spite of the difficulties in the measuring of growth, we do have a good picture of the patterns of growth of many organisms. **Growth curves** made up from measurements of growth throughout the life of an organism show when most growth takes place. Growth curves are very similar for most organisms, whether animals or plants, as figure 5.1.15 shows. After an initial relatively slow start there is a rapid period of growth until maturity is reached, when growth slows down and stops. This pattern is known as **continuous growth**.

Not all organisms undergo continuous growth. Insects grow in a series of **moults**. They shed one exoskeleton and then expand their bodies by taking in air or water and 'grow' whilst the new exoskeleton is soft. Once the new skeleton has hardened the air or water can be released, leaving room for the tissues of the insect to increase in size. This is known as **discontinuous growth** and gives a different growth pattern to that of organisms that grow continuously. If length is measured the arthropod appears to grow in a series of steps, as figure 5.1.15 shows.

Figure 5.1.15 Growth curves are usually measured by the linear dimensions. If dry mass is used the discontinuous growth curve becomes less pronounced.

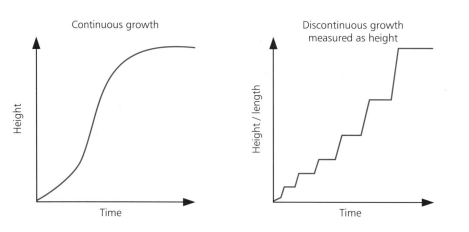

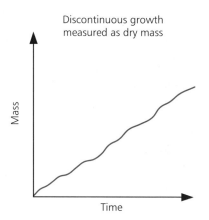

The development of the embryo is the time when the largest amount of growth occurs, in terms of percentage of body mass. At this time mitosis occurs at an immensely rapid rate. Growth continues until the organism reaches maturity. At this point growth slows down or stops completely in many organisms. Mitotic divisions do not stop, however, as cells are continually becoming worn out and being replaced. This continues until the onset of **senescence** or old age, when mitosis occurs less frequently and the dying cells begin to outnumber the new cells being formed. When this process reaches a certain point, death will occur.

Cancer – when normal cells run amok

Cancers have affected the living world, plants and animals alike, for millions of years. Signs of cancers have been found in the bones of dinosaurs and in Egyptian mummies. As people live longer, and infectious diseases have been brought increasingly under the control of drugs, so the toll of cancers on human health have become more apparent. But what is cancer?

If the answer to this question was fully understood, then cancer would be curable in all its forms. Sadly this is not yet the case, but scientists and doctors are rapidly increasing their understanding of cancer. Cancer cells carry out many cellular functions as normal, but the transformation of a normal cell into a cancerous cell brings about some important changes. The most important of these is uncontrolled growth. When we cut ourselves, the cells on the edge of the cut undergo rapid mitosis until the cut is healed. The rate of mitosis is then restored to its normal level. In cancer cells mitosis simply goes on and on because the normal regulatory systems fail to work. As cancer cells also live longer than ordinary cells the result is an enlarging tumour or other malignant condition which disrupts normal tissues, often to the point of killing the organism.

What causes cancers? Again, the answer is far from fully understood. Certain substances cause cancer – they are known as **carcinogens**. The tars in tobacco smoke, chemicals such as benzene, some dietary additives and many other substances may trigger the transformation of normal cells to cancer cells. Radiation, ultraviolet light from the Sun and X-rays can all have a similar effect. Some viruses are known to cause cancers in animals, and it is suspected that they play a role in some human cancers too. And finally, some people have genes which make them more vulnerable to cancer than others. **Oncogenes** which result in cancer, or **proto-oncogenes** which may become oncogenes if exposed to particular carcinogens, may prove to be the vital clues in the search for a cure – only time will tell.

SUMMARY

- **Asexual reproduction** involves one parent, is safe, certain and may produce many offspring which are genetically identical with the parent. **Sexual reproduction** is the meeting of **gametes** from two parents. It is less straightforward but produces genetic variety.

- Methods of asexual reproduction include **binary fission** – the splitting of an individual; **sporulation** – the production of asexual spores in organs called **sporangia**; **fragmentation** and **regeneration**; and **budding**, which is similar to fission except that the new organism is smaller than the parent.

- **Vegetative propagation** is asexual reproduction in flowering plants, and often involves **perennating organs** which store food and remain dormant in the soil. Methods of vegetative propagation include bulbs, corms, runners, rhizomes, stem tubers and root tubers.

- Asexual reproduction and growth result from cell division by **mitosis**, which produces two daughter cells with genetic material identical with that of the parent cell.

- Each species has a characteristic number of **chromosomes** in its body cells (46 in humans). This is the **diploid number**, made up of **homologous pairs** of chromosomes (23 pairs in humans). Chromosomes become visible during cell division and may be displayed as a **karyotype**.

- The **cell cycle** involves a period of mitotic division and a non-dividing state called **interphase**. Stages of interphase are G_1 when cell components are synthesised, S when cell components and DNA are replicated and G_2 when growth continues until mitosis occurs. The cell cycle can last for any period from about 24 hours to years.

- **Prophase** is the first stage of mitosis. The chromosomes become visible, made up of two **chromatids** joined at the **centromere**. The nuclear membrane breaks down and the nucleolus disappears. The centrioles (in animal cells) start to pull apart.

- In **metaphase** a **spindle** of microtubules has formed. The chromosomes jostle for position and line up on the **metaphase plate**.

5

- In **anaphase** the centromeres split and the chromatids separate to opposite poles of the cell.
- In **telophase** the spindle disappears and new nuclear membranes, nucleoli and centromeres form. The chromosomes disperse and the cytoplasm splits (**cytokinesis**).
- **Growth** is a permanent increase in the size of an organism. It involves **cell division**, **assimilation** and **cell expansion**.
- Growth may be measured by height or circumference or by mass, but none of these gives a very accurate picture. The **dry mass** is an accurate measure of the amount of biological material present but involves killing the organism.
- Organisms may show **continuous growth**, or **discontinuous growth** such as in insects which grow in a series of **moults**.
- Growth is greatest while the embryo develops. It continues to maturity when mitotic divisions are reduced and only replace worn-out cells. When this process slows down **senescence** occurs, followed by death.

QUESTIONS

1 a What are the advantages and disadvantages of asexual reproduction?
  b What are the advantages and disadvantages of sexual reproduction?
  c Produce a table summarising the main methods of asexual reproduction in both the animal and the plant kingdoms.

2 a What is the role of mitosis in growth?
  b Describe mitotic cell division in animal cells.

3 a What are the difficulties of measuring growth in organisms?
  b What is meant by:
    i continuous growth
    ii discontinuous growth?
  c Explain what happens when control over cell growth is lost or altered in some way.

5.2 Sexual reproduction

As we saw in section 5.1, sexual reproduction is the production of a new individual resulting from the fusing of two specialised cells known as **gametes**. The individuals resulting from sexual reproduction are *not* genetically the same as either of their parents, but contain genetic information from both. Sexual reproduction introduces variety – no two individuals (apart from identical twins, discussed in section 5.5) will have the same combination of genes.

As we have seen, asexual reproduction is a very successful process. A winning genetic formula is preserved and passed on to subsequent generations at minimum risk to the parental organism. It is often rapid and does not require any investment in specific reproductive organs. Whilst conditions remain stable all is well. However, populations with little or no genetic variability are vulnerable to new diseases or to environmental crises such as drought. Sexual reproduction, in spite of being more risky and expensive in terms of bodily resources, introduces genetic variety, and this variety enables organisms to adapt and cope with changing circumstances.

GAMETES

What are gametes?

The nucleus of a cell contains the chromosomes. As we saw in section 5.1, the vast majority of the cells of an organism are **diploid** – chromosomes occur in pairs. However, if two diploid cells combined to form a new individual in sexual reproduction, the offspring would have two pairs or four sets of

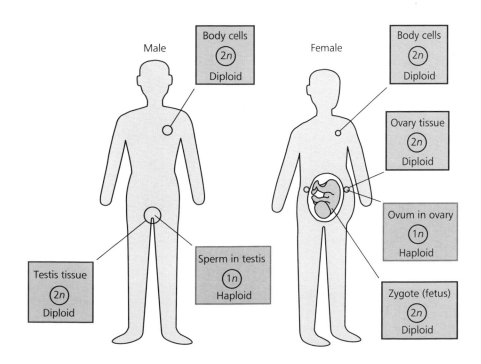

Figure 5.2.1 Diploid body cells, haploid gametes and the diploid zygote

chromosomes, twice the characteristic number for the species. Subsequent generations would each have double the previous number of chromosomes. To avoid this, **haploid** nuclei with half of the full chromosome number are formed, usually within the specialised cells called **gametes**. Sexual reproduction occurs when two haploid nuclei fuse to form a new diploid cell called a **zygote**, as figure 5.2.1 shows.

Types of gametes

For the gametes to fuse one of them at least must move to meet the other. In many organisms both gametes are motile and are released directly into the water, where they meet. Identical gametes such as these are known as **isogametes**. **Isogamy** is the production of these identical gametes. *Chlamydomonas* is an example of an isogamous organism.

In many other organisms the gametes are not identical. When the gametes differ from each other in size and structure they are known as **anisogametes**, and their production is known as **anisogamy**. The **female** gamete does not move. It is usually large and filled with food stores for the development of the embryo. The motile **male** gamete is much smaller and adapted for movement.

If one individual forms both male and female gametes it is said to be **hermaphrodite**. Plants in which both male and female gametes are formed on a single plant are called **monoecious**. Plants that form only male or only female flowers during their lifetime are known as **dioecious**. Examples are shown in figure 5.2.2.

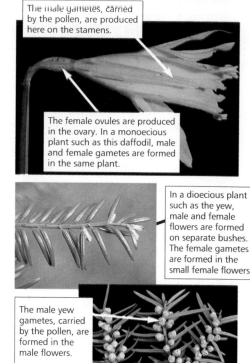

The male gametes, carried by the pollen, are produced here on the stamens.

The female ovules are produced in the ovary. In a monoecious plant such as this daffodil, male and female gametes are formed in the same plant.

In a dioecious plant such as the yew, male and female flowers are formed on separate bushes. The female gametes are formed in the small female flowers.

The male yew gametes, carried by the pollen, are formed in the male flowers.

Figure 5.2.2 A monoecious and a dioecious plant

Sexual reproduction in bacteria

For many years it was assumed that bacteria reproduced only asexually. Then in 1946 Joshua Lederberg, a student at Yale University, and his professor Edward Tatum, performed an experiment which showed that some sort of genetic exchange must occasionally take place in these organisms.

They grew two strains of bacteria. Strain A could grow on a minimal medium as long as the amino acid methionine and the vitamin biotin were added. The bacteria could synthesise everything else they needed. Strain B could grow on the minimal medium as long as the amino acids threonine and leucine were added. Neither strain could grow on the medium alone. But after mixing the two strains together, about one in every 10 million bacterial cells could grow on the minimal medium without any additions. This was only possible if these cells had received some genes from strain A and some from strain B. The electron microscope has since

shown clearly bacteria 'mating' and exchanging genetic information through a strand of cytoplasm known as the **sex pilus**. This has been vital in the development of **genetic engineering**, which will be discussed in more detail in section 5.7.

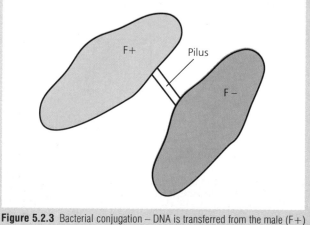

F+   Pilus   F−

Figure 5.2.3 Bacterial conjugation – DNA is transferred from the male (F+) cell to the female (F−) cell through the pilus which joins them.

Where are gametes formed?

Gametes are formed in special sex organs. In some animals and plants these are often temporary, forming only as they are needed. In larger animals they tend to be more permanent structures and may be referred to as **gonads**. In flowering plants the sex organs are the **anthers**, where the male gamete carried in the **pollen** is produced, and the **ovaries**, where the female **ovules** are formed. In animals the male gonads are the **testes**, responsible for production of **spermatozoa**, the male gametes. The female gonads are the **ovaries** and they produce **ova**. The male gametes are frequently much smaller than the female ones, but are usually produced in much larger quantities.

THE FORMATION OF GAMETES – MEIOSIS

When cells divide by mitosis the number of chromosomes in each of the daughter cells is the same as the number in the original parent cell. However, in the cell divisions which lead to the formation of gametes the chromosome number needs to be halved to give the necessary haploid nuclei. The formation of the gametes must therefore result from a different process of cell division. This **reduction division** is known as **meiosis**, and in most organisms it occurs only within sex organs. Meiosis occurs infrequently in a small number of cells, whilst mitosis occurs regularly in very many cells. Meiosis is of great biological significance – it is the basis of variation and thus also of evolution.

The stages of meiosis

In mitosis a single nuclear division gives rise to two identical diploid daughter cells. In meiosis two nuclear divisions give rise to four haploid daughter cells, each with its own unique combination of genetic material. As in the case of mitosis, the events of meiosis are continuous but are divided into defined stages for ease of description and understanding. The stages of meiosis are shown in figure 5.2.4. As in mitosis again, the contents of the cell, and in particular the DNA, are replicated whilst the cell is in interphase.

Figure 5.2.4 These diagrams show the stages of meiosis in a simplified cell which has only two pairs of chromosomes.

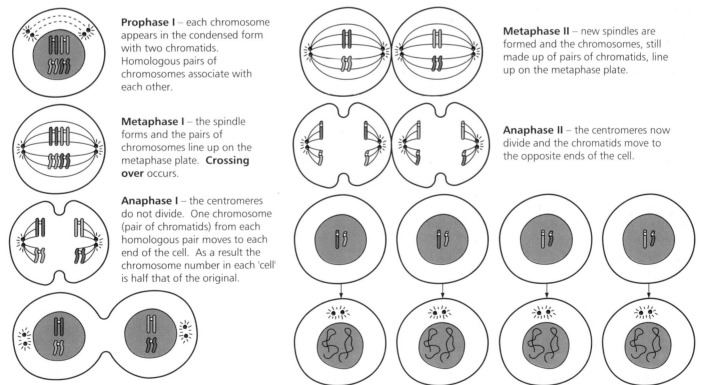

Prophase I – each chromosome appears in the condensed form with two chromatids. Homologous pairs of chromosomes associate with each other.

Metaphase I – the spindle forms and the pairs of chromosomes line up on the metaphase plate. **Crossing over** occurs.

Anaphase I – the centromeres do not divide. One chromosome (pair of chromatids) from each homologous pair moves to each end of the cell. As a result the chromosome number in each 'cell' is half that of the original.

Telophase I – the nuclear membrane reforms and the cells begin to divide. In some cells this continues to full cytokinesis and there may be a period of brief or prolonged interphase. During this interphase there is *no further replication* of the DNA.

Metaphase II – new spindles are formed and the chromosomes, still made up of pairs of chromatids, line up on the metaphase plate.

Anaphase II – the centromeres now divide and the chromatids move to the opposite ends of the cell.

Telophase II – nuclear envelopes reform, the chromosomes return to their interphase state and cytokinesis occurs, giving four daughter cells each with half the chromosome number of the original diploid cell.

Prophase I

The chromosomes have replicated to form chromatids joined by a centromere just as in mitosis. However, in meiosis the two members of the chromosome pair associate together. Imagine the original pair of chromosomes in the parent cell as a pair of shoes – the 'left shoe' from the mother and the 'right shoe' from the father. After replication in interphase there are two identical pairs of shoes. The two right shoes are joined by the shoelaces – the centromere – as are the two left shoes. In prophase I of meiosis the two right shoes and the two left shoes are paired, giving a structure made up of four 'shoes' or chromatids. It is at this stage that **chiasmata** may be formed, resulting in **crossing over** (this is shown in figure 5.2.6 and explained opposite). As in mitosis, the nuclear membrane breaks down, the nucleolus disappears and the centrioles pull apart to form the spindle.

Metaphase I

The homologous pairs of chromosomes line up on the metaphase plate of the spindle.

Anaphase I

One chromosome of each pair moves to one pole of the cell, whilst the other goes to the opposite pole. As the centromeres do not split, it is *pairs of chromatids* that move. To continue the shoe analogy, the laces are not untied, so two left shoes go one way and two right shoes the other.

Telophase I

The spindle fibres disappear and nuclear membranes form around the chromosomes. Cytokinesis begins, and there are half as many chromosomes in the new cells as there were in the original parent cell. In some cases the division is complete, but often prophase II begins almost immediately.

Prophase II

Dispersed chromosomes shorten and become visible again. They are still made up of two chromatids joined at the centromere as at the end of telophase I.

Metaphase II

A new spindle forms and the chromosomes (pairs of chromatids) line up on the metaphase plate.

Anaphase II

The centromeres finally divide (the 'shoelaces' are undone) and each chromatid moves to one of the poles of the cell.

Telophase II

The nuclear membranes reform as the chromosomes become invisible again. Cytokinesis occurs, resulting in four haploid daughter cells each with half the chromosome complement of the original parent cell. These daughter cells will then develop into gametes – spermatozoa, ova or the gametes in pollen grains – in the fullness of time.

The importance of meiosis

Having seen the stages of mitosis we can now look at the fundamental significance of the process. Firstly it reduces the chromosome number in gametes from diploid to haploid. But meiosis is also the primary means of generating new gene combinations.

Distribution of maternal and paternal chromosomes

The chromosomes from either parent are distributed in the new gametes completely at random. Taking the human situation as an example, each gamete receives 23 chromosomes. The parent cell which forms the gametes has 46 chromosomes, 23 from the original maternal gamete and 23 from the original paternal gamete, arranged in pairs. Any number of the 23 chromosomes which go into the gamete, from none to all 23, could come from either source. The most usual situation is a mixture of the two, as in figure 5.2.5. There are more than eight million potential chromosome combinations within the sperm or the egg.

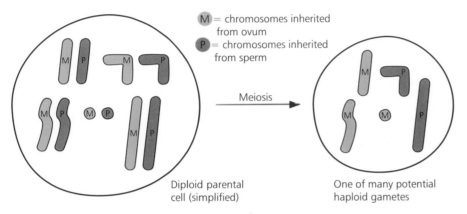

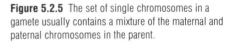

Figure 5.2.5 The set of single chromosomes in a gamete usually contains a mixture of the maternal and paternal chromosomes in the parent.

Recombination

Added to this is the phenomenon of **crossing over** or **recombination** shown in figure 5.2.6. Large, multienzyme complexes 'cut and stitch' maternal and paternal chromatids together during prophase I. The points where the chromatids break are called **chiasmata**. These are important in two ways. The exchange of genetic material produces new combinations of genes, as the recombined chromatids are unlike any other chromatids. Moreover, errors in the process lead to **mutation** and this introduces new genes into the genetic make-up of a species. We shall now go on to look at mutations in more detail.

Mutation

Mutations are changes in the chemical structure of an individual gene or in the physical make-up of the chromosome – in other words, changes in the arrangement of the genes on the chromosome. Mutations are the ultimate source of variation, but the vast majority of them are disadvantageous and so rapidly disappear. Many others are neutral – they neither improve nor worsen the chances of survival or reproduction. Very occasionally a mutation occurs which results in the production of a new and superior protein. The organism gains an advantage and so the mutation survives and remains in the gene pool of the species.

There are several different sorts of mutation. Some involve a change in the individual gene and are known as **point mutations**. If we consider genes as the letters of the alphabet and chromosomes as sentences, a point mutation is like changing a letter in one word. It may well still make an acceptable word, but the meaning will very probably be different. There are also a variety of mutations which involve changes in the position of genes within the chromosomes. These are known as **chromosomal mutations**. This is like rearranging the words within a sentence – if we are lucky they still make sense, but will probably not mean the same as the original sentence. Some examples of chromosomal mutations are shown in figure 5.2.7. And finally there are whole chromosome mutations in which an entire chromosome is either lost

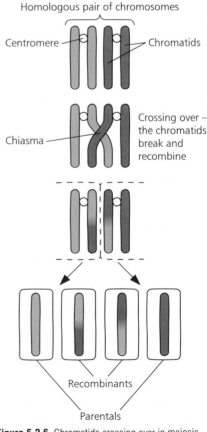

Figure 5.2.6 Chromatids crossing over in meiosis. Recombinant chromatids are unlike any other chromatid in the organism.

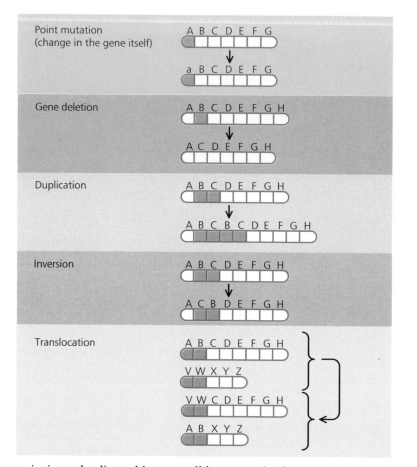

Figure 5.2.7 Several processes can alter the arrangement of the genes on the chromosomes. These are some of the most common types of chromosomal mutation. The relative size of the genes is very much exaggerated in the diagrams – there are in fact very many more genes on each chromosome.

during meiosis or duplicated in one cell by errors in the process – the loss of or repetition of a whole sentence in our analogy.

Mutations occur all the time, but broadly speaking they are only of major importance when they occur in the gametes, for they may then be passed on to future offspring. Such natural mutations which are passed on do not occur very frequently (see box below), but exposure to ionising radiation and certain chemicals increase the rate at which they appear. Anything which causes an increased frequency of mutation is called a **mutagen** and is said to be **mutagenic**. Because genetic mutations can be lethal, and because many severe human diseases are the result of genetic mutation, these mutagenic substances are usually best avoided.

How often do mutations occur?

The frequency with which mutations naturally occur (the **mutation rate**) was first demonstrated by Herman J. Muller at Columbia University. He devised a way of measuring the mutation rate in the fruit fly *Drosophila* by observing the occurrence of lethal genes carried on the sex chromosomes. Lethal genes show up because the ratio of males to females is affected – the males inheriting the lethal gene died and so there was a higher than usual proportion of female offspring. The mutation rate of each individual gene was between 10^{-5} and 10^{-6} mutations per gene per generation. He subsequently showed that exposure to X-rays increased the mutation rate, and that the greater the dosage of X-rays the higher the rate of mutation.

Mutations are relatively rare events – even in our complex genetic make-up, each one of us will contain on average just one new mutation. If we are 'normal' and healthy, that mutation is probably hidden in the recessive form in our genes, and may never be expressed unless by chance we meet up with and have children by another individual with the same mutated gene. (A mutation on the X chromosome could be expressed in male offspring even if the father does not have this mutation. There is more about this in sections 5.6 and 5.7.)

Gametogenesis in humans

Sperm cells in the male mammal are produced in the testes. Ova in the female are produced in the ovaries. Many millions of sperm are released every time a male mammal ejaculates. The numbers of eggs in a sexually mature female are usually numbered in thousands, and will eventually run out. Special cells (the **primordial germ cells**) in the gonads divide, grow, divide again and then differentiate into the gametes. How are the events of meiosis used to produce two such different sets of gametes?

Spermatogenesis – the formation of spermatozoa

Each primordial germ cell results in large numbers of spermatozoa. As there are enormous numbers of primordial germ cells, millions of spermatozoa (commonly known as sperm) are produced on a regular basis. Figure 5.2.8 shows this process.

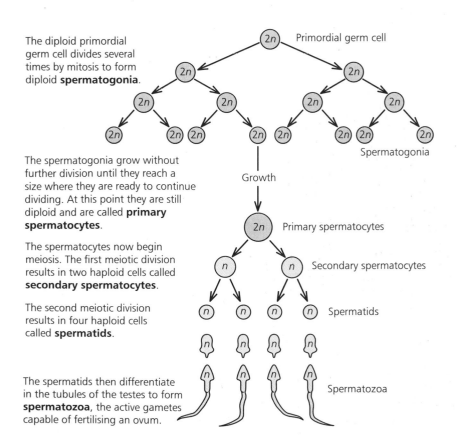

The diploid primordial germ cell divides several times by mitosis to form diploid **spermatogonia**.

The spermatogonia grow without further division until they reach a size where they are ready to continue dividing. At this point they are still diploid and are called **primary spermatocytes**.

The spermatocytes now begin meiosis. The first meiotic division results in two haploid cells called **secondary spermatocytes**.

The second meiotic division results in four haploid cells called **spermatids**.

The spermatids then differentiate in the tubules of the testes to form **spermatozoa**, the active gametes capable of fertilising an ovum.

Figure 5.2.8 The stages of spermatogenesis

Oogenesis – the formation of ova

Each primordial germ cell results in only one ovum. As a result the number of female ova is substantially smaller than the number of spermatozoa. As ova are much larger cells containing more cytoplasm than sperm, there is a much greater investment of resources in each one and therefore it would not make biological sense to produce too many. Figure 5.2.9 shows the process of oogenesis.

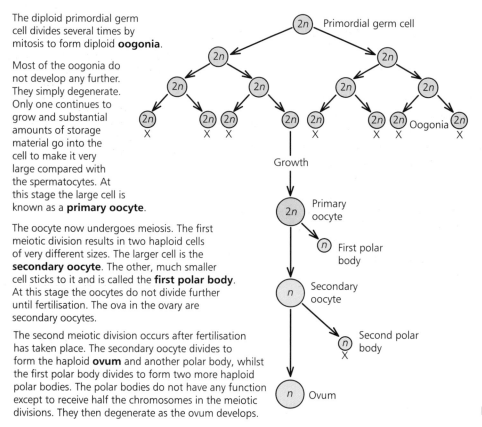

The diploid primordial germ cell divides several times by mitosis to form diploid **oogonia**.

Most of the oogonia do not develop any further. They simply degenerate. Only one continues to grow and substantial amounts of storage material go into the cell to make it very large compared with the spermatocytes. At this stage the large cell is known as a **primary oocyte**.

The oocyte now undergoes meiosis. The first meiotic division results in two haploid cells of very different sizes. The larger cell is the **secondary oocyte**. The other, much smaller cell sticks to it and is called the **first polar body**. At this stage the oocytes do not divide further until fertilisation. The ova in the ovary are secondary oocytes.

The second meiotic division occurs after fertilisation has taken place. The secondary oocyte divides to form the haploid **ovum** and another polar body, whilst the first polar body divides to form two more haploid polar bodies. The polar bodies do not have any function except to receive half the chromosomes in the meiotic divisions. They then degenerate as the ovum develops.

Figure 5.2.9 The stages of oogenesis

FERTILISATION

Asexual reproduction is a guaranteed method of passing on the genes from one individual into the next generation. For sexual reproduction to be successful, gametes must meet. If these gametes are coming from two different individuals, systems are needed to ensure the successful transfer of the male gamete to the female gamete. In plants, flowers either attract other organisms such as insects, birds or mammals to transfer the pollen from one plant to another, or they are arranged in such a way that the pollen is easily removed by the wind. In animals a wide variety of strategies have developed to ensure the meeting of gametes. They fall into two main categories – **external fertilisation** and **internal fertilisation**.

External fertilisation

This occurs outside the body, with the gametes shed directly into the environment. It is common only in aquatic species, because spermatozoa and ova are very vulnerable to drying and are rapidly destroyed in the air. Animals such as the coelenterates simply release copious amounts of male and female gametes into the sea and chance determines whether fertilisation takes place. Animals such as fish and amphibians have complex rituals which increase the likelihood of fertilisation by ensuring that the ova and sperm are released at the same time in close proximity to each other, as figure 5.2.10 illustrates.

In spite of strategies such as these, many of the gametes do not meet. External fertilisation can be very wasteful of resources, and is not an option for many of the organisms that have colonised the land.

Internal fertilisation

This involves the transfer of the male gametes directly to the female. Although this does not ensure fertilisation, it makes it much more likely. The way in which the sperm are transferred varies greatly, with many species producing packages of sperm for the female to pick up and transfer to her body. Many vertebrates have evolved a system whereby the male gametes are released directly into the body of the female during **mating** or **copulation**. This ensures that the ova and sperm are kept in a constantly moist environment and are placed as close together as possible, thus maximising the chances of fertilisation.

Fertilisation will not take place unless the spermatozoa meet a mature ovum. Many animals show complex behaviour patterns which help to ensure that mating only takes place at the time when the ovum is most likely to be fertilised, for example that shown in figure 5.2.11. In many cases the female is only receptive to mating when she is fertile. Chemicals called **pheromones** carry powerful messages through chemically based senses. Sexual behaviour is also often strongly linked to the seasons of the year, linking fertilisation with the optimum time for offspring to be born.

Figure 5.2.10 The female frog carries the male around for several days prior to the release of gametes. The male develops special grasping areas (**nuptial pads**) on his forelimbs to help him remain in place. It is vital that the sperm are squirted onto the eggs as soon as they are laid so that fertilisation can take place before the protective coating of the eggs absorbs water and swells to form a layer of jelly.

Figure 5.2.11 The courtship behaviour of the great crested grebe involves this very complex series of movements often referred to as a 'dance'. Only after the full series of movements has been completed by both birds will mating take place.

Inbreeding and outbreeding

Organisms which are dioecious, that is, have two separate sexes, must by definition have their gametes fertilised by another individual. They are termed **obligate outbreeders**. Organisms which are monoecious – produce male and female gametes in the same individual – have the potential to fertilise their own gametes. In spite of this, many monoecious organisms are outbreeders. However, **inbreeding** or self fertilisation can be advantageous. Organisms which live extremely isolated lives, such as tapeworms in the gut, are inbreeders by necessity. Although they fertilise their own ova, variety still arises due to meiosis in the production of the gametes. However, this variety is limited compared with that which arises from fertilisation by another genetically different individual. The biggest disadvantage of inbreeding is that any harmful or disadvantageous genes are likely to be expressed in the offspring and successive generations will become weaker and less able to survive.

Plants are at higher risk of inbreeding than animals, as a very large number of plant species are monoecious. However, most monoecious species have developed a variety of strategies to avoid self fertilisation except as a last resort.

The female gametes may mature before the male ones are ripe – this is **protogyny**. Equally the male gametes may mature before the female ones, in **protandry**. **Self sterility** may occur, in which case the male and female gametes of the same plant cannot fuse and fertilise each other: see section 5.3, page 360. The arrangement of the reproductive parts of a flower may be such that self pollination is most unlikely – for example, if the style is long so that the stigma is sited above the male anthers, then the likelihood of self pollination is substantially reduced.

Parthenogenesis – the redundant male

Parthenogenesis is not sexual reproduction, but it involves the same systems. In the females of some plant and animal species the ova can develop without fertilisation by the male. This may give rise to haploid offspring such as male honey bees (drones), or there may be an additional doubling of the chromosomes to give diploid offspring as occurs in aphids. Insects, lizards, some relatives of the lobsters and even turkeys (artificially) are capable of parthenogenesis.

Like asexual reproduction or inbreeding, parthenogenesis is very successful when conditions are stable. Greenfly reproduce by parthenogenesis in the spring, giving a very rapid rise in the population at a time when food supplies are good. In the autumn when variety is needed to help ensure survival over the winter, sexual reproduction occurs again.

So far we have looked at reproduction in a general sense – its purpose, and how cellular reproduction and gamete formation are brought about. In sections 5.3–5.5 we shall move on to consider in detail how sexual reproduction is brought about in both flowering plants and people.

SUMMARY

- The **gametes** are specialised cells which have half the normal number of chromosomes (23 in humans) – one chromosome from each homologous pair. They are **haploid**. In sexual reproduction two haploid nuclei fuse to form a diploid **zygote**.

- **Isogamy** is the production of identical gametes which are both motile. **Anisogametes** are different in structure and function. Usually the female gamete does not move, is large and contains food stores for the developing embryo. The male gamete is smaller and adapted for movement.

- **Hermaphrodite** organisms produce both male and female gametes. **Monoecious** plants form male and female gametes, while **dioecious** plants produce only male or only female flowers.

- Gametes are formed in specialised sex organs (**gonads**). In plants the anthers produce gametes carried in pollen and the ovaries produce ovules. In animals the testes produce spermatozoa and the ovaries produce ova.

- Cells in the gonads divide by reduction division or **meiosis** to produce haploid gametes. Meiosis results in four haploid daughter cells from one diploid parent cell.

- **Prophase I** is the first stage of meiosis. Homologous chromosomes, each made up of two chromatids joined at the centromere, pair up. The chromatids cross over (form **chiasmata**) which exchanges genetic material. The nuclear membrane breaks down, the nucleolus disappears and centrioles separate to form the spindle.

- In **metaphase I** the pairs of chromosomes line up on the metaphase plate.

- In **anaphase I** the chromosomes from each pair migrate to opposite poles of the cell.

- In **telophase I** the spindle disappears and nuclear membranes form around the chromosomes. Cytokinesis begins.

- In **prophase II** there is no further replication of DNA. The dispersed chromosomes shorten and nuclear membranes break down.

- In **metaphase II** a new spindle forms and chromosomes line up on the metaphase plate.

- In **anaphase II** the centromeres split and chromosomes migrate to opposite poles of the cell.

- In **telophase II** nuclear membranes reform, chromosomes disperse and cytokinesis occurs.

- Meiosis results in genetic variation because the parental chromosomes are distributed randomly in the gametes. **Recombination** (formation of chiasmata) exchanges genetic material, and **mutations** in the process increase variety further.

- **Point mutations** are changes in individual genes, while **chromosomal mutations** are changes in the positions of genes on a chromosome.

- **Fertilisation** is the meeting of gametes so that they can form the zygote. In **external fertilisation** the gametes are shed into the aquatic environment to meet by chance. In **internal fertilisation** the male gametes are transferred into the female, during mating in many land animals.

- **Obligate outbreeders** have their gametes fertilised by another organism, while **inbreeding** or **self fertilisation** involves the meeting of gametes from the same parent. To avoid self fertilisation in some flowering plants the female gametes may mature first (**protogyny**) or the male gametes may mature first (**protandry**).

QUESTIONS

1 a What are gametes?
  b How do gametes differ from normal cells?
  c Describe meiotic cell division.

2 a Where are the gametes formed in:
    i   the human male
    ii the human female?
  b Draw up a table comparing gametogenesis in the male and in the female.
  c Where does gametogenesis occur in plants?

  d What are:
    i   monoecious plants
    ii dioecious plants?

3 a What is mutation and why is it important?
  b What are the main problems to be overcome in reproduction for organisms living on the land rather than in the oceans?
  c What are the advantages and disadvantages of:
    i   inbreeding
    ii outbreeding?

5.3 Sexual reproduction in plants

The **angiosperms** or flowering plants reproduce by the production of flowers. As we have already seen in section 5.1, many flowering plants can reproduce asexually too. Evidence of sexual reproduction in plants is all around us – in the pollen which causes the misery of hay fever, in the flowers of the garden, the hedgerow and the rainforest, and in the fruits and vegetables which form a large part of our diet worldwide. What is the role of these various structures in the process of sexual reproduction?

FLOWERS

Flowers are some of the most beautiful manifestations of biological diversity. The range of colours, patterns, designs and scents is almost endless, as figure 5.3.1 illustrates. Flowers are collections of sex organs. Plants do not have sex organs as a permanent part of their structure – they are developed when conditions are appropriate. Many plants produce flowers which combine both the male and female reproductive parts. As mentioned in section 5.2, **dioecious plants** produce male flowers on one plant and female flowers on another. This prevents any problems of self fertilisation. **Monoecious plants** produce male and female gametes on the same plant. Flowers are very expensive in terms of the resources of the plant, so they are only produced at times when it is most likely that successful reproduction will result.

The male and female parts of this grass protrude beyond the dull-coloured petals.

The common garden tulip shows the bright colours and patterns frequently used to attract insects to flowers.

Only the tip of the female organs can be seen in this strange-looking flower of the hedgerows, the cuckoo pint.

Each *Gazania* flower is made up of a large number of very small flowers. It is therefore a **composite** head of flowers, with individual flowers maturing over several days. Different pollinators therefore fertilise each egg cell.

The structure of a flower

In spite of the great diversity shown in figure 5.3.1, most flowers are based on a very similar pattern. Flowering plants are divided into the monocotyledons and the dicotyledons – these terms are explained on pages 360–1. Both types

Figure 5.3.1 Flowers are sexual structures with an almost infinite variety of shapes, colours and smells.

of plants produce flowers, although those of the dicotyledons are generally larger and more colourful. In plants which produce separate male and female flowers, parts of the general design are present in the male flowers and other parts in the female flowers.

Flowers are a group of leaves modified for a particular function. A typical flower consists of four types of structure – **sepals**, **petals**, **stamens** and **carpels**. Their arrangement and functions are shown in figure 5.3.2.

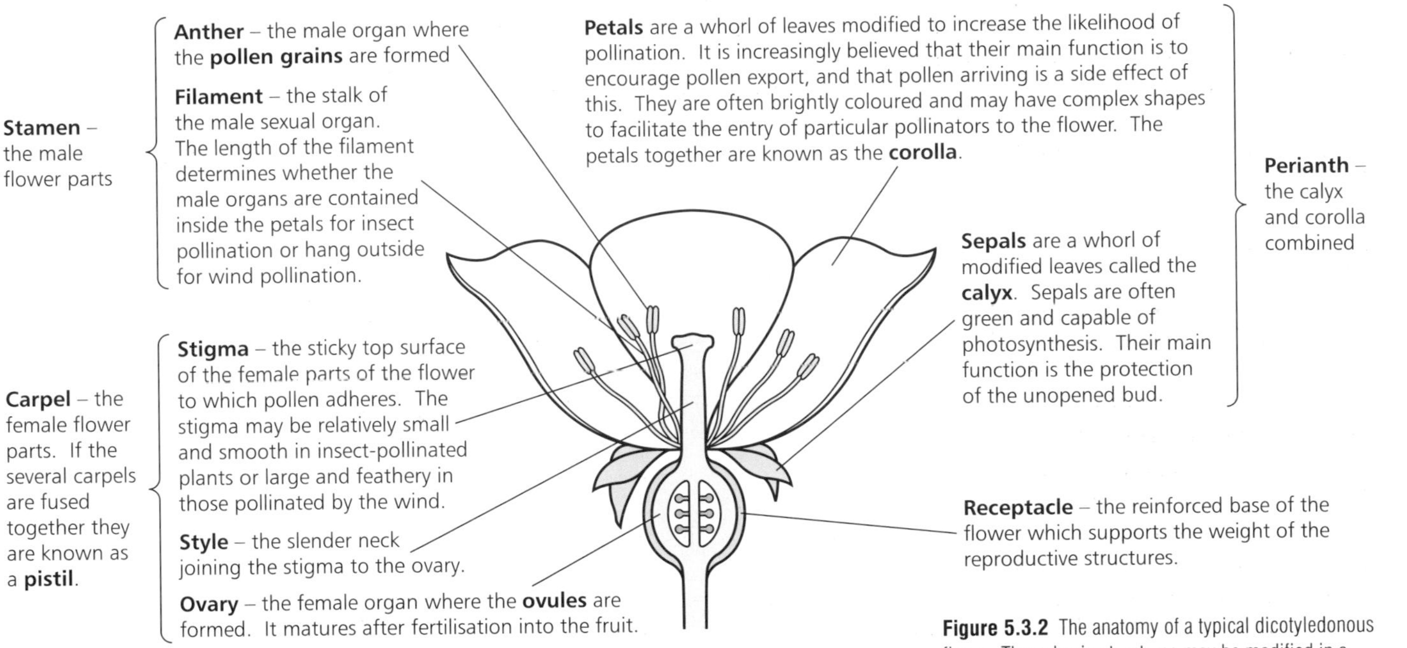

Anther – the male organ where the **pollen grains** are formed

Filament – the stalk of the male sexual organ. The length of the filament determines whether the male organs are contained inside the petals for insect pollination or hang outside for wind pollination.

Stamen – the male flower parts

Petals are a whorl of leaves modified to increase the likelihood of pollination. It is increasingly believed that their main function is to encourage pollen export, and that pollen arriving is a side effect of this. They are often brightly coloured and may have complex shapes to facilitate the entry of particular pollinators to the flower. The petals together are known as the **corolla**.

Perianth – the calyx and corolla combined

Sepals are a whorl of modified leaves called the **calyx**. Sepals are often green and capable of photosynthesis. Their main function is the protection of the unopened bud.

Stigma – the sticky top surface of the female parts of the flower to which pollen adheres. The stigma may be relatively small and smooth in insect-pollinated plants or large and feathery in those pollinated by the wind.

Style – the slender neck joining the stigma to the ovary.

Ovary – the female organ where the **ovules** are formed. It matures after fertilisation into the fruit.

Carpel – the female flower parts. If the several carpels are fused together they are known as a **pistil**.

Receptacle – the reinforced base of the flower which supports the weight of the reproductive structures.

Figure 5.3.2 The anatomy of a typical dicotyledonous flower. These basic structures may be modified in a wide range of ways to adapt the flower to particular pollinators.

Pollination

Sexual reproduction is the fusing of two gametes to form a zygote, as we saw in section 5.2. The gametes in a flowering plant are in the pollen and the ovules. For the gametes to meet, **pollination** must take place – the pollen must be transferred from the male organs to the female organs. As inbreeding is generally avoided, most plants require their pollen to be transferred to another plant, and to receive pollen from elsewhere in return. A variety of strategies have evolved to ensure that this happens, two major groups being those plants which depend on insects or other living organisms to transfer pollen and those which rely on a physical medium such as the wind. Flower structures vary greatly between these categories.

Those flowers which are pollinated by insects or other animals must initially attract these **pollinators**, and so coloured and patterned flowers are produced. Some have attractive scents and most also have rewards for the pollinators such as the sugary liquid **nectar** found in the base of many flowers. Flower structures are designed to maximise the chance of pollen from another plant reaching the stigma, and equally the chance of pollen from the anthers being removed and carried to another plant.

Flowers which rely on the wind as a pollinator do not need attractive structures. Many monocotyledons are wind pollinated, including all the grasses and cereal plants. The size and arrangement of the parts of the flower are modified to maximise the chances of pollen being blown from the stamens, and also of pollen from another plant of the same species being trapped on the stigma. The main structural modifications seen in these two types of flower are summarised in table 5.3.1.

Feature of flower	Insect-pollinated flowers	Wind-pollinated flowers
Sepals	May be green and photosynthetic or may be indistinguishable from the petals, brightly coloured to help attract insects	Often absent
Petals	Usually large and conspicuous, with colours and patterns which may be visible only in ultraviolet light. May be arranged or shaped to encourage only specific pollinators	Small, inconspicuous, often green or brown
Scent	Often attract specific pollinators – insects very sensitive to scents	Usually absent
Nectar	May be produced by nectaries at the bases of the petals so that the pollinator must enter the flower to reach the nectar	None produced
Stamens	Usually enclosed within the flower	Long filaments usually protrude from the flower with loosely attached anthers so pollen is easily blown off even in light wind
Carpels	Flat, lobed and sticky stigma usually enclosed within the flower	Stigma protrudes from the flower, and is usually feathery and sticky to trap pollen from the air
Pollen grains	Relatively large, often sticky	Very small, light and powdery, sometimes with air bladders

Table 5.3.1 A comparison of the main adaptations of insect-pollinated and wind-pollinated flowers

THE MECHANISM OF REPRODUCTION

Formation of the gametes

The formation of gametes in flowering plants is made more complex by the fact that plants have two phases to their life cycles. The **sporophyte** generation is diploid and produces spores by meiosis. The **gametophyte** generation which results is haploid and gives rise to the gametes by mitosis. In many plants such as mosses and ferns these two phases exist as separate plants, as we shall see in section 6.3. In seed-bearing plants, the two phases have been combined into one plant. The main body of the plant which we see is the diploid sporophyte. The haploid gametophytes are contained in the anther and the ovary. They are produced by meiosis from spore mother cells, as described below.

In plants the male gametes are contained in the pollen grains formed by the anthers, and the female gametes are contained in the ovules formed in the ovary. The gamete carried by the pollen is known as the **microgamete** and the ovule as the **megagamete**.

The formation of pollen – microgametogenesis

The anthers are analogous to the testes of animals. Meiosis occurs within the anthers resulting in vast numbers of pollen grains which carry the male gametes.

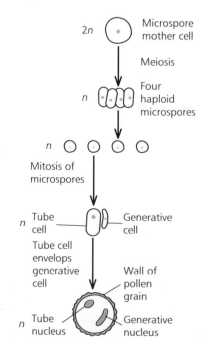

Figure 5.3.3 Microgametogenesis – the formation of the male gamete

Each anther contains four **pollen sacs** where the pollen grains develop. In each pollen sac there are large numbers of **microspore mother cells**. These cells are diploid and they divide by meiosis as shown in figure 5.3.3 to form haploid **microspores** which are the gametophyte generation. The gametes themselves are formed from the microspores by mitosis. A haploid tube cell and a haploid generative cell result from the mitosis of the microspore, and the tube cell then envelops the generative cell to form the pollen grain. This pollen grain contains two nuclei, the **tube nucleus** and the **generative nucleus**.

The pollen grains that are produced have extremely complex cell walls with three-dimensional patterns, as can be seen in figure 5.3.4. These pollen surface patterns are unique and species specific. They are produced by polysaccharides and proteins deposited in the cell wall. Pollen grains are extremely tough and resistant to decay so they remain in the soil for thousands of years. An expert examining an ancient sample of soil or dust can identify the species of plant growing at the time, and even give an idea of how abundant the different species were, from the distinctive pollen grains that survive.

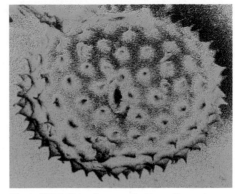

Figure 5.3.4 These fearsome looking pollen grains are from the common ragweed. They often cause the allergic reaction we call hay fever.

The formation of egg cells – megagametogenesis

The ovary of the plant is analogous to the animal ovary. Meiosis results in the formation of a relatively small number of egg cells contained within ovules (egg chambers). The site of egg development is within the ovary itself. Some plants, for example the nectarine, produce only one ovule, whilst others such as peas produce several.

The ovule is attached to the wall of the ovary by a pad of special tissue called the **placenta**. A complex structure of integuments (coverings) form around tissue known as **nucellus**, and in the centre diploid **megaspore mother cells** divide by meiosis to give rise to four haploid **megaspores**. Three of these degenerate leaving one to continue to develop. This megaspore undergoes three mitotic divisions which result in an **embryo sac** containing an egg cell (the **megagamete**), two **polar bodies** and various other small cells, some of which degenerate. This process is shown in figure 5.3.5.

Figure 5.3.5 Megagametogenesis – the formation of the egg cell – takes place in the ovary.

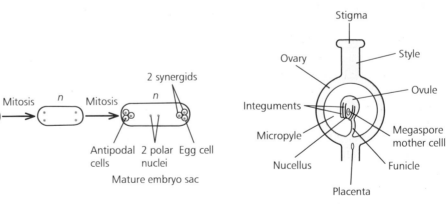

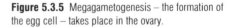

Pollination and fertilisation

Once pollen formation is completed the anthers ripen and their walls dry out and split. As a result the pollen is released for transfer to the stigma of another plant. As mentioned in section 5.2, self pollination is often avoided as far as possible by strategies such as a stigma positioned above the anthers, or female parts which mature either before or after the stamens. Most commonly the anthers mature first, described as **protandry**. Examples of protandrous flowers include white deadnettles and dandelions. **Protogyny** is the maturing of the stigmas first – an example is the bluebell.

Once a pollen grain has arrived on the stigma of a flower from another plant, transferred by either animals or the wind, **pollination** has taken place. The gametes must now fuse in the process of **fertilisation**.

The male gamete is contained within the pollen grain. The female gamete is embedded deep in the tissue of the ovary. To enable the two to meet an ingenious series of events takes place.

Molecules on the surfaces of the pollen grain and the stigma interact. If they 'recognise' each other as being from a compatible individual of the same species, the pollen grain begins to **germinate**. A **pollen tube** begins to grow out from the tube nucleus of the pollen grain, through the stigma into the style. It continues down towards the ovary, and the generative nucleus travels down it, dividing mitotically as it does so to form two **male nuclei**. Eventually the tip of the pollen tube passes through the **micropyle** of the ovule.

Once the tube has entered the micropyle, the two male nuclei are passed into the ovule so that fertilisation can occur. Angiosperms undergo what is known as **double fertilisation**. One of the male nuclei fuses with the two polar nuclei to form one **triploid** (with three sets of chromosomes) **endosperm nucleus**. The other male nucleus fuses with the egg cell to form the **diploid zygote**. This is shown in figure 5.3.6. At this point fertilisation has been completed and the development of the seed and the embryo within can begin.

The growth of the pollen tube is very fast due to the rapid elongation of the cells. If self pollination does occur in a plant which does not normally self pollinate, then **self incompatibility** (**self sterility**) means the pollen grain often does not develop or grows too slowly to bring about successful fertilisation. This is the result of self-incompatibility genes.

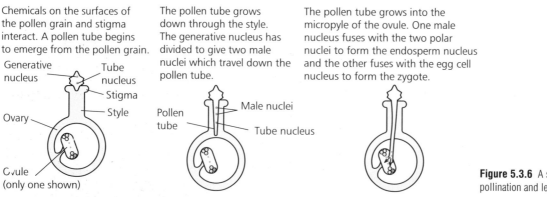

Chemicals on the surfaces of the pollen grain and stigma interact. A pollen tube begins to emerge from the pollen grain.

The pollen tube grows down through the style. The generative nucleus has divided to give two male nuclei which travel down the pollen tube.

The pollen tube grows into the micropyle of the ovule. One male nucleus fuses with the two polar nuclei to form the endosperm nucleus and the other fuses with the egg cell nucleus to form the zygote.

Figure 5.3.6 A summary of the events which follow pollination and lead to fertilisation of the ovum

Seed and fruit formation

Once the double fertilisation of the ovule has taken place, the next stage is the protection and dispersal of the embryo within the **seed**. Within the embryo sac two distinct processes take place which lead to the formation of the seed. First the endosperm nucleus divides repeatedly by mitosis and absorbs many of the nutrients produced by the parent plant. This gives rise to food-storage tissue known as the **endosperm**. Once this food source is well established, the zygote begins to undergo mitotic cell division to produce an embryo. The embryo plant consists of three main regions – the **plumule** (embryonic shoot), the **radicle** (embryonic root) and the **cotyledons** (embryonic leaves). It is the number of these embryonic leaves present which gives us the main division of the angiosperms into **monocotyledons** (with one seed leaf) and **dicotyledons** (with two seed leaves).

In monocotyledons the main food store is the endosperm and the cotyledon remains a very small part of the seed. In most dicotyledons the endosperm channels food into the cotyledons which become the main food store. By the time the seed is mature the endosperm has all but disappeared – the embryo with its food-swollen leaves takes up most of the seed, as shown in figure 5.3.7. Once the food store has been laid down and the embryo has developed the seed dehydrates, losing much of its wet mass, and with this loss of water the seed becomes dormant. The plant growth regulator abscisic acid (see section 4.5) appears to control this dormancy in seeds. The integuments of the ovule meanwhile become hardened to form the seed coat or **testa**.

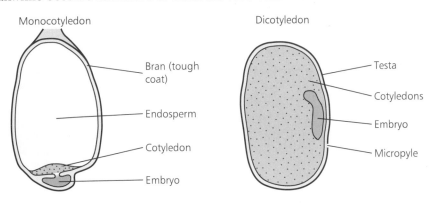

Figure 5.3.7 The internal arrangements of monocotyledonous and dicotyledonous seeds look different, but their basic parts are the same – an embryo ready to develop, a food store to supply the energy needed and a protective seed coat.

Once the seeds have formed and the seed coat has developed, the fruit begins to enlarge around the seed. In most cases the fruit develops from the wall of the ovary, although in some cases it is the result of an enlarged receptacle. Fruits come in many forms – the succulent, juicy cherry and grape, the leathery pods of the pea and bean, the tough husks of the coconut and the papery thin wings of the fruits of the sycamore and ash trees. All fruits are adaptations to aid in the spread or **dispersal** of seeds, as illustrated in figure 5.3.8.

The dandelion 'clock' is in fact a mass of fruits. Each pappus of hairs acts as a tiny parachute, enabling the seed to be carried long distances on currents of air.

Sweet and fleshy fruits like these blackberries rely on animals and birds eating them. The seeds then pass through the digestive tract and the action of the enzymes there prepares the testa for germination. Land sprayed with treated human sewage tends to yield a high proportion of tomato plants as tomato seeds pass through the human gut.

The need for seed dispersal

Any plant is in competition with other plants for soil nutrients, light and water. Tall plants with well established roots are at a great advantage over small unestablished plants in obtaining these resources. If the seeds of a plant simply fell off onto the ground beneath it, their chances of survival would be greatly reduced by competition with the parent plant and with each other.

The survival and successful germination of the seeds carrying the genetic material on into the next generation are very important. Resources are often used by the parent plant to produce fruits which not only protect the developing seed but also help to disperse the ripe seeds as far from the parent as possible. This increases the chance of successful growth of the seeds by reducing the likelihood of competition both with the mature plant and with other seedlings.

The extremely hard fruits of *Banksia serratifolia* can float in the sea off Australia. Once they have drifted ashore, they will resist opening until fire heats the nut wall and causes it to split.

Figure 5.3.8 Fruit adaptations showing various methods of dispersal of seeds

Methods of dispersal

Methods of seed dispersal range from the commonplace to the bizarre. Seeds that float in the air, pods that explode to expel the seeds and edible fruits are familiar to most of us. Tumbleweeds are less well known in Britain – the whole plant dies and is then uprooted and blown for miles in strong winds, scattering seeds as it rolls and tumbles along.

Dying with the dodo?

Theories about the calvaria tree of Mauritius show the vital role of dispersal. Calvarias used to be one of the most common trees on the island, but now only a few very ancient trees remain, in spite of the fact that the trees all produce their edible peach-like fruit each year. Why have no new trees developed? Some people believe that the calvaria trees relied on dodos for the dispersal of their seeds. The theory is that the passage of the seeds through the guts of these large birds native to Mauritius prepared the seeds for germination. Since the extinction of the dodos by sailors in 1681, calvarias have been unable to disperse their seeds – and now 300 years later the trees themselves face extinction.

Although this provides a neat and attractive explanation, and while dodos may indeed have played their part in calvaria seed dispersal when they were alive, it is actually far more likely that the ecology of the island has become so unbalanced with the introduction of increasing numbers of non-indigenous species that any seedlings produced cannot survive the competition.

GERMINATION, DEVELOPMENT AND GROWTH

Seed dormancy

The plant embryo survives in a dehydrated condition in the seed by becoming **dormant**. The respiration and metabolic rates drop to very low levels, and in this state the seed can survive very adverse conditions. In some species the seeds can survive in this state for many years. During a Second World War air raid in London, some dried plant specimens at the Natural History Museum were accidentally sprayed with water. The seeds contained in the plants germinated – and the specimens were known to be 250 years old!

The main function of the period of dormancy is to allow seeds to germinate when conditions are most suitable. They remain dormant until they are triggered to germinate by an environmental stimulus. This may be the disruption of the seed coat as it passes through an animal, exposure to light as the fruit rots away, exposure to the colder temperatures of winter in temperate climates or the washing of substances from the seed coat by rain in desert plants. Even with the appropriate environmental trigger, seeds will not germinate in the absence of water, oxygen and a suitable temperature. Once all these requirements are met, a period of intense activity starts within the seed and growth of the embryo occurs.

Germination

Germination or the development of the embryo plant from the seed may be seen as a race against time. The seed contains the embryo plant and a food store. Once the process of germination begins, the seedling must develop to become capable of obtaining water and minerals from the soil and carrying out photosynthesis before the food store of the seed runs out.

Taking in water

The first step in germination is for the seed to absorb water. This is taken in through the micropyle, as the rest of the testa is waterproof. The water is

Demonstrating conditions needed for germination

The conditions needed for seeds to germinate can be demonstrated by a simple series of experiments, as described in figure 5.3.9.

Cress seeds germinate readily on damp cotton wool at room temperature in the air. Dry seeds on dry cotton wool at room temperature in the air will not germinate – water is necessary for the process.

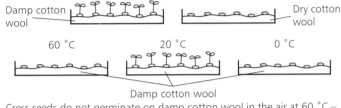

Damp cotton wool   Dry cotton wool

60 °C   20 °C   0 °C

Damp cotton wool

Cress seeds do not germinate on damp cotton wool in the air at 60 °C – the enzymes are denatured. At 20 °C they grow readily but not at 0 °C, although if this dish is subsequently warmed germination will occur. In the cold the enzymes are simply inactivated, not destroyed.

Cress seeds germinate readily on damp cotton wool at room temperature suspended in a sealed container over water. Those suspended over alkaline pyrogallol are deprived of oxygen and do not germinate, indicating that oxygen is required. As germination involves the rapid mobilisation and respiration of food stores to provide energy and materials for the growing embryo, this is not surprising.

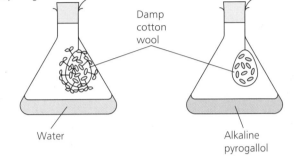

Damp cotton wool

Water   Alkaline pyrogallol

Figure 5.3.9 These simple experiments demonstrate the major requirements for the successful germination of seeds. If any of the elements is missing, germination does not take place.

absorbed by hydrophilic proteins within the seed and the large intake of water serves two major purposes. Firstly, the tissues of the embryo swell, and its enzymes can work again. Secondly, as a result of this swelling and the growth that occurs as the enzymes become active, the seed coat or testa splits.

The radicle emerges

The first structure to emerge from the ruptured seed coat is the **radicle**, the first short length of root. Once this has appeared the seed has germinated. The radicle, like all root tissue, is positively geotropic. This means that whichever way up the seed has fallen, the root will always grow downwards into the soil.

Development of the seedling

Up until the point where the radicle is growing down into the soil, germination is much the same in all plants. However, the way the seedling then continues to grow varies between monocotyledons and dicotyledons, and moreover dicotyledons may germinate in one of two different ways. In the developing seedling, the **hypocotyl** is the region beyond the radicle, that part of the seedling found *below* the cotyledons. The **epicotyl** is part of the developing stem *above* the cotyledons. The different processes involve different positions of these parts, as shown in figure 5.3.10.

Whatever the details of the development of the seedlings, the internal events are very similar. The food stores are mobilised to provide energy and materials for the rapidly growing and dividing cells. The rapid growth that is seen is largely the result of massive cell elongation. In monocotyledons the endosperm is used up and is then redundant. In many dicotyledons the cotyledons are brought up above the ground and once the food store inside them has been used up, they remain useful for some time as they photosynthesise until the seedling is well established with true leaves, when the cotyledons wither and drop off.

Primary growth – from seedling to mature plant

Once the seedling is well established, much growth is needed to give rise to a mature plant. In the growing regions of the stems and roots are regions known

In the germination of a monocotyledon such as maize the **coleoptile** leads the shoot through the soil and protects it. The food stores remain below the soil.

In many dicotyledons the cotyledons are brought above the ground, so that both the epicotyl and the hypocotyl emerge. This is known as **epigeal germination**.

Some dicotyledons develop leaving the cotyledons, hypocotyl and radicle below the ground. This is called **hypogeal germination**.

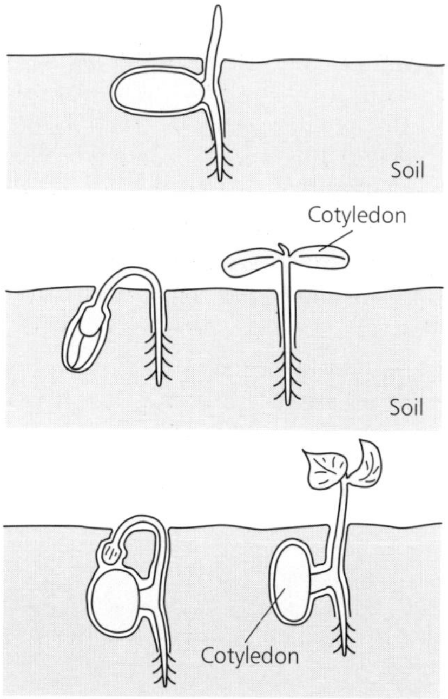

Figure 5.3.10 Whatever the method of germination and development, the race against time for the seed is completed when the root system is growing and the first true leaves are open and photosynthesising before the food store in the endosperm or cotyledons is used up.

as the **meristems**. It is here that rapid mitotic cell division occurs to generate new cells. These then undergo expansion and elongation by the uptake of water and the assimilation of new materials. The cells which result then differentiate into the different cell types needed within the stem or root. The growth which occurs at these growing root and shoot tips is known as **primary growth**.

Root growth

To make progress the growing root must probe through the soil. The delicate root cells are protected from damage by a tough **root cap**. There are three main zones in a rapidly growing root – the **meristematic region**, the **region of elongation** and the **region of differentiation**, shown in figure 5.3.11.

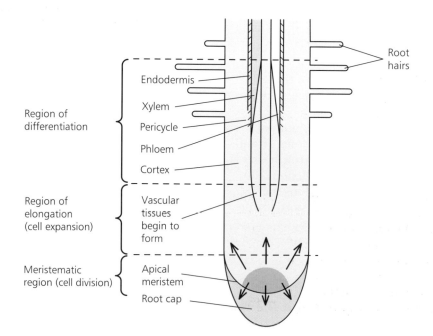

Figure 5.3.11 The three major growth zones in a young root. New cells arise by mitosis in the meristematic region. They increase in size and particularly in length in the region of elongation. Finally they differentiate into epidermis, vascular tissue, etc. in the region of differentiation. The development of root hairs is one of the earliest signs of differentiation.

Shoot growth

Stems grow from dividing cells behind the tip known as the **apical meristem**. The stem does not have the equivalent of a root cap, because there is not as much risk of mechanical damage growing through air as through soil. The meristematic zone in stems has two regions, the **tunica** and the **corpus**. As in the root, cells are formed and then elongate and differentiate into the tissues needed within the stem. The tunica and the corpus are involved in the production of regular swellings called **leaf primordia**. These subsequently develop into leaves. Within the leaf primordia are **axillary bud primordia**. Auxin from the main stem tip inhibits the development of these buds. As the stem lengthens and they become further from the auxin source, or if the **apical dominance** is removed by pinching off the auxin-producing growing tip, then these axillary buds develop into lateral shoots as shown in figure 5.3.12.

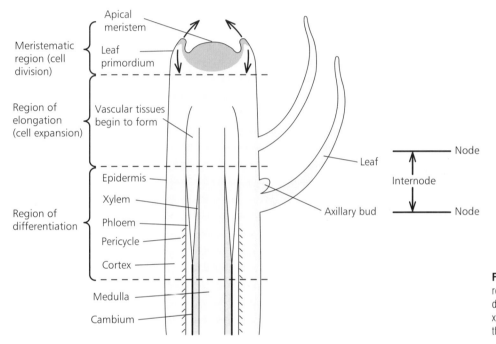

Figure 5.3.12 Cell division in the apical meristems results in stem growth. Most of the specialised tissues develop directly in the region of differentiation, except xylem and phloem – a procambial strand develops, and the xylem and phloem differentiate from this.

LIFE SPANS AND LIFESTYLES IN PLANTS

Angiosperms produce flowers and gymnosperms (conifers) produce cones, both forming seeds to reproduce. But they do not all grow, flower and set seeds in the same yearly pattern. There are three main divisions – **annuals**, **biennials** and **perennials**.

Annuals

Annual plants complete their life cycle within one year. The seed germinates and the plant grows and flowers, seeds develop and are dispersed and the parent plant dies, all in one growing season. The plant will survive the winter, or other adverse conditions such as drought, only as seeds. Annuals live for a relatively short time and much of their resources go into the production of flowers and seeds. Almost all the growth in these plants is primary growth. Many weeds, grains, garden flowers and vegetables are annuals. The major adaptive advantage of this lifestyle is that the plant can spread rapidly and produce many seeds in a short period.

Biennials

Biennial plants are less common than annuals. They complete their life cycles over two growing seasons. In the first season they germinate and grow, then the established plant overwinters. In the second season it produces flowers and seeds, and then dies. Parsley and foxgloves are typical examples.

Both annuals and biennials die after flowering because no more growth is possible. The apical meristem is converted into a flower, and once all the meristems have flowered the plant is doomed.

Perennials

Perennial plants may live for many years, and some such as trees have very large bodies which may survive for centuries. Sexual reproduction may not begin until a perennial plant is many years old. Although flowering does involve apical meristems, it takes place only on some of the lateral branches, and so growth can continue at the apical meristems of the rest of the branches.

Perennials can grow very large because of a process known as **secondary growth** or **secondary thickening** shown in figure 5.3.13. This occurs as a stem grows laterally, increasing its girth. This thickening of the stem involves production of secondary vascular tissues such as xylem and phloem and the laying down of lignified tissue, and enables the plant to withstand the added load of branches and leaves, and also to cope with the elements as it gets larger.

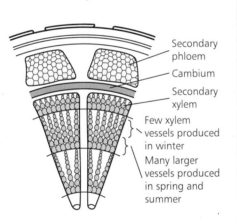

Secondary
phloem

Cambium

Secondary
xylem

Few xylem
vessels produced
in winter

Many larger
vessels produced
in spring and
summer

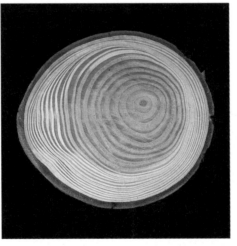

Figure 5.3.13 During autumn and winter the tree is becoming dormant and very few xylem vessels are produced by the cambium, giving narrow, dark rings. In spring and summer the cambium is reactivated and produces many larger xylem cells which gives the lighter regions in the rings. The photograph of a 24-year old pine tree shows how these rings appear in the wood.

Secondary thickening

The thickening of a stem or secondary thickening is brought about by two growth regions known as **lateral meristems**. One is called the **vascular cambium** and the other is the **cork cambium**.

The vascular cambium

The vascular cambium is the cylindrical layer of actively dividing cells that we saw when we considered the stem in section 2.2. As secondary thickening starts, a complete ring of cambium develops and its mitotic activity in turn results in complete cylinders of xylem and phloem, as shown in figure 5.3.14. The phloem cells die off and are regularly replaced, so there is only a narrow band of phloem around the edge of the stem. However, the lignified xylem vessels remain even when they have become blocked and non-functional, forming the wood which makes up the bulk of the stem. Because the activity of the vascular cambium is affected by environmental factors such as temperature and water availability the rings of growth recorded in the trunk of a tree reflect the environmental fluctuations over the years.

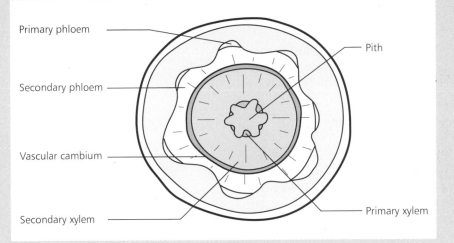

Figure 5.3.14 Apical meristems grow to increase the length of the roots or the height of the stem. Vascular cambium grows to increase the girth of the stem or root.

The cork cambium

The cork cambium is a layer of dividing cells found just below the epidermis, and the **cork** or bark which it produces replaces the epidermis as the stem gets larger and older. The cork cells become impregnated with suberin which makes them waterproof and prevents water loss. This also prevents the passage of oxygen and carbon dioxide into and out of the stem. Small openings called **lenticels** shown in figure 5.3.15 play the same role as stomata in leaves and allow gaseous exchange to occur.

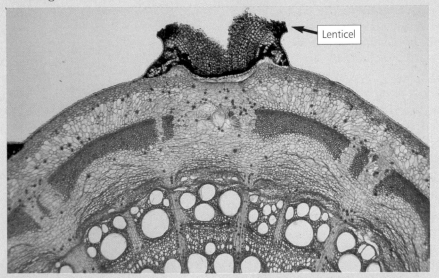

Figure 5.3.15 The waterproof cork produced by the cork cambium protects the stem from water loss. The lenticels perforate the layer to allow gaseous exchange.

Thus a variety of lifestyles and the strategies for asexual and sexual reproduction that we have seen enable plants to master an enormous range of habitats – their dominance over the surface of the land is such that the green colour they impart can even be seen from space.

Summary

- **Angiosperms** or flowering plants reproduce by the production of flowers. Flowers have **sepals** which protect the bud, **petals** which may be brightly coloured, **stamens** which are the male flower parts and **carpels** which are the female parts, all supported on the **receptacle**.

- Stamens are made up of the **anther** where pollen grains are formed supported on the **filament**. The **carpel** is made up of the **ovary** where ovules are formed, the **stigma** where pollen grains land and the **style** which joins the stigma to the ovary.

- **Pollination** is the transfer of pollen from one flower to another. **Insect-pollinated** flowers have attractive flowers and scents and may produce nectar. Their stamens and flat sticky carpels are usually enclosed in the flower, and their pollen grains are large and sticky. **Wind-pollinated** flowers have small inconspicuous flowers with long filaments and feathery stigmas protruding outside the flower. Their pollen grains are small and light.

- Flowering plants have a diploid **sporophyte generation** and a haploid **gametophyte generation**.

- **Microgametogenesis** is the formation of pollen. Each anther has four **pollen sacs** containing diploid **microspore mother cells** which divide by meiosis to produce haploid **microspores**. Mitosis of these results in two cells combining to form a **pollen grain** with two haploid nuclei, the **tube nucleus** and the **generative nucleus**.

- **Megagametogenesis** is the formation of egg cells in the ovule, which is attached to the ovary wall by the **placenta**. The **nucellus** is covered by **integuments** and contains diploid **megaspore mother cells**. These divide by meiosis to give four haploid **megaspores**. Three of these degenerate and one undergoes three mitotic divisions to give an **embryo sac** containing an **egg cell**, two **polar bodies** and other small cells.

- On pollination, molecules on the surface of the pollen grain and stigma interact to recognise each other. The pollen grain **germinates** – a **pollen tube** grows from the tube nucleus into the stigma, through the style to the ovary. The generative nucleus divides on its way down to form two male nuclei which pass through the micropyle of the ovule.

- Double fertilisation occurs – one male nucleus fuses with two polar nuclei to form the triploid **endosperm nucleus**. The other male nucleus fuses with the egg cell to give the diploid **zygote**.

- The endosperm nucleus divides by mitosis to give endosperm food stores. The zygote divides to form the embryo plant containing the **plumule** (shoot), **radicle** (root) and **cotyledons** (seed leaves – one in **monocotyledons** and two in **dicotyledons**).

- The seed dehydrates and becomes **dormant**. The integuments form the **testa**, a seed coat to protect the seed.

- The seed may be dispersed by various methods involving wind and water, or edible or hooked **fruits** developed from the ovary wall or the receptacle.

- The seed **germinates** after an environmental trigger in the presence of oxygen and water. Water is taken in through the micropyle causing the embryo tissues to swell and the testa to split. The radicle emerges first.

- In monocotyledons the **coleoptile** leads the shoot through the soil. In dicotyledons germination may be **hypogeal** or **epigeal**.

- Stems and roots have **meristems** where primary growth occurs.

- A root has a **root cap** to protect it and has a **meristematic region**, the **region of elongation** and the **region of differentiation**.

- A shoot has an **apical meristem** with two regions – the **tunica** and **corpus** which produce swellings called **leaf primordia**. These develop into leaves, and have **axillary bud primordia** which are inhibited by auxin from the apical meristem.

- **Annual plants** complete their life cycle in one year. **Biennials** germinate and grow in the first season and flower in the second season, after which they die. **Perennials** live for many years and may have **secondary growth** – a thickened stem which has lignified tissue.

QUESTIONS

1 a Draw an annotated diagram of the structure of a flower, relating structure to function.

b Describe some of the modifications to the basic flower pattern that have evolved to enable different agents to pollinate the flowers.

c What is the difference between pollination and fertilisation in a flower?

2 a Describe the fertilisation of an ovule.

b After fertilisation, a fruit containing the seeds will form. What is the function of this fruit and why is it necessary?

3 a Under suitable conditions seeds will germinate. Describe the main events of germination.

b After germination a period of growth and maturation occurs within the plant. Explain the main stages of this growth which result in a mature plant.

5

5.4 The mammalian sexual anatomy

Sexual reproduction is a process of paramount importance in the living world. It provides a way of introducing variety and change, allowing for the exploitation of new niches and the survival of organisms in difficult conditions. Sexual reproduction occurs in most groups of living things, and is carried out successfully by a wide variety of strategies. Of all the animal groups, mammals have perhaps the most sophisticated systems for ensuring both that fertilisation takes place and that the offspring survive the early stages of development, providing a ready supply of food for subsequent growth.

Reproduction in mammals

As members of the mammalian group we share many reproductive features with our nearest relatives. This means that as we look at the sexual anatomy and functioning of our own species we learn much of the reproductive physiology of others.

In mammals – including ourselves – it is very difficult to tell males from females in the early stages of embryo development. However, by the time of birth the external **genitalia** of the two sexes are quite obviously different, although otherwise the overall appearance is similar. The adult forms develop after a process of sexual maturation. The sexual anatomy of these adult males and females is very different. The differences are indicative of the distinct roles of male and female mammals in reproduction. What are these roles, and how is each anatomy designed to fulfil them? We shall look at the male and female in turn to answer these questions.

REPRODUCTION AND THE MALE MAMMAL

The male role

The male role is to provide the male sex cells or **spermatozoa**, commonly known as **sperm**, which are needed to fertilise the eggs within the female. Because most mammals live on land, it is important that these sperm are delivered inside the female reproductive tract at about the time of **ovulation** when a ripe egg is released. Spermatozoa cannot survive exposure to the air. As organisms evolved to colonise the land, systems developed by which the male gametes could be transferred to the female in a liquid environment. In mammals special glands provide a nutrient fluid containing the spermatozoa, known as **semen**.

In mammals the transfer of the gametes involves a specialised organ called the **penis** which, when engorged with blood and **erect**, is used to deliver spermatozoa into the female reproductive system. This takes place during **mating** or **copulation**, usually preceded by elaborate courtship displays to ensure that the female is both receptive and so will not attack the male, and fertile so that the spermatozoa are not wasted. In humans this element of the reproductive ritual has been, if not lost, at least substantially modified. Although it could be argued that in most societies sexual intercourse is still preceded by elaborate courtship displays (like buying rings, getting married, setting up home or other ritual behaviour), these displays are no longer linked to establishing whether or not a female is fertile – the sexual receptiveness of women is not linked to the time of ovulation.

For fertilisation to occur, vast numbers of spermatozoa need to be produced to allow for the enormous natural wastage which occurs, and to complete their task they need the right conditions – a moist environment, a supply of nutrients and the right temperature and pH.

Fulfilling the male role

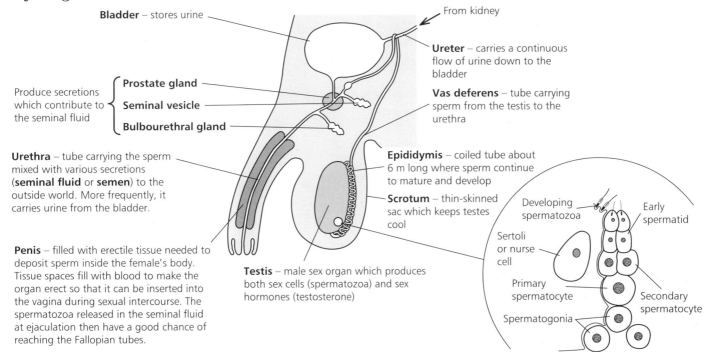

Bladder – stores urine

From kidney

Ureter – carries a continuous flow of urine down to the bladder

Produce secretions which contribute to the seminal fluid
- **Prostate gland**
- **Seminal vesicle**
- **Bulbourethral gland**

Vas deferens – tube carrying sperm from the testis to the urethra

Urethra – tube carrying the sperm mixed with various secretions (**seminal fluid** or **semen**) to the outside world. More frequently, it carries urine from the bladder.

Epididymis – coiled tube about 6 m long where sperm continue to mature and develop

Scrotum – thin-skinned sac which keeps testes cool

Penis – filled with erectile tissue needed to deposit sperm inside the female's body. Tissue spaces fill with blood to make the organ erect so that it can be inserted into the vagina during sexual intercourse. The spermatozoa released in the seminal fluid at ejaculation then have a good chance of reaching the Fallopian tubes.

Testis – male sex organ which produces both sex cells (spermatozoa) and sex hormones (testosterone)

Developing spermatozoa

Early spermatid

Sertoli or nurse cell

Primary spermatocyte

Secondary spermatocyte

Spermatogonia

Figure 5.4.1 The male sex organs need to produce vast numbers of gametes continuously and position them inside the body of a female. This diagram shows the important parts of the male reproductive system, including the links with the urinary system.

The male sex organs are the **testes**. In human males the testes are found outside the main body cavity in a thin-skinned sac called the **scrotum**, and the design of many other male mammals is similar. This external position makes these delicate organs vulnerable, but helps maintain the optimum temperature for the production of healthy sperm, 2–3 °C lower than normal body temperature. Thus for the species as a whole, the advantages of having the testes outside the body outweigh the disadvantages.

It is the presence of testes which results in the development of all the other characteristics of maleness. The testes have two main functions, controlled by hormones produced by the pituitary gland in the brain. **Follicle stimulating hormone (FSH)** stimulates the development of spermatozoa. **Luteinising hormone (LH)** stimulates the testis to make testosterone.

The testes as chemical controllers

The testes are made up of several different types of cells, most of which are involved in sperm production. However, the function of the **interstitial** or **Leydig cells** is to manufacture the male sex hormone **testosterone**. Luteinising hormone is released by the pituitary and picked up by receptors in the interstitial cells. It stimulates the synthesis of enzymes which in turn make testosterone. Luteinising hormone is identical in males and females, but in males it is sometimes called **interstitial cell stimulating hormone (ICSH)**, describing exactly what it does.

Testosterone is the main male hormone. It is partly responsible for the production of sperm and wholly responsible for the development and maintenance of the **secondary sexual characteristics** – those features which make the individual male. Before puberty the basic body shape of both boys and girls is very similar. In boys during adolescence (at the time of puberty) the brain starts to produce increased levels of ICSH and the testosterone level rises as a result. Sperm production begins and the entire physical appearance undergoes changes, some quite rapid, others taking years to complete. These physical changes all depend on a maintained level of testosterone. Other mammals show varying degrees of the same process, with adult males usually being easily recognisable from the females.

Male characteristics

Figure 5.4.2 illustrates the human male secondary sexual characteristics which develop at the period of life known as **puberty**.

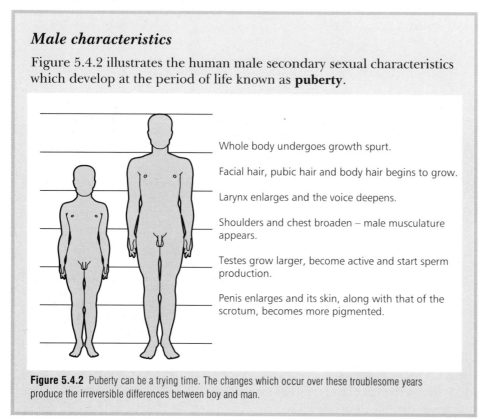

Whole body undergoes growth spurt.

Facial hair, pubic hair and body hair begins to grow.

Larynx enlarges and the voice deepens.

Shoulders and chest broaden – male musculature appears.

Testes grow larger, become active and start sperm production.

Penis enlarges and its skin, along with that of the scrotum, becomes more pigmented.

Figure 5.4.2 Puberty can be a trying time. The changes which occur over these troublesome years produce the irreversible differences between boy and man.

The testes as sperm producers

As figure 5.4.1 shows, the testes are extremely complex organs, consisting of a mass of seminiferous tubules. At the time of puberty they are 'kick-started' by a combination of follicle stimulating hormone from the pituitary and testosterone from their own interstitial cells. This sets in action the process of sperm production which normally continues throughout the life of the individual.

The testes contain germ cells which undergo the complex process of **gametogenesis** or **spermatogenesis** (see section 5.2, page 351), followed by a phase of differentiation called **spermiogenesis** before they become active fertile spermatozoa ready to fertilise an ovum.

Spermatozoa are produced in the epithelium of the seminiferous tubules. Many millions are made each day, and the maturation process takes about 70 days from the beginning of spermatogenesis to the end of spermiogenesis. **Sertoli** or **nurse cells** supply support, protection and nutrition to the maturing sperm. The glands along the system provide the nutrients and fluid that make up the seminal fluid released at ejaculation. The spermatozoa themselves make up only a tiny fraction of the 3 cm^3 of ejaculate – the remaining liquid helps to modify the pH of the vagina and maintain optimum conditions for the survival of the sperm.

The structure of a spermatozoon

The spermatozoa of most species are around 50 μm long. They have several tasks to fulfil. They must remain in suspension in the semen to facilitate transport through the female reproductive tract, and to carry out fertilisation they must be able to penetrate the protective barrier around the ovum and deliver the haploid genome into the cytoplasm of the egg. (This process is described in detail in section 5.5.) The close relationship between the structure of the human spermatozoon and its functions can be seen in the figure.

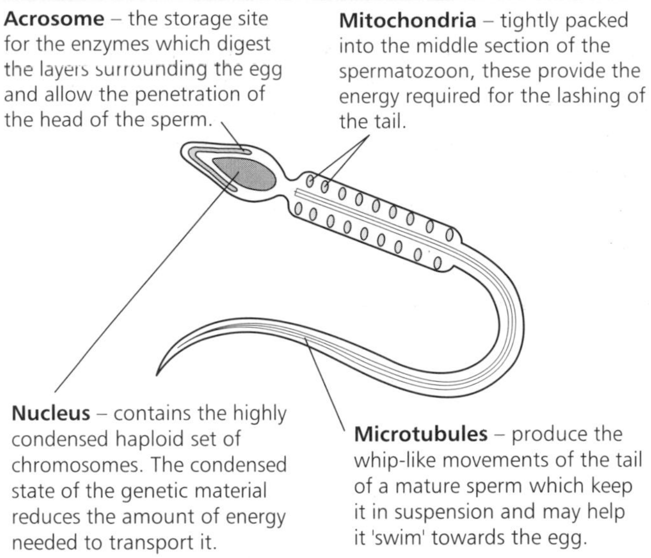

Figure 5.4.3 Millions upon millions of these motile gametes are produced in the lifetime of a human male, yet the average size of a family in the developed world is only around 2 children. Only one spermatozoon fertilises each egg. Biologists are currently considering several hypotheses about the role of the 'wasted millions'.

Acrosome – the storage site for the enzymes which digest the layers surrounding the egg and allow the penetration of the head of the sperm.

Mitochondria – tightly packed into the middle section of the spermatozoon, these provide the energy required for the lashing of the tail.

Nucleus – contains the highly condensed haploid set of chromosomes. The condensed state of the genetic material reduces the amount of energy needed to transport it.

Microtubules – produce the whip-like movements of the tail of a mature sperm which keep it in suspension and may help it 'swim' towards the egg.

REPRODUCTION AND THE FEMALE MAMMAL

The female role

The role of the female mammal in reproduction is to produce a relatively small number of large gametes or **ova**, to provide the developing embryo with food and oxygen, remove its waste products and, after delivering it into the world, provide it with a continued supply of food for a period of time.

Figure 5.4.4 The human female reproductive system produces relatively few, large gametes and provides a place for the growth and development of any that are fertilised – a strategy which has been enormously successful in evolutionary terms. These diagrams show the main parts of the female reproductive system.

Fulfilling the female role

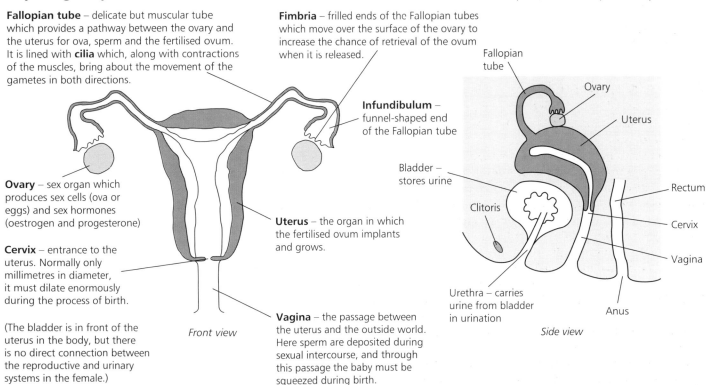

Fallopian tube – delicate but muscular tube which provides a pathway between the ovary and the uterus for ova, sperm and the fertilised ovum. It is lined with **cilia** which, along with contractions of the muscles, bring about the movement of the gametes in both directions.

Fimbria – frilled ends of the Fallopian tubes which move over the surface of the ovary to increase the chance of retrieval of the ovum when it is released.

Infundibulum – funnel-shaped end of the Fallopian tube

Ovary – sex organ which produces sex cells (ova or eggs) and sex hormones (oestrogen and progesterone)

Cervix – entrance to the uterus. Normally only millimetres in diameter, it must dilate enormously during the process of birth.

(The bladder is in front of the uterus in the body, but there is no direct connection between the reproductive and urinary systems in the female.)

Front view

Uterus – the organ in which the fertilised ovum implants and grows.

Vagina – the passage between the uterus and the outside world. Here sperm are deposited during sexual intercourse, and through this passage the baby must be squeezed during birth.

Fallopian tube

Ovary

Uterus

Bladder – stores urine

Clitoris

Rectum

Cervix

Vagina

Urethra – carries urine from bladder in urination

Anus

Side view

5

The female sex organs are the **ovaries**. These are found low in the abdominal cavity and are very closely associated with, but not attached to, the **uterus** with its **Fallopian tubes**.

It is the presence of the ovaries which results in the outward signs of femaleness. The ovaries have two main functions and they are controlled by chemical messages from the pituitary gland in the brain just like the testes. Follicle stimulating hormone stimulates the development of ripe **ova**, the female gametes, within structures called **follicles** in the ovaries. FSH also stimulates the production of the hormone **oestrogen**. Luteinising hormone (LH) brings about the production of the hormone **progesterone**.

The ovaries as chemical controllers

Of the two hormones produced by the ovaries, oestrogen is mainly responsible for the secondary sexual characteristics, those physical attributes associated with femaleness. The changes in the female body shape during puberty tend to be more dramatic than those in the male and on the whole take less time to be completed. The role of the female hormones in regulating the reproductive cycle is described opposite.

Female characteristics

Figure 5.4.5 shows the human female secondary sexual characteristics.

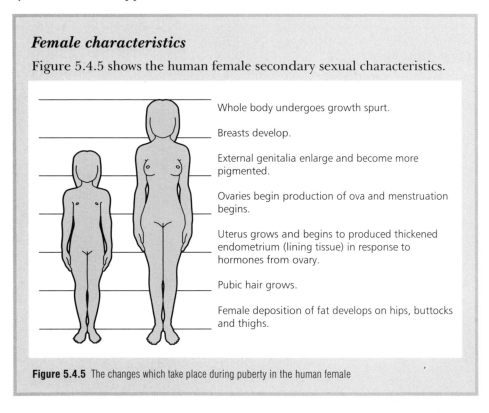

Whole body undergoes growth spurt.

Breasts develop.

External genitalia enlarge and become more pigmented.

Ovaries begin production of ova and menstruation begins.

Uterus grows and begins to produced thickened endometrium (lining tissue) in response to hormones from ovary.

Pubic hair grows.

Female deposition of fat develops on hips, buttocks and thighs.

Figure 5.4.5 The changes which take place during puberty in the human female

The ovaries as egg producers

The production of ova begins in the fetus long before birth. By the time of birth about 2 million (2×10^6) immature ova-containing follicles are present in the ovaries of a baby girl. Only about 450 of these ova will ever be released from mature follicles, and a tiny fraction of these will be fertilised and develop into new human beings.

At puberty, the menstrual cycle begins and is repeated approximately every 28 days for the years of a woman's reproductive life – that is, until she runs out of ova. In each cycle several follicles begin to develop and ripen, and from one a mature ovum is released. Those follicles which ripen but do not release ova simply shrivel and atrophy. When her ovaries are empty a woman enters what is called the **menopause**, and this marks the end of her ability to bear children.

The menstrual cycle

Once gamete production starts in men, it is a continuous process. In women and other female mammals the situation is very different. Mature ova are produced as the result of cyclical activity, so that females are only fertile for relatively brief but regularly spaced periods of time, compared with the constant fertility of the male. The way in which the reproductive cycle is organised varies, but in humans and their closest relatives it takes the form of the **menstrual cycle**.

The events of the cycle are controlled by the pituitary hormones and the hormones of the ovary itself. The menstrual cycle is measured from the first day of menstrual bleeding as this is an easy event to pinpoint. The old lining of the uterus breaks down and sloughs off. In the first stage of the cycle, follicles mature under the influence of follicle stimulating hormone from the pituitary gland (which seems to initiate the process). At the same time luteinising hormone, also from the pituitary, stimulates the cells of the developing follicle to produce oestrogen. Oestrogen suppresses the production of FSH by the pituitary through a negative feedback loop. Oestrogen also stimulates the lining of the uterus (the **endometrium**) to become thicker and more vascular in preparation for the arrival of a fertilised ovum. By about day 14 of the cycle oestrogen has reached a peak, FSH and particularly LH are released in relatively large amounts from the pituitary and **ovulation** occurs – a ripe ovum is released from a follicle and picked up by the fimbria to begin its journey towards the uterus.

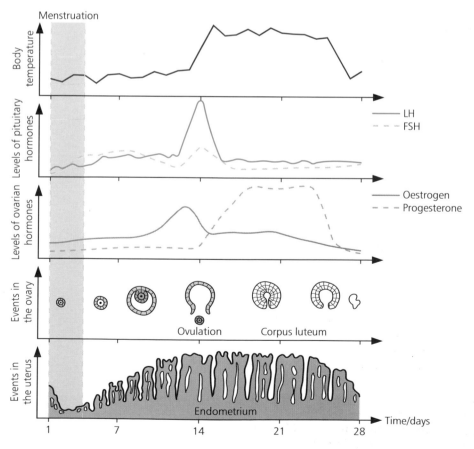

Figure 5.4.6 The menstrual cycle is an integrated sequence of complex events, the end result of which is a mature ovum ready for fertilisation and a uterus prepared to receive it every month.

After ovulation the cells of the empty follicle change to form a **corpus luteum**. The name literally means 'yellow body', from the yellow fat deposited in the empty follicle. The corpus luteum makes and releases the second ovarian hormone, progesterone. Progesterone maintains the endometrium in preparation for pregnancy and prevents the release of FSH and LH by the pituitary. However, if the ovum is not fertilised and pregnancy does not occur, then after about 11 days the corpus luteum atrophies and the level of progesterone drops. This has two effects – the lining of the uterus is shed in the menstrual bleeding and the levels of FSH and LH begin to rise again so that a new cycle begins. Figure 5.4.6 shows the events of the menstrual cycle.

The structure of the ovum

While the spermatozoa of most animals are very similar in size, eggs vary greatly in both their diameter and their mass. The human ovum is about 0.1 mm across, whilst the ovum contained in the ostrich egg is around 6 mm in diameter.

The main difference between eggs of various species is the quantity of stored food they contain. In species such as birds and reptiles, complex development of the young occurs before hatching. Large food stores are therefore contained within the egg to supply their needs. In mammals, the developing fetus is supplied with nutrients from the blood supply of the mother and so large food stores are unnecessary. Figure 5.4.7 shows a human ovum.

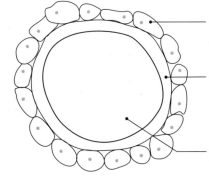

Follicle cells form a protective layer around the oocyte.

Zona pellucida – a clear jelly-like protective layer through which spermatozoa must penetrate

Oocyte containing chromosomes which still have to complete their meiotic divisions

Figure 5.4.7 The human oocyte is much larger than the spermatozoa and completely non-motile. It relies entirely on beating cilia and muscular contractions of the Fallopian tubes to move it along towards the uterus.

The sexual anatomy and the way it works to produce gametes in the right place at the right time is a complex system in both males and females. However the joining of these gametes and the forming and nurturing of the new life which results are even more complex, as section 5.5 will demonstrate.

SUMMARY

- The mammalian male gametes are **spermatozoa**. They cannot survive in air so a nutrient fluid called **semen** is secreted to carry them inside the female.

- Spermatozoa are introduced into the female through the **penis** when **erect** (engorged with blood) during **copulation**.

- Spermatozoa are produced under the influence of FSH and testosterone in the **testes** which are positioned outside the main body cavity in the **scrotum**, to keep them cool. Spermatozoa pass into the **epididymis**, a coiled tube, and to the penis via the **vas deferens**. Along this tube the **prostate gland**, **seminal vesicles** and **bulbourethral gland** produce seminal secretions. Semen passes out of the penis via the **urethra**.

- The **interstitial** or **Leydig cells** of the testis produce testosterone under the influence of LH from the pituitary gland. Testosterone brings about the development of male secondary sexual characteristics at puberty.

- **Spermatogenesis** takes place in germ cells in the testis followed by **spermiogenesis** or differentiation. **Sertoli** or **nurse cells** support, protect and feed the large numbers of developing spermatozoa. The sperm cell has a head containing the **acrosome** (a storage site for enzymes) and the nucleus, a middle section containing large numbers of mitochondria and a long tail containing microtubules which enable it to whip backwards and forwards.

- The female gametes are **ova** or eggs, produced in **follicles** in the **ovaries** under the influence of FSH from the pituitary. The **uterus** or womb contains the developing embryo and the ova reach this via the **Fallopian tubes** which have frilled ends called **fimbria** to catch the ripe ova that are released. The **cervix** is the entrance to the womb, and the **vagina** leads to the exterior.

- The ovaries produce the hormones **oestrogen** under the influence of FSH and **progesterone** under the influence of LH. Oestrogen brings about the development of female secondary sexual characteristics at puberty.

- One ovum is released from either one of the ovaries once every 28 days, in the **menstrual cycle**. Day 1 is the beginning of **menstruation**, the shedding of the **endometrium** or vascular uterus lining. A **follicle** starts to develop in the ovary under the influence of FSH. The endometrium is built up again under the influence of oestrogen. At about day 14 **ovulation** takes place under the influence of LH and the ovum released from the follicle travels down the Fallopian tube towards the uterus. The empty follicle becomes a **corpus luteum** and produces progesterone. If the ovum is not fertilised, the corpus luteum atrophies and the level of progesterone falls, causing the endometrium to be shed again, and the cycle is repeated. If the ovum is fertilised, the corpus luteum is maintained and progesterone prevents menstruation.

- Ova are larger than spermatozoa. The central **oocyte** is contained inside the jelly-like **zona pellucida** and the **follicle cells**.

QUESTIONS

1 a Draw the outline of an adult human male with details of the reproductive system, including the pituitary gland in the brain. Label it appropriately.
   b Using annotated arrows, show the organ of origin, the receptor sites and the effects of the main male reproductive hormones.
   c Draw the outline of an adult human female with details of the reproductive system, including the pituitary gland in the brain, and label the diagram.
   d Using annotated arrows, show the organ of origin, the receptor sites and the effects of the main female reproductive hormones.

2 Describe the similarities and differences between:
   a mammalian testes and mammalian ovaries
   b mammalian sperm and mammalian ova.

3 Produce a simple and clear explanation of the events of the menstrual cycle that could be used on a leaflet in an infertility clinic. Remember that in this situation patients need to understand the details of the processes, but may have little or no scientific background.

Reproduction in mammals depends on the production of haploid ova and spermatozoa within the bodies of sexually mature individuals. New life starts at **fertilisation**, with the union of the two germ cells. From that point onwards complex developments during the period of **gestation** or **pregnancy** result in the eventual birth of a fully formed, new individual. In this section we look into the events of fertilisation and beyond, while accepting that we can only scratch the surface of these incompletely understood areas of biology.

Figure 5.5.1 After ovulation the ovum is picked up from the surface of the ovary by the cilia on the infundibulum and passes into the Fallopian tube to meet the spermatozoa.

FERTILISATION

The arrival of the ovum

During the 28-day human menstrual cycle, a follicle ripens and bursts to release an ovum. This is picked up by the fimbria of the infundibulum of the Fallopian tube, as shown in figure 5.5.1. It is then moved along the tube towards the uterus by a cilial current and by the muscular contractions of the oviduct walls.

The human ovum begins to deteriorate only a few hours after ovulation, and is dead within 24 hours. For fertilisation to occur, spermatozoa must reach the ovum as soon as possible after ovulation, or must be already present in the Fallopian tube when the ovum is released. In most mammals the period of sexual receptiveness known as **oestrus** begins several hours before ovulation. This results in mating taking place before the ripe ovum leaves the ovary. However, in humans this link between mating and ovulation has been lost. Human females are, at least in theory, sexually receptive all the time and so there is no guarantee that sperm will be deposited at the optimum time for fertilisation to take place, in contrast to other mammals.

The journey of the spermatozoa

The movement of the spermatozoa from the testis to the site of fertilisation involves an epic journey. On leaving the testis the spermatozoa pass into the epididymis which consists of an extremely long, twisted tube. Here they are stored for a time. As the spermatozoa pass through the next tube, the vas deferens, they become **motile** or capable of moving themselves by the lashing of their tails. During sexual excitement the penis engorges with blood and becomes erect. This enables it to be positioned in the vagina of the female. The spermatozoa are mixed with secretions of the various glands to form **semen** and ejected into the female reproductive tract as the result of a brief series of muscular contractions known as **ejaculation**. Typically about 300–500 million sperm contained in 3–4 cm^3 of fluid will be deposited high up in the vagina.

The semen coagulates after ejaculation, which seems to help to prevent the loss of the fluid from the female reproductive tract. The sperm are then moved up through the cervix and the uterus into the Fallopian tubes very rapidly – spermatozoa reach the upper parts of the tubes between 5 and 30 minutes after intercourse has taken place. It was originally thought that they 'swam' but the current model is that the spermatozoa are moved mainly by contractions of the female reproductive tract. The movements of sperm which are observed seem to be more important for maintaining them in suspension in the semen and for the fertilisation of the egg than for movement. The difficulty of the journey undertaken is such that, in spite of the millions of sperm which start off, only a few hundred to a few thousand reach the ovum.

Copulation

Mating in most mammals is the result of instinctive behaviour patterns which occur when sex hormone levels reach a particular level and the environmental cues are appropriate. In humans the situation is more complicated, since mating (more commonly referred to as sexual intercourse or copulation in the human context) can take place at any time of year, regardless of the likelihood of ovulation occurring. Sexual arousal in the male causes the erection of the penis and the events already described. In the female, sexual arousal rather than simple hormone levels ensures the secretion of extra mucus in the vagina to make penetration by the penis easier. Also the **clitoris**, a tiny sexual organ similar in structure and sensitivity to the penis, gives sexual enjoyment when appropriately stimulated. It may be this level of voluntary rather than innate involvement of both partners which has freed human sexuality from the bonds of oestrus.

The female role in sperm movement

Once the sperm enter the female reproductive tract they are moved largely by the female system.

(1) Around the time of ovulation the vaginal mucus changes in pH in response to the changing levels of sex hormones. It is normally so acidic that it tends to kill sperm. At the fertile time it becomes more alkaline to prevent sperm damage.

(2) In a similar way, mucus which blocks the cervix preventing the entry of microorganisms becomes much less viscous, allowing the spermatozoa to move through it more easily.

(3) **Prostaglandins** (local hormones) in the semen and **oxytocin**, a hormone released from the female posterior pituitary gland during sexual intercourse, cause contractions in the uterus which move the semen towards the Fallopian tubes.

(4) Groups of cilia in the lower end of the Fallopian tubes appear to beat upwards, carrying the semen towards the ovum.

The meeting of the gametes

The female ovum is fully viable for only a few hours. The male sperm will survive a day or two in the female reproductive tract. There is little evidence to suggest that the sperm are attracted to the egg in any way. Yet, in spite of this, they frequently do meet and fuse. What happens when sperm meets egg?

During their passage through the female reproductive tract spermatozoa become fully activated and able to penetrate the egg. The ovulated ovum has not fully completed meiosis and is a secondary oocyte (see section 5.2, page 352). It is surrounded by a protective jelly-like layer known as the zona pellucida and also by some of the follicle cells. Many sperm cluster around the ovum, and as soon as the head of the sperm touch the surface of the ovum the **acrosome reaction** is triggered, shown in figure 5.5.2. Enzymes are released from the acrosomes which digest the follicle cells and the zona pellucida. One sperm alone does not produce sufficient enzyme to digest all the protective layers around the ovum, and the very large number of sperm initially released in ejaculation ensures that there are sufficient sperm present in the Fallopian tube to surround the ovum and digest its defences.

Eventually one sperm will wriggle its way through the weakened protective barriers and touch the surface membrane of the oocyte. This has several almost instantaneous effects. The oocyte undergoes its second meiotic division, providing a haploid egg nucleus to fuse with the haploid male nucleus. It is vital that no other sperm enter the egg, as this would result in triploid or

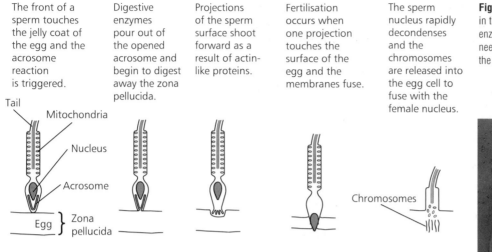

The front of a sperm touches the jelly coat of the egg and the acrosome reaction is triggered.

Digestive enzymes pour out of the opened acrosome and begin to digest away the zona pellucida.

Projections of the sperm surface shoot forward as a result of actin-like proteins.

Fertilisation occurs when one projection touches the surface of the egg and the membranes fuse.

The sperm nucleus rapidly decondenses and the chromosomes are released into the egg cell to fuse with the female nucleus.

Tail
Mitochondria
Nucleus
Acrosome
Egg } Zona pellucida
Chromosomes

Figure 5.5.2 The acrosome reaction plays a vital role in the fertilisation of the egg. It is thought that the enzymes from a number of acrosome reactions are needed to enable one sperm to eventually break through the protective layers.

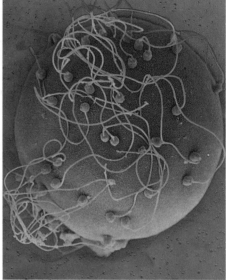

polyploid nuclei (with several sets of chromosomes), and the other events that follow fertilisation prevent this possibility. There are immediate changes in the electrical properties of the cell surface membrane of the ovum so that the inside, instead of being negative with respect to the outside, becomes positive. This alteration in charge prevents the entry of any further sperm. It is a temporary measure until a tough **fertilisation membrane** forms which then repels other sperm as the electrical charge returns to normal. Fertilisation is also referred to as **conception** in the case of humans, and is shown in figure 5.5.3.

The head of the sperm enters the oocyte, and the tail region is left outside. Once the head is inside the ovum it absorbs water and swells up, releasing its chromosomes to join those of the ovum and forming a diploid **zygote**. At this point fertilisation has occurred and a new genetic individual has been formed.

Figure 5.5.3 Sperm meets egg – and a new life begins. Once the head of a sperm has fused with the membrane of the egg a complex series of reactions takes place to ensure that no other sperm can enter, and the development of the new diploid zygote can begin undisturbed.

DEVELOPMENT OF THE EMBRYO

Formation of the embryo

If the ovum is fertilised, a complex series of events is set in motion which in humans will lead to the birth of (usually) one fully formed new individual. The first stage of this process is known as **cleavage**. Cleavage involves a special kind of mitosis, where cells divide repeatedly without a substantial interphase for growth between the divisions. A mass of small, identical and undifferentiated cells results which forms a hollow sphere known as a **blastocyst**. Cleavage occurs as the zygote is moved along the Fallopian tube towards the uterus, which in humans takes about one week. A large number of small cells result from one large egg cell.

Figure 5.5.4 By the time a woman misses a menstrual period and begins to suspect that she might be pregnant, the fertilised ovum has undergone a series of rapid mitotic divisions to produce the hollow ball of cells known as the blastocyst, and this has begun to implant itself in the lining of the uterus.

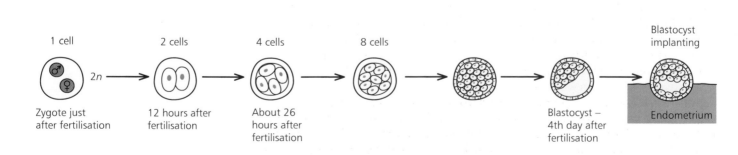

1 cell — 2n → 2 cells → 4 cells → 8 cells → Blastocyst – 4th day after fertilisation → Blastocyst implanting

Zygote just after fertilisation

12 hours after fertilisation

About 26 hours after fertilisation

Endometrium

After several days in the uterus the 50–100-cell blastocyst must embed itself into the lining of the uterus for the pregnancy to be recognised by the body and development to continue. Finger-like projections (**trophoblastic villi**) grow out from the outer layer (**trophoblast**) of the blastocyst. These embed themselves into the lining of the uterus, usually at the top end of the uterus, and eventually form the site of the placenta. This is known as **implantation**. It is usually completed about 11 or 12 days after fertilisation, and from this time onwards the woman is considered to be pregnant. Figure 5.5.4 illustrates the events leading up to implantation.

Formation of tissues and organs

Looked at very simply, an adult human being consists of an outer tube (the skin) and an inner tube (the gut) with various clusters of cells in between (the muscles, bones, liver, kidneys, etc.). The process which changes the simple ball of cells of the blastula into a complex three-dimensional organism with three layers is known as **gastrulation**.

Gastrulation leads to a three-layered embryo known as a **gastrula**. The layers are the **ectoderm** (outer layer), the **endoderm** (inner layer) and the **mesoderm** (middle layer) between them. Once this stage of development is reached the cells can differentiate and develop into organs and tissues. Only days after conception, cells are being committed or **determined** to become one type of tissue or another, as shown in figure 5.5.5. Even if these cells are then surgically moved around the embryo, as has been done in, for example, mice, they will still produce the predetermined tissue type, even if it is entirely inappropriate in the new setting. No one is yet entirely sure of the mechanism of this differentiation.

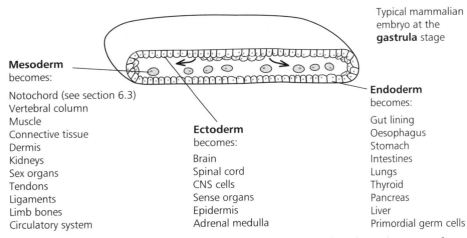

Typical mammalian embryo at the **gastrula** stage

Mesoderm becomes:

Notochord (see section 6.3)
Vertebral column
Muscle
Connective tissue
Dermis
Kidneys
Sex organs
Tendons
Ligaments
Limb bones
Circulatory system

Ectoderm becomes:

Brain
Spinal cord
CNS cells
Sense organs
Epidermis
Adrenal medulla

Endoderm becomes:

Gut lining
Oesophagus
Stomach
Intestines
Lungs
Thyroid
Pancreas
Liver
Primordial germ cells

Figure 5.5.5 Once the three main cell types have been established they rapidly become determined and form specific tissues.

Once determination has occurred the cells divide to produce tissues and organs. The rate of growth in a developing embryo is phenomenal. Rat pregnancies last just 22 days from fertilisation to birth. In that time the embryo increases from one to three billion cells, as the result of simple, steady mitosis. Even this level of growth pales into insignificance compared with the development of the blue whale. A whale ovum is less than a millimetre in diameter and weighs a fraction of a gram. The newborn calf that results from the fertilisation of such an egg is 7 metres long and weighs 2000 kg, representing a 20 million-fold increase in mass.

In the development of the human embryo the major organ systems are completed by around the 12th week of the 40-week pregnancy. By the end of this time the embryo is a tiny but recognisable human infant as shown in figure 5.5.6, although totally incapable of life outside the uterus. During these weeks the embryo is particularly vulnerable to any harmful substances which the mother may take in. 80% of the birth defects which occur in live births are caused by **teratogens**. These are environmental influences which affect the development of the embryo, usually at the early stages when organ systems are forming. Drugs such as thalidomide (which affects the growth of the limbs), nicotine from cigarette smoking and alcohol, diseases such as rubella (German measles) which affects the ears and brain and ionising radiation such as X-rays can all cause irreparable damage in the initial stages of pregnancy.

From around 12 weeks onwards the developing individual is known as a **fetus**. The remaining time in the uterus is used mainly for growth and maturation of the tissues and the fetus is much less vulnerable to the influence of teratogens.

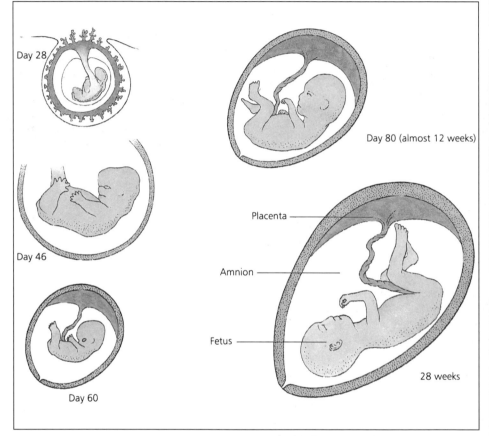

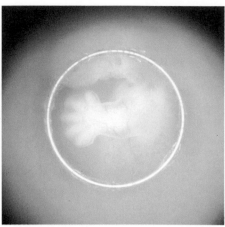

5 weeks

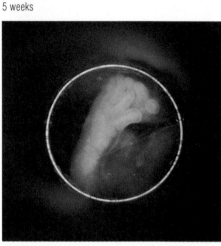

6 weeks

7–8 weeks

Figure 5.5.6 The human hand develops from a simple paddle to a sophisticated tool in just a few weeks, and the other systems of the body form similarly at the same time. Most of the period of gestation is spent in growth and maturation rather than in tissue differentiation.

Formation and functions of the membranes

The dividing mass of cells of the zygote in the early stages of gestation also forms the **extraembryonic membranes**. These membranes, shown in figure 5.5.7, play a vital role in the development of the fetus. They are present right up to the time of birth – the rupturing of the membranes and the subsequent release of the fluid they contain is a sign that the birth of the baby is imminent.

The membranes fold around the embryo to form the **amniotic cavity**. This is filled with **amniotic fluid** in which the embryo and fetus grows. The fluid supports the developing fetus, allowing it the freedom to move about easily. It provides a medium in which to practise swallowing and breathing movements and holds urine as it is voided. The amniotic fluid also cushions and protects the fetus from damage by external injury or impact to the mother – only a very violent blow would have any effect on the fetus. The inner membrane is the **amnion**, and the outer of the two is the **chorion**. After the initial implantation into the lining of the uterus by the blastocyst, it is the chorion that takes over the formation of the **placenta**, the area where the fetal and maternal blood exchange various substances (this is described below). Another cavity, the **allantois**, arises from the gut of the embryo and in humans this is destined to become the **umbilical cord** which links the embryo to the mother.

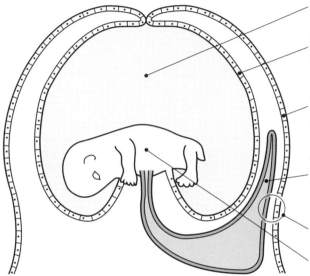

Amniotic cavity

Amnion – this lines the fluid-filled amniotic cavity in which the fetus grows and develops.

Chorion – outgrowths from this membrane extend into the lining of the uterus and form the placenta.

Allantois – this fuses with the chorion. It forms the umbilical cord which connects the fetus to the uterine wall.

Allantochorion – the placenta is formed in this region.

Embryo

Figure 5.5.7 The extraembryonic membranes in mammals are developed as part of the products of fertilisation – they are not produced by the maternal tissues. They play vital roles in the maintenance of the fetus during the period of development within the uterus. Babies are sometimes born still completely surrounded by these membranes – 'born in a caul'. This used to be regarded as a sign of good luck and an omen that the individual would never die by drowning. Now it is more likely to be regarded simply as a result of extremely tough membranes!

Formation and functions of the placenta

The surface of the chorion develops large numbers of finger-like projections into the endometrium, the maternal tissue lining the uterus. These form in the region shown in figure 5.5.7 as the allantochorion, and are known as the **chorionic villi**. They provide a vast surface area for the exchange of substances between the fetal and maternal blood. To make the exchange even easier the maternal blood vessels of the endometrium break down in the regions of the invading chorionic villi, so that the villi are bathed in maternal blood and the distance for diffusion of substances is only about 0.002 mm. A mature placenta towards the end of pregnancy is a disc about 20 cm across and 6 cm thick, with a surface area for exchange of 16 m^2 – equivalent to almost half a tennis court!

The site where the placenta forms is of great importance to the progress of the pregnancy. In the majority of cases it is in the upper third of the uterus, but sometimes the placenta forms lower down near the cervix (**placenta praevia**) when problems may arise at delivery. Without medical intervention the placenta would be delivered before the baby, which would then die from lack of oxygen. The mother would also be at risk from loss of blood from the site of the placenta. Routine ultrasound scans in pregnancy now pick up this condition so that the pregnancy can be monitored and a Caesarian section carried out.

What are the substances that are exchanged across this organ made up partly of tissue from one individual and partly of tissue from another? Oxygen and dissolved food substances pass from the mother to the fetus. Waste products of metabolism such as carbon dioxide and urea pass from the fetus to the maternal system for removal from the body. Some antibodies may also pass from the mother to the fetus, providing it with temporary immunity to certain diseases. The placenta also acts as a barrier, protecting the developing fetus from many microorganisms and drugs which cannot cross from the mother to the fetus. Figure 5.5.8 shows how exchange takes place across the placenta.

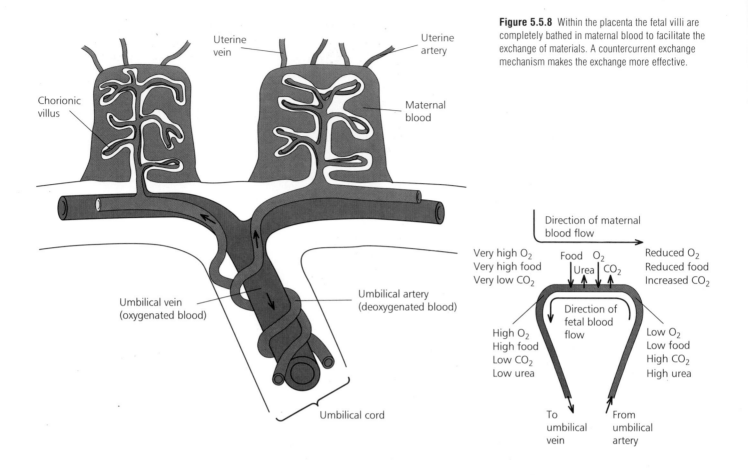

Figure 5.5.8 Within the placenta the fetal villi are completely bathed in maternal blood to facilitate the exchange of materials. A countercurrent exchange mechanism makes the exchange more effective.

Why the maternal and fetal blood are not mixed

It might be thought that the easiest way to supply the fetus with nutrients would be for maternal blood to flow through it. However, this would be disastrous for a number of reasons, detailed below.

(**1**) The maternal blood is under relatively high pressure compared with the fetal system which could damage the delicate tissues of the developing fetus.

(**2**) The maternal blood and that of the fetus may not be of the same blood group and so mixing would result in agglutination of one or the other.

(**3**) Most importantly, if the two bloods mixed the fetal blood would be recognised as foreign by the maternal blood. The maternal immune system would respond to and destroy the fetus. This is because half the genetic material in the fetus comes from the father and so the fetal cells are not identical to those of the mother. Surprisingly, the maternal system does not recognise and destroy the invading chorionic villi. They appear to be coated with special molecules which prevent the maternal blood from reacting adversely to them. The mechanism for this protection is not fully understood. When we understand how the fetus avoids rejection by the mother, then the treatment of skin grafts and organ transplants should be greatly improved if we can mimic this 'anti-rejection' message.

Multiple births

A **multiple birth** occurs when more than one fetus develops in the uterus at the same time. The most common type of multiple birth in humans is twins (two fetuses), which occur about once in every 90 births. Triplets (three fetuses) and quadruplets (four fetuses) may occur naturally but they are very rare.

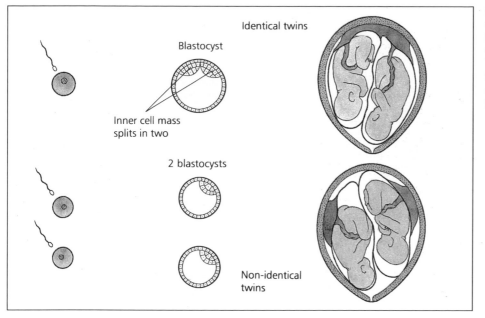

Figure 5.5.9 If twins are different sexes they are obviously the result of two fertilisations and non-identical. There is a genetic tendency to double-ovulate, which is why non-identical twins can run in families. The presence of one placenta rather than two suggests that identical twins result from the divisions of a single egg. No one is yet entirely sure what causes the formation of identical twins.

How do these relatively rare events come about? They are usually chance occurrences, happening either when more than one ovum ripens and is released at ovulation, or when a single ovum is fertilised but the zygote splits into two during the first two weeks of development. There is a genetic tendency to release more than one ovum at the fertile time of the menstrual cycle, but it can also happen randomly. The use of fertility drugs in the treatment of those with fertility problems often increases the likelihood of a multiple pregnancy by increasing the number of ova released, as does the technique known as *in vitro* fertilisation.

Twins may result from the fusing of one egg and one sperm (**identical** or **monozygotic twins**), or they may result from two quite different eggs and sperm (**non-identical** or **dizygotic**). The differences are shown in figure 5.5.9.

Hormonal control of pregnancy

As we saw in section 5.4 the female reproductive cycle is under the control of several hormones. When the cycle is interrupted by pregnancy, hormones again play a vital part. One of the most common causes of miscarriage – the spontaneous abortion of the embryo or fetus in the early stages of development – is insufficient production of hormones at the appropriate time.

Once the embryo has implanted it sends a hormonal message into the blood of the mother. This is **human chorionic gonadotrophic hormone (HCG)** and it stimulates the corpus luteum in the ovary to keep producing progesterone and so maintain the pregnancy. Some of this HCG is excreted in the urine of the mother, and this forms the basis for many of the pregnancy testing kits now available.

At the end of the first 12 weeks of pregnancy the production of HCG by the implanted embryo begins to decrease. This is a critical time for the pregnancy, because if the drop in the levels of HCG is matched by a corresponding drop in progesterone levels, then the lining of the uterus will be shed and the pregnancy lost in miscarriage. At the same time as the embryonic production of HCG falls away, the rapidly maturing placenta begins to manufacture and secrete large amounts of progesterone which carries on the role of maintaining the endometrium and inhibiting further menstrual cycles. In this way the pregnancy is maintained until the time of birth.

BIRTH

Initiation of birth

The most dangerous journey most of us will ever undertake is the short distance from the uterus of our mother to the outside world. It is a passage which, even today with advanced technologies, is fraught with difficulties and dangers for both mother and infant. At the end of the human gestation period, which lasts approximately 40 weeks, the fetus is outgrowing the uterus. The placenta becomes less efficient and space is at a premium. It is the increasingly cramped fetus rather than the mother which initiates the birth process.

Evidence for the role of the fetus in initiating birth

The importance of the fetal role in the birth process has been studied in a variety of ways, but some of the most interesting evidence came from simple observations of a natural phenomenon. Shepherds in Idaho noticed that certain pregnant sheep died with their lamb fetuses unborn but extremely large – up to two or three times the normal size. Careful observation showed that this occurred in sheep which had eaten large quantities of a certain weed (*Veratrum californicum*). When biologists investigated the weed they found that it contained a chemical which interferes with the pituitary and adrenal glands of the fetus. This in turn led to the discovery of the sequence of events which we now believe takes place to initiate and maintain birth.

The sequence of events which triggers the process of birth is believed to be as follows. The fetal pituitary secretes ACTH, which in turn stimulates the fetal adrenal glands to produce steroid compounds. These steroids affect the maternal cells of the placenta to produce substances called **prostaglandins**, which act as local hormones and cause the muscles of the uterus to contract. As the head of the fetus is pressed down onto the cervix, stretch receptors fire and send messages to the maternal brain causing the release of **oxytocin**. This also causes the uterine muscle to contract. The process is outlined in figure 5.5.10.

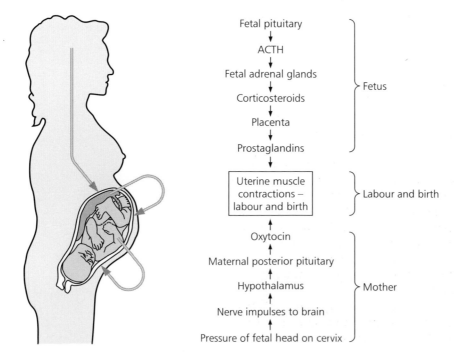

Figure 5.5.10 The fetus is believed to initiate the birth process. Hormones from the fetus, the placenta and the maternal brain all act together to bring about birth.

The stages of birth

To expel a fully developed fetus from the uterus into the outside world is no easy task. The process, known as **labour**, involves the contraction of the uterine muscles, which are the strongest in the body. The process is often lengthy and can be very painful.

There are three main stages. During the first stage of labour the uterine contractions increase in both frequency and strength. Their major function at this stage is to cause the dilation of the cervix from a gap of a few millimetres to the 10 cm needed to allow the passage of the head of the baby.

The second stage of labour occurs once the cervix is fully dilated. It involves the actual delivery of the baby. Both the involuntary muscles of the uterus and the voluntary muscles of the abdomen are used in a massive expulsive effort to push the fetus out from the uterus, through the cervix and out along the vagina.

The third and final stage of labour is the use of those same muscles to push the now redundant placenta out of the body after it has peeled away from the uterine wall.

The arrival of a healthy full-term baby is the hoped-for outcome of every planned pregnancy. It can be somewhat sobering to discover that it is not all that easy to achieve this target. From every 1000 human eggs exposed to sperm only around 310 live infants will result. Not all the eggs will be fertilised and many of those that are will be miscarried either before or after the pregnancy is recognised, due to hormone insufficiencies or abnormalities of the developing embryo/fetus. The result of this filtering is that the majority of fetuses which make it through the pregnancy to term are relatively free from abnormalities.

Changes at birth

As the fetus emerges at the moment of birth and the first breath is drawn into its lungs a new independent life begins. The transition from fetus to baby involves some very complex changes. Whilst in the uterus the fetus is supplied with all the food and oxygen it needs through the placenta. Waste products are removed in the same way. The body is cushioned and supported by the amniotic fluid. But after birth all these support systems are removed and the infant must perform their functions itself if it is to survive. What does this transition entail?

The heart and lungs

Some of the most critical initial changes occur in the cardiovascular system and lungs. In the uterus the lungs are filled with fluid and receive very little blood as they play no part in oxygenation. After birth they must immediately inflate with the first breath – this is made possible by the presence of **lung surfactant**, described in section 2.4. Changes must also take place in the circulation of the blood through the heart. The fetal heart differs from that of an adult in two main ways. The two sides of the fetal heart are not completely separate. They are joined by the **foramen ovale** and the blood in the two sides can mix relatively freely. Also, the pulmonary artery which leaves the right-hand side of the heart carrying blood to the lungs is joined to the main aorta by a small vessel called the **ductus arteriosus**. Up to 90% of the blood which enters the right atrium goes not to the lungs but into the aorta through the ductus, because the resistance of the fluid-filled lungs is much higher than that of the rest of the system. From the aorta blood flows to the placenta in the umbilical artery, where it will be oxygenated.

At birth this circulation must change to allow the blood to gain oxygen from the lungs rather than from the placenta. In the first few hours after birth the ductus arteriosus narrows and a flap descends to cover the foramen ovale.

Within minutes of birth 90% of the blood leaving the right atrium travels to the lungs, which have a much lower resistance now that they are inflated. This enables oxygenation of the blood to occur. Within a few weeks of birth both the ductus and the foramen have been permanently closed. If they do not close up properly, abnormal heart sounds and problems associated with ineffective oxygenation of the blood will be noticed. An incompletely closed foramen ovale, commonly called a 'hole in the heart', can be a life-threatening condition, but many people live long and healthy lives with a slight gap in their heart septum. The changes in the heart between fetus and child can be seen in figure 5.5.11.

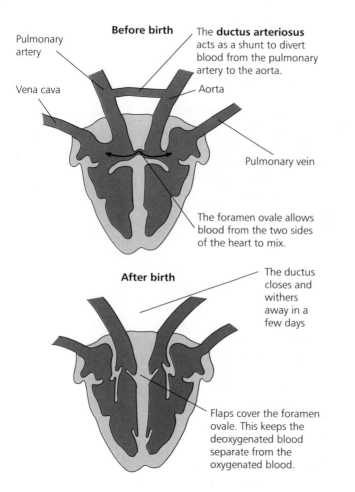

Before birth

Pulmonary artery

Vena cava

The **ductus arteriosus** acts as a shunt to divert blood from the pulmonary artery to the aorta.

Aorta

Pulmonary vein

The foramen ovale allows blood from the two sides of the heart to mix.

After birth

The ductus closes and withers away in a few days

Flaps cover the foramen ovale. This keeps the deoxygenated blood separate from the oxygenated blood.

Figure 5.5.11 The structure of the heart changes at birth.

The liver

The other major organ system which has a greatly increased workload after birth is the liver, which has to deal with the pigment bilirubin, one of the main breakdown products of fetal haemoglobin. The level of bilirubin in the blood increases to levels where it causes a yellowish tinge in the tissues of the newborn infant known as **physiological jaundice**. This is so common in newborn infants as to be normal, but the yellow colour may become very pronounced if the liver is slow to start bilirubin breakdown. If the bilirubin levels get too high, brain damage can occur. To prevent this babies are monitored carefully and treated by exposure to broad spectrum light if bilirubin levels rise too high. The light breaks down the bilirubin into substances which can be managed by the infant system until the liver becomes effective at breaking down haemoglobin.

Fetal development outside the uterus

The length of a full-term human pregnancy is around 40 weeks or 266 days, calculated from the first day of the last menstrual period a woman has before conception. Not all babies remain in the uterus for as long as this. Many pregnancies end a few days or even two or three weeks before the estimated date of delivery with the arrival of a normal-sized and healthy infant which is quite ready for independent life. Some babies, however, are born weeks or even months before they are due, and for these babies survival – even with the high levels of medical care now available – can be a struggle. Many of their body systems are still very immature, and maturation processes which should be occurring during uterine life have to happen whilst carrying out their functions in the baby.

Some of the major difficulties of premature babies and the ways in which they may be overcome are summarised below.

- Lack of a stable environment – the uterus provides a constant environment for fetal development. An incubator attempts to mimic this stability. The wires and tubes that are attached to the baby ensure that monitoring can take place with minimal disturbance and the temperature and humidity of the air is carefully controlled and kept constant.
- The ductus arteriosus of the heart may not close so that the blood does not circulate properly to the lungs for oxygenation. This closure can be carried out surgically to allow the circulatory system to function properly.
- Lung surfactant may not be present in the lungs at birth and may not be synthesised for some time after the birth. This makes ventilating the lungs very difficult (see section 2.4). Surfactant has been synthetically produced and can be sprayed into the lungs to enable them to be inflated with minimal damage.
- The respiratory centres in the brain are not mature and babies may forget to breathe or respond inappropriately to high or low levels of carbon dioxide. Air containing higher than average levels of oxygen may be supplied or if necessary the infant may be artificially ventilated.
- Feeding may be difficult – the infant may be too weak to suckle, and premature babies cannot coordinate the sucking, swallowing and breathing reflexes. Food can be delivered by tubes directly to the gut or, if the gut is very immature or the infant very ill, nutrients may be passed straight into the blood.
- The normal adult temperature control mechanisms are absent – this is overcome by careful monitoring of the temperature of the baby and provision of a constant suitable external temperature. In addition infants may be wrapped in materials such as cling film and foil to reduce heat loss, and woolly hats placed on their heads (up to a third of the body heat loss is through the head).
- Excess water loss can occur as the skin is not properly waterproofed – this can be overcome by humidifying the air around the infant and by wrapping it in cling-film like materials which trap a moist layer of air (see figure 4.7.8, page 317).

In spite of these and many other problems which have to be overcome, many premature babies survive the ordeal of continuing fetal development postnatally, and go on to live healthy lives unaffected by their somewhat unorthodox start.

CARE OF THE INFANT

Reproduction represents a considerable investment of biological resources, particularly for the female of the species. Many aquatic organisms which release their gametes directly into the water provide no care for their offspring. In other animals the parental involvement in the development of the offspring is greater, and the level of parental care once the offspring have hatched or been born is generally greater too. There are examples of fish, amphibians and reptiles which take great care of their young, and birds are well known for their extensive parenting, both in incubating eggs and feeding the offspring after hatching. Mammals show an even greater level of infant care.

Producing milk

Even the most prolific mammalian breeders produce a relatively small number of offspring. However, the mammals have evolved a method of providing almost guaranteed food for their offspring. This maximises their chances of survival and so minimises the wastage of resources. This evolutionary breakthrough is the production by specialised sweat glands of a protein-rich liquid to feed the offspring. The modified glands are the **mammary glands**, and it is the possession of these that provides the mammals with their name. The protein-rich liquid is more commonly known as **milk**.

In humans breast feeding provides infants with the ideal food – the balance of fats, proteins, carbohydrates, minerals and vitamins is exactly appropriate to the needs of a growing human baby (see figure 5.5.12). The balance of the different components varies with how long the baby has been suckling, the weather and the age of the baby to ensure that all the infant needs are met. Until the age of around six months, no additional feeding is necessary. In the developed world many people begin additional feeding earlier than this, though physiological rejection mechanisms in the infant have to be overcome before this can be done. As well as providing a perfectly balanced diet, the major advantage of breast milk is that it provides the infant with immunity to those diseases to which the mother has acquired antibodies. This protection can be very important in maintaining the health of the infant.

The World Health Organisation has produced some very strongly worded directives advising that all babies should be breast fed if at all possible. Unfortunately in the developed world there is often great pressure for women to bottle feed, using modified milk from cows to provide the nutritional needs of their babies. Whilst this is frequently done successfully, the immunological advantages of breast milk cannot be mimicked. Of even greater concern is the spread of bottle feeding to the developing world. Here the water used to make

Figure 5.5.12 Milk production from the mammary glands is almost unaffected by the levels of nutrition of the mother, thus ensuring that the infant is adequately fed under most circumstances. Not content with human milk, we have attempted to modify the milk of cows to feed our offspring, in spite of the fact that the balance of protein, fats, minerals and vitamins in the two milks are very different as they are tailored to different needs.

up infant feeds may be infected by a variety of diseases, the feed is often not concentrated enough as the milk powder is expensive and the lack of immunological protection is particularly important.

> ### Different milks for different mammals
>
> The milk mammals produce for their young is uniquely adapted to the particular demands of the growth patterns of the species. For example, newly born whales and seals have to cope with living in very cold conditions. They need to develop a thick layer of blubber as rapidly as possible and therefore they need a very high calorie intake. The milk of female whales and seals is extremely high in fats – it is very, very thick and creamy and has the necessary high calorie content.
>
> The young of grazing herbivores such as antelope and cows need to grow very rapidly to keep up with other members of the herd and to escape predators. The milk of these animals is relatively low in fats but high in the protein needed to produce new tissues.
>
> Human milk is relatively low in both protein and fat. Human babies need to grow quite slowly to allow their brains to develop and assimilate the enormous amount of complex behaviour that allows human societies to function. Thus the milk of each mammal is tailored to meet the requirements of its growing offspring.

CONTRACEPTION

As we have seen, humans appear to be the only mammal in which copulation and reproduction are no longer synonymous with each other. The human sexual drive or **libido** is not seasonal. Whilst tiredness, ill-health and many other factors may influence the frequency of copulation between a couple, in theory intercourse may occur at any time throughout the year. However, many people do not wish to risk pregnancy whenever they have sex, and for centuries various methods of **contraception** (ways of preventing conception) have been suggested. These have included vinegar-soaked sponges placed in the vagina before intercourse, mixtures of camel dung and various herbs similarly placed, and reusable condoms made from animal intestines.

At the present time methods of contraception fall into several categories. **Natural methods** are based on an understanding of the menstrual cycle and accurate predictions of the point of ovulation. Their major advantage is that they are accepted even by those religious groups which condemn other methods of controlling family size. **Physical** or **barrier methods** of contraception involve physical barriers which prevent the meeting of the ovum and the spermatozoa. **Chemical methods** involve chemicals which work in a variety of ways to do the same thing. **Sterilisation** is the ultimate form of contraception. By cutting or tying the tubes along which egg or sperm travel, conception is rendered almost impossible. This has the additional benefit of not having to remember to use contraception, the major cause of failure in the other methods. These methods of contraception are summarised in table 5.5.1.

Contraception is largely used by people in the developed world. A combination of factors, including a desire to have a large family for fear of children dying and therefore lack of support in old age, lack of medical supervision, expense and religious condemnation, mean that in the developing world contraception does not as yet play a major role.

5

Method of contraception	How it works	Advantages/disadvantages
Rhythm (natural) **method** 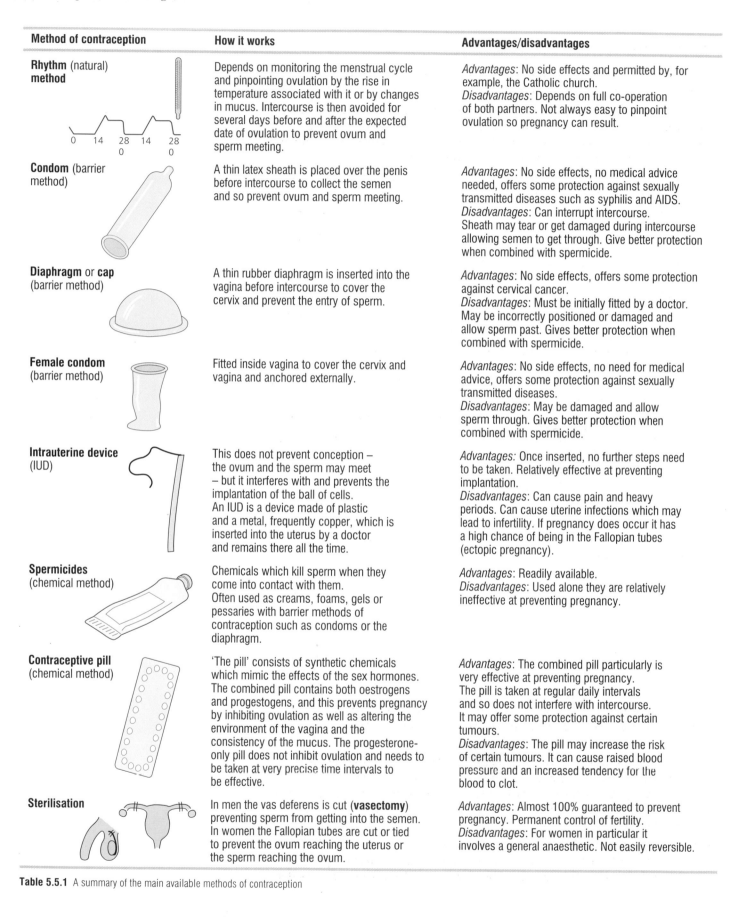	Depends on monitoring the menstrual cycle and pinpointing ovulation by the rise in temperature associated with it or by changes in mucus. Intercourse is then avoided for several days before and after the expected date of ovulation to prevent ovum and sperm meeting.	*Advantages*: No side effects and permitted by, for example, the Catholic church. *Disadvantages*: Depends on full co-operation of both partners. Not always easy to pinpoint ovulation so pregnancy can result.
Condom (barrier method)	A thin latex sheath is placed over the penis before intercourse to collect the semen and so prevent ovum and sperm meeting.	*Advantages*: No side effects, no medical advice needed, offers some protection against sexually transmitted diseases such as syphilis and AIDS. *Disadvantages*: Can interrupt intercourse. Sheath may tear or get damaged during intercourse allowing semen to get through. Give better protection when combined with spermicide.
Diaphragm or **cap** (barrier method)	A thin rubber diaphragm is inserted into the vagina before intercourse to cover the cervix and prevent the entry of sperm.	*Advantages*: No side effects, offers some protection against cervical cancer. *Disadvantages*: Must be initially fitted by a doctor. May be incorrectly positioned or damaged and allow sperm past. Gives better protection when combined with spermicide.
Female condom (barrier method)	Fitted inside vagina to cover the cervix and vagina and anchored externally.	*Advantages*: No side effects, no need for medical advice, offers some protection against sexually transmitted diseases. *Disadvantages*: May be damaged and allow sperm through. Gives better protection when combined with spermicide.
Intrauterine device (IUD)	This does not prevent conception – the ovum and the sperm may meet – but it interferes with and prevents the implantation of the ball of cells. An IUD is a device made of plastic and a metal, frequently copper, which is inserted into the uterus by a doctor and remains there all the time.	*Advantages:* Once inserted, no further steps need to be taken. Relatively effective at preventing implantation. *Disadvantages*: Can cause pain and heavy periods. Can cause uterine infections which may lead to infertility. If pregnancy does occur it has a high chance of being in the Fallopian tubes (ectopic pregnancy).
Spermicides (chemical method)	Chemicals which kill sperm when they come into contact with them. Often used as creams, foams, gels or pessaries with barrier methods of contraception such as condoms or the diaphragm.	*Advantages*: Readily available. *Disadvantages*: Used alone they are relatively ineffective at preventing pregnancy.
Contraceptive pill (chemical method)	'The pill' consists of synthetic chemicals which mimic the effects of the sex hormones. The combined pill contains both oestrogens and progestogens, and this prevents pregnancy by inhibiting ovulation as well as altering the environment of the vagina and the consistency of the mucus. The progesterone-only pill does not inhibit ovulation and needs to be taken at very precise time intervals to be effective.	*Advantages*: The combined pill particularly is very effective at preventing pregnancy. The pill is taken at regular daily intervals and so does not interfere with intercourse. It may offer some protection against certain tumours. *Disadvantages*: The pill may increase the risk of certain tumours. It can cause raised blood pressure and an increased tendency for the blood to clot.
Sterilisation	In men the vas deferens is cut (**vasectomy**) preventing sperm from getting into the semen. In women the Fallopian tubes are cut or tied to prevent the ovum reaching the uterus or the sperm reaching the ovum.	*Advantages*: Almost 100% guaranteed to prevent pregnancy. Permanent control of fertility. *Disadvantages*: For women in particular it involves a general anaesthetic. Not easily reversible.

Table 5.5.1 A summary of the main available methods of contraception

SUMMARY

- In most mammals except humans, **oestrus** is the period of sexual receptiveness in the female, which occurs just before ovulation.

- The spermatozoa are deposited high in the vagina by **ejaculation** during **copulation**. They are helped up through the cervix and uterus to the Fallopian tubes by contractions of the female reproductive tract. Sperm movements keep them in suspension in the semen.

- Many spermatozoa cluster round the ovum and as their heads touch its surface the **acrosome reaction** is triggered. Enzymes are released which digest the follicle cells and zona pellucida.

- One sperm touches the oocyte membrane and the oocyte undergoes its second meiotic division, producing a haploid egg nucleus to fuse with the haploid male nucleus. The oocyte membrane becomes positive on the inside to prevent any more sperm entering. A **fertilisation membrane** forms to repel further sperm as the electrical charge returns to normal.

- The head of the sperm enters the oocyte and its chromosomes fuse with the haploid egg nucleus to form the zygote.

- The zygote undergoes **cleavage**, repeated mitotic divisions to form a **blastocyst** or hollow ball of cells. During **implantation** this becomes embedded in the endometrium and **trophoblastic villi** grow out from the outer layer or **trophoblast**.

- **Gastrulation** leads to a three-layered embryo, the **gastrula**. The outer **ectoderm**, the middle **mesoderm** and the inner **endoderm** will each differentiate to produce particular types of organ or tissue.

- From around 12 weeks the human embryo is called the **fetus** and its major organ systems are completed.

- The **extraembryonic membranes** (**amnion** and **chorion**) fold around the embryo to form the **amniotic cavity**, filled with **amniotic fluid**. The chorion forms the **placenta** and the allantois the **umbilical cord** which connects the fetus to the placenta.

- To make the placenta the chorion forms many **chorionic villi** which protrude into sinuses in the endometrium and are bathed in maternal blood. Oxygen, food and some antibodies pass from mother to fetus and carbon dioxide and urea from fetus to mother. The placenta is a barrier to some microorganisms and drugs.

- **Identical** or **monozygotic twins** result from a single egg and sperm, while **non-identical** or **dizygotic twins** result from two different eggs and sperms.

- Pregnancy is first maintained by progesterone from the corpus luteum under the influence of **human chorionic gonadotrophin** from the embryo. After about 12 weeks the placenta takes over secreting progesterone to maintain the pregnancy.

- Birth is initiated by fetal ACTH which stimulates the fetal adrenals to produce steroids. These cause the maternal placenta to secrete **prostaglandins** which cause contractions of the uterus. Stretch receptors in the cervix stimulate the brain to secrete **oxytocin** which also stimulates contractions.

- In the first stage of labour contractions increase in strength and frequency and the cervix becomes dilated. In the second stage the baby is pushed out through the cervix and vagina. In the third stage the placenta is expelled.

5

- The baby's lungs inflate immediately after birth aided by **lung surfactant**. The sides of the fetal heart are connected by the **foramen ovale** which closes over after birth. The fetal pulmonary artery is joined to the aorta by the **ductus arteriosus** which bypasses the lungs, and this narrows after birth.

- The fetal liver starts breaking down fetal haemoglobin, increasing the level of bilirubin and causing **physiological jaundice**.

- The **mammary glands** of the mammals produce protein-rich **milk** to feed the young, which also confers maternal antibodies.

- Conception may be prevented by various **contraceptive** measures. These may be:

  natural methods – the rhythm method

  barrier methods – the condom, diaphragm and female condom

  the **IUD** which prevents implantation

  chemical methods – spermicides which are best used with barrier methods, and the contraceptive pill which prevents ovulation

  sterilisation – irreversible surgical techniques which involve cutting the Fallopian tubes or vas deferens.

QUESTIONS

1 a Describe the journey of the human ovum from the ovary to the point of fertilisation.
  b How does the sperm arrive in the Fallopian tube?
  c Describe the events of fertilisation.

2 Produce a leaflet suitable for the antenatal clinic of your local hospital explaining the main stages of fetal development, the role of the placenta in supporting the fetus and some of the factors which may have an adverse effect on fetal development.

3 a Before birth, a fetus has all its requirements supplied by the maternal system. Once a baby is born it has to immediately adapt to independent life. Describe the changes which occur in a baby in the first few minutes after birth.
  b Discuss the advantages and disadvantages of different feeding methods for babies in the developed countries and the developing world.

5.6 Genetics

The fruit, vegetables and farm animals of several hundred years ago were very different from those we see today, as figure 5.6.1 illustrates. These differences have been brought about by a process of **selective breeding** – plants and animals with the desired features are bred to give offspring which also exhibit them.

For a long time these crosses were carried out without any real understanding of the mechanisms behind the breeding process. But over the last hundred years or so the underlying mechanisms of inheritance have been worked out and understood, and the science of **genetics** has emerged.

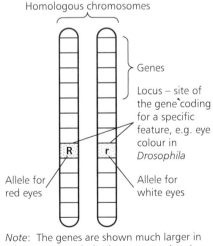

Ornamental cabbages add an extra dimension to the herbaceous border.

Figure 5.6.1 Selective breeding has taken the cabbage plant of the eighteenth century in a variety of directions to suit a variety of demands. The mechanisms by which such changes come about are now well understood.

A nineteenth century cabbage was a relatively unspecialised plant.

The familiar large round cabbage from the supermarket shelf fulfils the requirements of the British shopper.

THE BASIS OF INHERITANCE

The physical and chemical characteristics which make up the appearance of any organism are known as its **phenotype** – for example, the shape of a cabbage, the scent of a flower and the colour of hair. This phenotype is the result of the genetic information handed on from parents to their offspring (the **genotype**) along with the effect of the environment in which an organism lives. The shape of a cabbage will depend on the levels of soil nutrients and sunlight as well as on the genetic make-up of the individual. For many characters, the phenotypic variation between individuals results largely from genetic differences between individuals. These genetic differences are due in part to the shuffling of genes which occurs during the process of meiosis, and in part to the inheritance of genes from two different individuals in sexual reproduction. How does the process of inheritance work?

As we have seen in sections 5.1 and 5.2, the cells of any individual organism contain a species-characteristic number of chromosomes. These occur as matching **homologous pairs**. One of each pair is inherited from the female parent and one from the male. Along each chromosome are thousands of **genes**, and the genes coding for particular characteristics or **traits** are always found in the same position or **locus** on a particular pair of chromosomes. A gene is a segment of DNA coding for a specific RNA molecule to give a particular characteristic. A gene may be one of two or more different **alleles**, each allele producing different alternative forms of the characteristic. An allele is a particular sequence of bases which may or may not result in a functional protein. For example, the fruit fly *Drosophila melanogaster* is very commonly used as a laboratory animal for genetic experiments. The gene for eye colour is always found at the same locus on the chromosomes, but either the allele for red eyes or the allele for white eyes (different sequences of base pairs) may be present at the locus. This is shown in figure 5.6.2.

Homologous chromosomes

Genes

Locus – site of the gene coding for a specific feature, e.g. eye colour in *Drosophila*

Allele for red eyes

Allele for white eyes

Note: The genes are shown much larger in relation to the whole chromosome than is actually the case.

Figure 5.6.2 In sexual reproduction it is the alleles passed on which determine the genotype and hence eventually the phenotype of the offspring.

GENETIC CROSSES

Monohybrid crosses

If both of the alleles coding for a particular characteristic on a pair of homologous chromosomes are identical, then the organism is **homozygous** for that characteristic – it is a **homozygote**. Homozygotes are referred to as **true breeding**, because if two similar homozygotes are crossed all the offspring will be the same. The offspring of two homozygous parents are called the F_1 (**first filial generation**). If those offspring are crossed the next generation, the F_2 (**second filial generation**), will also have the same genotype and phenotype and this will continue on through the generations.

If the two alleles coding for a characteristic on a pair of homologous chromosomes are different, the organism is **heterozygous** for that feature and is referred to as a **heterozygote**.

Some alleles are described as **dominant**. This means that their effect is expressed even in the presence of another, **recessive** allele. Thus if one allele is dominant over another the individuals will show the dominant feature whether they are homozygotes or heterozygotes, as figure 5.6.3 shows. The dominant allele might result in the production of a particular protein – for example, a pigment molecule or an enzyme – whilst recessive alleles might result in a lack of that protein or in a different, non-functional protein. Recessive alleles are only expressed in the phenotype when they are present in the homozygous form. Dominant genes are usually represented by a capital letter, recessive genes by the lower case version of the same letter. When one genetic trait is considered at a time the cross is referred to as a **monohybrid cross**.

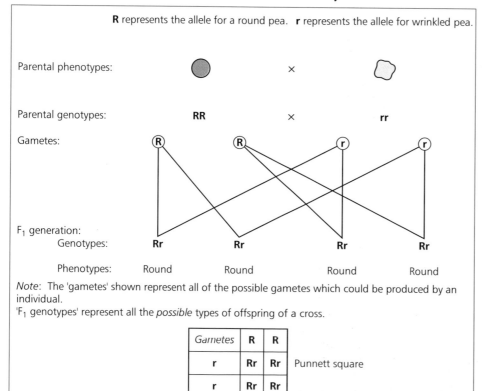

R represents the allele for a round pea. **r** represents the allele for wrinkled pea.

Parental phenotypes:

Parental genotypes: **RR**   ×   **rr**

Gametes: Ⓡ   Ⓡ   ⓡ   ⓡ

F_1 generation:
Genotypes: **Rr**   **Rr**   **Rr**   **Rr**

Phenotypes: Round   Round   Round   Round

Note: The 'gametes' shown represent all of the possible gametes which could be produced by an individual.
'F_1 genotypes' represent all the *possible* types of offspring of a cross.

Gametes	R	R
r	Rr	Rr
r	Rr	Rr

Punnett square

Figure 5.6.3 This representation of a cross between a pea plant producing round peas with a plant producing wrinkled peas shows how to present a typical genetic cross. The Punnett square is not always needed in a monohybrid cross such as this, but becomes very important later when dihybrid crosses are considered. The cross also demonstrates the typical heterozygous genotype but dominant phenotype in the offspring from a cross between a homozygous dominant individual and a homozygous recessive individual.

Heterozygotes are not true breeding. If two heterozygotes are crossed then the offspring may include homozygous dominant, homozygous recessive and heterozygous types, as figure 5.6.4 shows.

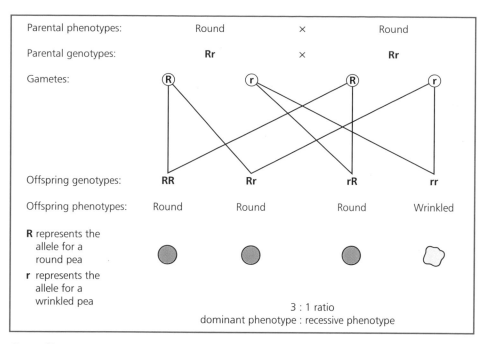

Parental phenotypes: Round   ×   Round

Parental genotypes: **Rr**   ×   **Rr**

Gametes: R   r   R   r

Offspring genotypes: **RR**   **Rr**   **rR**   **rr**

Offspring phenotypes: Round   Round   Round   Wrinkled

R represents the allele for a round pea

r represents the allele for a wrinkled pea

3 : 1 ratio
dominant phenotype : recessive phenotype

Figure 5.6.4 The phenotypes of heterozygotes such as these parental peas show no evidence of their underlying genotype. However, when they are crossed, the occurrence of homozygous recessive forms reveals the hidden alleles.

Sampling errors

Considering genetic crosses in theory gives rise to *predicted ratios* for the proportions of different types of offspring. These ratios are seen in real-life genetic experiments, but they are unlikely to correspond exactly to the predictions. The ratios are only approximate due to a variety of factors, for example the death of some offspring before they can be sampled, or inefficient sampling techniques – for example, it is very easy to let a few *Drosophila* escape. Also, chance plays a large role in reproduction – the joining of particular gametes is a completely random affair. Thus **sampling error** must be taken into account when a cross is considered, and the smaller the sample, the larger the sampling error. Therefore the results of a genetical experiment with thousands of offspring will be much more predictable than those of an experiment with under 100 results.

The ratios of genotypes and phenotypes expected from genetic crosses are known as **Mendelian ratios** after Gregor Mendel, the Austrian monk who first developed the study of genetics (described later in this section).

Test crosses

As can be seen from figures 5.6.3 and 5.6.4, individuals which are homozygous dominant or heterozygous will have identical phenotypes. In practical terms this can present all sorts of difficulties for breeders of plants and animals. A breeder needs to know that stock will breed true – in other words, that it is homozygous for the desired feature. If the feature is inherited through recessive genes then the genotype can be seen from the phenotype. However, if the required feature is inherited through a dominant gene the physical appearance does not show whether the organism is homozygous or heterozygous. To find out the genotype of an individual showing the phenotype corresponding to a dominant allele, it is crossed with a homozygous recessive individual. The recessive genes have no effect on the phenotype of the offspring unless they are in the homozygous state, so this type of cross can reveal the parental genotype. If there are offspring with the recessive characteristic the parent must have been heterozygous, as figure 5.6.5 shows. Such a cross is known as a **test cross**.

5

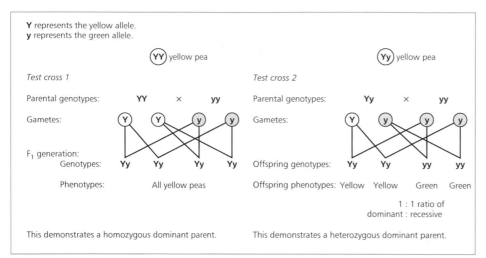

Y represents the yellow allele.
y represents the green allele.

(YY) yellow pea

(Yy) yellow pea

Test cross 1

Parental genotypes: YY × yy

Gametes: Y Y y y

F₁ generation:
Genotypes: Yy Yy Yy Yy

Phenotypes: All yellow peas

This demonstrates a homozygous dominant parent.

Test cross 2

Parental genotypes: Yy × yy

Gametes: Y y y y

Offspring genotypes: Yy Yy yy yy

Offspring phenotypes: Yellow Yellow Green Green

1 : 1 ratio of
dominant : recessive

This demonstrates a heterozygous dominant parent.

Figure 5.6.5 Test crosses are carried out to discover the unknown genotype of a parent. The colour of the pea seed is inherited as a dominant allele **Y** for yellow seeds or a recessive allele **y** for green seeds. Thus a plant producing yellow seeds could have the genotype **YY** or **Yy**. The phenotypic ratios of the progeny of the test cross reveal whether the yellow-seed producing parent was homozygous or heterozygous.

Codominance

When a pure-breeding red-flowered antirrhinum is crossed with a pure-breeding white-flowered antirrhinum, we might expect either all red- or all white-flowered offspring, depending upon which allele is dominant. In fact, pink flowers are the result – a 'mixture' of the two phenotypes. This is due to **codominance** between the alleles. Both alleles result in the production of functional proteins that interact in a specific way. Neither is dominant over the other, hence the term codominance. In antirrhinum the red-flower allele and the white-flower allele are codominant and so heterozygotes display pink flowers. This type of interaction between alleles is sometimes referred to as **incomplete dominance** as both alleles are represented, yet neither is completely dominant to the other.

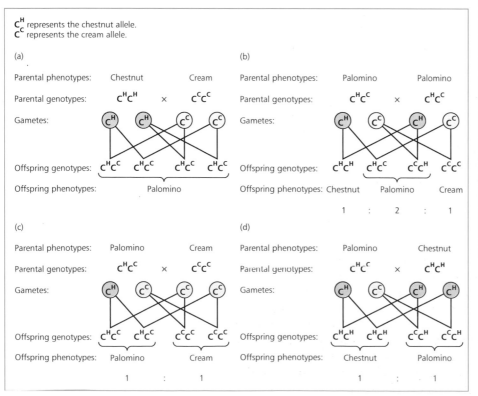

C^H represents the chestnut allele.
C^C represents the cream allele.

(a)

Parental phenotypes: Chestnut Cream

Parental genotypes: $C^H C^H$ × $C^C C^C$

Gametes: C^H C^H C^C C^C

Offspring genotypes: $C^H C^C$ $C^H C^C$ $C^H C^C$ $C^H C^C$

Offspring phenotypes: Palomino

(b)

Parental phenotypes: Palomino Palomino

Parental genotypes: $C^H C^C$ × $C^H C^C$

Gametes: C^H C^C C^H C^C

Offspring genotypes: $C^H C^H$ $C^H C^C$ $C^C C^H$ $C^C C^C$

Offspring phenotypes: Chestnut Palomino Cream

1 : 2 : 1

(c)

Parental phenotypes: Palomino Cream

Parental genotypes: $C^H C^C$ × $C^C C^C$

Gametes: C^H C^C C^C C^C

Offspring genotypes: $C^H C^C$ $C^H C^C$ $C^C C^C$ $C^C C^C$

Offspring phenotypes: Palomino Cream

1 : 1

(d)

Parental genotypes: Palomino Chestnut

Parental genotypes: $C^H C^C$ × $C^H C^H$

Gametes: C^H C^C C^H C^H

Offspring genotypes: $C^H C^H$ $C^H C^H$ $C^C C^H$ $C^C C^H$

Offspring phenotypes: Chestnut Palomino

1 : 1

Figure 5.6.6 The photographs show, from the top, a chestnut, a cream and a palamino horse. The attractive golden colour with the white mane and tail which characterises the palomino horse is much sought after by some breeders. However, it is a difficult colour to obtain because it does not breed true. These genetic diagrams show the results of crosses between a) a chestnut horse and a cream horse, b) two palomino horses, c) a palomino horse and a cream horse, and d) a palomino horse and a chestnut horse.

When true-breeding cream horses are bred, cream foals result. When true-breeding chestnut horses are bred, chestnut foals result. But the result of a cross between a cream horse and a chestnut horse is a palomino – a horse with a very distinctive gold colouring and a white mane and tail. This again is due to codominance between the alleles and a specific interaction between the resulting proteins.

The example of the palomino horses is unusual – a heterozygote for codominant alleles usually shows a condition much closer to a simple mixture of the two codominant forms rather than the very distinctive coat colour shown in figure 5.6.6.

Both the human ABO blood groups (met in section 1.6) and the MN blood groups are examples of inheritance involving codominant alleles. In the ABO system there are three possible alleles – **A**, **B** and **O**. **A** and **B** are both dominant to **O**, but are codominant to each other so an individual with the genotype **AB** has the blood group AB. The MN blood group system is another way of tissue typing human blood, and just like the ABO system is dependent on the presence or absence of antigens on the surface of the red blood cells. In the MN system, homozygous **MM** individuals have blood group M. Homozygous **NN** individuals have blood group N. Heterozygotes carrying **MN** have blood group MN. Codominance may be considered as a possible explanation of the more unexpected phenotypic results of some genetic crosses.

Experimental species

To carry out genetic experiments, organisms are needed that are relatively easy and cheap to raise. They need to have short life cycles so that the results of crosses and/or mutations can be seen quickly. They also need to produce large numbers of offspring so that the results of any crosses can be checked statistically against the predicted ratios. For example, an expected 3:1 ratio of phenotypes in the offspring of a cross is unlikely to show itself if only four offspring are produced. With 400 offspring the likelihood of seeing the expected ratio is increased and 4000 offspring would be better still.

Commonly used organisms include the fruit fly *Drosophila melanogaster*, pea plants, fungi such as *Aspergillus* and a variety of bacteria. Specially bred 'fast plants' are now being used increasingly, particularly in schools and colleges. These are members of, for example, the cabbage family which have been bred to germinate, mature, flower and set seed in a matter of weeks. This means that they can be used in much the same way as *Drosophila*, without the problems of specimens getting stuck in the nutrient medium, dying or escaping!

Multiple alleles

Most of the traits we have considered so far are inherited as genes with a pair of possible alleles. However, this is not always the case. Sometimes there will be several possible alleles (**multiple alleles**) – even as many as 100 – all resulting in a slightly different phenotype for the same trait, and with varying degrees of dominance over each other. Figure 5.6.7 shows an example. Of course, however many alleles there are, any one diploid individual will possess only two of them.

The human ABO blood group is determined by three alleles. I^O is recessive, and a homozygote is blood group O. I^A and I^B are both dominant to I^O, and are codominant to each other. Thus a genotype $I^A I^B$ gives an AB blood group,

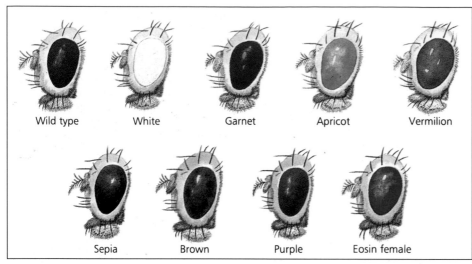

Wild type   White   Garnet   Apricot   Vermilion

Sepia   Brown   Purple   Eosin female

Figure 5.6.7 The normal *Drosophila* eye is red, but a whole range of mutant alleles can produce a great variety of other eye colours. These multiple alleles are in fact sex linked. This is explained later in the section.

whilst either I^AI^A or I^AI^O gives blood group A and I^BI^B or I^BI^O gives blood group B. Figure 5.6.8 shows some genetic crosses involving human blood groups.

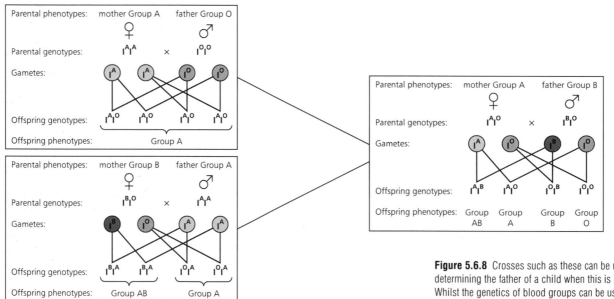

Figure 5.6.8 Crosses such as these can be used in determining the father of a child when this is in doubt. Whilst the genetics of blood groups can be used to show that a certain man is *not* the father of a child, it can never be used to prove that someone definitely *is* the father. Whilst possessing a particular blood group may make him a possible father, several million other men will also have the same blood group and so are, in theory at least, equally likely to be the father!

Lethal genes

Manx cats originate on the Isle of Man. They are well known for the physical peculiarity of lacking a tail, and they are very difficult to breed. When a Manx cat is mated to a normal long-tailed cat approximately half the offspring will be Manx and half will have normal tails. This indicates that the Manx cat is heterozygous for a dominant mutant gene which causes the lack of tail. However, if two Manx cats are mated the expected Mendelian ratio for a heterozygote cross (3 dominant phenotypes to each 1 recessive) does not occur. The ratio is always 2:1. Why do Mendelian ratios not apply to Manx cats? The answer is that the homozygous form of the mutated Manx gene is **lethal**. Kitten embryos which inherit it fail to develop the entire back end of the body and usually die and are reabsorbed before birth. This means that one quarter of the expected progeny of the cross do not appear in the actual offspring, giving the distorted 2:1 ratio, as shown in figure 5.6.9.

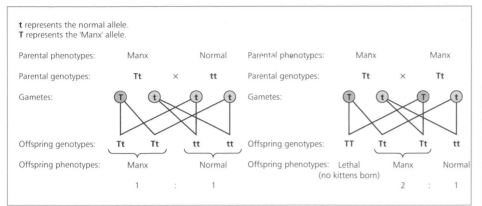

t represents the normal allele.
T represents the 'Manx' allele.

Parental phenotypes:	Manx		Normal	Parental phenotypes:	Manx		Manx
Parental genotypes:	**Tt**	×	**tt**	Parental genotypes:	**Tt**	×	**Tt**
Gametes:	(T) (t)		(t) (t)	Gametes:	(T) (t)		(T) (t)
Offspring genotypes:	**Tt** **Tt**		**tt** **tt**	Offspring genotypes:	**TT**	**Tt** **Tt**	**tt**
Offspring phenotypes:	Manx		Normal	Offspring phenotypes:	Lethal (no kittens born)	Manx	Normal
	1	:	1			2	: 1

Figure 5.6.9 The gene which causes the phenotype of the Manx cat is dominant. However, the lethal aspect of the gene acts as if recessive – when it is present in the homozygous form the kittens do not survive. It is an example of a single gene affecting several traits. True-breeding stock of Manx cats has never been achieved.

Lethal genes usually occur as the result of a chance mutation. If they persist in a wild population they must be recessive for the lethal characteristic. This would enable them to be passed on in the heterozygous form, only causing problems when two heterozygotes breed giving rise to a homozygote which would then die as a result. A dominant lethal mutation would usually be wiped out instantly as all the offspring which inherited the mutation would die before reproducing. Most lethal genes take effect before the embryo is fully developed or in early life, although exceptions to this are seen where the lethal genetic disorder only manifests itself at a stage when reproduction has already occurred and the gene has been passed on.

The remarkable physical differences between these two cats from the same litter of kittens are due to the difference in a single gene.

A brief history of genetics

For centuries the theories developed to explain the inheritance by offspring of characteristics from their parents made the same fundamental error. The assumption was made that the characteristics of the parents in some way blended or fused so that the distinct characteristics of each were lost. Followed to its logical conclusion, this theory means that the offspring from mating a pure-breeding black dog with a pure-breeding white dog would always be grey. Not only is this rarely the case, but even if grey puppies do result they may produce black or white offspring at a later date. The birth of Gregor Mendel in 1822 was to herald the end of these ideas.

Mendel became a monk at the Augustinian monastery at Brunn, and here he formulated his ideas about the particulate nature of the hereditary material – what we now call genes. He was educated in maths, physics and statistics as well as botany, which may well explain his rigorous approach and analysis of his results. His abbot supported his research and had a large greenhouse built to grow the peas with which he worked. Much of his work was carried out using the features of peas shown in figure 5.6.10.

In 1866 Mendel presented his results and the two fundamental laws of heredity he had deduced from them. However, as has often been the case with great ideas, he was ahead of his time and the work was poorly received and almost ignored. Bear in mind that no one knew of the existence of chromosomes at this time, let alone genes, so Mendel's ideas were presented to a scientific community with no framework ready to receive them. Along with this, his statistical work was strange to many biologists of the time, and the full impact of his work was not recognised until 16 years after his death.

Anther

Stigma

Both the male and female parts of the pea flower are held within a hood-like petal so that self fertilisation frequently occurs. Mendel opened the bud of one flower before the pollen matured and fertilised the stigma with pollen from another chosen flower. In this way he could control the cross.

Mendel used seven clearly differentiated, pure-breeding traits of the pea plant for his experiments. They are shown here in both their dominant and recessive forms.

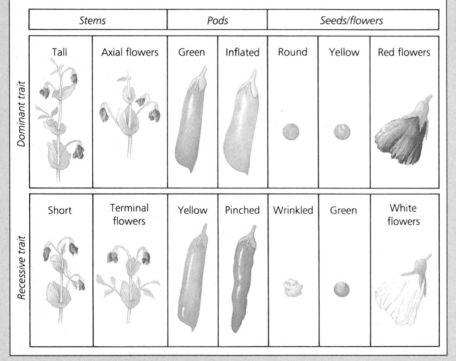

	Stems		Pods		Seeds/flowers		
Dominant trait	Tall	Axial flowers	Green	Inflated	Round	Yellow	Red flowers
Recessive trait	Short	Terminal flowers	Yellow	Pinched	Wrinkled	Green	White flowers

Figure 5.6.10 Pea plants were Mendel's main experimental organism, and by focusing on particular traits he was able to build up a mass of statistics about the various crosses he made.

By 1900 chromosomes had been discovered and meiosis had been seen and described. Hugo de Vries in Holland and Karl Correns in Germany discovered Mendel's work and duplicated his results. To their credit, they gave him the recognition he deserved. Then in 1902 it was suggested that Mendel's units of inheritance might be found on the chromosomes, and in the early 1900s Thomas Morgan and his team in the USA worked with *Drosophila* to gain evidence for this idea. From this point onwards the study of genetics snowballed. Today a complete unravelling of all the human DNA (the **genome**) is well on the way to becoming reality – and also the capability to alter and engineer parts of it, for good or ill.

Dihybrid crosses

In all the genetic crosses we have looked at so far, we have considered a single trait inherited on a pair of alleles. However, hundreds and thousands of genes go to make up the genotype of any one individual, and they are all passed on at the same time. **Dihybrid crosses** involve the inheritance of two different genes at the same time. Although still a very long way from the complexity of real events, this takes us one step closer to the real situation.

As we have seen, pea plants possess a variety of traits which are inherited through individual pairs of alleles. Mendel set up experiments to determine how two characters interact during inheritance, using the shape and colour of peas. Round peas (**R**) are dominant to wrinkled peas (**r**). Yellow peas (**Y**) are dominant to green peas (**y**). A dominant parent (round and yellow, **RRYY**) crossed with a recessive parent (wrinkled and green, **rryy**) gives rise to offspring which all display the dominant phenotype characteristics (round and yellow). A test cross with a homozygous recessive and a self cross between two of the offspring both give rise to four different phenotypes – round and yellow, round and green, wrinkled and yellow and finally wrinkled and green. Two of these – round and yellow, wrinkled and green – are the same as the original parents (**parental phenotypes**). The other two phenotypes – round and green, wrinkled and yellow – are different from the parents and have a **recombinant phenotype**. The ratios are different for the two types of cross. Figure 5.6.11 shows how such a dihybrid cross works in peas, and also demonstrates the value of a Punnett square in dealing with these more complex crosses.

The explanation for this situation is that the two sets of alleles are separated into the gametes quite independently of each other. In other words, four types of pollen and four types of egg are produced. Imagine the alleles as two pairs of shoes (pair 1 and pair 2, as shown in figure 5.6.12). If they are used to make

Figure 5.6.11 In a dihybrid cross each of the two alleles of one gene may combine randomly with either of the alleles of another gene. As a result of this independent assortment, four possible phenotypes result when heterozygotes are involved, whether selfed or in a test cross.

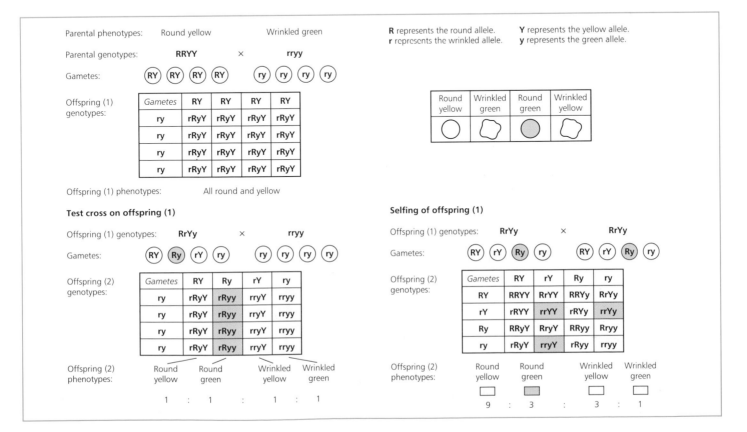

a gamete (a single pair of shoes) it must contain one shoe from each pair. The possible combinations are two left shoes, two right shoes, a left shoe from pair 1 with a right shoe from pair 2 or a right shoe from pair 1 with a left shoe from pair 2. Thus two of the pairs resemble the original shoes with a left and a right shoe, and two of the pairs are new combinations. In the same way, when gametes are formed in a dihybrid cross they segregate quite independently of each other and so recombine in all the possible ways.

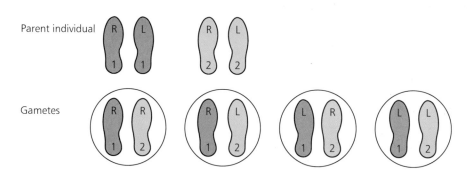

The shoes represent the alleles. If the parent is homozygous for both genes then all the possible gametes will carry the same information.
If the parent is heterozygous for one of the genes then there will be two possible types of gamete.
If the parent is heterozygous for both genes then four possible gametes result.

Figure 5.6.12 Alleles are separated independently when the gametes are formed.

Another example of dihybrid inheritance occurs in the fruit fly *Drosophila*. Grey bodies are dominant to ebony bodies, and long wings are dominant to the very short vestigial wings. A cross between two heterozygotes as seen in figure 5.6.13 shows again the use of a Punnett square to determine the results of dihybrid inheritance.

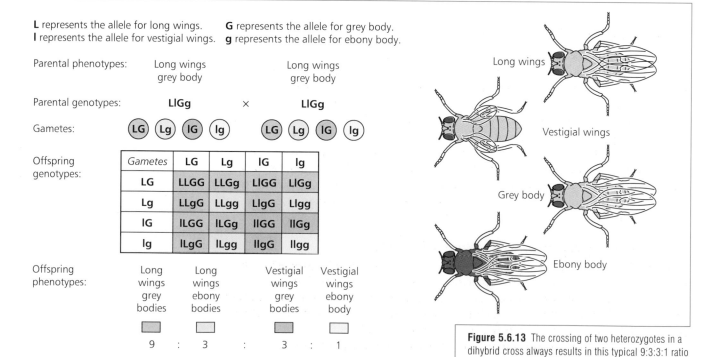

Figure 5.6.13 The crossing of two heterozygotes in a dihybrid cross always results in this typical 9:3:3:1 ratio of parental and recombinant offspring.

Using probability to predict the outcomes of complex crosses

As crosses become more complex, the use of simple diagrams to predict the possible offspring and their ratios becomes more difficult. However, it is possible to use probability to work out very simply what genotypes might be expected for any given cross. This only applies when the two characteristics are inherited independently of each other – in the case of gene linkage (see later in this section) these calculations need modification.

In the cross between the heterozygous grey-bodied long-winged *Drosophila* shown in figure 5.6.13, the probability of certain phenotypes occurring is worked out as follows:

- *Grey-bodied long-winged flies* – the grey allele is dominant, so $\frac{3}{4}$ of the F2 offspring will be grey. Similarly long wings are dominant, so $\frac{3}{4}$ of the offspring will have long wings. Therefore the likelihood of any fly inheriting a grey body and long wings is $\frac{3}{4} \times \frac{3}{4} = \frac{9}{16}$.
- *Grey-bodied vestigial-winged flies* – as above, grey bodies have a $\frac{3}{4}$ chance of appearing in the offspring. The recessive character for vestigial wings has a $\frac{1}{4}$ chance of occurring in the phenotype. Therefore the probability of grey-bodied vestigial-winged flies appearing is $\frac{3}{4} \times \frac{1}{4} = \frac{3}{16}$.
- *Ebony-bodied long-winged flies* – the probability of ebony bodies appearing is $\frac{1}{4}$ as they are recessive, and that of long wings is $\frac{3}{4}$ as they are dominant. Thus the probability of this phenotype being inherited is $\frac{1}{4} \times \frac{3}{4} = \frac{3}{16}$.
- *Ebony-bodied vestigial-winged flies* – each of the vestigial characters has a $\frac{1}{4}$ chance of occurring in the offspring, so the probability of the double recessive phenotype being seen is $\frac{1}{4} \times \frac{1}{4} = \frac{1}{16}$.

Notice that this provides the same 9:3:3:1 ratio of phenotypes as was seen from the Punnett square.

MENDEL'S LAWS OF GENETICS

The contribution of the monk Gregor Mendel to genetics cannot be overstated (see box 'A brief history of genetics'). With his work on pea breeding he founded the subject, and arrived at two fundamental laws of heredity. He worked with pure-breeding strains of peas and carried out large numbers of crosses, followed as necessary by test crosses to determine whether individuals were homozygous or heterozygous. His results showed ratios which coincided remarkably well with the theoretical ones. It has been suggested that Mendel worked out in advance what he expected to happen and then either he or his helpers 'managed' the actual results to achieve the desired ratios. Even if this was the case, the ideas were correct and the laws, although oversimplifying matters, still apply today to all diploid organisms.

The law of segregation – Mendel's first law

The first law which Mendel presented is known as the **law of segregation**. It was the result of work with monohybrid crosses. The law states that in a diploid organism one unit or allele for each trait is inherited from each parent to give a total of two alleles for each trait. The **segregation** (separation) of each pair of alleles takes place when the gametes are formed.

Mendel himself knew nothing of the process of meiosis, but this is the modern-day explanation for his proposed segregation, as during meiotic division chromosome pairs are split and haploid cells produced. This idea of independent units of inheritance, some dominant to others, which are maintained throughout the life of an individual and do not fuse to form a homogeneous mass, was Mendel's real breakthrough.

The law of independent assortment – Mendel's second law

The **law of independent assortment** states that different traits are inherited independently of each other. What this means is that the inheritance of a dominant or recessive allele for one characteristic, such as grey or ebony bodies, has nothing to do with the inheritance of alleles of other genes, such as those for wing length or eye colour. Considering that Mendel formulated these laws so early in the history of our understanding of inheritance it is remarkable that they are still so relevant today, although we now recognise that the second law in particular has many exceptions. This is the result of gene linkage, discussed below.

Polygenic traits

When we consider monohybrid and dihybrid crosses we are looking at genetic traits which are inherited as the result of a single gene. However, many traits are **polygenic**. This means that they are controlled by several interacting genes, rather than one single gene. (This is distinct from multiple alleles for *one* gene seen in *Drosophila* eye colours.) Human eye colour is an example of a fairly simple polygenic situation – colour is controlled by two genes. Height, weight and intelligence involve the interaction of many more genes, and so the consideration of their inheritance is far more complex than the simple situation with round and wrinkled peas.

Gene linkage

In dihybrid inheritance we have seen how the alleles of the two genes segregate quite independently of each other. However, the situation is not always as predictable as this. When *Drosophila* with normal broad abdomens and long wings are crossed with flies displaying recessive narrow abdomens and vestigial wings, the first generation produced is as expected, with all the offspring showing the dominant phenotype but possessing the heterozygote genotype. However, when these first generation flies are crossed, the normal 9:3:3:1 ratio of parental and recombinant types does not occur, as figure 5.6.14 shows. Instead there is a 3:1 ratio of dominant:recessive phenotypes, which is what we would expect in a monohybrid cross. The explanation of this apparent discrepancy is that the genes are **linked** – that is, they are sited on the same chromosome and inherited as if they were one unit.

If genes are completely unlinked, approximately equal numbers of recombinants and parental type gametes are formed. But if genes are linked then they are inherited to a greater or lesser degree as if they were a single gene. When genes are closely linked recombinants rarely if ever occur. If the genes are more loosely linked then the number of recombinants which results will be higher. The tightness of the linkage is related to how close the two or more genes are on the chromosome – genes that are very close together are less likely to be split during the crossing over stage of meiosis than genes which are further apart (see section 5.2, page 349).

Cross over values

If the linked genes are very close together on the chromosome they may be so tightly linked that they are effectively never split up during meiosis, and so the

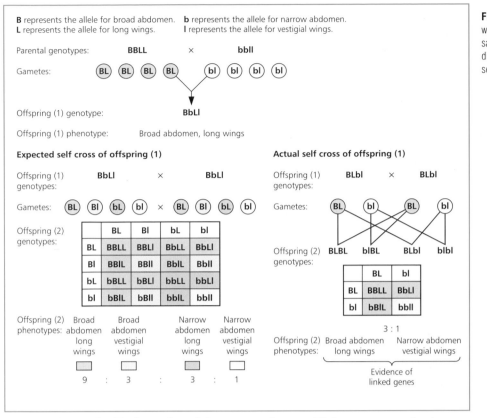

Figure 5.6.14 These two genes are inherited as if they were one unit because they are closely linked on the same chromosome. The normal Mendelian ratios for a dihybrid cross only occur when the genes are on separate chromosomes.

gametes formed will always be of the parental types. If the genes are further apart, crossing over between them is more likely to occur. Although in the majority of cases they will be passed on as a parental unit, sometimes they will be mixed and recombinant gametes will be produced, which will in turn be reflected in the offspring. This gives us a way of working out how close together various genes are on a chromosome – in fact, a genetic map of a chromosome can be built up by studying linked genes. To produce a genetic map the **cross over value** (**COV**) has to be worked out:

$$\text{Cross over value} = \frac{\text{number of recombinant offspring} \times 100}{\text{total number of offspring}}$$

Thus closely linked genes producing only small numbers of recombinant offspring will result in low cross over values. Genes which are further apart produce larger numbers of recombinants and have higher cross over values. This can be used to build up a chromosome map as shown in figure 5.6.15 and table 5.6.1. The clue that linkage is involved in a dihybrid cross is the absence of the expected 9:3:3:1 ratio. Whilst small differences in numbers may be put down to experimental error, larger discrepancies, with a heavy weighting to parental types, indicate that linkage between the genes is involved.

Genes	Cross over value
X and Y	25%
Y and Z	5%
X and Z	20%

Table 5.6.1 To create a genetic map the cross over values are converted into arbitrary **map units**. The first cross over value positions genes X and Y relative to each other. However, the position of gene Z can only be decided by reference to its cross over values in relation to both gene X and gene Y. This technique can be very slow and laborious as most chromosomes carry thousands of genes, but it is vital in the process of altering individual genes or groups of genes.

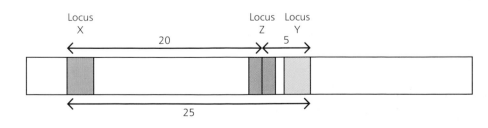

Figure 5.6.15

Sex linkage

The chromosomes in diploid organisms occur as homologous pairs. In organisms such as mammals where there are clear differences between the males and females, sex is determined by the chromosomes. As mentioned in section 5.1, all the pairs of chromosomes except one carry information about the general body cells and their biochemistry. They are known as the **autosomes**. The final pair of chromosomes carry information about the sex of the individual and are known as the **sex chromosomes**. In mammals the female has two large X chromosomes. Thus all of her eggs contain an X chromosome and she is referred to as **homogametic**. The male has one X chromosome and one much smaller Y chromosome. Thus half of the sperm will contain X chromosomes and half will contain Y chromosomes – the male is **heterogametic**. In humans there are 22 pairs of autosomes and one pair of sex chromosomes. In some organisms, for example birds, it is the male which is homogametic.

However, not all the genes on the sex chromosomes are concerned with matters related to sex. The small human Y chromosome seems to carry little but male sex information and a gene for very hairy ears, but the X chromosome carries a variety of other genes coding for traits such as clotting factors in the blood and the ability to see in colour. Genes which are carried on the X chromosome are said to be **sex linked**. Any recessive or mutant genes passed on the X chromosome from a mother to her sons will be expressed in the phenotype as there is no corresponding gene on the Y chromosome to mask them. Because of the importance of the X and Y chromosomes in sex linkage, they are shown on genetic diagrams.

Sex linkage in *Drosophila* was discovered by Thomas Morgan. In these flies eye colour is affected by sex, as shown in figure 5.6.16. In humans sex linkage leads to a variety of conditions known as **sex-linked diseases**, some of which will be considered in section 5.7.

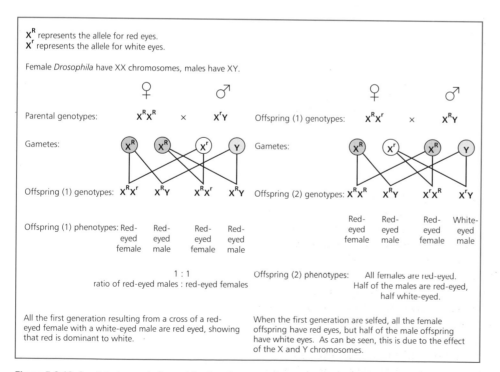

Figure 5.6.16 Sex-linked genes in *Drosophila* affect the eye colours which may be found in males and females.

Inactivation of X chromosomes

All the cells of a female mammal other than those of the ova contain two X chromosomes. However, during the early stages of development one of these X chromosomes is inactivated and coils up into a mass known as the **Barr body**.

Which X chromosome is inactivated in each embryonic cell seems to be a random event, but all the cells which are descended from that cell will have the same inactivated X chromosome. This means that females are made up of mosaics of cells, some with one X chromosome inactivated and some with the other. Tortoiseshell cats are a good example of this. Some of the X chromosomes carry the allele for black fur and some for orange. The mixture of colours on the cat reflects the mosaics of cells with different X chromosome inactivation. The situation is the same in all mammals, including humans. An example of human X inactivation is shown in figure 5.6.17.

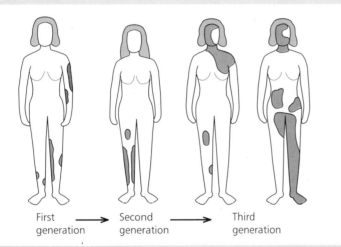

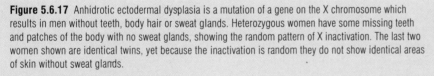

First generation → Second generation → Third generation

Figure 5.6.17 Anhidrotic ectodermal dysplasia is a mutation of a gene on the X chromosome which results in men without teeth, body hair or sweat glands. Heterozygous women have some missing teeth and patches of the body with no sweat glands, showing the random pattern of X inactivation. The last two women shown are identical twins, yet because the inactivation is random they do not show identical areas of skin without sweat glands.

Nature versus nurture

Some genetic traits result in very distinct phenotypic features such as eye colour or number of digits. However, the phenotype of an organism is not the result of its genotype alone. Genetically identical plants grow very differently when exposed to varying amounts of light and soil nutrients. The ability of an animal to achieve its full genetic potential in terms of growth depends heavily on the amount of food available to it. Thus in looking at the genetic make-up of an individual, the role of the environment in shaping the apparent phenotype has to be considered (nature versus nurture). It is important during any genetic experiment that all the organisms are subjected to the same conditions so that as far as possible any differences between them can be seen as the results of genetic differences alone. The role of the environment in shaping an organism becomes even more important when we consider human development – an aspect which will be explored more fully in section 5.7.

5

SUMMARY

- The **phenotype** of an organism is made up of its observable characteristics. Its **genotype** is the genetic information in its chromosomes. The phenotype depends on the genotype and also on environmental conditions.

- A **gene** is a segment of DNA coding for a particular trait or characteristic, and is found at a particular position (the **locus**) on a specific homologous pair of chromosomes. Each homologous chromosome may carry different codings for this gene called **alleles**, and these produce alternate characteristics.

- An organism with the same allele for a particular gene on both homologous chromosomes is **homozygous** for that gene or characteristic.

- Homozygotes are **true breeding**. If two homozygous parents with the same characteristic are crossed and their offspring subsequently crossed, all generations of offspring have the same characteristics as each other and as the parents.

- The offspring of a cross between homozygous parents make up the **first filial generation** (F_1). The offspring all have the same phenotype. Crossing these offspring results in the **second filial generation**, and so on.

- An organism that has two different alleles for a particular gene on its homologous chromosomes is **heterozygous** for that gene or characteristic.

- A **dominant** allele expresses itself in the heterozygous state, masking the **recessive** allele. Recessive alleles are only expressed in homozygotes. Dominant alleles code for a particular protein, while recessive alleles usually code for its absence or an inactive protein.

- A **monohybrid cross** is a cross in which one genetic trait is considered. The predicted ratios of genotypes and phenotypes expected from a cross are known as **Mendelian ratios**, and the smaller the sample the larger the sampling error and the larger the error will be in applying these ratios.

- The Mendelian ratio for the phenotypes of a monohybrid cross between two heterozygotes is 3 dominant:1 recessive.

- A **test cross** is carried out to test whether a characteristic coded for by a dominant allele is the result of a homozygous or heterozygous genotype. The individual is crossed with a recessive homozygote. If the individual is heterozygous the offspring will have the ratio 1 dominant:1 recessive, while all will be dominant if it is homozygous.

- **Codominance** or incomplete dominance occurs when both alleles code for different proteins. The heterozygote has a characteristic in between those of the homozygotes. A heterozygote crossed with a homozygote results in offspring in the ratio 1 heterozygote:1 homozygote.

- **Multiple alleles** may result in several different phenotypes in different individuals.

- **Lethal genes** result in the death of individuals with particular gene combinations, for example homozygous recessive individuals may die before birth. This gives a ratio of 2:1 instead of the expected 3:1 for the first generation of a cross between two heterozygotes.

- **Dihybrid crosses** consider the inheritance of two traits at the same time. Offspring may show the same combination of traits as the parents (**parental phenotypes**) or they may show different combinations (**recombinant phenotypes**).

- The expected ratio for the second generation of a cross between two different homozygous parents is 9:3:3:1. A ratio of 3:1 in the second generation shows that the genes are **linked** (on the same chromosome).

- Mendel's **law of segregation**, in modern language, states that in a diploid organism one allele for each trait is inherited from each parent to give a total of two alleles for each trait. The pairs of alleles are separated during gamete formation (meiosis).

- Mendel's **law of independent assortment** states that different traits are inherited independently of each other.

- Exceptions to the law of independent assortment occur in **polygenic traits**, which are controlled by several interacting genes.

- Another exception to the law of independent assortment is genes being linked. This can happen to a greater or lesser extent, depending on how close together the genes are on the chromosome. Inheritance of linked genes can indicate how close together the genes are and hence the mapping of chromosomes, by working out the **cross over value**:

$$COV = \frac{\textbf{number of recombinant offspring} \times \textbf{100}}{\textbf{total number of offspring}}$$

- Another exception to the law of independent assortment is **sex linkage**. Human males have one X and one Y chromosome, while females have two X chromosomes. Recessive alleles carried on the X chromosome will be expressed in the male since there is no other X chromosome to carry the dominant allele and mask them.

QUESTIONS

1 As any gardener knows, F_1 hybrid seeds do not breed true. Thus plants raised from these seeds cannot be used as stock for the next year.
  a What is meant by the terms:
    i F_1 hybrid
    ii breeding true?
  b Why do F_1 hybrids not breed true if they are self-fertilised?
  c What is the advantage of using these hybrids?

2 a Human ABO blood types are determined by three alleles at one locus, I^A, I^B and I^0. I^A and I^B are codominant. What does this mean?
  b I^0 is recessive. Is it possible for one individual to carry all three alleles? Explain your answer.
  c What gametes will someone with blood type AB produce?
  d What gametes will someone with blood type 0 produce?
  e Produce a genetic diagram to show the possible outcomes of a couple with AB and B blood groups.

3 In *Drosophila melanogaster*, kidney bean-shaped eye is recessive to round-shaped eye and orange eye colour is recessive to red eye colour.
  A fly, homozygous for both round-shaped eye and red colour, was crossed with a fly having kidney bean-shaped orange coloured eyes. The offspring (F_1) were allowed to interbreed and the investigator expected to find a 9:3:3:1 segregation for eye shape and colour in the F_2 generation. Instead, the following progeny were produced.

red colour, round-shaped eye	520
orange colour, kidney bean-shaped eye	180
red colour, kidney bean-shaped eye	54
orange colour, round-shaped eye	45

  a Account for the basis for the expected 9:3:3:1 segregation in the F_2 generation. (**6 marks**)
  b Explain the observed result. (**6 marks**)
  (**Total 12 marks**)
  (O&C June 1991)

5.7 Human genetics

Although all human beings are members of the same species, there is tremendous variation between individuals. Height, weight, skin colour and many other features show an enormous range. Most families have particular genetic traits which can be traced back through the generations, such as unusually shaped noses, red hair or dimples. Some families have genetic traits which cause disorders such as colour blindness, haemophilia or Huntington's chorea (a disease characterised by involuntary muscular movements and mental retardation). The genetic principles which we have looked at so far also apply to our own genetics. However, like those of other complex organisms, many human features are polygenic, making working out the details of a human genotype considerably more difficult than understanding the genotype of *Drosophila*.

Figure 5.7.1 Within the human species the variety is enormous – and the genetics behind this poses a complex puzzle.

STUDYING HUMAN GENETICS

Practical problems

A major problem in understanding human inheritance is that people are not available for experimentation. Apart from the obvious social and moral objections, humans are very poor laboratory animals! We have long life cycles, and only produce a very small number of offspring. The effects of any experiments would take a very long time to become apparent.

One of the main tools of genetic investigation is, as we have seen, the setting up of deliberate crosses between individuals – but humans do not reproduce to please geneticists. Secondly, it is frequently the second generation which reveals most about a particular cross. In humans a true F_2 generation almost never occurs, as it would be the result of a union between a brother and sister. Such relationships frequently result in children born with abnormalities due to the increased likelihood of two undesirable recessive genes coming together. Brother–sister marriages are taboo in most human societies. Finally, the sample size in human families is too small for statistical analyses – the average family in the UK is now less than two children per couple.

Bearing these problems in mind, how are human genetics studied? Much of the work done on human inheritance has involved studies of a genetic disorder. The production of a family tree or pedigree is one technique where a particular trait can be traced through several generations of a family and deductions made about the mode of inheritance. A **karyotype** or photograph of the chromosomes like that seen in figure 5.1.9 may be produced and used to uncover abnormalities involving whole chromosomes. Biochemical analysis of specific metabolic pathways can reveal the presence of mutant genes which cause severe genetic diseases. These techniques, along with the application of the Mendelian principles of genetics already considered earlier, and increasingly intricate gene mapping, are rapidly increasing our understanding of the human **genome** (the total number of genes). We shall now go on to consider these techniques, along with some disorders studied using them.

Family trees

The analysis of pedigrees or family trees can be very useful for deciding how a trait is inherited. Sex-linked diseases show up in the males rather than the females. Dominant genes express themselves in every generation whereas recessive genes may skip one or more generations, being passed on in the recessive states. A **carrier** is a healthy individual who possesses an unexpected recessive allele which would result in a genetic disease in the homozygous state or if inherited by a male. The use of a family tree allows potential carriers of a trait to be identified.

Achondroplastic dwarfism

An example of a family tree for a condition known as **achondroplastic dwarfism** is shown in figure 5.7.2. This form of dwarfism affects the long bones of the body which do not grow to normal size, although in every other way the affected individuals are normal. The trait is dominant – it is exhibited in every generation and it does not appear to be sex linked, as both female and male dwarfs are seen. Just as in the genetics of the Manx cat already considered in section 5.6, the homozygous state of this gene is lethal and the fetuses die before birth. The genotypes of the individuals in the family tree can therefore be deduced.

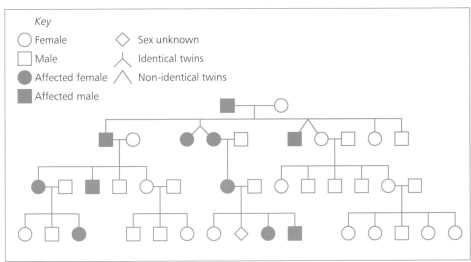

Figure 5.7.2 Achondroplastic dwarfism is a genetic condition which has existed for a very long time, as this portrait from the early seventeenth century shows. From a family tree such as this involving a dominant allele, the genotypes of the individuals can be worked out with considerable accuracy.

Red-green colour blindness

Family trees are also useful for demonstrating the inheritance of some of the better known human sex-linked diseases – in particular red-green colour blindness and haemophilia. **Colour blindness** is due to a recessive mutation of a gene on the X chromosome. Around 8% of Caucasian males are affected by this, but only 1% of females. The gene does not markedly affect the chances of survival of an individual, and because the homozygous form is not lethal colour blindness does occasionally occur in women. This is explained in figure 5.7.3.

Haemophilia

Colour blindness is usually an inconvenience but nothing more. **Haemophilia** is a much more severe sex-linked trait in which one of the proteins needed for the clotting of the blood (factor 8) is missing. The mutated gene is carried on the X chromosome. The homozygous form is unlikely to arise since haemophiliac men rarely have offspring, but if it does it is lethal when the affected female reaches puberty if not before. In an untreated haemophiliac the slightest injury can lead to death through excessive bleeding – even exercise can result in internal bleeding of the joints. Haemophilia can now be treated by regular transfusions of factor 8, and although this is expensive it can allow haemophiliacs to lead a more or less normal life.

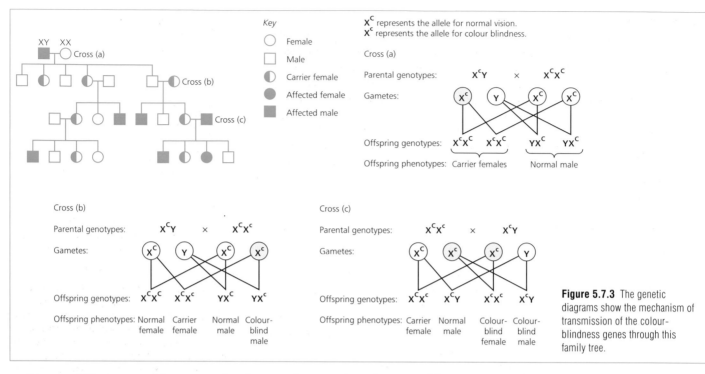

Cross (a)

Parental genotypes: X^cY × X^cX^c

Offspring genotypes: X^cX^c X^cX^c YX^c YX^c

Offspring phenotypes: Carrier females   Normal male

Cross (b)

Parental genotypes: X^cY × X^cX^c

Offspring genotypes: X^cX^c X^cX^c YX^c YX^c

Offspring phenotypes: Normal female, Carrier female, Normal male, Colour-blind male

Cross (c)

Parental genotypes: X^cX^c × X^cY

Offspring genotypes: X^cX^c X^cY X^cX^c X^cY

Offspring phenotypes: Carrier female, Normal male, Colour-blind female, Colour-blind male

Figure 5.7.3 The genetic diagrams show the mechanism of transmission of the colour-blindness genes through this family tree.

Haemophilia is caused by a mutation known since ancient times, and long before Mendel some idea of the genetics of the disorder was grasped. The primary source of Jewish law, the Talmud, specifies that if a boy dies of bleeding after circumcision, his younger brothers should not undergo the ritual and his male cousins on the mother's side were also exempt. This shows an understanding that blood which does not clot runs in families, and is inherited from the mother rather than the father. Haemophilia plagued the royal families of Europe after a mutation in the earliest stages of the embryonic development of Queen Victoria, as figure 5.7.4 shows.

Figure 5.7.4 Haemophilia in the royal families of Europe – as a result of the wide-ranging intermarriages between the various royal families, haemophilia spread from the original mutation in Queen Victoria to most of the royal houses of Europe.

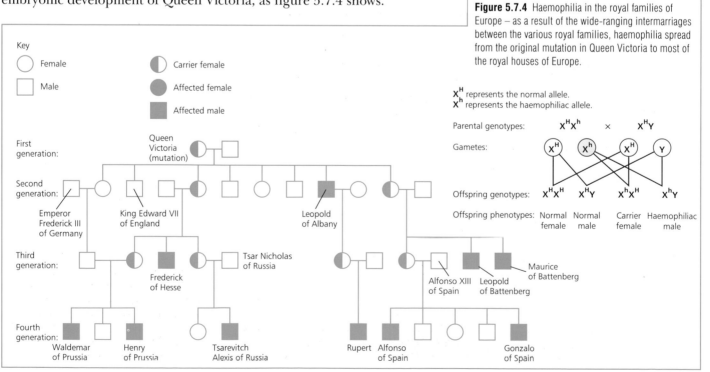

Karyotypes

Karyotypes (photographs of the chromosomes from a single diploid cell) are used for showing up chromosome abnormalities such as **Down's syndrome**. In Down's syndrome the chromosomes do not separate fully at meiosis and the offspring contains an extra autosome. The likelihood of this occurring increases with maternal age, as shown in figure 5.7.5. Known as trisomy on chromosome 21, Down's syndrome results in mental retardation, heart defects, a short stocky body and characteristic eye folds and lack of muscle tone. Karyotyping is often used to determine whether a fetus has chromosome abnormalities so that the pregnancy may be terminated if the parents wish. Chromosome abnormalities are not usually passed from generation to generation as individuals are often sterile or do not live long enough to reproduce.

Biochemical analysis

Phenylketonuria

Biochemical analysis of certain biochemical pathways can uncover evidence for genes which cause severe problems. The metabolism of the amino acid phenylalanine is prone to errors. One of the most common is known as **phenylketonuria** (**PKU**), which is inherited as a recessive allele. Newborn infants who have inherited this condition have high levels of phenylalanine in their blood, but are not damaged. As the breakdown products of the amino acid build up, severe mental retardation results. Babies in the UK are routinely tested for PKU in the first few days of life. If affected they can be given a diet almost free of phenylalanine so avoiding the problems of the defective metabolism and the child will develop more or less normally.

Tay-Sachs disease

Biochemical analysis has also been used to detect a genetic disorder known as **Tay-Sachs disease**. Particularly prevalent in the Ashkenazi Jews from central Europe, this is the result of a recessive allele which, when present in the homozygous form, prevents the production of the enzyme **hexosaminidase A**. This enzyme is required for lipid metabolism and without it the nervous system degenerates. This degeneration begins a few months after birth so that the infant becomes blind and deaf, suffers seizures and dies before reaching the age of five. Heterozygous carriers show no obvious symptoms, but it has been found that their blood levels of hexosaminidase A are reduced. Thus by testing the blood, carriers in the 'at risk' group can be identified, as shown in figure 5.7.6. This information can be used to choose partners or have fetal screening as the identified carriers wish, in an attempt to eradicate this dreadful genetic disease.

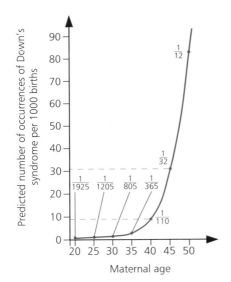

Figure 5.7.5 When chromosome 21 is 'sticky', two copies will be delivered to one gamete and none to another. If the first of these two gametes is subsequently fertilised the individual resulting will be trisomic for chromosome 21 and will have Down's syndrome. If the second is fertilised the zygote cannot survive and fails to develop. The likelihood of this 'stickiness' occurring increases with the age of the mother, particularly beyond the age of 40.

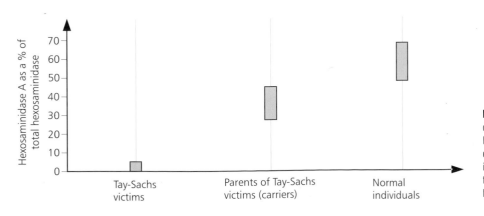

Figure 5.7.6 A biochemical assay of Tay-Sachs disease shows that hexosaminidase A levels are very low in sufferers of the disease, and that carriers of the disease have a level between that of victims and normal individuals. This can be used to screen people before they marry or have children. Identifying carriers may help to eradicate this tragic condition.

Pleiotropic genes and sickle cell anaemia

Many genes are **pleiotropic**. This means that one gene affects more than one phenotypic feature. A well researched example of pleiotropy in humans is a gene affecting the production of haemoglobin. The allele **Hb^A** allows for the formation of normal haemoglobin. The mutant form **Hb^S** causes a disease known as **sickle cell anaemia** which usually results in an early death. This mutant allele is recessive and has at least five readily observed phenotypic consequences if it is inherited. These include:

- red blood cells which are sickle shaped instead of the normal biconcave disc, as shown in figure 5.7.7
- severe and eventually lethal anaemia – a greatly reduced number of red blood cells
- pain in the abdomen and joints
- an enlarged spleen
- resistance to malaria in heterozygotes.

Figure 5.7.7 Normal and sickled red blood cells. Sickle cell anaemia is the result of a homozygous recessive gene. In the homozygous state it is usually lethal. However, in the heterozygous form it confers resistance to malaria, and so the gene is maintained in the population.

All these phenotypic traits are in fact the result of a change in a single amino acid in the structure of the haemoglobin molecule. The abnormal form tends to join together in long strands which distort the shape of the red blood cells, producing the typical sickle shape. The spleen removes old and damaged red blood cells, and it becomes enlarged as a result of dealing with the sickled cells. This in turn leads to the severe anaemia because many red blood cells are removed from the circulation and destroyed. The sickled cells in the circulation block tiny capillaries and so starve nearby cells of oxygen, causing the pain in joints and abdomen.

The final effect of the mutation is that in the heterozygous form it confers resistance to malaria. This might seem to have little to do with the shape of the haemoglobin, and the way in which the two are connected has taken much unravelling by doctors, biochemists, population geneticists and electron microscopists. The heterozygous genotype **Hb^{A}Hb^S** results in individuals who produce both normal and abnormal haemoglobin. Under normal body conditions their red blood cells are disc shaped. When malarial parasites are present in the blood cells the cells cannot cope with reduced oxygen tension, and they sickle as the oxygen level falls. These mis-shaped cells are then removed by the spleen, incidentally getting rid of the malarial parasites at the same time. Normal shaped, unaffected cells are unharmed. The result of this is individuals who might have slightly enlarged spleens or suffer a degree of anaemia – but who are resistant to malaria. The sickle cell gene is found in populations almost exclusively where malaria is also prevalent. The advantages it gives in terms of malarial resistance ensure that it is retained within the genetic mix of the population in spite of the problems it causes in the homozygous form, as demonstrated in figure 5.7.8.

Nature versus nurture – the role of twin studies

As we saw in section 5.6, the phenotype of an individual does not depend solely on the genes it inherits, but also on the environment in which it develops and lives. Many human traits are polygenic, and added to this the environment within the uterus, during childhood and at adolescence affects the eventual phenotype of the individual. Thus deciding whether nature or nurture determines a certain phenotypic characteristic in humans is very difficult indeed. To help unravel some of these strands, twin studies are useful.

In twin studies the differences between identical (monozygotic) twins are considered. Many identical twins brought up together show remarkable physical similarities, but these could be due to being brought up in the same

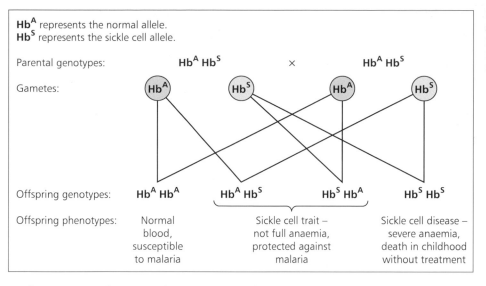

Hb^A represents the normal allele.
Hb^S represents the sickle cell allele.

Parental genotypes: **Hb^A Hb^S**   ×   **Hb^A Hb^S**

Gametes: (Hb^A)   (Hb^S)   (Hb^A)   (Hb^S)

Offspring genotypes: **Hb^A Hb^A**   **Hb^A Hb^S**   **Hb^S Hb^A**   **Hb^S Hb^S**

Offspring phenotypes: Normal blood, susceptible to malaria   Sickle cell trait – not full anaemia, protected against malaria   Sickle cell disease – severe anaemia, death in childhood without treatment

Figure 5.7.8 Homozygotes for the sickle cell allele die as a result of it. Many of those homozygous for the normal gene will die or be seriously debilitated by malaria. Sickle cell heterozygotes usually survive both conditions.

environment at the same time. However, in some cases identical twins are separated at birth to be adopted by different families. If separated twins are considered and compared with non-separated identical twins, non-identical twins and ordinary siblings then some idea can be gained of traits which are largely determined by genes and those which are strongly affected by the environment. The data in table 5.7.1 show that height appears to have a strong genetic component and is influenced relatively little by environmental factors. On the other hand, mass seems to be much more affected by environmental factors such as learned eating habits.

Trait	Identical twins reared together	Identical twins reared apart	Non-identical twins	Non-twin siblings
Height difference	1.7 cm	1.8 cm	4.4 cm	4.5 cm
Mass difference	1.9 kg	4.5 kg	4.6 kg	4.7 kg
IQ score difference	5.9	8.2	9.9	9.8

Table 5.7.1 The differences between twins, based on 50 pairs each of identical twins reared together, non-identical twins and non-twin siblings, and 19 pairs of identical twins reared apart.

Our understanding of the genetic basis of human traits is still far from complete. Genetics has been shown to play a part in some remarkably complex patterns of behaviour.

GENETIC ENGINEERING

What is genetic engineering?

Genetic engineering is a term used to describe the skills which have been developed enabling molecular biologists to move genes from one chromosome to another. They have made possible the creation of new genomes resulting in the production of altered organisms such as the giant mice shown in figure 5.7.9 and in bacteria which express human genes. Genetic engineering involves locating a desirable gene, isolating it and then moving into the genome of another organism. The proteins produced in response to the new piece of DNA will either have a useful effect in the host cell or may be harvested for use elsewhere.

Figure 5.7.9 One of these mice is quite normal. The other has the human growth-hormone gene in the chromosomes of its cells, inserted using the techniques of genetic engineering.

How is genetic engineering carried out?

The techniques used in genetic engineering are complex. Special enzymes called **restriction endonucleases** chop up DNA strands, cutting them at specific sites. Other enzymes known as **DNA ligases** act as 'genetic glue' and join pieces of DNA together. The required fragment of DNA cut from the chromosome of one organism is pasted into another piece of DNA which will carry it into the host cell. **Plasmids**, the circular strands of DNA found in bacteria, are frequently used as these vectors. Once the plasmid is incorporated into the host nucleus it becomes part of the new **recombinant DNA** of the engineered genome.

What is the value of genetic engineering?

Genetic engineering as a technique has or may have immense value to human beings in three main ways:

- Genes may be placed into microbes and the microbes will then produce large quantities of a desired substance. Examples of successes in this area include the production of human insulin by bacteria, reducing the need for insulin from other animals to be used.
- Engineered genes may be placed in domestic species of plants or animals to improve traits such as their growth rates or the protein quality produced. It may enable more plants to produce first class protein, or reduce the fat content of the protein-rich muscle tissues in animals. Engineered genes in crop plants may soon enable the plants to synthesise chemicals which will act as inbuilt pesticides, removing the need for extensive crop spraying.
- Gene therapy involves putting engineered genes into people to cure genetic diseases. As we have seen throughout this section, the human race is plagued by a large number of genetic diseases such as sickle cell anaemia, haemophilia, cystic fibrosis and thalassaemia. They affect about 1% of all children, placing a heavy burden of both care and cost on the individual families and on the society in which they live. The first few tests of gene therapy are currently being undertaken with genes to combat immune insufficiency and to alleviate cystic fibrosis. The day is getting ever closer when engineered genes will be inserted routinely into the relevant somatic (body) cells of an affected individual, allowing normal functioning of the cells to begin.

The genetic disorders of the human species may be inherited in a wide variety of ways. The lethal gene causing Huntington's chorea is a dominant gene that is only expressed later in life. Albinism and sickle cell anaemia are the result of recessive alleles on the autosomes, and haemophilia caused by a mutation on the X chromosome – these are but a few examples. With the advance of screening techniques such as those described for Tay-Sachs disease, and the increasing use of gene manipulation and genetic engineering, we may hope to see a substantial reduction in the human suffering caused by these genetic diseases within the next 50 years. Perhaps one day they, like smallpox, will become a thing of the past.

Genetic engineering – the final step?

With time and technical advances, the insertion of healthy genes into the body cells of individuals suffering from genetic diseases may soon become an everyday procedure. However, this only alleviates the problem for each individual, and the therapy would need to be repeated in each succeeding generation. To eliminate the disease completely it would be necessary to engineer the germ cells so that the altered genome was passed on to subsequent generations. This is seen as a very major step – interfering with the inheritable material of an individual could open a Pandora's box of ethical and moral controversies. Some individuals might well demand 'intelligence' genes or 'good-looking' genes, and some societies might wish to engineer social classes... the implications for this area of biology are immense. At the moment any engineering of the germ cells is forbidden, but once the technology for engineering is in place, society has some very big decisions to make about what is to be done with it.

SUMMARY

- **Family trees** are used to trace human genetic disorders through the generations and identify potential **carriers** who may be heterozygous for a particular harmful trait.
- **Achondroplastic dwarfism** is a dominant trait which is lethal in the homozygous state.

- **Red-green colour blindness** is caused by a recessive mutation on the X chromosome and so affects more males than females, though recessive homozygotes (colour-blind women) do occur.

- **Haemophilia** is a sex-linked recessive trait carried on the X chromosome in which a protein needed for blood clotting is missing. The homozygous recessive condition in females is rare and lethal.

- **Down's syndrome** is trisomy on chromosome 21 (three chromosomes instead of two) and is detected by **karyotyping** (photographing the chromosomes). An affected individual suffers from mental retardation, heart defects, a characteristic body shape and eye folds and a lack of muscle tone.

- **Phenylketonuria** is a recessive allele that causes defective metabolism of the amino acid phenylalanine. The condition is detected by **biochemical analysis**.

- **Tay-Sachs disease** is caused by a recessive allele that prevents the production of the enzyme hexosaminidase A, required for lipid metabolism. Without this enzyme the nervous system degenerates. Carriers can be identified by biochemical analysis as they have reduced blood levels of hexosaminidase A.

- A **pleiotropic gene** affects more than one phenotypic feature. The recessive mutant allele for **sickle cell anaemia** causes sickle-shaped red blood cells, severe anaemia, pain in the abdomen and joints, an enlarged spleen, and resistance to malaria in the heterozygous state. These all result from a change in a single amino acid in the haemoglobin molecule. The trait is selected for in areas where malaria is prevalent.

- **Twin studies** give information about which aspects of a phenotype are caused by the genotype, and which by environmental influences. Studies on identical twins separated at birth and so subjected to different conditions have shown, for example, that height is largely genetically determined, unlike mass.

- In **genetic engineering**, a desirable gene is removed from its chromosome by enzymes called **restriction endonucleases**. It is joined into the genetic material of another organism, for example into a bacterial plasmid, by enzymes called **DNA ligases**. The resulting recombinant DNA is inserted into the host cell. The host cell may produce a substance, for example human insulin, in large quantities. The recombinant DNA may confer improved traits to domestic plants or animals, such as inbuilt pesticide activity. In **gene therapy** the recombinant DNA is used to cure genetic diseases in people.

QUESTIONS

1 A fairly common sex-linked trait in humans is colour blindness. The allele for normal vision is dominant. In the United States 6% of males are colour blind, compared with fewer than 1% of females. Explain why there are more colour-blind males than females. Does every colour-blind male have a colour-blind parent? Does every colour-blind female? Use genetic diagrams to explain your answer.

2 Discuss the difficulties of investigating human genetics and comment on some of the techniques used.

3 Select two different human genetic diseases and discuss the impact genetic engineering might have on such diseases in the future.

1 Write an essay on gene expression, including reference to animal and plant examples where appropriate. **(20 marks)**
(ULEAC January 1991)

2 a Describe and account for the differences between fetal and adult circulatory systems. **(8 marks)**
b Fetal and adult haemoglobin are different. Describe their different properties and relate these to their functions. **(4 marks)**
c Explain the possible problems for the fetus if the mother is Rhesus negative and the father is heterozygous for the Rhesus factor. **(8 marks)**
(NEAB June 1991)

3 When a female fruit fly with red eyes and a grey body was crossed with a male with brown eyes and yellow body all of the 47 offspring had red eyes and grey bodies. These offspring were then used to set up a backcross in order to test the validity of Mendel's law of independent assortment of genes and the phenotypes of their offspring were carefully recorded.
a i Explain what is meant by independent assortment of genes.
ii What evidence is there that these parents were pure breeding for these characteristics?
iii Explain how a backcross would be set up. **(9 marks)**
b What phenotypes would you expect from interbreeding the F_1 if the two genes in question were located:
i on separate autosomes
ii on the same chromosome
iii on the X chromosome?
Explain your answers. **(10 marks)**
c Name *one* statistical test that could be applied to test the significance of these results. **(1 mark)**
(NEAB June 1991)

4 Describe the similarities and differences between each of the following pairs:
a mitosis and meiosis **(8 marks)**
b mammalian sperm and mammalian ovum. **(6 marks)**
(From ULEAC June 1990)

5 a Explain the differences between the members of *each* of the following pairs of genetical terms and give *one* example of *each* term to illustrate your example.
i complete and incomplete dominance
ii continuous and discontinuous variation
iii chromosomal mutation and crossing-over
iv polyploidy and haploidy. **(12 marks)**
b Crosses between ginger female cats and black male cats produce only tortoiseshell females and ginger-coloured males. A single gene controls expression of colour in cats.
i Give a reasoned explanation of these results and show the genotypes of the parents, their gametes and the offspring produced in these crosses.

ii Is it possible to have tortoiseshell male cats? Explain your answer. **(8 marks)**
(NEAB June 1989)

6 This question is concerned with reproduction in flowering plants.
a Outline any *three* methods of asexual reproduction. **(6 marks)**
b Describe the role of nuclear division in sexual and asexual reproduction. **(6 marks)**
c One population of a plant reproduces by sexual means but a second population of the same species reproduces exclusively by asexual means. Describe the advantages and disadvantages of these methods of reproduction and comment on their possible evolutionary consequences. **(8 marks)**
(NEAB June 1989)

7 Explain the following terms, using an example in each case to make the meaning clear:
a alleles
b phenotype
c independent assortment
d polygenes
e polyploidy.

8 Write an essay on mechanisms for outbreeding, and the advantages and disadvantages of outbreeding, in flowering plants. **(20 marks)**
(ULEAC January 1992)

9 In African Swallow-tail butterflies, the male is the homogametic sex (XX) and the female is heterogametic (XY). Males and females differ in appearance as shown below.

Male   Female

a By means of a genetic diagram using the heads below show how reproduction in butterflies leads to approximately equal numbers of males and females in the next generation.

	Male XX	Female XY
Gametes		
Offspring		

(2 marks)

b In butterflies one of the two cells arising from the first division of the zygote gives rise to all of the left side of the body. The other cell of the zygote gives rise to all of the right side. In some rare cases, the first division of a male (XX) zygote may leave one cell with only one X chromosome instead of two. The resulting individual will be male on one side of its body and female on the other side. This is shown diagramatically below.

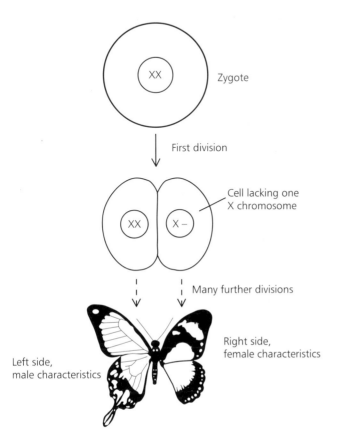

i What type of division occurs in zygotes and embryonic cells? **(1 mark)**

ii Suggest how a cell may be produced during the first division of the zygote which lacks an X chromosome. **(2 marks)**

iii What information is given by the outcome of such a loss about the relative importance of the X and Y chromosomes in sex determination in butterflies? **(3 marks)**

c African Swallow-tail butterflies are preyed on by birds. Female Swallow-tails are found in three distinct forms (see following figure). These resemble (mimic) three different, unrelated species which birds will not eat because they are distasteful. Swallow-tail butterflies are not distasteful to birds.

i Outline a possible genetic mechanism that would restrict these mimic forms to females of the species. **(3 marks)**

ii Suggest how resemblance to a distasteful species may be advantageous to animals which are not themselves distasteful to predators. **(2 marks)**

iii Most organisms showing this kind of mimicry resemble only one distasteful species. What additional advantage may African Swallow-tail butterflies receive by mimicking three different distasteful species? **(2 marks)**

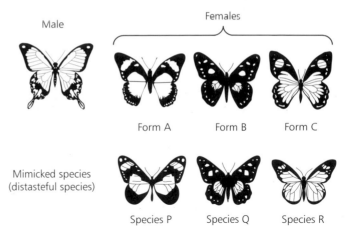

iv Comment on the fact that mimicry protects female African Swallow-tail butterflies but not males. **(2 marks)**

v If the species labelled Q were to become extinct in a certain area, what changes might occur in the local population of African Swallow-tail butterflies? Explain your answer.
(3 marks)
(Total 20 marks)
(ULEAC June 1990)

10 Read through the following passage and then answer the questions which follow.

The Siberian tiger is an endangered species; there are probably no more than 300 roaming in the wilds of Siberia. An international zoo programme has been set up to preserve the species until they can be gradually re-introduced to the wild once the world's human population has been stabilised. There are nearly 1000 Siberian tigers in captivity around the world, but they are all descended from about 20 'founder' animals. Part of the zoos' dilemma is to decide which descendants recorded in the international Siberian tiger 'studbooks' will be allowed to breed to keep the genetic stock evenly balanced.   **10**

The Bali race of tiger is already extinct while the Bengal race of tiger in the Indian sub-continent is vulnerable. To keep stocks in zoos and to breed selectively is now the only way of ensuring survival. Zoos could use contraceptive implants to prevent unwanted tigers from breeding but then precious tiger 'spaces' costing £1500   **15** each a year in food alone would be occupied. At present there are 3–4000 places for tigers in zoos around the world.

[Adapted from a report in *The Guardian*]

a The Siberian and Bengal tigers are referred to as 'races' or sub-species of a single species. State *one* reason to justify their description as races rather than as separate species.   **(1 mark)**

b Explain briefly how each of the following could have operated to produce separate races:

i *separate gene pools*

ii *evolution in different environmental conditions*

iii *natural selection.*   **(6 marks)**

c Charles Darwin noted that the members of a species produce more offspring than are required to maintain the numbers of that species.

  i State *one* example of how this is being used to advantage in the preservation of the tiger races.

  ii What factor is most likely to prevent such an increase in a wild population? **(2 marks)**

d State *two* disadvantages of a breeding programme based on a small number of founder animals. **(2 marks)**

e Explain the value of:

  i keeping a 'stud book' (line 9)

  ii gradually re-introducing the tigers into the wild only when the world's human population has been stabilised. **(4 marks)**

f **i** What position would you expect the tiger to occupy in a food chain?

  ii Relate this position to the statement in the passage that it is expensive to feed a captive tiger (lines 15–16) **(3 marks)**

(Total 18 marks)

(O&C June 1991)

11 The diagrams below show two cells, X and Y, from the same organism. The nucleus of one cell is dividing by mitosis, the other by meiosis.

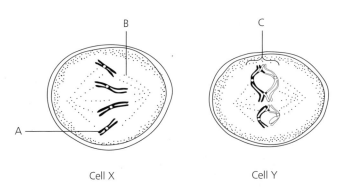

Cell X   Cell Y

a Name the structures labelled A, B and C. **(3 marks)**

b **i** Which cell is dividing by meiosis? Give *two* reasons for your answer. **(3 marks)**

  ii State *two* sites in which meiosis occurs in a flowering plant. **(2 marks)**

c Outline the roles of **i** mitosis and **ii** meiosis in the lives of flowering plants. **(4 marks)**

(Total 12 marks)

(ULEAC June 1991)

12 Squirrel monkeys are unusual in the way in which their colour vision is genetically controlled.

In squirrel monkeys, production of the pigments responsible for red and green vision is controlled by two alleles of a single sex-linked gene on the X chromosome represented by X^R and X^G. There is no such gene on the Y chromosome. Red can be seen by a squirrel monkey only if the X^R allele is present, and green can be seen only if the X^G allele is present.

a A male squirrel monkey of genotype $X^G Y$ mates with a female of genotype $X^R X^G$.

  i Construct a genetic diagram to show the possible genotypes of the offspring, and indicate their ratios. **(3 marks)**

  ii For this cross comment on the ability of the parents, their male offspring, and their female offspring, to see colour. **(6 marks)**

b Squirrel monkeys live in trees and feed largely on fruit and insects. Suggest *three* reasons why the ability to distinguish between red and green may be of survival value to squirrel monkeys. **(3 marks)**

c It has been suggested that, during the course of evolution, a mutation allowing the X^R and X^G alleles to co-exist on a single X chromosome in squirrel monkeys would be of great selective advantage, especially to males.

  i Explain why such a mutation would be advantageous. **(2 marks)**

  ii Outline the chromosomal events by which a mutation of this kind might occur. **(3 marks)**

  iii Explain how selection might spread an advantageous mutation of this sort through a population. **(3 marks)**

(Total 20 marks)

(ULEAC June 1991)

13 a The diagram below shows part of a flower with germinating pollen grains.

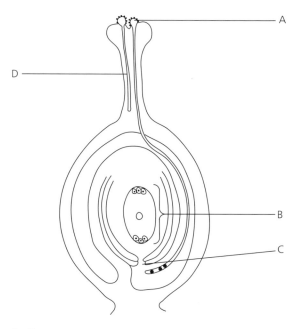

  i Name parts A, B and C. **(3 marks)**

  ii Comment on the likely fate of part D. **(2 marks)**

b State *three* ways in which pollen grains of different species may be adapted for different methods of pollination. **(3 marks)**

(Total 8 marks)

(ULEAC January 1991)

14 The diagram shows a red blood cell from a person suffering from sickle cell anaemia.

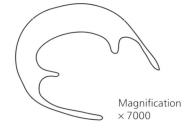

Magnification × 7000

a Suggest *two* reasons why this cell may be less efficient than a normal red blood cell in transporting oxygen. **(2 marks)**

b Sickle cell anaemia is caused by an allele of a single gene. People who are homozygous for the allele for sickled cells are severely anaemic and may die at a young age.

People who are heterozygous for this are healthy but have some red blood cells which will sickle in extremely low oxygen partial pressure.

   i What does the information above suggest about the inheritance of sickle cell anaemia (i.e. the dominance of the allele)? **(1 mark)**

   ii One of the children of a healthy couple has sickle cell anaemia.

    Using appropriate symbols to represent the normal and sickle cell alleles, write down the genotype of the parents, the child with sickle cell anaemia and the possible genotypes of the couple's healthy children. **(3 marks)**

c The allele for sickle cell anaemia arose as a mutation which changed one amino acid in a polypeptide chain forming haemoglobin. Explain how a change in the gene leads to a change in amino acid sequence. **(4 marks)**

(Total 10 marks)

(ULEAC January 1991)

15 a Name the structures indicated by the label lines on the diagram of the flower below. **(4 marks)**

b Define *pollination*. **(2 marks)**

c The flower shown in the diagram is wind pollinated. List *five* ways in which an insect pollinated flower could be expected to differ. **(5 marks)**

d After fertilisation, into what do each of structures 4, 5, 7 and 8 shown on the diagram develop? **(4 marks)**

e What does the term *double fertilisation* in flowering plants mean? **(2 marks)**

f List *four* features found in flowers which increase the chances of cross-fertilisation (outbreeding). **(4 marks)**

(Total 21 marks)

(O&C June 1991)

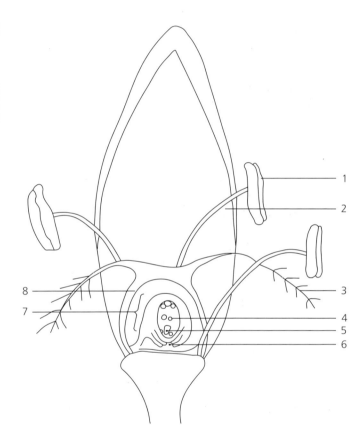

POPULATION
BIOLOGY

Integrated pest management – food for the future?

Producing enough food for an ever-growing population is one of the biggest problems facing the human race. The world population is fast approaching over 6 billion people – an enormous number of individuals all needing food. The difficulty of food production on this scale is not eased by the fact that up to one-third of the crops we grow worldwide are spoiled by pests. Animals (particularly insects) and plants can wreak untold damage to stored food, to crops, to agricultural land and to health. The battle to control them is all-out war.

Chemical control

For many years the main weapons in the arsenal turned on crop pests were chemical. From early sprays of tar and copper sulphate to the sophisticated modern cocktails applied ever more frequently, chemical warfare has been very successful in reducing crop damage. The current high crop yields which have led, in the developed world at least, to reasonably priced food available all year round, owe much to chemical pest control. However, for a variety of reasons, there is now a move towards alternative methods of controlling the pests which are the scourge of agriculture worldwide. One of the most promising of these is **biological pest control**.

Biological control

Biological pest control involves using one living organism to control the activities of another. In the developed world we have become increasingly concerned about the long-term effects that chemical pesticides and herbicides may have on us as we eat our well-sprayed food. We have also become progressively 'greener' over recent years, with more and more people expressing concern over the future of the Earth and our effect on it. Substituting biological control for chemical intervention therefore seems like a very good idea.

The developing world cannot yet afford such concerns – the main struggle for many developing nations is to be able to feed all the hungry mouths. But in these countries too the cost of chemical control and the increasing resistance of pests to the expensive chemicals is adding another powerful voice to the arguments in favour of biological control as an integrated part of pest management.

So how does biological control work? In a natural **ecosystem** (a community of interdependent living things) a balance is set up. This balance is between the plants that provide the primary source of food (the **producers**), the animals that eat the plants (the **primary consumers**) and the animals that eat the animals that eat the plants (the **secondary consumers**). Chemical pesticides destroy

this delicate balance. Biological pest control attempts to deal with the pest without destroying this balance. There are three different approaches, discussed below.

1 Encouraging natural enemies

If the natural enemies of a pest species can be conserved and encouraged, they may well control the pest as effectively as any spray, and far more cheaply. A good example of the effectiveness of this approach is the control of rice pests in Indonesia. In the 1970s the development of high-yielding strains of rice and increasing use of fertilisers and pesticides allowed two rice crops to be grown each year instead of one. Unfortunately it also led to an enormous growth in the population of the brown planthopper (*Nilaparvata lugens*) which is a devastating pest of rice plants. Farmers were spraying up to eight times a season to try to reduce the damage done by this pest, with huge Government subsidies to help with the cost. It was then shown that spraying had caused the problem in the first place!

The sprays had wiped out all the natural predators of the brown planthoppers, particularly spiders, and yet had only limited effect on the pest itself. An **integrated pest management** (IPM) system was introduced. The Indonesian Government reduced the subsidies on chemical sprays and banned the use of 57 insecticides on rice. A nationwide training programme was set up to help farmers conserve predators such as spiders. It only costs about £10 to train each farmer – but there are about 7 million farmers in Indonesia alone! In spite of this, by the third year following the ban, pesticide use had been reduced by 90% with large savings in

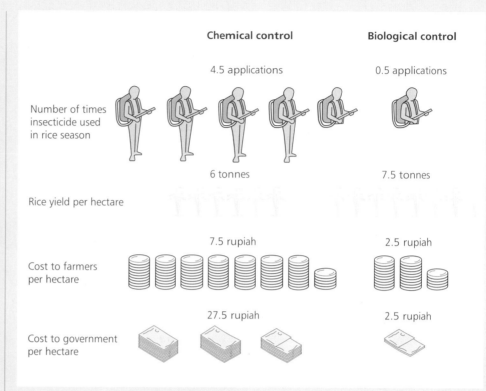

Figure 1 The figures speak for themselves! These charts show the effect on rice farming in Indonesia of introducing an integrated pest management system.

cost for both the farmers and the Government – and the yields of rice were increasing. Spraying is now only considered as a last resort. Similar IPM programmes for rice are being successfully introduced in Bangladesh and India.

At Southampton University there is a team investigating the possibility of biological control of the grain aphids which damage our wheat crops by sucking the sap. The larvae of hoverflies are voracious predators of aphids. The plan is to have strips of 'weeds' in and around fields of growing grain. The 'weeds' will be flowering plants that attract female hoverflies, which need protein from pollen to successfully produce eggs. The plan is that the flies will move out into the crop after feeding and lay their eggs near the aphids.

Once the larvae hatch they attack and destroy the aphids with ever-

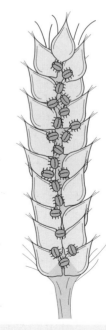

Figure 2 The attentions of these tiny insects can severely reduce the yield of wheat. It is hoped that weeds and hoverflies between them will control the problem in years to come – resulting in more food for all.

6

increasing enthusiasm. The older larvae can each eat between 30 and 40 aphids a day. The cost of using land for flowering plants to attract hoverflies rather than to grow crops should be more than regained by savings on chemical pesticides. If the idea is shown to work, it would be yet another example of successful IPM – and would also add extra colour to the countryside!

2 Biological pesticides

We are not generally used to being told what we should wear in the evening. But thanks to *Simulium posticatum*, down in Dorset everyone in certain areas has been advised to wear trousers (preferably with cycle clips as well) and thick socks before going out on early summer evenings! *Simulium posticatum* is a tiny black fly, known locally as the Blandford fly. It infests the river Stour and has a vicious bite – hence the trousers and cycle clips! A blister and swelling form around the site of the bite and some people suffer fevers, swelling of the lymph glands and dizziness. A bite on the face can close an eye!

A method of biological control was tried on a small tributary of the Stour. A biological pesticide was introduced to the water in the form of *Bacillus thuringiensis israelensis* (BT). This bacterium kills the larvae of the Blandford fly in the river without damaging other forms of life or polluting the water. BT is a biological pesticide which has a variety of forms. It has been used internationally with a great deal of success against many of the appalling fly-borne diseases of the developing world. A particular success story has been its use to clear thousands of square miles of West Africa of the organism which carries river blindness, a disease that has affected almost every family in the past. Best of all, in removing river blindness, the African waterways have not been chemically polluted and the remaining fauna is unharmed.

3 Exotic enemies

The third, and perhaps best known, form of biological control is the release of an exotic species from one country to control a pest in another. The introduction of the vedalia beetle from Australia onto the citrus groves of California saved the Californian citrus growers from destruction by the cottony cushion scale, a pest that was completely

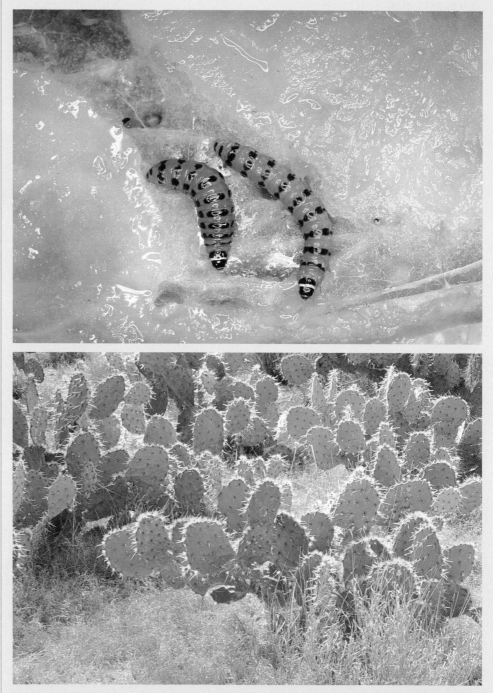

Figure 3 In 1920 these large cacti were taking over the Australian countryside in much the same way as bracken is spreading in Britain today. The introduction of an Argentinian moth controlled the situation. In Britain it is hoped that the caterpillar of the *Conservula cinisigna* moth will perform a similar function in controlling the spread of bracken.

destroying the fruit trees. This happened a century ago and since then exotic species have played their part in the battle to preserve our food from the multitude of pests who would like to share it with us.

Nowadays the introduction of a new exotic species to a country is only carried out after extensive research to make sure that it will do the job required and nothing else. It would be catastrophic if a beetle introduced to control a plant pest turned out to prefer a diet of corn or potatoes and laid waste to acres of food crops!

Biological control of weeds by the introduction of herbivorous insects has been successful in many parts of the world. Australia in the 1920s was threatened with a take-over by the prickly pear cactus *Opuntia*. It was controlled by the introduction of the Argentinian moth *Cactoblastis cactorum* which cleared over 60 million acres. When biological control of weed pests is achieved it is permanent, very cost effective and environmentally friendly.

Exotic and alien organisms have been used for biological control for over a century now. To begin with, introductions were not screened thoroughly and problems arose, but lessons have been learned and several thorough steps now have to be taken before the release of an exotic organism can be considered.

How is a biological control system developed?

When a new system of biological pest control is considered, a great deal of background work is necessary to discover whether the pest organism is a suitable target for this type of control. Research is expensive, and an initial study is needed to determine whether the money will be well spent.

Then a range of possible control organisms are collected. As far as possible, these need to be species specific and unlike any naturally occurring organism in the country affected by the pest. Laboratory trials demonstrate the effectiveness or otherwise of the control organism, the specificity of its action and its effect on the environment as a whole. Only after the successful completion of all of these tests and meeting international safety guidelines will a biological pest control organism be released.

What does the future hold?

People are increasingly aware of the value of integrated pest management systems, where natural predators are preserved and chemical sprays used as a last resort. This puts management of the land back in the hands of individual farmers. It is cheaper both for the individual farmer and for governments freed from subsidising the purchase of expensive chemicals, and the lowered doses of pesticides make the prospect of eating the food more pleasant too.

Genetic engineering and pest control

The use of biological pesticides and the careful importation of exotic species are both full of promise for the years to come in many areas of the world. But one area associated with biological control which does raise some important questions is the use of genetically engineered organisms in pest control.

What is the aim of manipulating genes to produce biological control agents? The work is very much concentrated on pathogens. Some of the

gene manipulation is designed to increase the virulence of existing pathogens – creating 'super-pathogens' which are much more effective at attacking the pest. Other engineering work is designed to increase or change the range of organisms affected by the pathogen.

The other thrust of genetic engineering is to introduce genes from insect pathogens into plants, so that when an insect feeds on a plant, it also takes in a dose of toxin. Genes from *Bacillus thuringiensis* (BT) have been introduced into various crop plants, including cotton, potatoes and tomatoes. Although very effective, a problem has arisen. Following the repeated exposure of the insects to the toxin, for the first time resistance to a biological control method has been found. This poses profound difficulties, because BT is potentially one of the most useful organisms for biocontrol when targeted against very specific pests. It would be a tragedy if this potential were lost due to the injudicious use of engineered crop plants.

This area of biological control obviously opens up many avenues of exciting research. The answer to many pest problems may well lie with engineered organisms. But it does seem important that stringent safety measures are applied when introducing totally new and alien genetic material into the environment. Up until now biological control has an excellent safety record, with over 4000 introductions of exotic agents against insects and 1000 against weeds worldwide, and with no major accidents (although the control has not always worked as well as expected!). In fulfilling all the hopes for a pest-free, well-fed world, long may that record continue.

6.1 The origins and diversity of life

Throughout this book the way in which living organisms are built up and function has been examined in some detail and the wide variety of living organisms has been indicated. So far we have focused on individual organisms, whereas most of this final section deals with interactions between individuals and groups of living things. Over the years many attempts have been made to classify the great variety and diversity of living organisms into ordered groups, and as well as the groups themselves we shall consider how organisms have arrived at their present state (their **evolution**), and how they interact with their environments and with other organisms (their **ecology**).

Figure 6.1.1 The diversity of life – the biosphere contains millions of different organisms which are interdependent in many ways, and can colonise even the most daunting of habitats.

THE ORIGINS OF LIFE

For some, the origins of life on this planet are a matter for religious faith rather than scientific thought. But for many others the events of the beginning of life on the Earth offer the challenge of a puzzle which can never be completely solved. Theories about the origins and evolution of life are still being discussed, although amongst most scientists a consensus has been reached on the most probable pattern of events, which is described here. Around 4500 million years ago there was no life on Earth. To date more than four billion species of living organisms have appeared, most to disappear again in the mists of time and others to persist until the present day.

The formation of monomers

Before life emerged on the planet, the organic molecules which are the building blocks of life had to be formed. In the 1920s two scientists – the Russian Aleksandr Oparin and the Briton J. B. Haldane – developed a revolutionary theory that the energy of ultraviolet light, heat, other radiation or massive lightning discharges could have catalysed the formation of small organic molecules or **monomers**. In the 1950s these predictions were tested experimentally by Harold Urey and Stanley Miller using the apparatus shown in figure 6.1.2, and large quantities of amino acids and simple sugars were produced.

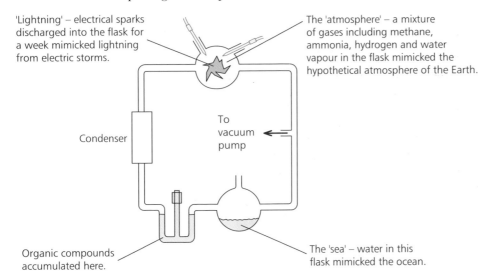

'Lightning' – electrical sparks discharged into the flask for a week mimicked lightning from electric storms.

The 'atmosphere' – a mixture of gases including methane, ammonia, hydrogen and water vapour in the flask mimicked the hypothetical atmosphere of the Earth.

Condenser

To vacuum pump

Organic compounds accumulated here.

The 'sea' – water in this flask mimicked the ocean.

Figure 6.1.2 The successful production of organic compounds by Miller and Urey in this apparatus gave support to the theory of chemical evolution.

Current evidence suggests that the atmosphere of the early Earth was in fact much more similar to the present atmosphere, though without the oxygen, than was hypothesised by Miller and Urey. By using various combinations of gases up to 100 monomers have been produced – almost all the building blocks of living cells.

From monomers to cells

The chemical mix of monomers which, in theory, formed in the earliest stages of the development of life moved on with the formation of polymers. These long-chain molecules could have been produced using the energy of the Sun or the heat of the Earth as an energy source. Alternatively, ATP is one of the chemicals produced in a Urey–Miller chamber, and this could have supplied the chemical energy for polymerisation.

Polymers may form aggregates, groups of molecules which begin to exhibit some of the characteristics of living things – for example, aggregates of lipids will show membrane-like properties. From then on, through a series of steps open to hypothesis, what we consider to be 'life' emerged. It seems to have been the inevitable consequence of a variety of physical and chemical processes, though no one will ever know for certain when, where and how life on Earth began.

The earliest cells

The oldest remains of living organisms are fossil rod-shaped structures found in rocks in Australia which have been dated as 3.5 billion years old. The first cells were almost certainly anaerobic and heterotrophic. They would have fed on the organic chemicals which formed spontaneously in the atmosphere and waters of the Earth. Mutations occurred which resulted in cells with the ability to exploit another energy source – sunlight. These cells would have been the first autotrophs. From single-celled organisms such as these early microbes all subsequent life on Earth has evolved. The evidence for this evolution will be considered in more detail in this section. In section 6.3 we shall look at the immense variety of life now present on the Earth as a result of the evolutionary process.

THEORIES OF EVOLUTION

The diversity of life

The range of organisms on the Earth is vast (see section 6.3). There are microorganisms, so small that their existence only became known with the development of the microscope. They are capable of life in almost every available niche, and yet we are usually unaware of them except when they cause disease. Fungi of different types decompose dead and dying organic material. Trees form forests, produce oxygen and provide timber. Birds fill the air, fish and many other diverse creatures fill the seas, lakes and rivers. Animals and plants of every description are part of our lives, and we in turn are part of the great diversity of living things. How has this diversity of life come about?

The background to evolutionary theory

For many years the commonly held belief was that the Earth and all that is on it were spontaneously created by God over a period of six days, as described in the first chapter of Genesis in the Bible. The implication of this belief was that all of the species present on the Earth had been there from the beginning and were unchanging. However, by the end of the eighteenth century some scientists were beginning to put forward a scientific explanation for the existence of life on Earth. Simple observation of the development of farm animals seemed to indicate that living things could and did change from generation to generation. In 1809 (the year Darwin was born) Jean Baptiste de Lamarck presented a new evolutionary theory. He accepted the idea that life forms evolve, and proposed that the driving force behind this evolution was the **inheritance of acquired characteristics**. In other words, organisms change physically as they struggle to meet the demands of their environment, and these advantageous changes are passed on to their offspring. His classic illustration was the evolution of the giraffe as he saw it, as shown in figure 6.1.3.

Figure 6.1.3 'Use it or lose it' and 'Use it and it may improve' were the key ideas of Lamarck's theory. His idea was that over a series of generations the neck got longer and longer as giraffes stretched to reach the topmost leaves.

We now know that Lamarck's theory is not correct, because activities such as stretching to feed do not affect the germ cells or the gametes. However, research such as the work on learning behaviour in flatworms (section 4.3, page 267) has led to recent discussion that there may be unexpected Lamarckian elements in the process of evolution after all.

Darwin's theory of evolution

Charles Darwin set off on the surveying ship HMS *Beagle* when he was in his early twenties. The ship circumnavigated the world, taking five years to do so. During that time Darwin read, amongst other things, books on geology by Lyell speculating that, on the basis of the fossil record, animal and plant species arose, diversified and then died out. Darwin also spent hours observing the wide range of living organisms, both plant and animal, in the various countries where they stopped. His most famous observations were made on the Galapagos archipelago (group of islands). On returning to England Darwin read an essay by Thomas Malthus proposing that populations grow faster than their food supplies and so organisms are forced into a 'struggle for existence'. There followed a long period of thought, during which Darwin made copious observations on breeding pigeons and other domestic animals, until eventually he wrote his now famous work *On the Origin of Species by Natural Selection, or the Preservation of Favoured Races in the Struggle for Life*, published in 1859. A large part of Darwin's book argues that **evolution** has taken place. He then explains a possible mechanism by which evolution has come about, and that is **natural selection**.

The main argument put forward by Darwin was as follows. Living organisms which reproduce sexually show great variety in their appearance (or phenotype as we would now describe it). Also, more offspring are produced than survive so that there is always a struggle for existence, a competition between members of the same species. Those organisms which inherit characteristics that put them at an advantage in this struggle are most likely to survive and reproduce, passing on the desired feature to their offspring. Organisms which possess features which put them at a disadvantage will be more likely to die out and thus fail to reproduce themselves. This can be summed up in the phrase 'the survival of the fittest', where fitness is the ability of an organism to survive and reproduce in the environment in which it is living. He called this process **natural selection**, and the long-term changes in organisms which occur as a result of natural selection are **evolution**. Alfred Wallace, another British naturalist, proposed a very similar theory at the same time as Darwin, but Darwin's ideas were backed up by a far greater bank of research and observations and so it was he who received most of the accolades (and criticisms) which went with the theory.

What is a species?

A species is a specific group of closely related organisms which are all potentially capable of interbreeding to produce fertile offspring.

Figure 6.1.4 Darwin's ideas in many ways reflected the changing mood of his times. But for some of the more traditional thinkers and the hierarchy of the Church, his ideas were heretical and damaging to the status quo.

Darwin speaks

Charles Darwin was a modest and retiring man who spent most of his time at home with his wife and children. Yet his mind was continually active and he produced one of the most revolutionary biological ideas of all time – the theory of evolution by natural selection has affected all subsequent studies. In the summary of his work *On the Origin of Species* he writes:

> *As more individuals are produced than can possibly survive, there must in every case be a struggle for existence, either one individual with another of the same species or with the individuals of a distant species, or with the physical conditions of life. . . . Can it therefore be thought improbable seeing that variations useful in some way to each being in the great and complex battle of life, should sometimes occur in the course of thousands of generations? If such do occur, can we doubt (remembering that many more individuals are born than can possibly survive) that individuals having any advantage, however*

slight, over others would have the best chance of surviving and of procreating their own kind? On the other hand, we may feel sure that any variation in the least degree injurious would be rigidly destroyed. This preservation of favourable variations and the rejection of injurious variations, I call natural selection.
(Charles Darwin, *On the Origin of Species*, 5th edition 1896)

Darwin's book was an immediate success – every copy of the first edition was sold the day it was published. When we remember that Darwin's ideas were formulated with no knowledge of genetics, his perceptiveness and vision become more remarkable. He is rightly remembered as one of the truly great figures in biology.

Neo-Darwinism

Throughout the twentieth century our knowledge of genetics, molecular biology, ecology and palaeontology have grown at an astonishing rate. As a result of some of this knowledge, Darwin's theory has been slightly modified to give what is called **neo-Darwinism**. Neo-Darwinism considers that the evolution of organic forms occurs as a result of the **differential survival and fertility** of organisms with different genotypes (and therefore phenotypes) within an environment. In other words, a disadvantageous trait need not mean that those individuals are wiped out, simply that they are less reproductively successful than others, for evolution to occur. It also indicates that what constitutes an advantageous or disadvantageous trait will differ with the environment. We have only to think of the sickle cell allele considered in section 5.7 to recognise the truth of this.

EVIDENCE FOR EVOLUTION

Where Darwin leads...

The importance of Darwin's ideas in the study of ecology and the changing populations of organisms on Earth today will be considered in section 6.6. Here we shall go on to look at the picture of evolution which has emerged over the years since Darwin published *On the Origin of Species*, and the various strands of evidence which support the ideas.

The evolution of the different phyla – reptiles, fish, mammals, angiosperms, etc. – can be seen as the result of many small changes resulting in new species. These major changes or **macroevolution** occurred over millions of years, in a time scale measured by geologists going back to the beginnings of the planet Earth. The most important developments are summarised in figure 6.1.5. The information in a diagram such as this comes from a variety of sources, relying very heavily on information from the fossil record (see page 434). As you can see, human beings have been on the Earth for a very short time. If the history of the Earth was represented by one year, the first bacteria appeared around March, and the first photosynthesisers arrived in the sea in May. In contrast the first vertebrates, primitive fishes, appeared around 20 November. Amphibians crawled onto the land around 30 November. The reptiles became dominant around 7 December and the mammals became common by 15 December. The human genus *Homo* did not arise until 11 p.m. on 31 December!

There are many strands of evidence which can be applied to evolution. These include palaeontology (fossil studies), the distribution of animals across continents, comparative anatomy, embryology, biochemistry and cell biology. We shall look at each in turn.

Era	Period/epoch		Millions of years ago	Geological events	Biological events – plants	Biological events – animals
C E N O Z O I C	QUATERNARY	Holocene		Cooling after middle of period. Glaciations. Continents separate.	Gymnosperms and angiosperms become widespread. Grasslands and forests expand.	Modern *Homo sapiens* arises 0.135 million years ago.
		Pleistocene	2			
	TERTIARY	Pliocene Miocene Oligocene Eocene Palaeocene	65			Mammals diversify, primates arise.
M E S O Z O I C	CRETACEOUS			Last expansion of shallow seas. Rocky mountains arise and climate cools.	Angiosperms arise and diversify.	Major extinction event. Most large reptiles and ancient birds extinct.
	JURASSIC		136	Very stable climate.		Teleost fish diversify. Dinosaurs dominant. Modern crustaceans present.
	TRIASSIC		190	Land high, widespread deserts.	Gymnosperms and ferns dominant.	Dinosaur ancestors. First mammals and birds.
P A L A E O Z O I C	PERMIAN		225	Cold climate slowly warming.	Conifers appeared.	Mammal-like reptiles common. Major extinction of the invertebrates and amphibia.
	CARBONIFEROUS		280	Lowland forests, coal swamps, mountain building.	Forests widespread – coal deposits formed.	First reptiles. Amphibians diversify.
	DEVONIAN		345	Continents rising. Appalachian mountains forming. Cooler climates.	First forests. Vascular plants and seeds present.	Fish diversify. First insects, sharks and amphibians.
	SILURIAN		395	Land uplifting slowly. Still extensive seas.	Green, red and brown algae common.	First land arthropods. Jawed fish arise.
	ORDOVICIAN		430	Warm shallow seas extensive. Warming continues.	First vascular land plants probably appeared.	Second major extinction event. Jawless fish and molluscs diversify. Large invertebrates present.
	CAMBRIAN		500	Warm climate. Huge equatorial shallow seas.	Algae dominant.	First major extinction event. Trilobites common. Evolution of many phyla.
A R C H A E O Z O I C	PRECAMBRIAN		570 1500		Algae abundant. Cyanobacteria diversified. Eukaryotes present.	Worm-like animals and cnidarians.
			2500	Atmosphere oxidising. Shallow seas.	Origin of life. Prokaryotic heterotrophs, chemical evolution.	
	Origin of Earth		3000 3500 4000	Earth's crust hardens. Sea, atmosphere and rocks develop? Formation of the Earth?		

Figure 6.1.5 A summary of the main events in the evolution of life on Earth

Palaeontology – the fossil record

Fossils are the remaining traces of organisms that lived millions of years ago. Many of them are the hard body parts left behind when an organism dies – bones, shells, carapaces and tough plant cell walls – whilst others are tracks, burrows, nests, eggs and faeces left behind in sediments and transformed into rocks. In a few rare cases fossils of entire animals with the skin, tissues and organs still intact have been found in ice or in amber. In general the fossil record depends on the chance preservation of organisms millions of years ago, the chance survival of the fossil-bearing rocks, and the chance discovery of those fossils in an accessible area by people who recognise what they are, and so it gives only a tantalising, fragmented picture of the evolution of living organisms.

The main problems with the fossil record are firstly that there is no fossil record of the vast majority of the early organisms of the Earth, or indeed of many of those that followed. Soft-bodied creatures have few fossilisable parts and thus a substantial part of the early history is poorly represented. Secondly, although certain lines of descent such as that of the modern horse (see figure 6.1.6) are very well documented, other vital evolutionary milestones such as possible links between various fish and the amphibians are lost without trace.

Thirdly, there is the difficulty of aging fossil finds. Radioisotopic dating is the method most used, with radiocarbon dating being the most common. This measures the amount of radioactive carbon 14 in a sample, and with knowledge that the half-life is 5730 years, the age of the fossil can be calculated. Unfortunately this half-life is very short in evolutionary terms, and so other elements such as uranium and potassium are used. Dating a single

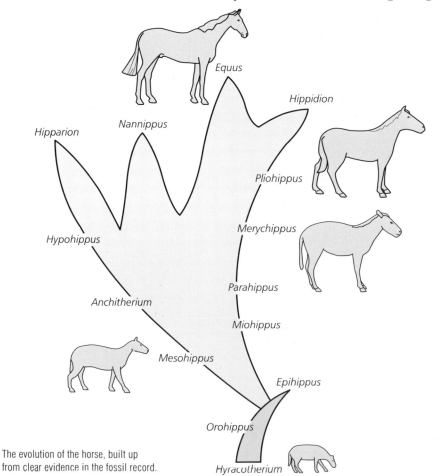

The evolution of the horse, built up from clear evidence in the fossil record.

The fossil record can provide us with a glimpse of some of the animals and plants which lived on Earth before the advent of humans.

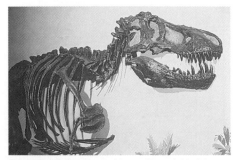

Figure 6.1.6 Fossils provide us with useful insights into early life but the record is limited and fragmented – we need evidence from other sources.

fossil is only part of the problem. When fossils are found together in a large geological formation they may not all originate from the same time – rocks move and a fossil bed may encompass thousands or millions of years.

Thus fossils give us incontrovertible evidence of life forms which have long since died out, and show us, in some cases, the evolutionary pathways to modern animals and plants. But there is a limit to what fossils can tell us and so other strands of evidence are needed.

Distribution studies

The animals and plants of the world are as likely to show surprising differences as they are to be similar. The flora (plants) and fauna (animals) of the two continents of the northern hemisphere are broadly similar, but the organisms of the southern continents of Africa, Australia and South America are not only different from those of the north, but also markedly different from each other. A look at the events in figure 6.1.7 shows that throughout the history of the Earth, the structure of the great land masses has changed. During the Permian period all the land was joined into one supercontinent referred to as Pangaea. At this stage there was therefore no geographical separation of the different groups of organisms. As time progressed the southern continents became separated from the northern ones and from each other, as shown in figure 6.1.7. As a result the same ancestral fauna evolved in different directions to fill the available niches. This is known as **adaptive radiation**.

The Australian marsupials

Australia shows a particularly good example of this adaptive radiation. The land link between Australia and the other southern land masses was broken relatively late, when vertebrates were well established and the first steps towards the true mammals – the monotremes and the marsupials – were already in place. However, with isolation from the rest of the world, true placental mammals never reached Australia until they were introduced at a much later date by people. Marsupials filled all the available 'mammal' niches as there was no competition from the better adapted placentals.

Figure 6.1.7 The differences and similarities between groups of animals can be explained by considering at which point they became evolutionarily isolated. Once isolation occurred the animals and plants continued to evolve independently, undergoing adaptive radiation.

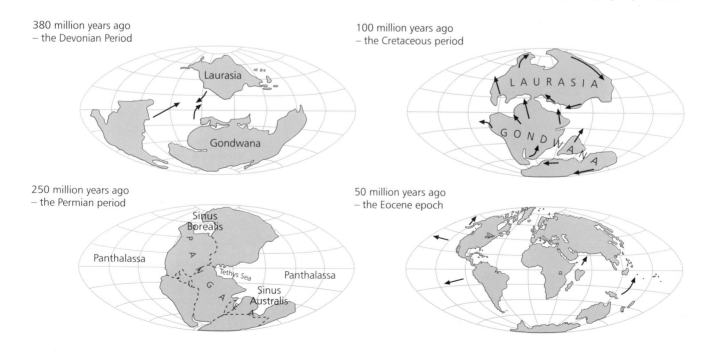

Darwin's finches

The effects of geographical isolation are shown very clearly by oceanic islands. A good example is the Galapagos archipelago which had a profound influence on Darwin when he observed its unique fauna and flora. Organisms which have migrated to the islands early in their history have developed quite independently of their mainland relations, and often in a variety of ways. For example, the birds known as 'Darwin's finches' (incidentally, these were not mentioned in *On the Origin of Species*) are 12 separate species of finch which are all assumed to have evolved from one original type. In the absence of other birds they adapted to fill all the available niches, evolving a wide range of sizes and beaks suited to a variety of foods.

Parallel evolution

Geographical isolation does not always result in very different animals. Porcupines have evolved independently in both Africa and South America. It seems that, more than 70 million years ago before the two land masses separated, the porcupines shared an ancestor resembling a large furry rat. In similar natural surroundings, what occurred was an example of **parallel evolution**. The two types of porcupine are remarkably similar in all their body features, including the sharp hollow spines which have evolved for defence, as shown in figure 6.1.8.

Comparative anatomy

The basic idea of evolution is that all living things evolved from one common ancestor, and that some animals and plants are more closely related than others. For example, the common ancestor of humans and chimpanzees is closer in time to us than the common ancestor of humans and rabbits, and that in turn is closer than the common ancestor of a person and a fish, or a person and a sea anemone. Evolutionary relationships can be deduced by similarities between organisms, and this is the study of **comparative anatomy**.

Convergent evolution

Interesting patterns of evolution are revealed in this way. Some organisms have features which at first sight are similar, for example, vertebrates, insects and octopuses all have large well developed eyes and demonstrate good vision. However, these organisms lack a recent common ancestor. The structures are **analogous** – that is, they have the same function but have evolved independently. Another example is the wings of birds and insects. In this situation, where anatomical structures evolve independently to perform the same function, **convergent evolution** has occurred, as illustrated in figure 6.1.9.

New World American porcupine

Old World porcupine

Figure 6.1.8 Seventy million years of separate development have resulted in these two remarkably similar animals in an example of parallel evolution.

Figure 6.1.9 Convergent evolution is very striking in the bodies of these very different animals which have all evolved for rapid swimming. The dolphin (a mammal), the penguin (a bird), the shark (a fish) and the Ichthyosaurus (an extinct reptile) all evolved quite separately yet arrived at a similar shape to avoid drag.

Divergent evolution

On the other hand, sometimes structures which are fundamentally closely related evolve into very different forms which, without careful consideration of their comparative anatomy, might be taken as totally unrelated. The wing of a bird and the flipper of a whale might seem far removed from each other and from the leg of a horse. However, these are all **homologous structures** – they have the same underlying biological structure even though they may perform different functions. This is the result of **divergent evolution**, where related organisms evolve to fill a diverse range of available niches, as illustrated in figure 6.1.10.

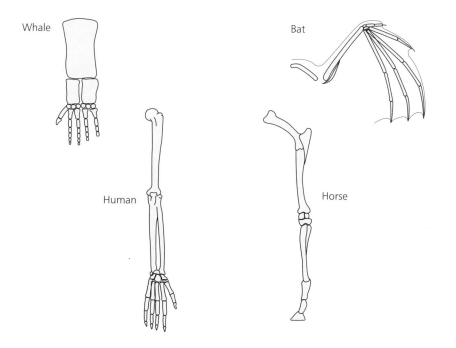

Figure 6.1.10 In mammals the same basic pentadactyl limb has evolved to perform a wide variety of tasks. This is a well documented example of the homologous structures which can result from divergent evolution.

Comparative anatomy can be used to chart the progress of evolution by comparisons of the hearts, eyes or any other major feature of the various phyla. It can confirm or deny relationships if structures can be shown to be analogous or homologous, and is a very useful tool.

Embryology

Sometimes comparing the embryos of animals or plants can reveal their true relationships. For example, as is discussed in the classification of the chordates in section 6.3, many vertebrates and particularly the mammals only show their chordate features in the embryo stage. Equally, annelid worms and molluscs are very different in their appearance as adults, yet their embryos show striking similarities and in evolutionary terms they must be quite closely related. It has been thought in the past that during embryonic development each organism 'retraces its evolutionary pathway'. It is certainly true that, for example, embryonic mammals pass through a 'fish-like' stage with gills and a tail with distinct muscle blocks, but they certainly do not pass through all the stages from an anaerobic prokaryote to a large multicellular, heterotrophic eukaryote! An old adage that 'ontogeny (embryo development) mimics phylogeny (the evolutionary pathway)' is far from the truth, although a careful consideration of comparative embryology certainly has a place in determining evolutionary relationships.

Cell biology and biochemistry

The detailed study of the structure of cells which has been made possible by the development of the electron microscope has revealed that many of the basic structures of cells, including the organelles, are common to most living organisms. This reinforces the idea that living things have a common ancestor. Added to this, biochemical analysis of many of the fundamental chemicals of life – DNA, RNA, proteins, etc. – shows that they too are almost universal. However, whilst many of these chemicals are broadly similar, differences are revealed when the molecules are broken down to their constituent parts. It has been proposed that the more similarities there are between the biochemistries of different organisms, the more closely related they are in terms of evolution. So far these biochemical relationships have reinforced or clarified relationships proposed on the basis of structural similarities.

Some examples of such biochemical findings are as follows:

- The vertebrates and the echinoderms appear, from the evidence of comparative anatomy and embryology, to come from one line of ancestors and the annelids, molluscs and arthropods from another. Biochemical evidence supports this. It shows that the **phosphagens** required to provide the phosphate group for the synthesis of ATP are of two different sorts. **Phosphocreatine** occurs almost exclusively in the muscle tissue of vertebrates and echinoderms whilst **phosphoarginine** occurs in the other groups.
- Blood pigments are important in a wide number of animal groups. Analysis has shown that any one group contains only one type of blood pigment – vertebrates and many of the invertebrates have **haemoglobin**, polychaete worms have **chlorocruorin** and molluscs and crustaceans have **haemocyanin**.
- Sequence analysis of the amino acids in particular proteins has allowed relationships within a phylum to be mapped out – for example, mammalian relationships have been investigated by analysis of fibrinogen.
- The cytochromes in the electron transfer chain can also be analysed and those in related species show remarkable similarities.

Extinction

Extinction is the permanent loss of all members of a species. Well known examples include the dinosaurs, sabre-toothed tigers and the dodo. What is perhaps less well known is that extinction has been the fate of the vast majority of species that have ever evolved. Of the estimated four billion species that have evolved since life appeared on Earth, only a few million are still in existence today. Individual species are continually disappearing for a range of reasons. At the present time humans are responsible for many extinctions by habitat destruction and pollution. But this is overshadowed by the **mass extinctions** which appear to have occurred at times in the history of the Earth – for example, at the end of the Permian period 96% of all the then-living marine invertebrate species became extinct. At least five of these mass extinctions appear to have occurred. Even species that are well adapted under normal circumstances may be wiped out as their adaptations become irrelevant under the extremes of environmental stress which seem to be responsible for such extinctions. These major eradications of large numbers of species seem to have cleared the way for the evolution of new and opportunistic groups, and thus to have had a major effect on the pattern of life on Earth.

Figure 6.1.11 The mass extinction of the dinosaurs took place over a very long period of time. Removal of large numbers of species of plants or animals opened the way for the relatively rapid evolution of new species to fill the available niches.

Is evolution the whole story?

Although the majority of the scientific community of the developed world accept the ideas of Darwin as the basis for the evolution of life on Earth, the discussions continue. There will always be some who *believe* in Creation – as they have every right to do, for belief and science are very different things and belief requires no evidence. Many other scientists accept the evidence available for evolution, yet still believe in God. Yet others *believe* in evolution, and regard God as a figment of the human imagination. And many individuals have different views on the details of the process by which life has evolved on Earth. Whilst these diverse viewpoints continue to exist, discussion and indeed argument will always be maintained.

SUMMARY

- The organic monomers that make up the large polymeric compounds in living things could have been produced by the action of lightning or radiation from the Sun on the early sea and atmosphere, as demonstrated by Miller and Urey. The monomers may have been polymerised by the action of the Sun, or using energy from ATP.

- The earliest known cells (3.5 billion years) were anaerobic and heterotrophic, taking in organic molecules from the ocean. These cells subsequently evolved to form all the life now present on Earth.

- Lamarck's evolutionary theory proposed that evolution takes place by **inheritance of acquired characteristics**. This is now largely disproved.

- Darwin's theory says that due to the variety among offspring within a species, those organisms best suited to their environment are most likely to survive and reproduce because of the **struggle for existence** – more offspring are produced than ever survive into adulthood. The favourable characteristics are thus passed on to the next generation. Changes in the environment will bring about a change in the characteristics by selecting for favourable characteristics within the variation normally present. **Evolution** is brought about by **natural selection**.

- **Neo-Darwinism** considers that evolution occurs as a result of the **differential fertility** of organisms with different phenotypes – those that reproduce the most successfully will be selected for, and advantageous features vary according to the environment.

- The history of the Earth is divided into four main **eras** – the Cenozoic, Mesozoic, Palaeozoic and Archaeozoic. These are subdivided into **periods** and **epochs**. The evolution of plants and animals can be traced through these periods using the fossil record. There were several mass extinctions of animals, and humans appeared relatively recently in the evolution of the Earth.

- **Palaeontology** is the study of **fossils** – the remains of the hard body parts of living organisms, or the remains of their imprints (tracks, burrows, etc.) preserved in rocks, ice, amber, etc. There are few fossils of soft-bodied creatures. The finding of fossils is largely dependent on chance, and fossils are difficult to date accurately.

- **Distribution studies** show how organisms have evolved differently when separated geographically. The continents were originally joined in one supercontinent called **Pangaea**. As the continents drifted apart the same ancestral forms evolved differently in **adaptive radiation**. **Parallel evolution** is the separate evolution of similar organisms from the same ancestral forms.

- **Comparative anatomy** uses the similarities between organisms to deduce how recently they had a common ancestor. **Convergent evolution** results in structures that are **analogous** – they perform the same function but have evolved independently. **Divergent evolution** results in features that have similar anatomical structures but perform different functions – they are **homologous**.

- **Embryology** can reveal relationships between organisms that are not obvious in the adults.

- **Cell biology** and **biochemical analysis** can reveal differences and similarities between organelles and chemicals within cells, confirming whether they had a recent common ancestor.

QUESTIONS

1 Explain the scientific hypothesis for the beginning of life on Earth.

2 Charles Darwin introduced his theory of evolution and the mechanism of natural selection by which it occurred in the nineteenth century. Discuss the evidence now available to back up or refute his theories.

6.2 Population genetics

The complexity of the evolutionary events leading to the development of the vast array of modern species can seem overwhelming. Looked at from the viewpoint of individuals in populations the mechanisms of evolution become easier to understand.

POPULATIONS

Defining terms

Before we begin to look at population genetics, we need to know what is meant by the term 'population'. In biological terms, a **population** is a group of individuals of the same species occupying a particular **habitat**. The habitat of an organism is the place where it lives, and takes into account both the physical and biological elements of the surroundings. This will be discussed in more detail in section 6.4.

In population genetics the gene is taken as the unit of evolution. The genetic make-up of a population evolves over time. The sum total of all the genes in a population at a given time is known as the **gene pool**. For all the genes contained within the genome of a population the gene pool will run to millions or even billions of genes. However, at any one time it is usually the gene pool for a particular trait which is considered. This tends to be a more manageable number of genes of around several thousand. At any point in time a population of organisms will have a particular gene pool, with different alleles occurring with varying frequencies. Evolution may be considered as a permanent change in **gene frequencies** within a population. What does this mean and how is it measured?

Gene frequencies

The **frequency** of a particular allele in a population is a statement of how often it occurs. This is usually expressed as a decimal fraction of 1. Surprisingly, the frequency with which a gene occurs within the population has little bearing on whether it is a dominant or recessive allele, as figure 6.2.2 illustrates. Take an imaginary gene with two possible alleles **X** and **x**. If all of the individuals in a breeding population of 100 are heterozygous, then the frequency of each allele is 100/200 or 0.5. For example, tests were done on a sample of white Caucasians living in the USA to determine the frequencies of the alleles for the MN blood groups, L^M and L^N (see section 5.6, page 399). These turned out at L^M 0.539 and L^N 0.461.

There is a general formula which can be used to represent the frequency with which the dominant and recessive forms of an allele occur in the gene pool of a population. The frequency of the dominant allele is represented by the letter p and the frequency of the recessive allele is represented by the letter q:

$$p + q = 1$$

The frequency of the dominant allele plus the frequency of the recessive allele will always equal 1. Thus for the MN blood group:

$$0.539 + 0.461 = 1$$

Evolution involves a change in the allele frequencies within a population.

Figure 6.2.1 When the frequencies of the genes within a population change, favouring one phenotype over another, then evolution can be seen to be taking place. These photographs show the white and blue forms of the lesser snow goose.

This simple equation is of very limited use as it stands, because it is almost impossible to measure the frequencies of alleles or the ways in which they change. What can be observed readily in a population is the distribution of phenotypes. In 1908 the British mathematician G. H. Hardy and the German physician W. Weinberg independently developed an equation for stable gene frequencies within a population, which can be solved using observable phenotypes.

The Hardy-Weinberg equilibrium – an ideal case

The algebraic equation developed by Hardy and Weinberg describing stable gene frequencies is:

$$p^2 \quad + \quad 2pq \quad + \quad q^2 \quad = \quad 1$$

p^2	$2pq$	q^2	= 1
Frequency of homozygous dominant individuals in population	Frequency of heterozygous individuals	Frequency of homozygous recessive individuals	Total

Figure 6.2.2 The frequency of a phenotype in a population is the result of the frequency of particular alleles and that is independent of the dominant or recessive nature of the allele. Thus blue-eyed, blond-haired children (both features the result of recessive alleles) are not uncommon in this country, and in Scandinavia they are the most common phenotype.

One of the main problems in finding gene frequencies is that it is not possible to distinguish between the homozygous dominant and the heterozygote based on the appearance of the individuals. However, using the Hardy-Weinberg equation these gene frequencies can be calculated from the number of homozygous recessive individuals in the population. The number of homozygous recessive individuals gives us q^2. From this, q is readily obtained by finding the square root. The result gives the frequency of the recessive allele and by substituting this figure into our initial formula of $p + q = 1$, the frequency p of the dominant allele can be found.

The Hardy-Weinberg equation was developed to describe the situation in a stable equilibrium, where the relative frequencies of the alleles and the genotypes stay the same over time. The implication is that in the absence of any factors which affect the equilibrium, gene frequencies will remain constant within a population from generation to generation. However, this is only true for an ideal population. An ideal population is one in which:

- there is random mating – there are no factors at work to cause the choice of mates to be non-random
- the population size is large
- the population is isolated so that there is no exchange of genetic material with other populations
- there are no mutations
- no natural selection takes place – in other words, all alleles have the same level of reproductive advantage or disadvantage.

In the real world these ideal conditions rarely if ever exist, and it is the upsetting of the gene-pool equilibrium which results in evolution. We shall examine each of these conditions in turn.

Using the Hardy-Weinberg equation

In the case of the MN blood groups, the frequency of the NN phenotype in American Caucasians is 0.21. Thus q^2 is 0.21 giving q as $\sqrt{0.21}$ or 0.461. The frequency of the $\mathbf{L^N}$ allele is 0.461, and this can then be used to find the frequency of the $\mathbf{L^M}$ allele:

$$p + 0.461 = 1$$
$$p = 1 - 0.461 = 0.539$$

By substituting these values into the Hardy-Weinberg equation, the frequency of homozygous MM individuals [(0.539)^2 or 0.29] and heterozygous MN individuals (2 × 0.539 × 0.461 or 0.5) can be calculated. As the L^M and L^N alleles show incomplete dominance, these theoretical frequencies can be confirmed by phenotypic observations.

THE MECHANISMS OF EVOLUTION

Upsetting the Hardy-Weinberg equilibrium

The Hardy-Weinberg law is never really valid as the ideal conditions it demands cannot be met in real situations. Mutation occurs constantly in any population and many other factors also influence the gene pool. However, the equation is invaluable as a way of measuring evolutionary change, and also as a way of demonstrating that evolution is a constantly occurring feature of the living world. Deviations from the Hardy-Weinberg equilibrium show that species, far from being unchanged since time immemorial, are in a continuous state of evolutionary flux. The factors which result in deviations from the hypothetical equilibrium state are the factors which bring about a long-term change in the gene pool and so drive evolution forward.

Non-random mating

One of the most important requirements for a gene pool to remain in equilibrium is for **random mating** to occur. Random mating means that the likelihood that any two individuals in a population will mate is independent of their genetic make-up – for example, the pollen from a wind-pollinated grass flower is likely to blow to any one of thousands of other similar grass flowers. **Non-random mating** occurs when some feature of the phenotype affects the probability of two organisms mating.

   Animals in which the male displays in some way to attract the female do not show random mating. The male peacock with the most impressive tail, the stag with the largest antlers and most effectively aggressive nature and the male stickleback with the brightest belly and most available nest will all appear to be more attractive than average to the females of the species. As a result they will be more likely to have the opportunity to mate and pass on their genes, ensuring that their offspring in turn are likely to carry these attractive characteristics. In plants and some animals self fertilisation or other forms of inbreeding mean that mating is not random and this affects the genetic ratios.

   Within human populations non-random mating is the normal state. In every human culture value judgements of one sort or another are used in the selection of a partner. The selection may be made by the family or social groups to which the individuals belong, in which case the professional and social standing of the prospective partners and their families will be weighed in the balance. Equally the selection may be made by the individuals concerned and called 'falling in love', but even this seemingly random process is governed by choices based as much on society and peer group pressure as on the perceived attractiveness of the proposed partners. In different populations different traits are regarded as desirable and so the gene pool shifts, as figure 6.2.3 illustrates.

A large population

The Hardy-Weinberg equation is only valid if it is applied to a large population, that is, one with several thousand individuals. This is because the

Figure 6.2.3 In the general human population albinism occurs in one person out of every 30 000. However, albinos are highly esteemed in certain Native American tribes and as a result the trait occurs in one out of every 240 individuals. This is a remarkable shift in the gene frequency within the population and shows the effect of non-random mating on the gene pool equilibrium.

maintenance of genetic equilibrium depends on random assortment of the genes. Compare the situation with tossing a coin. In theory, every time a coin is tossed it has a 50/50 chance of landing with the 'heads' side upwards. In practice, if a coin is tossed 20 times it could land 'tail' side up every time. If it is tossed 200 times, then at least some 'heads' will almost certainly occur. By the time 2000 tosses have been carried out the number of 'heads' and 'tails' resulting will be nearing the expected 50/50 ratio.

A similar situation exists in populations. If an allele occurs in 10% of the population and that population has only 10 individuals, then only one will carry the allele. Should a calamity befall that individual, the gene will be lost from the population completely. In a population of 5000 individuals, 500 will carry the allele in question and the likelihood of all of these being destroyed is remote.

A clear example of the importance of large populations in the maintenance of stable genetic frequencies occurs in the reverse situation, that is, when a small number of individuals leave the main population and set up a separate new population. Any unusual genes in the founder members of this new population will be amplified as the population grows. This is known as the **founder effect**, which is demonstrated clearly amongst the Amish, an American religious sect which exists in three isolated communities. One of the groups has a high frequency of a very rare genetic disorder known as Ellis-van Creveldt syndrome, since one of the founder members carried the gene. More cases of Ellis-van Creveldt syndrome have been found in this one small population than in the whole of the rest of the world (for more details see box below).

The founder effect

When Mr and Mrs Samuel King immigrated to Pennsylvania in 1744 as members of a small group of 200 souls who founded an Amish community, they can have had no idea of the legacy they took with them. One of them was heterozygous for Ellis-van Creveldt syndrome. This rare genetic disease results in a type of dwarfism – the limbs are shortened, there may be extra digits and as a result of associated problems the victim usually dies in the first year of life. The Kings produced many children, who in turn were also particularly prolific. This raised the frequency of the allele within the population way above its normal level. Purely by chance at least one other of the founding group also possessed this recessive gene in the heterozygous form. As a result of interbreeding in this isolated Amish community, one in 14 individuals in the population now carry the gene. This results in a distressingly high number of affected births – in 1964 there were 43 cases of Ellis-van Creveldt syndrome in a population group of 8000.

Gene flow – a lack of isolation

One of the constraints on a population in order to maintain the Hardy-Weinberg genetic equilibrium is that it should exist in isolation. There should be no migration of organisms either into or out of the population. Of course, this is very rarely the case in the living world. Insects carry pollen from one population of flowers to another, and the wind can carry it for miles. Male animals frequently leave their familial groups and go in search of other populations to find a mate. Many simple organisms release their gametes directly into the water to be carried great distances before fertilisation occurs. In all of these cases migration of either the whole organism or the genetic material into or out of the population takes place. As a result of this, **gene flow** occurs, tending to make the individual populations more alike. In extreme cases of isolation such as the Amish described above, this gene flow effect is missing.

Gene flow in black Americans

When the ancestors of many black Americans arrived in America from Africa around 300 years ago, they almost certainly had the rhesus **R** allele at the modern-day frequency of the African population – around 0.630. The present-day frequency of this allele in the black American population is 0.446. The frequency of the **R** allele in the white American population is only 0.028. The change in the frequency of the allele in the black population is thought to be the result of gene flow between black and white American populations. This gene flow has been calculated as about 3.6% of the genome per generation, over the last 10 generations. This rate is statistically significant and shows that gene flow can have a considerable effect on the gene pool of a population and therefore on evolution.

Mutations

For the Hardy-Weinberg equation to apply to a population, no mutations must occur within that population. This is because mutations involve changes in the inheritable material, that is, the alleles are altered. As we have seen in section 5, a mutation-free situation is an unrealistic expectation as mutations occur within a population all the time. Mutations in the somatic cells of animals will not be passed on to their offspring. They may or may not affect the individual themselves. The only animal mutations which are passed on to the next generation (and so affect the gene pool) are mutations in the cells of the germ line – in other words, those cells which will form the eggs and sperm. In plants there is no germ line – any of the shoots of a plant may develop into a flower bud – and so any mutation may become part of the gametes.

Although mutations occur continuously, they do not happen very rapidly. In a single generation each gene has between a 1 in 10^4 and a 1 in 10^9 chance of mutation. For human beings with a structural gene pool of around 10^6 this probably means that each of us contains only one new mutation. The vast majority of these will be recessive and are unlikely to be expressed, but occasionally mutations arise which confer benefits to an individual and may eventually become entrenched within the gene pool.

NATURAL SELECTION

Survival of the fittest

To quote Stephen Jay Gould, a leading authority on evolution: 'The essence of Darwin's theories is his contention that natural selection is the creative force of evolution – not just the executioner of the unfit.' Natural selection acts at the level of each individual organism. Some individuals carry alleles that result in phenotypes which provide them with some sort of advantage. These phenotypes can vary from the very obvious traits which make them more likely to be selected as mates or less likely to be eaten by predators, through to more subtle features such as forms of enzymes which allow them to digest food more effectively. Any individual with advantageous alleles will be more likely to survive and reproduce – in other words, they will be selected for over other less well adapted individuals. Thus natural selection means that the best adapted organisms are more likely to survive and reproduce, and as a result the frequencies of those advantageous alleles within the population increase, as illustrated in figure 6.2.4.

Natural selection and the Hardy-Weinberg equilibrium

For the Hardy-Weinberg equation to apply to a population, no natural selection must occur. But within any natural population selection is always

Figure 6.2.4 The ptarmigan is a member of the grouse family found in the Scottish Highlands. After the autumn moult the plumage regrows almost white. This is the result of natural selection – those birds which produced lighter feathers for the winter were less likely to be seen and eaten by predators during the snowy winter months than darker birds and so stood a greater chance of surviving to the spring and reproducing those advantageous alleles. Over many years this has become the normal state for the ptarmigan and almost all birds become white in the winter as the frequency of the allele controlling this feature has risen within the population. Similar colour changes can be seen in the Arctic hare.

occurring as some phenotypes are more advantageous than others and are therefore selected for. Natural selection works in several different ways to affect gene frequencies within a population, and we shall consider the main methods below.

Stabilising selection

Natural selection is often thought of as the selection of some new and advantageous features into a population. This is far from the truth – often natural selection acts to preserve what is already present in a population. This **stabilising selection** has resulted in many organisms which have changed little from the relatively early days of their history, for example ferns, sharks and lungfish. Wherever the environment of an area has been stable for a long period of time, then the effect of stabilising selection can be seen, as figure 6.2.5 shows.

Directional selection

This is the 'classic' type of natural selection, involving a change from one phenotypic property to a new one more advantageous in new circumstances. There are many excellent examples of this type of evolution in progress. One of the best known is the story of the peppered moth *Biston betularia* which underwent natural selection in response to environmental changes caused by industrial processes (**industrial melanism**). *Biston betularia* is a creamy speckled moth found in British woodlands. In the eighteenth century the relatively rare black specimens resulting from a random dominant mutation were very popular with butterfly collectors of the day. Their dark colour made them easily visible against the pale bark of the trees for both human collectors and birds looking for a meal, and so the dark allele was selected against and its frequency in the population remained low.

Then in the mid-nineteenth century the Industrial Revolution changed the face of much of Britain and Europe. The soot and smoke from the factory chimneys darkened the bark of the trees and the surfaces of buildings. As a result the dark melanic form of *Biston betularia* was at a selective advantage and the frequency of the allele within the population began to increase as more and more of the light-coloured moths fell prey to predators.

Now that there is considerably less industry in Britain, and the remaining industries produce much less pollution, trees and buildings are becoming cleaner and paler again and the selection pressure is beginning to move back in favour of the paler moth. Figure 6.2.6 shows the forms of the peppered moth.

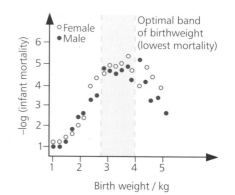

Figure 6.2.5 This graph shows the negative log of human infant mortality plotted against birth weight. The higher the vertical axis, the lower the death rate. The evidence demonstrates the effect of stabilising selection on the birth weight of babies. Babies with very low birth weights (often also born very early) have a high death rate. Very large babies may well endanger the mother at the time of birth and may also have some metabolic disorders. They too have a relatively high death rate. Those babies born in the 2.7–4 kg (6–9 lbs) range are most likely to survive and reproduce themselves, and this is by far the most common birth weight range.

Figure 6.2.6 The two main forms of *Biston betularia* on the bark of a clean tree and a polluted tree. In the 1950s H. B. D. Kettlewell of Oxford University set up several experiments to examine the selection of the two forms. In an unpolluted area with clean trees he released equal numbers of light and dark moths. 12.5% of the light moths were subsequently recaptured, but only 6% of the dark ones were retrieved. In another series of observations birds were seen to eat 26 light moths and 164 dark ones as they rested on a light tree trunk. In an industrial area 40% of the dark moths were recaptured but only 19% of the light ones. Of the moths picked off the blackened trees from equal releases, 43 were light and 15 were dark. In recent years the move has been back in favour of the paler moth and the frequency of its allele in the population has begun to increase again.

Another example of directional selection is the oysters of Malpeque Bay, Prince Edward Island, Canada. In 1915 the oyster fishermen of Malpeque Bay began to notice that amongst their usually healthy catches there were a few diseased oysters, small and flabby with pus-filled blisters. By 1922 the oyster beds had been all but wiped out by this Malpeque disease. However, a small number of the millions of offspring produced by each oyster each year carried an allele giving resistance to the disease. Not surprisingly the frequency of this gene in the population increased rapidly – by 1925 a small oyster harvest was again possible and by 1940 the beds were as prolific as ever, as figure 6.2.7 shows. However, there was now a rather different gene pool, the new one containing a high frequency of disease-resistant alleles.

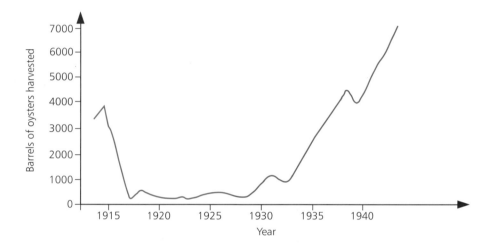

Figure 6.2.7 Oyster yields from Malpeque Bay, 1915–40. Disease devastated the populations but as a result of the increased selection of the disease-resistant allele within the population, healthy oyster beds developed.

Directional selection occurs whenever environmental pressure is applied to a population. It is frequently seen in populations of insects and plants that are regarded as pests and sprayed with chemical insecticides or herbicides. The chemicals may have a devastating effect initially, but directional selection ensures that within relatively few generations resistant individuals become more common within the population. The introduction of the rabbit disease myxomatosis into this country almost wiped out the rabbit population 40 years ago, but rabbits are now common once more. Many of them carry alleles which render them immune to myxomatosis – the frequency of such alleles in the rabbit gene pool has increased enormously.

Diversifying selection

Diversifying selection is another variation of directional selection. The difference is that the outcome is an increase in the diversity of the population rather than a trend in one particular direction. This occurs when conditions are very diverse and small subpopulations evolve different phenotypes suited to their very particular surroundings. Darwin's finches are a good example of diversifying selection.

Balancing selection

Here natural selection maintains variety, keeping an allele within the population even though it might seem to be disadvantageous. One clear example of **balancing selection** we have already come across is the sickle cell allele for human haemoglobin. Although the homozygous form of the allele is usually lethal, the heterozygous form gives protection against malaria and so the allele remains at a relatively high frequency within the population. This is known as **heterozygote advantage** or **hybrid vigour.**

The effect of this hybrid vigour is clearly seen in domestic dogs and cats. Pure breeds of dogs and cats have frequently been highly inbred over the generations to arrive at and maintain the particular features required for the breed. However, this also concentrates 'disadvantageous' genes and highly inbred animals are frequently relatively low in overall resistance to disease and also particularly prone to particular problems – Labradors tend to get arthritis, spaniels have ear problems, etc. On the other hand cross-bred cats and mongrel dogs, whilst of a far less predictable phenotype, tend to have much higher disease resistance and a much reduced tendency to succumb to inherited diseases – they are full of hybrid vigour.

Natural selection then can take a variety of forms. It may result in populations with a smaller gene pool or it may produce populations with such increased variety that two new populations or even species result. Along with all the other factors we have considered, natural selection occurs constantly in the living world, bringing about evolution. Some of the ways it does this are shown in figure 6.2.8. In section 6.3 we shall consider in more detail some of the great variety of living organisms which have resulted from the evolutionary process over millions of years.

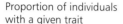

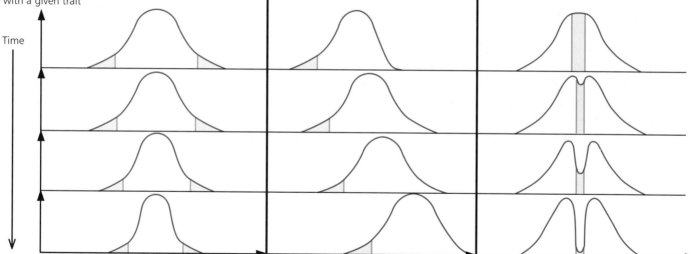

Stabilising selection tends to reduce variety so that only the phenotypes best adapted to a particular set of circumstances are maintained within the population.

Directional selection leads to a change in the phenotypes of a population in a particular direction, making them better suited to their environment.

Diversifying selection increases the variety within a population – if the selection pressure is strong two non-overlapping populations may result which could end up as two new species.

Figure 6.2.8 The effects of stabilising, directional and diversifying selection on populations. The shaded areas represent the alleles selected against, and their loss results in the next graph down. Natural selection, acting in these different ways, is a major factor in the evolution of life on Earth.

SUMMARY

- A **population** is a group of individuals of the same species occupying a particular **habitat**. The habitat includes the geographical place where an organism lives and its interactions with other organisms.

- The **gene pool** is the sum total of the genes in a population at a given time. **Evolution** is the permanent change in **gene frequencies** within a population.

- If p represents the frequency of a dominant allele and q the frequency of a recessive allele, then $p + q = 1$. p and q are impossible to measure because heterozygotes and homozygous dominants have the same phenotype.
- The **Hardy-Weinberg equation** can be solved to give the genotypes by observing the number of individuals with the recessive phenotype (homozygous recessives): $p^2 + 2pq + q^2 = 1$. The Hardy-Weinberg equilibrium only occurs where the gene frequencies are **stable**, i.e. the population is not evolving.
- Factors which result in evolution in a population are:

  non-random mating, where the phenotype affects the chances of mating

  small isolated populations, where unusual genes have a greater effect than would normally be the case due to the **founder effect**

  gene flow – migration to or interbreeding with different populations

  mutations – changes in alleles.
- **Natural selection** results in an increase in the proportion of advantageous alleles in a population.
- **Stabilising selection** maintains advantageous characteristics already present in a population.
- **Directional selection** results in a change to a new phenotype better suited to a changing environment.
- **Diversifying selection** results in different subpopulations with their own phenotypes suited to different habitats.
- **Balancing selection** maintains variety, where a disadvantageous allele is kept in a population, e.g. by being advantageous in the heterozygous state.

QUESTIONS

1 Define the following terms and give examples where appropriate:
   a adaptation
   b gene pool
   c population
   d allele frequency
   e genetic equilibrium
   f founder effect.

2 a What conditions must operate to maintain the Hardy-Weinberg equilibrium?
   b Do these conditions occur in nature?
   c What conclusions can you draw from your answers to **a** and **b**?

3 What is meant by the term 'natural selection'? Explain how it is implicated in the evolution of new species.

6

ORDERING THE PRODUCTS OF EVOLUTION

Along with the great diversity of life on Earth goes a diversity of names. The same organism will have a wide variety of names not only in different countries but also within different regions of the same country and even depending on whether it is a child or an adult who is speaking. When biologists from different countries discuss a particular organism, they need to be sure exactly which organism is being referred to. This problem is solved by using names given to organisms as a result of their place in a system of **classification**. The names are in Latin, which avoids the problem of choosing and thus favouring any single living language.

Taxonomy – categorising living things

Attempts to categorise living things in an orderly way have been made since before the time of the Greek philosopher Aristotle. However it was not until the work of the Swedish botanist Carolus Linnaeus in the eighteenth century that real progress towards a universal classification system was made. The aim of a classification system is to represent the true ancestral relationships of living things. Such a system must be based on homologous features (see section 6.2). If analogous features were used then all wriggly worm-like organisms (including worms, slugs and snakes) would be grouped together, as would bats, birds and flying insects. A valid classification must be based on careful observation and the use of structures which genuinely demonstrate a common ancestry. Linnaeus achieved such a classification and nomenclature system which is still relevant today in spite of the fact that he was working before the time of Darwin.

Linnaeus published a book called the *Systema Naturae* in which he organised and named all the animals and plants known at the time. He introduced a system known as the **binomial system of nomenclature** by which all organisms were assigned two Latin names. The first name indicates the **genus** to which the organism belongs (a genus is a group of very similar species). The second term indicates the **species** of the organism. A species is a specific group of closely related individuals which are all capable of interbreeding to produce fertile offspring. All the members of a species are potentially capable of interbreeding, although the geographical spread of the species may substantially reduce the chances of members from one environment meeting members from another. The full binomial (two-part) name of an organism always includes both genus and species, for example *Homo sapiens* (human being), *Equus caballus* (domestic horse), *Zea mays* (maize or sweetcorn) and *Lumbricus terrestris* (earthworm).

When Linnaeus devised his classification system the number of known species was much smaller than it is today. The only way of differentiating between them was by their physical appearance or **morphology**. However, since the eighteenth century many more species have been discovered and there are many more ways of identifying differences between organisms.

Taxonomic groups

The discovery of many more organisms has led to the development of broader groups. Similar genera are arranged into **families**; similar families are placed in the same **order**; similar orders go in the same **class**; similar classes into a

phylum and finally, the largest division of the living world, similar phyla are placed into the same **kingdom**. This arrangement is illustrated in table 6.3.1, along with the classifications of several common organisms. Named groups within the classification hierarchy (such as the phylum Chordata or the order Hymenoptera) are known as **taxa** (singular **taxon**).

Taxon	Human	Honeybee	Corn	Mushroom
species	*Homo sapiens*	*Apis mellifera*	*Zea mays*	*Agaricus campestris*
genus	*Homo*	*Apis*	*Zea*	*Agaricus*
family	Hominidae	Apidae	Poaceae	Agaricaceae
order	Primates	Hymenoptera	Commelinales	Agaricales
class	Mammalia	Insecta	Monocotyledoneae	Basidiomycetes
phylum	Chordata	Arthropoda	Angiospermophyta	Basidiomycota
kingdom	Animalia	Animalia	Plantae	Fungi

Table 6.3.1 The classification of organisms still follows the basic hierarchical system devised by Linnaeus in the eighteenth century, although the classification criteria used today are different from those on which Linnaeus relied.

Classification today

Modern classification systems are no longer based purely on simple morphological observations such as the number and type of limbs possessed by an organism. There have been two major developments since Linnaeus's time which have altered the way in which organisms are compared in taxonomy. One of these is Darwin's theory of evolution, which we have already considered in some detail. The other is the development of new techniques in physiology, embryology and biochemistry which have revealed similarities between living organisms which the most careful external observer could not hope to discover.

One of the most vexed issues of taxonomy has concerned the broadest level of classification of organisms into kingdoms. From the original two kingdoms Plantae and Animalia, proposed by Linnaeus and maintained for many years, the general consensus at present is that there should be five kingdoms, as detailed below. Not all biologists accept this however, and some still refer to the original two kingdoms only. Even five kingdoms do not provide an entirely appropriate niche for all living things – more kingdoms may yet need to be created.

The kingdoms

The five kingdoms in most common use for classification purposes are:
- **Prokaryotae** – the single-celled prokaryotic organisms (bacteria)
- **Protoctista** – often single-celled eukaryotic organisms including protozoa and some of the algae
- **Fungi** – eukaryotic organisms with non-cellulose cell walls, from the single-celled yeasts to the more complex toadstools
- **Plantae** – eukaryotic photosynthetic organisms including multicellular algae
- **Animalia** – non-photosynthetic multicellular organisms with nervous coordination.

When biologists set out to identify organisms which they have not met before, they rely largely on observation as a first step in the classification process. The following summaries of the kingdoms of the living world highlight the major features enabling the identification of the organisms within each kingdom.

PROKARYOTAE

The prokaryotic cells which make up bacteria are smaller and simpler than those of any other type of organism, as described in section 1.1 (see box 'Eukaryotic and prokaryotic cells', pages 16–17). They obtain their nutrition in a wide variety of ways – it has been said that bacteria exploit every energy-yielding reaction in nature.

Bacteria are unicellular, although the cells may be grouped together to form chains or clumps. They occur in a variety of shapes and sizes, as shown in figure 6.3.1. Some feed autotrophically by chemosynthesis or photosynthesis, others are heterotrophs. Bacteria may therefore be classified as photosynthetic or non-photosynthetic. Bacteria usually reproduce by binary fission, although sometimes a simple form of sexual conjugation occurs. Some bacteria are pathogenic (cause disease), for example *Streptococcus*, others are beneficial to humans, for example nitrifying bacteria and gut bacteria.

The photosynthetic bacteria include the cyanobacteria, which used to be called the blue-green algae. They are pigmented and occur as unicells or filamentous organisms, as shown in figure 6.3.2.

'Black smoker' bacteria

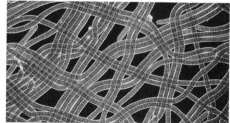

Cocci e.g. *Staphylococcus*

Diplococci e.g. *Diplococcus*

Streptococci e.g. *Streptococcus*

Bacilli e.g. *Escherichia coli*

Vibrios e.g. *Vibrio*

Spirilla e.g. *Spirillum*

Figure 6.3.1 Some of the many different forms of bacteria which colonise almost every known habitat on Earth. Within the last few years scientists have found bacteria in deep vents almost 3000 m below the surface of the ocean. Conditions in the mineral-rich water in these 'black smokers' are extreme – temperatures of around 350 °C and high pressures. 'Black smoker' bacteria stop reproducing if the temperature falls to 100 °C! These bacteria challenge many of the accepted theories about the biochemistry of enzymes and cells, and also illustrate the ability of bacteria to take advantage of environments in which nothing else can survive.

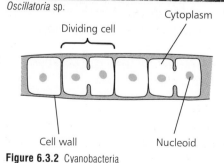
Oscillatoria sp.

Cytoplasm

Dividing cell

Cell wall

Nucleoid

Figure 6.3.2 Cyanobacteria

Bacteria and disease

Robert Koch and Louis Pasteur discovered and refined our knowledge of the link between bacteria and disease early in the twentieth century. Many human diseases are the result of bacterial infections – typhoid, cholera, dysentery, food poisoning, plague, diphtheria, TB, meningitis, pneumonia, tonsillitis, gonorrhoea and tetanus to mention but a few. The bacteria may be spread from person to person through faecal contamination, exhalation droplets (for example in a sneeze), animal bites, sexual contact or by direct entry into wounds.

The symptoms of bacterial diseases may result from:
- the destruction of the host cells and tissues by the bacteria
- irritation by bacterial waste products
- exaggerated immune responses to the presence of foreign cells
- reactions to bacterial toxins.

Most bacterial diseases can be treated with antibiotics or other drugs. These interfere with various aspects of the biochemistry of the bacterial cell – for example, interrupting protein synthesis or disrupting the structure of the membrane. Alternatively, many bacterial diseases can be avoided by vaccination programmes. Unfortunately, bacterial strains are developing immunity to the antibiotics which have been used to control them in the last few years. To avoid a return to the days when a bacterial infection in a wound, after childbirth or in a young child meant probable death, research is continually probing for new antibiotics or other drugs to control these mutant forms of an old enemy.

PROTOCTISTA

The protoctists consist of unicellular eukaryotic organisms which have a variety of methods of nutrition and reproduction. These organisms come in an enormous range of shapes and life styles – from amoebae to diatoms, including plant-like cells that swim, animal-like cells that cannot move, cells shaped like bells or fans or shells, and cells grouped into colonies. The number of phyla within the kingdom varies from expert to expert up to 45, but only a small sample will be considered here. Within the protoctista are organisms which make up the greater part of the plankton of the oceans, and a few which cause some of the most serious diseases to affect the human race.

Phylum: Rhizopoda

These single-celled amoebae are common in both fresh and salt water as well as in the soil. They move and capture their prey by means of temporary **pseudopodia**. Most Rhizopoda are naked, but a few secrete a shell (the radiolarians), as shown in figure 6.3.3. One group of Rhizopoda are parasitic – for example, dysentery in humans is caused by *Entamoeba hystolytica*, a gut parasite.

Phylum: Zoomastigina

The 'animal flagellates' have one or more flagella at some stage of their life cycle. They do not contain chlorophyll. They are heterotrophs, absorbing their food through the cell membrane or taking it in by phagocytosis. An example is *Trypanosoma*, a blood parasite causing the disease African sleeping sickness, shown in figure 6.3.4.

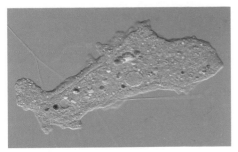
Amoeba

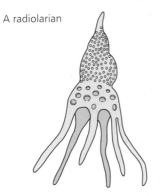
A radiolarian

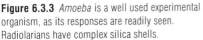

Figure 6.3.3 *Amoeba* is a well used experimental organism, as its responses are readily seen. Radiolarians have complex silica shells.

Shining seas, red tides

Two well known phenomena of the sea are caused by protoctists known as **dinoflagellates**. At night the sea may seem to give off a greenish glow, and where waves break or a swimmer splashes, greenish sparks may be seen. This is due to species of dinoflagellates containing enzyme systems which result in bioluminescence. When sufficient of the organisms are present in any particular area of the sea they cause these green sparkles.

At times, offshore waters turn a rusty red. These red tides contain toxins which can poison fish, shellfish and even people. They too are the result of large **blooms** of certain species of dinoflagellates. So far, the exact trigger for the appearance of these unwelcome blooms is unknown.

Phylum: Apicomplexa

The Apicomplexa or sporozoans all move by wriggling. They reproduce by producing spores, and most importantly they are parasitic, using animals as their hosts. *Plasmodium*, the blood parasite responsible for the human disease malaria, is perhaps the best known sporozoan. Its complex life cycle is shown in figure 6.3.5.

Phylum: Ciliophora

Ciliophora are unicellular organisms covered by cilia. These cilia have basal bodies which are connected by a complex network of fibres so that the cilia all beat together in complex rhythms. The cilia are frequently used for locomotion.

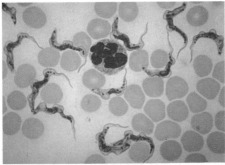

Figure 6.3.4 *Trypanosoma* is an example of an animal flagellate.

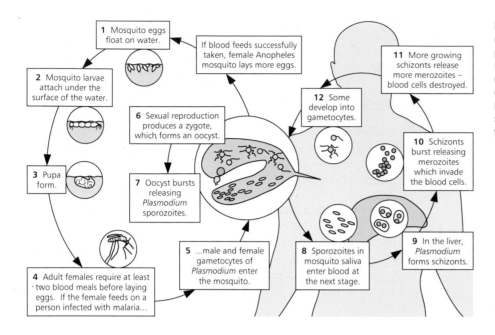

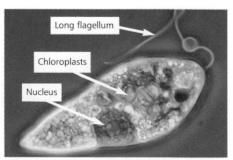

Figure 6.3.5 The *Plasmodium* parasite is responsible for malaria, a disease which affects hundreds of millions of people worldwide and kills several million each year. The life cycle of the parasite involves two hosts, the *Anopheles* mosquito and people. The stages of the life cycle which occur within the human body involve the lysis of large numbers of red blood cells and the concomitant disease symptoms of fevers, sweating, shaking, long-term weakness, anaemia and eventually death.

The Ciliophora have one particularly unusual feature. They have two nuclei, a larger **macronucleus** which controls all the normal functioning of the cell, and a much smaller **micronucleus** containing the chromosomes, which is vital for sexual reproduction. Figure 6.3.6 shows a ciliophoran.

Phylum: Euglenophyta

This phylum includes the 'plant flagellates' which are predominantly plant-like. They contain chlorophyll and feed by photosynthesis. Euglenophyta have flagella used for locomotion at some stage of their life cycle. Phytoplankton (plant plankton) are shown in figure 6.3.7.

Phylum: Oomycota

These are important because they are the cause of a variety of serious plant diseases, including potato blight (*Phytophthora infestans*) which resulted in the Irish potato famines in 1845 and 1847. Oomycetes have **hyphae** similar to those of the fungi (see page 455) containing few cross-walls. They reproduce asexually by structures called **conidia** and by flagellated motile **zoospores**. Their method of sexual reproduction gives them their name – the prefix 'oo-' refers to eggs. Oomycetes produce large, immobile egg cells which are fertilised by the male gametes, as shown in figure 6.3.8.

Phylum: Chlorophyta

The green algae may be unicellular or filamentous. They are photosynthetic, and the main photosynthetic pigment is chlorophyll. Both chlorophyll *a* and chlorophyll *b* are present in the green algae, as they are in the higher plants – the green algae are the most likely ancestors of the land plants. Most green algae are found in fresh water or damp places, with only a few such as the sea lettuce inhabiting the sea. The classic example is *Spirogyra*, a single chain of identical cells found in freshwater ponds, shown in figure 6.3.9.

Phylum: Rhodophyta

The red algae are mainly marine algae found in tropical waters. Their photosynthetic pigments are the red **phycoerythrin** and the blue

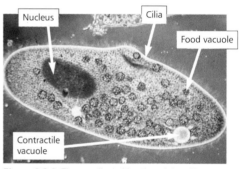

Figure 6.3.6 The coordinated beating of the cilia on the surface of ciliates such as *Paramecium* allows directional movement to take place.

Figure 6.3.7 Organisms such as this *Euglena* – plant-like in their method of nutrition and animal-like in their way of moving about – caused great problems for early biologists, who could not decide whether they were plants or animals. The obvious solution is to place them in a kingdom of their own.

phycocyanin as well as chlorophyll *a*, but they have no chlorophyll *b*. These pigments give them characteristic red-purple colours, and also enable them to absorb the blue and green wavelengths of light which filter through to deeper water. As a result they can live in water more than 150 m deep – five times deeper than green or brown algae. Red algae are used to produce agar, used as a nutrient medium for many laboratory cultures, and carrageenan, which is used as a stabilising agent in products as diverse as paints, ice-cream and cosmetics. A red alga is shown in figure 6.3.10.

Phylum: Phaeophyta

The brown algae are the giants of the algal world. They are the most complex with considerable cell differentiation. The main photosynthetic pigments are chlorophylls *a* and *c* and the brown **fucoxanthins**. They are generally familiar as seaweeds. The brown algae frequently show differentiation into a **holdfast** for clinging to rocks or the sea bed, a **stipe** or stem equivalent and **fronds** or leaf equivalents. It appears some vessel cells may be present which allow the transport of the products of photosynthesis – necessary in species such as kelp, shown in figure 6.3.11, which may grow up to 100 m long!

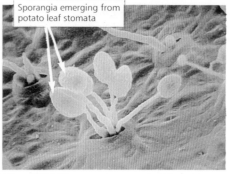

Figure 6.3.8 The oomycetes are protoctists, but are similar to fungi in many ways. *Phytophthora infestans*, illustrated here, caused the deaths of over a million people in the infamous Irish potato famines. Fear of further infestations led a third of the remaining population to flee the country and emigrate to the USA and Canada. Through this and other similar examples, oomycetes have had a major influence on human populations.

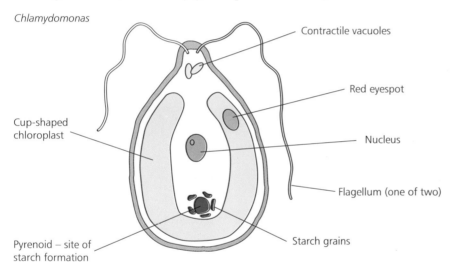

Chlamydomonas

- Contractile vacuoles
- Red eyespot
- Nucleus
- Flagellum (one of two)
- Starch grains
- Pyrenoid – site of starch formation
- Cup-shaped chloroplast

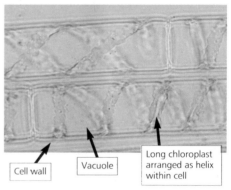

Figure 6.3.9 *Spirogyra* (above) is a typical green alga, present mainly in fresh water. *Chlamydomonas* (left) is a unicellular chlorophyte.

Cell wall   Vacuole   Long chloroplast arranged as helix within cell

FUNGI

For many years the fungi were classified as an unusual sort of plant. Now the great differences between fungi and true plants are recognised and they have a kingdom of their own. Fungi are eukaryotic and frequently multicellular. They are heterotrophic, digesting food extracellularly and then absorbing the nutrients. They play a vital role within ecosystems as **decomposers**.

In most fungi the body structure is made up of thread-like **hyphae** which have walls usually made of chitin. These hyphae form a tangled network known as the **mycelium**. Reproduction is usually by the production of spores or by simple sexual conjugation. The kingdom of the fungi contains a wide variety of organisms and some of the main divisions are considered here. Fungi are of immense importance in the ecology of the Earth. In their role as decomposers they prevent the build-up of the bodies of dead animals and plants by digesting them and returning nutrients to the soil. They provide the human race with food and with vital drugs such as penicillin. Conversely, fungi cause major human problems. They destroy vast quantities of crops and food worldwide, as well as causing a variety of diseases in both people and domestic animals.

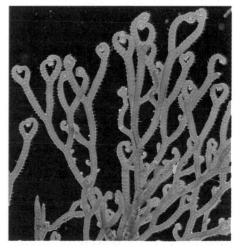

Figure 6.3.10 The more complex forms of the red algae photosynthesise at depths beyond the reach of others due to their cocktail of photosynthetic pigments.

Phylum: Zygomycota

The zygomycetes are all terrestrial and are largely saprophytic. The main body of the fungus, the mycelium, is haploid. They reproduce asexually by non-motile spores which form in a tough spore case known as the **sporangium**. They reproduce sexually by conjugation between neighbouring hyphae as shown in figure 6.3.12. Typical examples are the pin-moulds such as *Mucor* which form on rotting fruit and vegetables. Zygomycetes are also very important for their role in the symbiotic relationships which they form with the roots of the vast majority of other plants. These associations between roots and fungi are known as **mycorrhizae**. The masses of fungal hyphae increase the surface area of the roots for the absorption of water and nutrients, and the saprophytic activity of the fungus also increases the availability of these mineral nutrients. The plant in turn supplies the fungus with the products of photosynthesis.

Figure 6.3.11 The Phaeophyta are the 'trees' of the algal world, with complex structures needed to support the functions of large organisms.

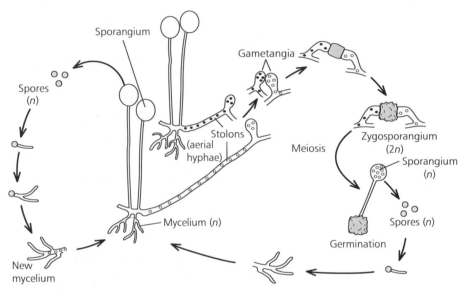

Asexual reproduction involves the formation of sporangia which contain haploid sporangiophores produced by mitosis. When conditions are appropriate the spores are released and carried by air currents to new sites.

In sexual reproduction two hyphae meet and form a zygosporangium which is diploid. After a period of dormancy the sporangium undergoes meiosis, and when it subsequently germinates a haploid sporangium is produced. The spores from this go on to produce new haploid mycelia.

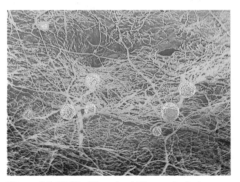

Figure 6.3.12 The life cycle of *Mucor*, a typical zygomycete fungus

Phylum: Ascomycota

The ascomycetes form the largest group of fungi. Asexual reproduction involves the formation of spores called **conidia**. These are produced in large quantities and are frequently the cause of human 'hay fever'-type allergies. Sexual reproduction involves conjugation followed by the production of **ascospores** inside a sac-like **ascus**.

The yeasts, which are useful to us in the fermentation reactions which produce alcoholic drinks and leaven bread, are rather atypical members of the ascomycetes. Yeasts are unicellular fungi which usually reproduce asexually by budding. Occasionally, however, two cells fuse and form an ascus and ascospores.

Some ascomycetes form toxins which cause hallucinations and can be fatal – LSD is derived from an ascomycete which attacks rye. Ascomycetes also include the *Penicillium* moulds which have played a vital role in reducing the impact of bacterial diseases. These fungi have also been used for centuries to ripen and impart flavour to many cheeses such as Camembert and Roquefort.

Dutch elm disease

Throughout Britain, Europe and North America the sight of an elm tree has become increasingly rare. This is due to the spread of a fungal disease called Dutch elm disease. The spores of *Ceratocystis ulmi* are carried from tree to tree by bark-boring beetles and the fungus rapidly spreads through the tree and kills it. Traditional methods of control – quarantine of timber, removing and destroying infected trees and spraying with expensive pesticides (for the beetles) and fungicides (for the fungus) – have had only limited success and elm trees have become steadily rarer.

Two new approaches may halt this fungal killer. Chemical messages or **pheromones** have been isolated and used to create 'trap trees'. The beetle carriers are attracted by the pheromones and are then caught on large sheets of sticky paper attached to the tree. Also, biological control is being tested. *Pseudomonas syringae*, a bacterium which produces a natural fungicide, has been shown to kill the fungal spores so effectively that the infection rate is reduced from 100% to 2%. To wipe out Dutch elm disease, strains of the bacteria need to be developed which are resistant to cold winters and which spread easily to the topmost branches of the elm trees, where the infection usually starts. With luck, the fungus will be overcome before elm trees disappear from our landscapes altogether.

Figure 6.3.13 Within the last century, the fungus *Ceratocystis ulmi* has almost removed the elm from our landscapes.

The attraction of the truffle

Many fungi are used as human food, but the most highly prized and expensive of all is the truffle. Truffles are the fruiting bodies of members of the ascomycetes. Known as the Tuberales, these rare fungi live in mycorrhizal symbiosis with the roots of oak and beech trees. The fruiting bodies are formed underground. The spores are only released as the truffle slowly decays, or if a burrowing animal breaks it apart. Truffles usually have strong and pungent smells, ranging from rich garlicky odours to the smell of sewer gas, depending on the species. Only a small number are edible.

To harvest these elusive fungi, European truffle hunters have traditionally used pigs to sniff out and dig up the fruiting bodies. A well trained truffle pig will dig down to a metre to seek out truffles. What stimulates a pig to be this persistent? It seems unlikely that it is the taste of the truffles. Researchers have discovered that truffles produce a highly volatile pig sexual attractant, usually found in the testes of boars. Androstenol is released in the saliva during pre-mating behaviour. It seems that truffles have evolved to attract sows, ensuring that the spore case is broken open and the spores dispersed. Androstenol is also produced by the human testes and released in the sweat glands of male armpits. Perhaps this explains why people are prepared to pay high prices for such an unlikely delicacy!

Phylum: Basidiomycota

The basidiomycetes are the best known of the fungi, including mushrooms, toadstools, puffballs and bracket fungi. Most of the basidiomycetes form large conspicuous fruiting bodies from a mass of hyphae buried within the soil or in a plant body such as a tree. Whilst the hyphae are haploid, the fruiting body is the result of sexual conjugation and consists of diploid hyphae. Meiosis results in haploid **basidiospores** which are formed in microscopic club-shaped **basidia**, as shown in figure 6.3.14.

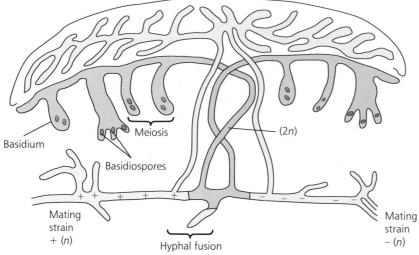

A typical basidiomycete. The fruiting body is the result of sexual reproduction and so is diploid.

The classic toadstool of fairytales is the fly agaric, a basidiomycete producing a toxin which is a powerful hallucinogenic chemical and can be fatal. Reindeer are very fond of eating fly agaric.

Figure 6.3.14 Basidiomycetes

PLANTAE

The plant kingdom encompasses a variety of organisms which range from the simple liverworts to the giant redwoods. They are multicellular eukaryotic organisms relying on photosynthesis for their nutrition. A typical plant cell (see section 1.1) has a cellulose cell wall, sap vacuole and chlorophyll *a* along with other photosynthetic pigments and plastids (structures such as chloroplasts). Plants reproduce in a variety of ways, both asexually and sexually (see sections 5.1 and 5.3). Plant reproduction is characterised by **alternation of generations** – the haploid **gametophyte** generation and the diploid **sporophyte** generation. So far botanists have identified around 300 000 living plant species, with over 80% of these being flowering plants.

Phylum: Bryophyta

The bryophytes are the simplest land plants. They do not have vascular tissues and so are unable to transport material around the body of the plant. They are small – the largest species is less than 60 cm tall – and are found in damp places. A large percentage of bryophytes live in tropical rainforests.

The life cycle shows a clear alternation of generations. The haploid, gamete-forming gametophyte is usually the more prominent phase with the diploid spore-forming sporophyte attached to and dependent on it. The gametophyte usually shows differentiation into **stem**, **leaves** and **rhizoids** for anchorage. Bryophytes are limited in their colonisation of the land by their dependence on water for their swimming sperm, their lack of vascular tissue for transport and lack of stomata to prevent water loss.

Class: Hepaticae

The liverworts are made up of a flat, lobed gametophyte **thallus**. On the thallus forms the **antheridium**, a sperm-producing chamber, and the **archegonium** which produces eggs. Sometimes both these structures form on a single gametophyte, sometimes they form on two different thalli. Water is needed for the sperm to swim to the egg and fertilise it. The fertilised egg develops into a diploid sporophyte dependent on the thallus. Meiosis occurs within the mature sporophyte to produce haploid spores. When released, these develop to form new haploid gametophytes. This life cycle is shown in figure 6.3.15.

Class: Musci

The mosses are similar to the liverworts, but the gametophyte is always differentiated into simple stems and leaves rather than a thallus. The rhizoids are multicellular rather than unicellular. Mosses form velvet-like carpets wherever the air is moist and clean. Like the liverworts, mosses produce sperm and eggs in antheridia and archegonia, and depend on water from rain, streams or other sources to carry the male gamete to the female one, as shown in figure 6.3.15.

Figure 6.3.15 The bryophytes are limited in their ability to colonise different ecological niches due to their constant requirement for water and a moist environment.

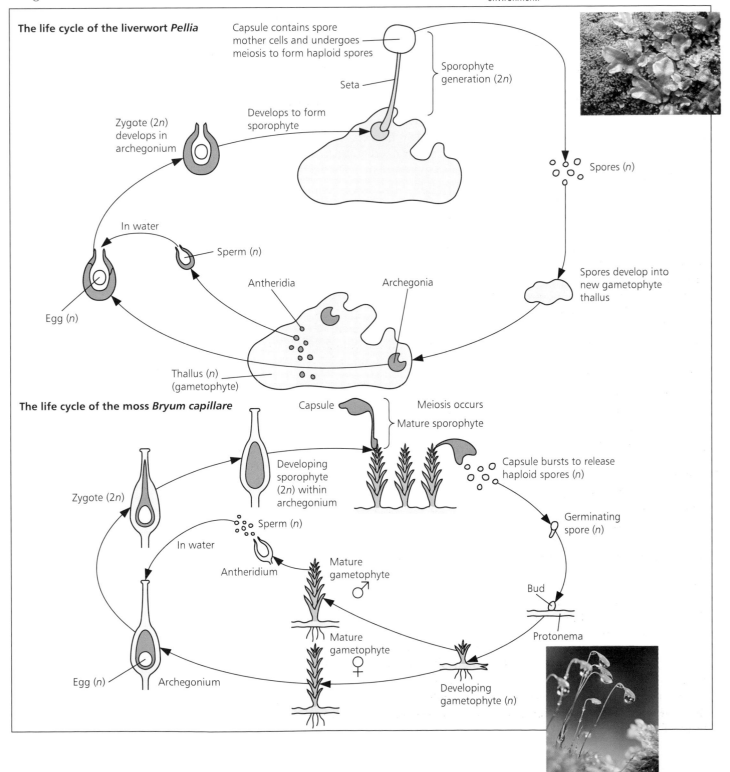

The life cycle of the liverwort *Pellia*

Capsule contains spore mother cells and undergoes meiosis to form haploid spores

Sporophyte generation (2n)

Seta

Zygote (2n) develops in archegonium

Develops to form sporophyte

Spores (n)

In water

Sperm (n)

Spores develop into new gametophyte thallus

Antheridia

Archegonia

Egg (n)

Thallus (n) (gametophyte)

The life cycle of the moss *Bryum capillare*

Capsule

Meiosis occurs

Mature sporophyte

Developing sporophyte (2n) within archegonium

Capsule bursts to release haploid spores (n)

Zygote (2n)

Sperm (n)

In water

Germinating spore (n)

Antheridium

Mature gametophyte ♂

Bud

Mature gametophyte ♀

Protonema

Egg (n)

Archegonium

Developing gametophyte (n)

Phyla: Lycopodophyta, Sphenophyta and Filicinophyta

In these phyla **true leaves**, **stems** and **roots** occur. The two generations exist as separate plants, but the sporophyte is the larger and dominant plant, with the gametophyte reduced to a small, simple prothallus.

The Lycopodophyta are the club mosses, the Sphenophyta the horsetails and the Filicinophyta are the ferns. Horsetails are an ancient group, dating back to the Carboniferous swamps. They have a spore-bearing cone at the top of the stem. The club mosses are small-leaved plants with sporangia arranged between the leaves. By far the largest group is the ferns.

Filicinophyta (ferns)

Ferns usually have large, lacy, prominent fronds which are true leaves. The dominant sporophyte generation is diploid. At maturity sporangia form in clusters called **sori** on the undersides of the fronds. Meiosis takes place within these sporangia to give rise to haploid spores. The spores are discharged into and dispersed by the air. A spore then develops into a small heart-shaped **prothallus** with no vascular tissue. Antheridia and archegonia form and give rise to male and female gametes. Because the sperm need water to swim to the egg, ferns are most commonly found in moist environments. The fertilised egg gives rise to a new diploid sporophyte, as shown in figure 6.3.16.

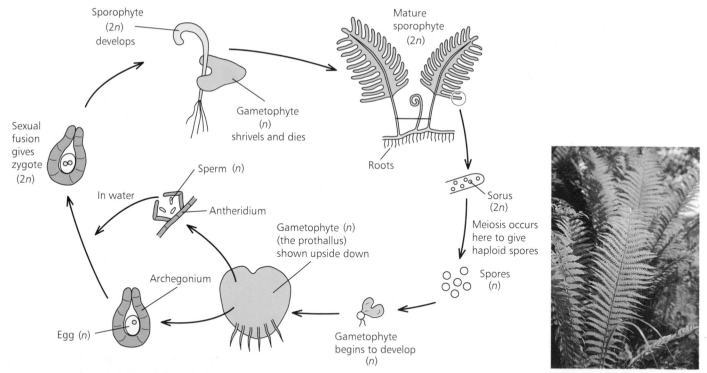

Figure 6.3.16 The life cycle of *Dryopteris*, a typical fern

The seed-bearing plants

The seed-bearing plants include the vast majority of plant species. They have enormous economic and ecological significance. They provide the human race with much of its food, drink, cloth, building material, drugs and more. Fossil evidence suggests that the seed-bearing plants have been the most widespread and abundant land plants for 250 million years. In seed-bearing plants the gametophyte generation is part of the physiology of the sporophyte generation which is completely dominant (for life cycles see section 5.3). There are two phyla of seed plants, shown in figure 6.3.17 and described opposite.

Phylum: Coniferophyta

The conifers or 'naked seed plants' include pine trees, spruces and cedars. They are characterised by narrow needle-like leaves, no vessels in the xylem tissue (only tracheids, lignified single cells) and most of all by their reproductive structures which are carried on spore-bearing leaves arranged into **cones**. Conifers are usually the dominant vegetation in cold and mountainous regions.

Phylum: Angiospermophyta

The flowering plants have their reproductive structures carried in flowers. The ovules occur in an ovary and after fertilisation this develops into a fruit. Angiosperms also have vessels in their vascular tissue, making it much more effective at transporting substances. The two classes of the angiosperms are the monocotyledons and the dicotyledons.

Class: Monocotyledoneae

The embryo has a single seed leaf (cotyledon). Leaves generally have parallel veins and the stem contains scattered vascular bundles. As there is no cambium (see section 5.3, pages 366–7), no secondary thickening takes place and so monocotyledons do not generally reach great sizes. Palm trees are one of the better known exceptions to this rule.

Class: Dicotyledoneae

The embryo has two seed leaves (cotyledons). The leaves usually have a network of veins and the stem contains a ring of vascular tissue. Cambium is present so secondary thickening can take place and some dicotyledons reach great sizes.

Rubus fructicosus (bramble)

Pinus (Scots pine)

Figure 6.3.17 The seed-bearing plants are the largest and most successful plants on the Earth.

ANIMALIA

The majority of the species on Earth are the complex eukaryotic organisms we know as animals. There are around 2–12 million known species of animals alive today. They feed heterotrophically and reproduce mainly sexually. Here we shall consider the main animal phyla.

The multicellular state has allowed great diversity of form to evolve, but also imposes certain limits. As a result of surface area:volume constraints the cells of most animals are very similar in size. This means that the major difference between a flea and an elephant is the number of cells, not the size of the cells. As animals get larger, more powerful muscle systems are needed to move them around. Fewer of the cells are in contact with the external environment and so respiratory and transport systems must become more sophisticated. Communications systems need to be more highly developed in larger animals to supply information to all areas of the body. Thus as we consider increasingly more complex animals, it is necessary to bear in mind the developments in these systems which make the increasing sophistication possible.

Phylum: Porifera

The sponges are sessile – they are permanently attached to a surface on the sea bed. They are hollow filter feeders, and the body cavity is connected to the external environment by pores. The body cavity is lined by flagellated cells which create currents of water through the sponge, as shown in figure 6.3.18. There is little coordination or control. They range in size from a few millimetres to 2 m and are supported by a series of calcareous spicules. As far as is known the sponges are an evolutionary dead end and have no other close living relatives. Some examples of sponges were shown in figure 1.6.3, page 73.

Phylum: Cnidaria

The cnidarians include some exceptionally beautiful creatures and also some very poisonous ones. Sea anemones, hydra, jellyfish and coral are among the members of this phylum. They have a sac-like body cavity called the **enteron** with only one opening and a **diploblastic** or two-layered structure to the body wall. The outside layer is called the **ectoderm** and the inner layer is the **endoderm**. The body plan of these animals demonstrates **radial symmetry** – they look circular when viewed from above or below and body structures radiate in all directions from the centre, as shown in figure 6.3.19.

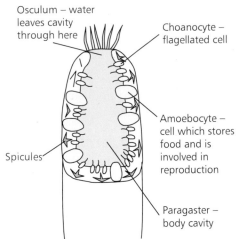

Figure 6.3.18 An evolutionary dead end, sponges are a very ancient form of life well adapted to their environment.

Figure 6.3.19 Though relatively unspecialised, with no organs to perform specific tasks, the cnidarians achieve a much higher level of organisation than the sponges.

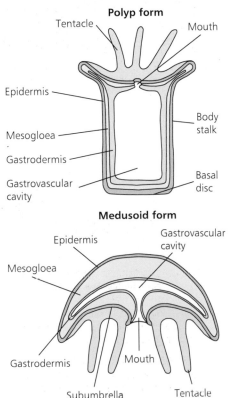

This brightly coloured sea anemone is an example of a polypoid cnidarian

This jellyfish is an example of a medusoid cnidarian

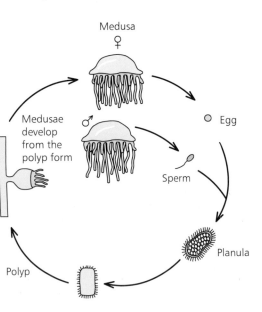

Most cnidarians exist in two forms, as **polyps** and **medusae**, and these forms alternate in the life cycle. This is an example of alternation of generations in the animal kingdom. Most cnidarians are carnivorous, trapping and killing their prey by means of special stinging cells on the tentacles known as **nematoblasts** (see figure 4.1.1, page 232). A simple nerve net over the entire body coordinates movement and response. These animals do not have a front and back, and they do not move in a single direction.

Phylum: Platyhelminthes

The flatworms show a much greater level of organisation. They possess a front end where the mouth, major sense organs and the main integrating region of the nervous system is sited. This anterior moves forward first and so is the first to encounter new situations. Such front-end specialisation is known as **cephalisation**. Along with the development of the front end, flatworms show **bilateral symmetry**. This means that the right and left sides of the animal are mirror images. Platyhelminthes are also the first major group to have organs rather than simply specialised tissues.

Flatworms range from 1 mm to 30 cm in length. They are dorsoventrally flattened, have a mouth but no anus and many blind-ending branches to the gut. They develop from three embryonic layers. **Ectoderm** gives rise to the surface layers, **mesoderm** to muscle fibres, excretory organs known as **flame cells** and the reproductive parts, whilst the **endoderm** lines the gut. There are no respiratory or circulatory systems – flatworms are so thin and flat that diffusion can fulfil these needs. The majority of flatworm species are parasitic. They are **hermaphrodite** – they possess both male and female reproductive organs, although self fertilisation is rare.

There are three classes of flatworms, shown in figure 6.3.20 and described below.

Class: Turbellaria

Turbellarians are the free-living flatworms commonly found in streams, for example *Planaria*. Some marine flatworms are vividly coloured. They usually possess cilia for movement.

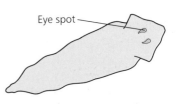

Eye spot

Planaria, a typical turbellarian

This tiger flatworm shows the bilateral symmetry of the Platyhelminthes.

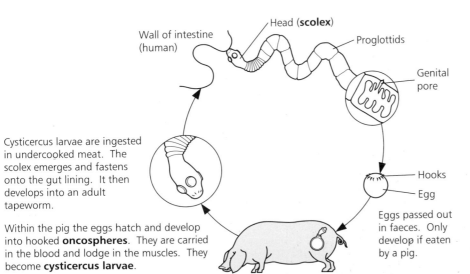

Head (**scolex**)

Wall of intestine (human)

Proglottids

Genital pore

Cysticercus larvae are ingested in undercooked meat. The scolex emerges and fastens onto the gut lining. It then develops into an adult tapeworm.

Within the pig the eggs hatch and develop into hooked **oncospheres**. They are carried in the blood and lodge in the muscles. They become **cysticercus larvae**.

Hooks

Egg

Eggs passed out in faeces. Only develop if eaten by a pig.

Cestode tapeworms like *Taenia* in the guts of domestic animals and humans can grow to many metres in length and cause considerable wasting of the body, as the tapeworm absorbs most of the digested nutrients in the gut.

Trematode liver flukes can cause great damage to the tissues of their host. They have a wide variety of adaptations to avoid rejection by their hosts, including in some cases the ability to change the immunological markers on their cell surfaces. Most flukes have life cycles involving at least two hosts.

Figure 6.3.20 The Platyhelminthes as a group are largely parasitic, although some are free living. They show cephalisation and bilateral symmetry.

Class: Trematoda

Trematodes are all parasitic and are known as flukes. They may be endo- or ectoparasites. They have no cilia but possess suckers and other adaptations for attaching themselves to their host. An example is *Schistosoma*, a blood fluke causing the disease schistosomiasis or bilharzia.

Class: Cestoda

Cestodes are all endoparasites known as **tapeworms**. They have no cilia and no gut as they absorb digested food directly from their host's gut. They have adaptations such as suckers and hooks for attachment to the host and tough outer coverings to avoid attack by host enzymes. The body is divided into sections known as **proglottids**. An example is *Taenia*.

The scourge of schistosomiasis

Trematode flatworms of the genus *Schistosoma* have a life cycle which involves aquatic snails and human beings. Infected snails release a larval stage into water, for example in a river, and these penetrate the skin of people entering the water. The larvae develop into adults within the body of the human host and eventually produce eggs, some of which are passed out of the body in the urine or faeces. If these eggs are released into water they hatch and develop into a different larval form which infects snails. Development then proceeds within the snails before the cycle begins again.

Not all the eggs are passed out of the human host, however. Those that remain become lodged in the tissues. It is the reaction of the body to these eggs which causes the symptoms of the disease **schistosomiasis**, which affects around 200 million people in the world at any one time. Symptoms include damage to the bladder, kidneys, liver and spleen as well as the intestines. Repeated internal bleeding leads to weakness and anaemia and for around 200 000 people a year it leads to death.

River blindness and elephantiasis

Nematode worms are an important and useful part of the soil fauna. However, many members of the phylum are debilitating parasites of animals and people, as the examples in figure 6.3.21 illustrate.

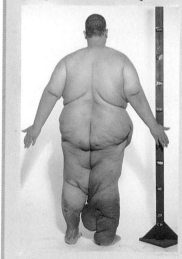

The larvae of certain nematode worms are passed to humans by the bite of infected mosquitoes. They develop as adult worms in the lymph vessels, blocking the vessels and causing the swelling which gives **lymphatic filariasis** its more common name, **elephantiasis**.

The infective nematode larvae which cause river blindness are transmitted to people through the bite of infected *Simulium* blackflies. Adult worms develop under the skin and reproduce, releasing large numbers of larvae known as **microfilariae**. These migrate into the skin and eyes, where they cause intense itching and damage which leads to blindness.

Figure 6.3.21 Elephantiasis affects in the region of 90 million people throughout the world, while around 17.5 million people are affected by river blindness, and in some areas most families have at least some members blinded. However, the damage caused by this nematode worm is gradually being halted by a programme controlling the blackfly vector and treating infected individuals with the drug ivermectin which kills the nematode larvae. In years to come the grip of the nematodes may finally be loosened and the disease eradicated.

Phylum: Nematoda

Roundworms are thought to be the most numerous animals in the world. It has been estimated that there are around 5 billion roundworms in the top 7 cm of an acre of soil. Nematodes have narrow bodies pointed at both ends, and they have a mouth at the front and an anus at the back. They are round in cross-section – hence their name – and have a cavity, the **pseudocoelom**, between the body wall and the gut. The phylum contains many important parasites, such as *Ascaris* which infects the guts of both humans and pigs, and the family Filariidae which cause several appalling tropical diseases affecting the lives of up to 1 billion people in Africa and Asia.

Phylum: Annelida

The segmented worms have a body divided into regular segments with structures and organs repeated along the body. The segments are separated from each other by **septa** – thin sheets of membrane – and as a general rule each segment has bristle-like **chaetae** used for locomotion. **Nephridia**, specialised organs for excretion and osmoregulation, are also present in each segment. There is a closed circulatory system. The development of segmentation in this phylum seems to be an important step towards the evolution of larger animals, giving the potential for increased body size, efficient movement and specialisation of particular areas for particular tasks. The annelids have a true body cavity or **coelom**, as shown in figure 6.3.22.

There are three classes of Annelida, described below.

Class: Polychaeta

Polychaetes have many chaetae which are attached to outgrowths of the body wall called **parapodia**. They have single sexes and are all marine, for example *Nereis*, the carnivorous ragworm and *Arenicola*, the burrowing lugworm.

Class: Oligochaeta

Oligochaetes have relatively few chaetae on each segment. They are hermaphrodites but not self fertile. They are found in freshwater and terrestrial habitats and the common earthworm, *Lumbricus terrestris*, is a good example (see figure 2.6.2 on page 149).

Class: Hirudinea

The leeches have no chaetae. They possess suckers at both the anterior and posterior ends for locomotion and attaching to their prey, for example *Hirudo*, the medicinal leech which is still used in some circumstances for human blood-letting.

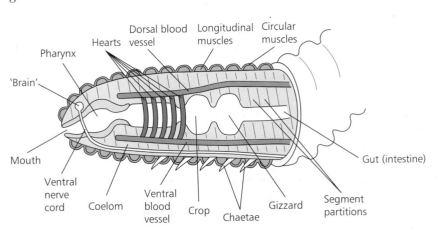

Figure 6.3.22 The annelids have a coelom, true gut, excretory organs and effective muscles, so paving the way for larger and more highly organised animals.

Phylum: Mollusca

The molluscs encompass a wide range of lifestyles and include the most intelligent of the invertebrates. Some molluscs have shells, and molluscs may live in the sea, in fresh water or on land. The main features of the molluscs include a soft muscular foot on the ventral side of the body, and a dorsal visceral hump containing the organs of the body which is often protected by the shell. Respiration is carried out by **gills** in the **mantle cavity**. There are several classes of molluscs, but the most common and important are listed below, and illustrated in figure 6.3.23.

Class: Gastropoda

The gastropods or 'stomach-foots' have a large flat muscular foot and often a coiled shell which is the result of the rotating of the visceral hump as it develops. A further result of this is that the anus and gills of the animal are positioned over its head! Many species have a rasping **radula** for feeding. The common garden snail *Helix*, the whelk *Buccinium* and the common garden slug *Arion hortensis* are all examples of this group.

Scallop, a pelycopod

Class: Pelycopoda

The bivalves are molluscs which have shells secreted in two halves. They have sheets of gills (they are sometimes called **lamellibranch molluscs**) and these are frequently used as filters for feeding, as bivalves do not have radulas. Examples include oysters, scallops and mussels such as *Anodonta* the freshwater mussel and *Mytilis* the marine mussel.

Class: Cephalopoda

The cephalopods or 'head-foots' have a foot which is reduced and incorporated in the head, and 8–10 tentacles. The members of this class are not typical molluscs. They include squids (the largest invertebrate on Earth is the giant squid), nautiluses and octopuses. The squids and octopuses have well developed eyes and octopuses in particular have large brains with areas not unlike cerebral hemispheres. They can be taught to run mazes and distinguish shapes – they are the most intelligent invertebrates.

The banded snail *Cepea nemoralis*, a gastropod

Phylum: Arthropoda

The arthropods are the most varied animals on the Earth, with around 3–12 million different species, though some estimates are as high as 30 million. They have made use of a wide range of available ecological niches – the only stumbling block to their total dominance of the planet has been their inability to reach great sizes, as the exoskeleton limits growth.

Arthropods have jointed appendages, both legs and antennae. One of the most distinctive features of the arthropods is their rigid exoskeleton, the result of a chitinous cuticle which is hardened to various degrees. Muscles are attached to the exoskeleton, which is shed periodically in the process of moulting (ecdysis) to allow for growth. The body cavity is a blood-filled haemocoel. They show bilateral symmetry and a degree of segmentation, although the segments are often modified for special functions. Some of the groups have compound eyes. The enormous number of arthropod species are arranged in a number of classes, four of which will be considered here.

Nudibranch mollusc, a marine gastropod

Superclass: Crustacea

The crustaceans are mainly aquatic arthropods, and the superclass contains several classes. Crustaceans have two pairs of sensory antennae and jaw-like mouthparts called **mandibles**. The dorsal surface of the body is usually

A blue-ringed octopus – a cephalopod, this is one of the smallest but most poisonous of the octopuses.

Figure 6.3.23 The molluscs – a phylum containing the largest and the most intelligent of the invertebrates.

covered with a tough protective **carapace** which in some cases completely covers the ventral surface as well. Examples include *Carcinus* (shore crab), woodlice and shrimps.

Class: Chilopoda

The centipedes are carnivorous arthropods with a pair of poison claws. The body is flattened dorsoventrally. They have one pair of legs per segment.

Class: Diplopoda

The millipedes are herbivores with rounded bodies and no poison claws. They have two sets of legs on each body segment.

The Chilopoda and the Diplopoda, along with two other small classes, used to be classified all together as the Myriapoda (many legs).

Class: Arachnida

The arachnids are the spiders, scorpions, mites, ticks and harvestmen. They have several simple eyes, and they have a **cephalothorax** and an **abdomen**. They contain poison glands. Most have eight legs.

Class: Insecta

The insects include around 900 000 named species, which have adapted successfully to life on land, in fresh water and in the air. Their bodies are divided into **head**, **thorax** and **abdomen**. The legs and wings are carried on the thorax. They have a single pair of antennae and well developed compound eyes. Gas exchange occurs within a tracheal system and they have Malpighian tubules for excretion. Insects are resistant to water loss and have efficient muscles. They also have life cycles which in general involve changes in form (**metamorphosis**) (see figure 6.3.24). This allows them to exploit different food sources and to avoid competition between adults and immature forms. Many insects exhibit extremely complex behaviour patterns on both social and individual levels which enable them to survive and reproduce in a wide variety of situations.

Figure 6.3.24 The insects are the most diverse group in the animal kingdom. Their complex life cycles are part of the key to their success.

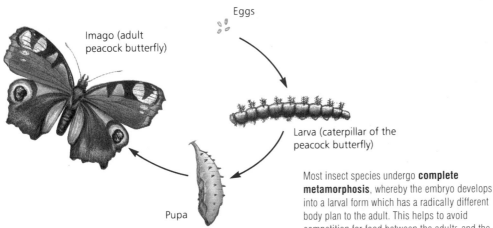

Most insect species undergo **complete metamorphosis**, whereby the embryo develops into a larval form which has a radically different body plan to the adult. This helps to avoid competition for food between the adults and the offspring.

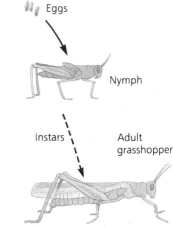

A minority of the insect species including grasshoppers and cockroaches have a development pattern known as **incomplete metamorphosis**. The embryo develops into a **nymph**, an immature version of the adult form. The nymph undergoes a series of moults through stages called **instars** until the adult finally emerges with fully mature organ systems.

Phylum: Echinodermata

The sea urchins, starfish and brittle stars make up the echinoderms. The skin contains many spines. Although they appear very simple, echinoderms have a mouth (on the lower side), a gut and an anus (on the upper side). They are all marine, and move around using **tube feet**. The adults are **pentaradiate**, which means they have five arms, but the larval stages are bilaterally symmetrical. Examples include *Asterias* the common starfish, *Echinus* the common sea urchin and *Paracucumana tricolor*, a brightly coloured sea cucumber known as a sea apple.

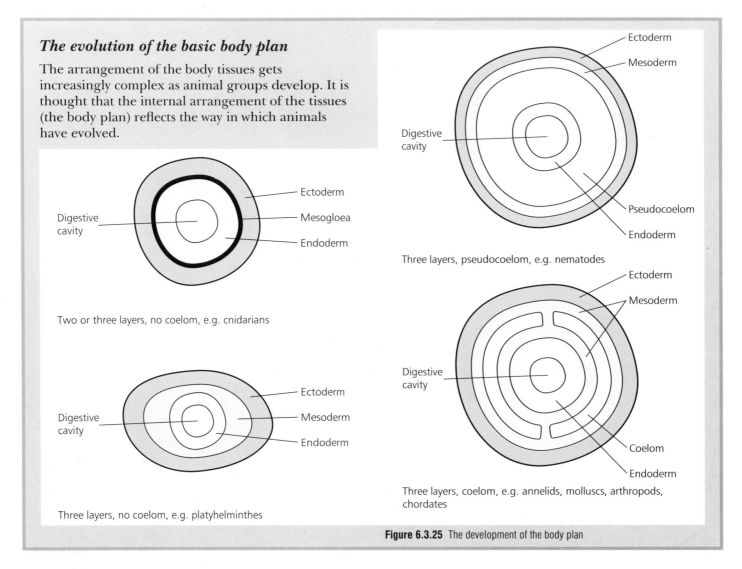

The evolution of the basic body plan

The arrangement of the body tissues gets increasingly complex as animal groups develop. It is thought that the internal arrangement of the tissues (the body plan) reflects the way in which animals have evolved.

Two or three layers, no coelom, e.g. cnidarians

Three layers, no coelom, e.g. platyhelminthes

Three layers, pseudocoelom, e.g. nematodes

Three layers, coelom, e.g. annelids, molluscs, arthropods, chordates

Figure 6.3.25 The development of the body plan

Phylum: Chordata

The invertebrates make up more than 99.9% of the animals alive on Earth today. Yet most people, when asked to name ten animals, would suggest ten from the remaining 0.1% – the vertebrates. The vertebrates are in fact one subphylum of the **chordates**. Chordate features include **gill slits**, a **notochord** (a dorsally situated strengthening rod), a **hollow nerve cord** and a **tail and associated muscle blocks**. All chordates, including ourselves, possess these features at some point during their life cycle. The chordate phylum is divided up into two subphyla, the non-vertebrate chordates and the vertebrate chordates.

The non-vertebrate chordates

Class: Hemichordata

The hemichordates make up a small class of invertebrates that appears to be closely related to the vertebrate chordates. The hemichordates have gill slits, and feed as the result of a current of water passing through the mouth and out over the gills. This is also the basic feeding mechanism of the chordates. Hemichordates also possess a dorsally situated nerve cord. An example is the acorn worm *Balanoglossus* shown in figure 6.3.26.

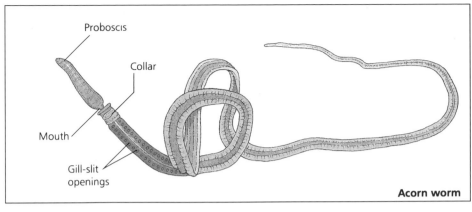

Acorn worm

Figure 6.3.26 An ancient creature not dissimilar to this hemichordate may well have given rise to the vertebrate chordates and so, indirectly, the human race.

Class: Urochordata

Most of the urochordates are sea squirts, sessile filter feeders which sit on the bottom of the sea as shown in figure 6.3.27. They are included in the phylum because the free-swimming larval stages possess a notochord in the tail.

Class: Cephalochordata

The cephalochordates consist of only one type of organism, the lancelet *Amphioxus*, shown in figure 6.3.27. A small marine animal, the lancelet has a notochord which extends into its head. It also possesses gill slits, a hollow nerve cord and a tail with associated muscle blocks. It is clearly a chordate, although as the lancelet notochord contains muscle-like fibres not seen anywhere else in the phylum it was probably an early evolutionary offshoot of the vertebrate line.

The vertebrate chordates

The true vertebrates possess all of the chordate features at some stage of their development. In addition, the notochord occurs only in the embryo form and is replaced in the adult by a **vertebral column**, with the brain developed and enclosed in a protective case called the **cranium**. The fossil record suggests that the early jawed fish were the first of the modern vertebrates to evolve, and that they gave rise to all the other vertebrate groups. There are five main classes of vertebrates, and we shall look briefly at each in turn.

Class: Chondrichthyes

The cartilaginous fishes include the sharks, skates and rays. They show several of the chordate features even into adulthood, with gill slits, tails with distinctive muscle blocks and hollow nerve cords. Their skeleton made of **cartilage** is light and strong. It is combined with large oil-storing livers so that many cartilaginous fishes have almost neutral buoyancy and need to expend very little energy to keep afloat. As a result many of them are very fast and efficient swimmers and the majority of the class exploit this and are carnivorous. They have five pairs of gill slits. The class is almost entirely marine.

Class: Osteichthyes

The modern bony fishes also have gill slits, tails with associated muscle blocks and hollow nerve cords. They have four pairs of gill slits which are covered by a flap called the **operculum**. They also usually have an **air sac** to control their buoyancy in the water. Much of the success of the bony fish is due to their ability to colonise both fresh and salt water. Within the bony fish is a small group of fleshy-finned fish. Extinct *Rhipidistans* of this type in Devonian times were probably the ancestors of the land vertebrates. The few modern forms include the lungfish and coelacanths.

Sea squirt

Lancelet

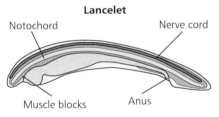

Figure 6.3.27 Primitive chordates – were creatures like these our dim and distant ancestors? It seems more likely that these were evolutionary 'dead ends' and that they and the vertebrates all came from a single ancestral type lost in the passage of time.

Class: Amphibia

The amphibians were the first vertebrates to colonise the land. They have simple sac-like lungs (which are not very efficient) and smooth, moist skin which is used as a supplementary respiratory surface. Their life cycle includes metamorphosis, as shown in figure 6.3.28, and they need water for successful reproduction as fertilisation is external and the larval form (tadpole) is aquatic. Gills are only present in the larval forms. Examples include frogs, toads, newts and salamanders.

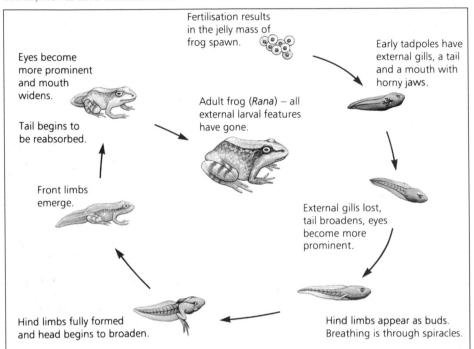

Fertilisation results in the jelly mass of frog spawn.

Early tadpoles have external gills, a tail and a mouth with horny jaws.

Eyes become more prominent and mouth widens.

Adult frog (*Rana*) – all external larval features have gone.

Tail begins to be reabsorbed.

External gills lost, tail broadens, eyes become more prominent.

Front limbs emerge.

Hind limbs fully formed and head begins to broaden.

Hind limbs appear as buds. Breathing is through spiracles.

Figure 6.3.28 Although the amphibians are no longer a large group on Earth, their ancestors took the first steps towards life on land for other animals to follow.

Class: Reptilia

The reptiles are mainly terrestrial animals which were for many millions of years the dominant species on Earth. They have dry skin with **scales** and their gas exchange takes place exclusively in the **lungs**. They have developed **internal fertilisation** and lay eggs on land which are enclosed within a leathery shell. Some reptiles retain the eggs within the body and give birth to fully developed young. These reproductive developments have freed the reptiles from the need to return to the water to breed, and so have enabled them to colonise dry and hot environments. The gill slits which are a chordate feature are seen only in embryonic development. Examples include snakes, crocodiles and lizards, as shown in figure 6.3.29.

Class: Aves

Birds are thought to have evolved from the reptiles or even the dinosaurs, and indeed they have many features in common. Instead of scales birds have **feathers** over most of the body. The forelimbs have been adapted as **wings** which in most cases are used for flight. The sternum or breastbone has been enlarged into a big keel shape for the attachment of the wing muscles, particularly in those birds which fly. The jaws are toothless and are covered by a horny beak. Reproduction involves the production of a well developed egg with a hard shell and in many cases considerable parental care is invested in the raising of the young. Birds have light skeletons and are endothermic. Of the four main chordate features, only the hollow nerve cord remains in the adult bird.

Figure 6.3.29 Snakes, crocodiles and chameleons have very different body forms and habits, yet they are all members of the reptiles. In spite of their obvious differences, the most important aspects of their biology – embryonic development, reproduction and breathing methods – are the same.

Class: Mammalia

A true mammal is distinguished by the production of milk for its young in **mammary glands**, a high internal body temperature and the presence of **hair**. Most mammals also produce live young which have developed for a time within the body of the mother in a structure called the **uterus**. As with the birds, chordate features are only plainly visible in embryonic development when gill slits, tails with segmental muscles and notochords may be seen as shown in figure 6.3.30. The mammals include organisms of an enormous range of sizes, from tiny shrews to elephants and whales. Their sophisticated temperature control and reproductive capabilities have made it possible for them to live in almost all of the Earth's environmental niches. Human beings are just one example of this highly developed class of animals.

The mammals are in fact divided into three subclasses based on the sophistication of the reproductive processes. These are:

- **Subclass: Monotremes** include the duck-billed platypus and the spiny anteaters. They are between the true mammals and the reptiles. The fertilised egg and early embryo are kept in the uterus for about two weeks. A reptilian type of egg is then built round the embryo and the egg is laid and incubated for a short time. A primitive newborn then emerges from the egg and suckles milk from the mother's mammary glands.
- **Subclass: Marsupials** are pouched mammals. Here the embryo grows for three to four weeks in the uterus before being born whilst almost still a fetus. The blind and tiny infant crawls from the vaginal opening through the fur of the mother to the **marsupium** or pouch. Here it latches on to the teat of a mammary gland and feeds, grows and develops until large enough for life outside the pouch. The length of time for which a marsupial can develop in the uterus is limited by the lack of a placenta in most species. Marsupial mammals include the kangaroo and the opossum.
- **Subclass: Eutherians** or placental mammals make up the majority of the mammals. The formation of a placenta which provides nourishment for a fetus and prevents immunological rejection by the maternal tissues allows the developing fetus to reach a large size and a high level of development before birth. In terms of evolutionary success this is of enormous significance. The mother remains freely mobile during pregnancy without the need to incubate eggs, and the mammary glands provide a guaranteed source of food for the offspring after birth. In view of all these factors the rapid rise of the placental mammals as a group is easily understood.

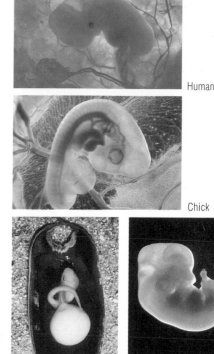

Human

Chick

Shark   Rat

Figure 6.3.30 Adult people, rats, birds and fish – all chordates – are easy enough to tell apart. In the early embryos of these species the similarities are much more striking than the differences, and only an expert would tell at a glance which was which.

Figure 6.3.31 A kangaroo and a cow are of comparable size, yet there is an enormous discrepancy between the size of their newly born infants. A newborn kangaroo is about the size of a bean and weighs around 1 g, whilst a newborn calf weighs around 45 kg. The comparatively large size of the calf is largely due to the role of the placenta in its development.

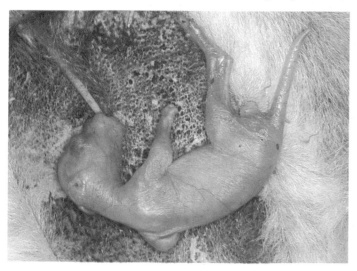

SUMMARY

- Organisms are classified according to the **binomial system of nomenclature**, in which each organism is given a **genus** and **species** name. Organisms are grouped together in further **taxa** (groups) – **families**, **orders**, **classes**, **phyla** and **kingdoms**. These classifications are given in table 6.3.2.

Kingdom	Phylum	Class	Example
Prokaryotae	(Non-photosynthetic bacteria)		*Staphylococcus*
	(Photosynthetic bacteria)		*Oscillatoria*
Protoctista	Rhizopoda		*Amoeba*
	Zoomastigina (flagellates)		*Trypanosoma*
	Apicomplexa (sporozoans)		*Plasmodium*
	Ciliophora (ciliates)		*Paramecium*
	Euglenophyta		*Euglena*
	Oomycota		*Phytophthora infestans*
	Chlorophyta (green algae)		*Spirogyra*
	Rhodophyta (red algae)		*Chondrus*
	Phaeophyta (brown algae)		*Laminaria* (kelp)
Fungi	Zygomycota		*Mucor*
	Ascomycota		*Penicillium*
	Basidiomycota		*Agaricus* (mushroom)
Plantae	Bryophyta	Hepaticae (liverworts)	*Pellia*
		Musci (mosses)	*Bryum capillare*
	Lycopodophyta (club mosses)		*Selaginella*
	Sphenophyta (horsetails)		*Equisetum*
	Filicinophyta (ferns)		*Dryopteris*
	Coniferophyta (conifers)		*Pinus* (Scots pine)
	Angiospermophyta (flowering plants)	Monocotyledoneae	*Triticum* (wheat)
		Dicotyledoneae	*Rubus fructicosus* (bramble)
Animalia	Porifera (sponges)		*Spongia* (commercial sponge)
	Cnidaria		*Aurelia* (jellyfish)
	Platyhelminthes (flatworms)	Turbellaria	*Planaria*
		Trematoda (flukes)	*Schistosoma*
		Cestoda (tapeworms)	*Taenia*

Kingdom	Phylum	Class	Example
	Nematoda (roundworms)		*Ascaris*
	Annelida	Polychaeta	*Arenicola*
		Oligochaeta	*Lumbricus terrestris* (earthworm)
		Hirudinea (leeches)	*Hirudo* (medicinal leech)
	Mollusca	Gastropoda ('stomach-foot')	*Helix* (snail)
		Pelycopoda (bivalves)	*Mytilis* (marine mussel)
		Cephalopoda ('head-foot')	*Loligo* (squid)
	Arthropoda	Crustacea (superclass)	*Carcinus* (crab)
		Chilopoda (centipedes)	*Lithobius* (centipede)
		Diplopoda (millipedes)	*Lulus* (millipede)
		Arachnida	*Scorpio* (scorpion)
		Insecta	*Locusta* (locust)
	Echinodermata		*Echinus* (sea urchin)
	Chordata	Chondrichthyes (cartilaginous fishes)	*Scyliorhinus* (dogfish)
		Osteichthyes (bony fishes)	*Clupea* (herring)
		Amphibia	*Rana* (frog)
		Reptilia	*Lacerta* (lizard)
		Aves (birds)	*Columba* (pigeon)
		Mammalia	*Homo* (human)

Table 6.3.2

QUESTIONS

1 a What features distinguish prokaryotes from eukaryotes?
 b In what ways are some protoctista like plants, some like fungi and some like animals?

2 a Why are the fungi known as the great decomposers?
 b Compare the vascular and non-vascular groups of land plants, indicating the major differences in their structures and lifestyles.

3 a What are some of the innovations of reptiles which freed them from dependence on a very moist environment?
 b The three lines of modern-day mammals are the monotremes, the marsupials and the placentals. What features distinguish each group?

6.4 Ecology

In our society the term 'ecology' often carries emotive meaning. It may conjure up pictures of brave (or foolhardy) conservationists in small rubber dinghies sailing in front of ships carrying toxic waste. It may be associated with earnest biologists, frequently sporting bushy beards and sandals and eating a vegetarian diet. Ecology is intimately linked in the public mind with the best and the worst of our efforts to undo some of the damage human beings have inflicted on our planet. These images are far from the true definition of ecology. The word comes from the Greek *oikos* meaning 'house', and ecology is the study of the interactions that determine the distribution and abundance of organisms within a particular environment. Information gained from the study of ecology may be useful to those involved in conservation, but ecology and conservation are not the same.

ECOSYSTEMS

What is an ecosystem?

An **ecosystem** is a life-supporting environment. It includes all the living organisms interacting together, the nutrients cycling through the habitats in the system, and the physical and chemical environment in which the organisms are living. An ecosystem consists of a network of several habitats and the communities associated with them (see box below). Particular emphasis is placed on the energy flows and links between the various components of the ecosystem. The largest ecosystem is the **biosphere** – the life-supporting seas, soils and atmosphere of the planet Earth and the organisms which live upon it, as illustrated in figure 6.4.1.

Figure 6.4.1 The largest ecosystem is the biosphere. It has many parts, each of which can be considered as an ecosystem in its own right. These range from the frozen Arctic wastes to the richness of a tropical rainforest, from the depths of an underground cave to a lush grassy meadow. All the ecosystems within the biosphere are dependent on the input of energy from the Sun.

The terminology of ecology

Figure 6.4.2 Within a diverse habitat such as a tropical rainforest, many different populations of organisms live interdependently as part of a complex community. These populations occupy all the available niches, unless circumstances change either to create a new niche or to empty an old one, when different species may enter the ecosystem.

Like any other area of scientific study, ecology has its own specific terms. Here are definitions of some of the most common terms you will meet.

- **Habitat** – the place where an organism lives. An organism's habitat can be considered as its address. Examples of habitats include a freshwater pond, a deciduous woodland or a rocky seashore. Many organisms live and move only in a small part of a particular habitat, such as a tree in a woodland or a rock on the seashore, and these are referred to as **microhabitats**.
- **Population** – a group of organisms, all of the same species, living together in a particular habitat. The three-spined sticklebacks in a particular pond and the skin mites in your mattress are examples of populations.
- **Community** – the total of all the populations of animals and plants living in a habitat at any one time. For example, in a habitat such as a rock pool there will be populations of seaweeds of various types, of sea anemones, of shrimps, of small fish such as gobies, and of crabs to name but a few. This collection of populations makes up a community.
- **Niche** – the ecological niche of an organism is difficult to define. It can best be described as the role of the organism in the community, its way of life. If the habitat is the address of the organism, the niche describes its profession. A niche can be broken down into specific elements, for example the **food niche** or the **habitat niche**.
- **Biome** – one of the world's major ecosystems – grasslands, forests, deserts and oceans are a few examples.
- **Abiotic factors** – these make up the non-living elements of the habitat of an organism. Abiotic factors include the amount of sunlight and rainfall experienced by a habitat, the water currents in an aquatic habitat, and all the chemical nutrients available to organisms along with the pH of a habitat. Abiotic factors can determine the ability of an organism to succeed within a particular habitat.
- **Biotic factors** – these are the living elements of a habitat which affect the ability of a group of organisms to survive there. For example, the presence of suitable prey species will affect the numbers of predators in a habitat.

Biomes – the major ecosystems

The biosphere is, as we have said, the largest ecosystem associated with the planet Earth. However, the biosphere is so large that it is very difficult to study as a whole. The biosphere is divided into smaller ecosystems, each readily visible around the planet. These major ecosystems or biomes are frequently further subdivided many times for ease of study. The major biomes around the world are shown in figure 6.4.3, and a summary of their features follows.

Tropical rainforests

Tropical rainforests have become part of our everyday language through high-profile international conservation efforts. Tropical rainforests occur at low latitudes where the rain falls abundantly all year long and the temperature is warm. The high humidity, warmth and sunlight provide ideal growing conditions so that trees grow very tall, supporting many species of orchids, vines and fungi. Animal species tend to be concentrated either in the canopy 40 m or so above the ground, or on the forest floor, which is relatively clear of

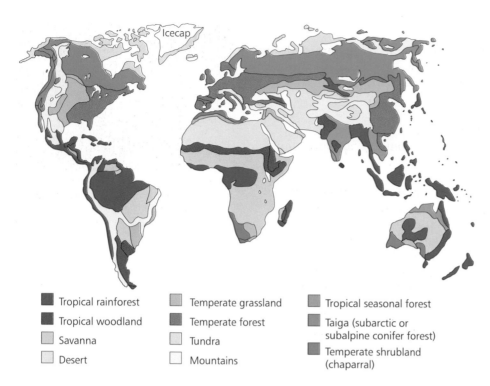

Icecap

◼ Tropical rainforest   ◼ Temperate grassland   ◼ Tropical seasonal forest

◼ Tropical woodland   ◼ Temperate forest   ◼ Taiga (subarctic or
                                                    subalpine conifer forest)
◻ Savanna   ◻ Tundra
                                                 ◼ Temperate shrubland
◻ Desert   ◻ Mountains   (chaparral)

Figure 6.4.3 The major biomes of the world. Their distribution is a result of abiotic factors – the geology of the area along with global wind, water and weather patterns.

vegetation because very little light penetrates the dense canopy. There is a great richness and diversity of species in tropical rainforests.

Tropical seasonal forests

Tropical seasonal forests occur where the climate is rather drier than in areas of rainforest. The trees may lose their leaves when the drier seasons come, so the rate of growth is lower than in the rainforests, but these forests are nevertheless a very diverse ecosystem.

Savannas and tropical woodlands

Savannas are the great tropical grasslands, well known in Africa. The driest savannas consist almost entirely of grasses, though with increasing rainfall they support increasing numbers of trees until they merge into **tropical woodlands**. Savannas support enormous herds of grazing animals and their predators, as well as many invertebrate species. Tropical woodlands are made up largely of thornwoods, bushes and trees such as the *Acacia* protected with long spines. They grow rapidly during the short rainy seasons but then frequently lose their leaves and grow very little during the dry seasons.

Deserts

Deserts are areas of very little rainfall. Some are almost entirely barren, while others have seasonal rainfall and so can support some vegetation. Relatively few species can cope with the extremes of a habitat such as a desert and most have very specific adaptations to allow them to succeed in harsh conditions.

Temperate grasslands, shrublands and forests

Temperate grasslands are similar to savannas but occur in cooler climates. Examples are the American prairies, the Eurasian steppes and the pampas of Argentina. They usually have rich, deep soil and are often used for human agriculture. **Temperate shrublands** grow where there are hot, dry summers and cool, wet winters. **Temperate forests** include deciduous and evergreen forests but contain far fewer species than a tropical rainforest.

Figure 6.4.4 The grasslands of the savannas support large herbivores and large carnivores in clearly defined predator–prey relationships.

Taiga and tundra

Taiga is evergreen forest which is found in subarctic and subalpine conditions. The number of tree species found is usually very low – two or three – but there is a greater variety of animal life, frequently including animals such as wolves, elk, moose and jays.

Tundra is found in the Arctic and on the world's highest mountains. It consists of treeless plains with grasses, lichens, sedges and mosses as the predominant species. Many migrants use the tundra during the short summers.

Aquatic biomes

As well as land biomes, there are several aquatic biomes. **Freshwater biomes** include **lentic** (lake and pond) communities and **lotic** (stream communities). **Marine biomes** include **shorelines** both sandy and rocky, **marshes**, **coral reefs**, **open ocean** communities and **benthic** communities (bottom dwellers).

This brief review of some of the major ecosystems shows something of the variety of habitats existing on the Earth. The make-up of a habitat is affected both by the climate – the wind, rainfall and sunlight it receives – and by the populations of living organisms which make that habitat their home. For any ecosystem to survive there must be a through-flow of energy and a renewal of resources. We shall consider these vital aspects of an ecosystem next.

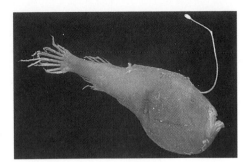

Figure 6.4.5 The biome of the deep ocean is one of the least understood. Many of the species show strange adaptations to their extreme environment, and there are probably many other populations as yet undiscovered in this major ecosystem.

THE EVOLUTION OF ECOSYSTEMS

How ecosystems evolve

The major biomes just described are the results of processes occurring over long stretches of time, changing the original bare rock of the Earth into the ecosystems of today. This change has been brought about by **succession**, a process by which communities of animals and plants colonise an area and then over time are replaced by other, more varied organisms. Succession can be observed everywhere – the species composition of the community of any habitat is gradually changing all the time.

New communities are usually started by plant species known as **opportunists** or **pioneers**. These plants interact with the abiotic environment and thus change the habitat. Other organisms, both plants and animals, then enter the community which passes through stages known as **seres** until finally a stable community is reached. This is known as the **climax**. There are two main types of succession, **primary succession** and **secondary succession**, which we shall go on to look at in more detail.

Primary succession

Primary succession occurs from the starting point of bare rock. This type of succession is seen after the eruption of a volcano or a landslide, or after the emergence of a new volcanic island such as Surtsey. The first organisms are lichens, algae and mosses which can penetrate the rock surface and trap organic material to begin the formation of humus. Other pioneer species such as grasses and ferns then take advantage of the pockets of soil formed, and the death and decay of several generations of these pioneers adds to the accumulating soil, as shown in figure 6.4.6. More water and nutrients gradually become available for plant roots so that other species can survive and the succession moves on. Gradually the diversity of species increases and larger plants can be supported. An increasingly diverse assortment of animals also colonise the environment, changing along with the dominant plant types.

6

Eventually a climax community is reached, where the species are regarded as being stable until the environment changes again.

A specialised form of primary succession is aquatic succession when a lake or pond silts up and a new terrestrial habitat is formed. Although in the formation of the biosphere primary succession must have been of prime importance, today it is relatively rare as volcanic eruptions and earthquakes are not everyday events in most parts of the world.

Volcanic eruptions result in bare rocks on which a primary succession starts.

The appearance of mosses and lichens on bare rock is the beginning of a primary succession.

A few years after the emergence of bare rocks an increasing diversity of plant species is beginning to colonise the new habitat.

Figure 6.4.6 Primary succession is the gradual progression from bare rock to a full and diverse ecosystem such as a woodland, coral reef or tundra. The first stages are illustrated here.

Secondary succession

Secondary succession is the evolution of an ecosystem from bare existing soil, as illustrated in figure 6.4.7. It occurs as rivers shift their courses, after fires and floods and after disturbances caused by humans. The sequence of events is very similar to that seen in primary succession, but because the soil is already formed there are much larger numbers of plants and animals present at the beginning of the succession.

Figure 6.4.7 The stages of a secondary succession from bare earth to an oak woodland are shown here. The time scale is very approximate – it will depend on many factors including rainfall and temperature – but the types of plants found at different stages can be seen. The types of plants found at the end of a secondary succession vary greatly. Simply digging a patch of earth and leaving it allows the beginnings of a secondary succession to be observed.

Age/ years		1	2	3 – 20	25 – 100	150+
Types of plant communities	Bare earth grass, weeds	Grass stage		Grass and shrubs	Young forest – pines, some young hardwoods	Mature forest – mainly oak

Succession in the sand dunes

Observing succession is not easy, because it occurs over very many years. However, Gibraltar Point on the east coast of England is one of several places where sand dunes hold a complete record of the stages of the succession. The oldest dunes, those furthest from the sea, are in the late stages of the succession. Nearest to the sea are the youngest, newly formed dunes which are in the very earliest stages of the ecological

Figure 6.4.8 Fires frequently denude areas of heath, moor and forest of their vegetation. Pioneer plants such as this rosebay willow herb are adapted to germinate rapidly and take advantage of the newly created habitat.

succession process. The gradual change from sand to a more mature soil, and the differences in the plants and animals making up the communities populating each habitat, can readily be observed in figure 6.4.9.

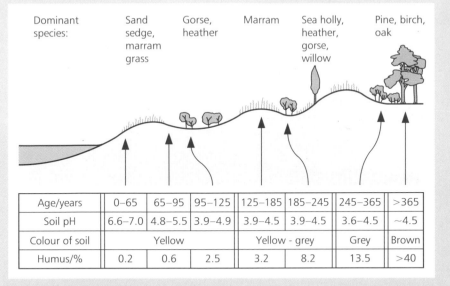

Age/years	0–65	65–95	95–125	125–185	185–245	245–365	>365
Soil pH	6.6–7.0	4.8–5.5	3.9–4.9	3.9–4.5	3.9–4.5	3.6–4.5	~4.5
Colour of soil	Yellow			Yellow - grey		Grey	Brown
Humus/%	0.2	0.6	2.5	3.2	8.2	13.5	>40

Figure 6.4.9 Studying sand dunes shows how the plants growing at each stage alter the environment and make it suitable for different plants in the later stages.

ENERGY TRANSFER IN ECOSYSTEMS

The idea of energy transfer

In the 1920s a young Oxford biologist called Charles Elton began to study the relationships between the animals in the Arctic tundra of Bear Island (off the northern coast of Norway) and the scarce food resources there. The relatively few available plant food species probably made his study easier and more effective than studying a more diverse habitat. The main carnivores he observed were the arctic foxes, which ate birds such as sandpipers and ptarmigan. These birds, mainly summer migrants, ate the leaves and berries of the tundra plants or, in some cases, ate insects which in turn fed on the plants. Elton referred to this series of feeding interactions between organisms as a **food chain**.

A model for a food chain

Elton then proposed a general model for a food chain. The first stage involves the trapping of solar energy by a plant in photosynthesis. Some of the energy from the Sun is converted into stored chemical energy within the structure of the plant cells and for this reason plants are known as **producers**. This energy is then passed on from the plant to the herbivore (known as a **primary consumer**) which eats it. The energy within the herbivore is, in its turn, passed on to the carnivore (a **secondary consumer**) which eats the plant eater, and this continues along the chain. At the end of every chain are the **decomposers**, bacteria and fungi which break down the remains of animals and plants and return the mineral nutrients to the soil. From here they will be absorbed again and recycled into new plant material. Each link in the food chain represents a specific **trophic** (or feeding) **level**, as shown in figure 6.4.10.

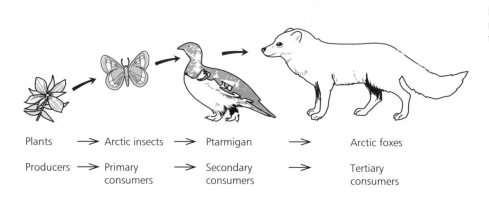

Plants ——→ Arctic insects ——→ Ptarmigan ——→ Arctic foxes

Producers ——→ Primary consumers ——→ Secondary consumers ——→ Tertiary consumers

Figure 6.4.10 A food chain is the simplest representation of the feeding relationships and therefore the energy flow within an ecosystem.

Food webs within an ecosystem

The description of a food chain makes sense and is relatively simple to understand. It is easy to think of all the organisms in a habitat taking part in a whole range of food chains. However, further thought suggests that this must be an oversimplification. Few animals eat only one single food – giant pandas eating only bamboo and koala bears feeding only on eucalyptus are two of a small number of examples. Most animals have a variety of food sources and exist not in simple food chains but as part of complex interconnected **food webs**. This is precisely what Elton went on to observe on Bear Island and a representation of some of the feeding relationships he observed is shown in figure 6.4.11.

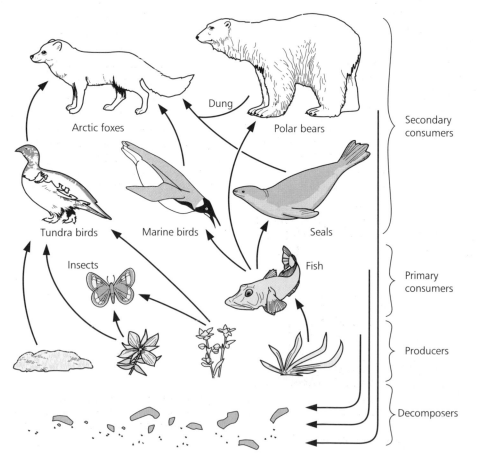

Figure 6.4.11 A food web is a complex system of interrelated food chains. This example of a food web from a tundra biome is still very much simplified, but begins to show how energy moves through the system.

In situations involving a simple food chain, such as the giant panda with its diet almost exclusively of bamboo, the ecosystem is very easily destroyed. Any event which reduces the availability of the bamboo will also threaten the pandas, making them very vulnerable to habitat destruction. When an organism is part of a complex food web a change in any one component, whilst potentially affecting the balance of the ecosystem, is far less likely to have catastrophic effects and so the system will be more stable, as illustrated in figure 6.4.12.

Ecological pyramids

Since Elton first began to consider the transfer of energy through ecosystems, much work has been done in a wide variety of habitats to look at the feeding relationships between the members of the various communities. A common observation is that in most communities there appear to be more plants than herbivores, more herbivores than small carnivores, more small carnivores than large carnivores, and so on. These observations are represented at their simplest as a **pyramid of numbers**, as in figure 6.4.13.

However, in many instances a simple pyramid of numbers is not helpful in demonstrating the flow of energy through an ecosystem. For example, a single rose bush will support a very large population of aphids which in turn will be eaten by a smaller population of ladybirds and hoverfly larvae, which are the prey of a few birds. A more realistic picture is gained by considering a **pyramid of biomass**. This shows the combined mass of all the organisms in a particular habitat measured in grams. Either wet or dry biomass may be used – dry mass eliminates the inaccuracy of variable water content but involves destroying the material. To avoid destruction of the habitat a small sample is usually taken and measured and the total biomass of the population then calculated. It is much more time consuming to produce a pyramid of biomass than a simple pyramid of numbers, but it gives much more accurate information about what is happening in an ecosystem, as figure 6.4.14 shows.

Even pyramids of biomass do not always supply us with the information we need about a particular ecosystem. For example, when a sample of water from the English Channel is analysed and the biomass of the organisms found, there appears to be a greater mass of zooplankton than of the phytoplankton on which it feeds. The reason for this is that the sample represents a picture of the water only at the moment the sample is taken – it is in effect a still photograph of the ecosystem. What is really needed is a moving film of the system, and a repeated sampling procedure shows that the phytoplankton reproduce much more rapidly than the zooplankton. Thus although the total population of phytoplankton at any one time (the **standing crop**) is smaller than that of the zooplankton, the turnover of phytoplankton is much higher and so the biomass over a period of time is greater. An analogy is a bowling

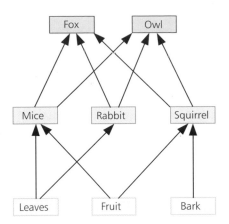

Figure 6.4.12 A web such as this simplified one for an English deciduous woodland represents a relatively stable system. For example, when rabbits were decimated by myxomatosis, owls did not die out but survived by eating more of their other prey species. Whilst squirrels and mice were more heavily predated as a result, they also gained the benefit of reduced competition for plant resources so they in turn could reproduce more successfully. Thus the change in the rabbit population did not, in the long term, inflict any major damage on the rest of the community.

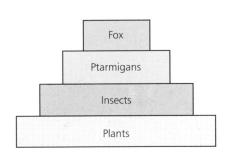

Figure 6.4.13 For the simplified food chain showing the feeding relationships on Bear Island a pyramid of numbers such as this can be built up. The number of individuals falls as we move further along the chain.

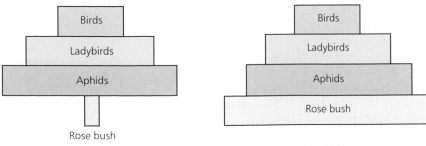

Pyramid of numbers

Pyramid of biomass

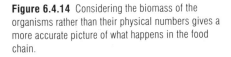

Figure 6.4.14 Considering the biomass of the organisms rather than their physical numbers gives a more accurate picture of what happens in the food chain.

green. The smooth closely cropped grass does not seem to represent a very large standing crop of grass. But at the height of the growing season a bowling green will be cut several times a week, and only if all the piles of grass clippings are taken into account will the true productivity of the grass be seen. A pyramid made up from observations taken over time is called a **pyramid of energy** and this gives the most accurate picture of energy flows within an ecosystem, as figure 6.4.15 shows.

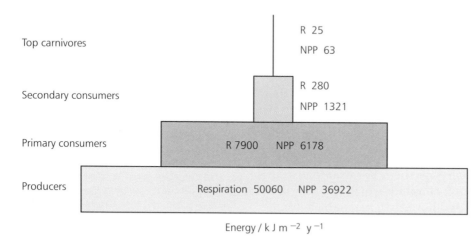

Top carnivores — R 25 NPP 63

Secondary consumers — R 280 NPP 1321

Primary consumers — R 7900 NPP 6178

Producers — Respiration 50060 NPP 36922

Energy / $kJ\ m^{-2}\ y^{-1}$

Pyramid of energy for a pond ecosystem in Florida

R = respiration, NPP = net primary production (see text)

Figure 6.4.15 A pyramid of energy such as this is difficult to produce, as it makes allowances for energy gains and losses over a period of time, but it gives the most accurate picture of the energy flows within a food chain or web.

Energy relationships within ecosystems

The transfer of energy through an ecosystem leads to the accumulation of biomass as new plant and animal material is produced. There are two kinds of production in an ecosystem – **primary production** is the synthesis and storage of organic molecules as a result of photosynthesis, and **secondary production** is the processing and storage of organic material by herbivores and carnivores. All the consumers in a food chain ultimately depend on the photosynthesisers for energy. The rate at which producers convert the Sun's energy into organic material determines the flow of energy within an ecosystem.

As we have seen in section 3.1, only a very small percentage of the energy from the Sun is transferred into plant material. Similarly, not all the organic material produced by photosynthesis is available to secondary consumers. Only about 10% of the material is passed on, the remainder being dispersed in a variety of ways, mainly by respiration and metabolism. Thus the net rate of production can be described as:

$$\text{Net primary production} = \text{gross primary production} - \text{respiration and metabolism}$$

The **net primary production** or **NPP** of different biomes has been estimated and is shown in table 6.4.1. The enormous productivity of the tropical rainforests can clearly be seen. Although only about 5% of the surface of the Earth is covered by tropical rainforest, it yields about 28% of the total NPP. For this reason alone the continuous massive deforestation presently taking place is a major cause for concern (see box 'Deforestation'). The human population of the world is growing at a very rapid rate, as we shall see in section 6.6, and even at its current level humans are consuming up to 40% of the NPP of the planet.

Habitat	Mean net primary production/ g m^{-2} y^{-1}
Terrestrial	
Tropical rainforest	2200
Temperate evergreen forest	1300
Temperate deciduous forest	1200
Savanna	800
Taiga	800
Cultivated land	650
Temperate grassland	600
Tundra and alpine communities	140
Desert and semidesert scrub	90
Extreme desert	3
Aquatic	
Algal beds and coral reefs	2500
Estuaries	1500
Upwelling zones	500
Continental shelf	360
Lake and stream	250
Open ocean	125

Table 6.4.1 Net primary production of a variety of ecosystems (*after R.H. Whittacker*, Communities and Ecosystems, *2nd edition, Macmillan 1975*)

Deforestation – the destruction of the rainforests

Tropical rainforests are the most productive and diverse of the land biomes. They have formed over millions of years and although they cover only a relatively small percentage of the surface of the Earth, they play a vital role both in maintaining species diversity and in absorbing carbon dioxide from the atmosphere for photosynthesis. During the latter part of the twentieth century a combination of world overpopulation, poverty and greed have driven people to ever-increasing destruction of these vital ecosystems. Rainforests have been destroyed to provide hardwoods such as mahogany for furniture and for other artefacts to supply the developed world. They have also been cut down and burned to make way for agriculture, both for growing crops and for rearing beef cattle. By 1980 around 44% of all the tropical rainforest had been destroyed, and in the 1990s this destruction is continuing at the rate of around 35 acres every minute – if it continues at this rate, in about 90 years' time there will be no tropical rainforests left.

Why does this deforestation continue? Apart from demands for tropical hardwoods as timber, the need for agricultural land might be expected to be finite. However, the result of deforestation is only a temporary gain in agricultural land. After a few years of supporting crops or cattle the land structure breaks down. Without the roots of the massive trees to hold the soil together, rain and wind cause extensive erosion and the land rapidly becomes barren. At this point another massive tract of forest has been destroyed.

Figure 6.4.16 To provide the developed world with goods such as mahogany toilet seats and cheap beefburgers, rainforest destruction is occurring at a frightening rate. In less than 100 years' time there may be no rainforests left, and the effects of this both on the gene pool of the planet and on its climate are as yet unknown. Only concerted conservation efforts by all parties – both the owners of the forests and the consumers of the developed world – can prevent the long-term loss of this irreplaceable biome.

What effects could deforestation have? The projected loss of around a million species by the year 2010 (that represents about 25% of the estimated diversity of life on Earth) would be a catastrophe in biological terms. Previous mass extinctions have occurred over millions of years rather than a few decades. Of equal importance is the effect of the loss of vast acreages of tropical rainforest on the world climate. The forests fix substantial amounts of carbon dioxide in the process of photosynthesis. If the forests are destroyed, not only will the carbon dioxide remain in the atmosphere, but burning the trees will release even more. Carbon dioxide is a greenhouse gas. A substantial increase in carbon dioxide levels in the atmosphere is likely to lead to global warming, and the resultant possible melting of the polar icecaps and shifting patterns of wind and rainfall could have catastrophic effects on agriculture throughout the world.

Energy budgets in ecosystems

As in any budget, the outgoings from each trophic level of an ecosystem need to be balanced by the incomings. This means that all the energy entering one trophic level must be accounted for when considering the transfer to the next trophic level. The energy bound up in plant material, the NPP, is transferred to heterotrophs. At this and every subsequent step, relatively little of the energy (around 10%) is used for producing new animal material. What happens to the rest?

As we have seen, the energy used to make new animal is known as secondary production. Much of the energy passed on from plants to consumer or from one consumer to another is used for respiration, and most of this is consequently transferred to the atmosphere as heat.

Yet more of the energy will be lost as metabolic waste products and heat in the urine of the animal, and as undigested and therefore unused material in the faeces, as illustrated in figure 6.4.17. Thus relatively little of the Sun's energy that reaches the Earth is used to produce plant material, and increasingly small amounts of that energy find their way into the individuals of a food chain or web. This is why the biomass and energy content of a food chain decreases along the chain.

However, the role of the decomposers must not be forgotten in this energy budget. They break down the bodies of animals and plants when they die, thus releasing and utilising much of the energy held within them. Decomposers also feed on and digest the faeces produced by living animals and thus make use of this energy too.

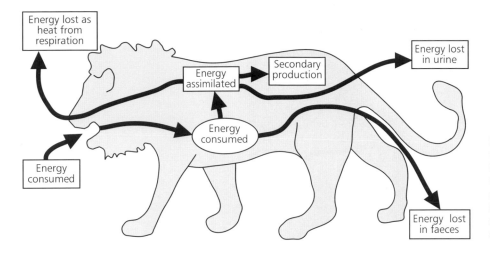

Figure 6.4.17 The fate of the food eaten by a consumer – a lion is shown but the situation is similar in any herbivore, omnivore or carnivore. Much of the food consumed is assimilated by the lion, but only a small proportion is used to produce new lion tissue. Some of the food is never assimilated at all and is simply lost as faeces – this proportion will be significantly larger in a plant-eater than in a carnivore.

The relatively inefficient transfer of energy through food chains and webs limits the number of trophic levels. Usually by the fourth or fifth level at most, harvesting the energy becomes too inefficient to be biologically worthwhile. This is why, for example, the killer whale at the top of a food chain (killer whale → seal → large fish → small fish → shrimp → phytoplankton) has no real predators of its own – an animal large enough to capture and kill a killer whale would need to expend so much energy hunting that it could not realistically survive.

Biological magnification – a by-product of food chains

Biological magnification occurs when materials present only in minute amounts in the environment accumulate in increasing quantities in the members of a food chain or web. Examples are seen when chemicals added to the environment by humans accumulate in other organisms and cause damage.

DDT is an extremely effective pesticide, but it is not readily excreted. In many animals it is stored in the body fat. Insects killed by DDT and zooplankton living in contaminated water are eaten by small fish, which in turn are eaten by larger fish. Finally the larger fish are eaten by top carnivores, fish-eating birds such as ospreys and herons.

When DDT was introduced as a pesticide in this country it was observed that the numbers of large fish-eating birds began to decline. The tissues of the birds were analysed and found to contain high concentrations of DDT, particularly in the fat, muscles and livers. The levels of DDT were not high enough to kill the birds, but were affecting their ability to reproduce. The eggs laid had very thin shells which broke readily, so many fewer offspring than normal were hatched and raised. Although the chemical was accumulating in the tissues of the entire food chain, it was only in the top consumers that it reached high enough levels to have an effect. These effects were noticed in the 1960s and 1970s. Since that time the use of chemicals such as DDT has been banned in most of the developed world. However, because they are very effective pesticides they are still in use in much of the developing world, where the priorities for disease vector control are very much higher.

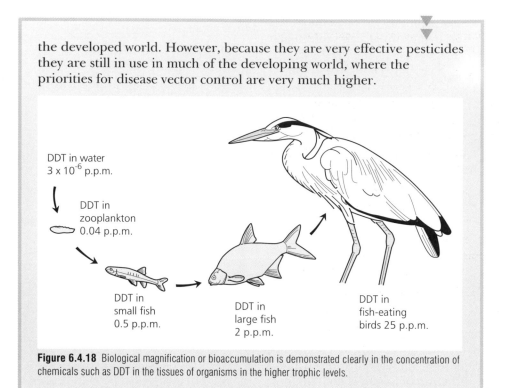

DDT in water
3×10^{-6} p.p.m.

DDT in zooplankton 0.04 p.p.m.

DDT in small fish 0.5 p.p.m.

DDT in large fish 2 p.p.m.

DDT in fish-eating birds 25 p.p.m.

Figure 6.4.18 Biological magnification or bioaccumulation is demonstrated clearly in the concentration of chemicals such as DDT in the tissues of organisms in the higher trophic levels.

Summary

- **Ecology** is the study of the interactions that determine the distribution and abundance of organisms within a particular environment. An **ecosystem** is a life-supporting environment consisting of organisms, nutrients and the physical and chemical environments. The biosphere is the largest ecosystem.

- A **habitat** is the place where an organism lives. A **population** is a group of organisms of the same species living in a habitat. A **community** is all the different populations living in a habitat. The ecological **niche** of an organism describes its role in the community. **Abiotic factors** make up the non-living elements of a habitat, and **biotic factors** its living elements.

- A **biome** is one of the world's major ecosystems.

  The major biomes of the world include:

  tropical rainforests   temperate grasslands

  tropical seasonal forests   temperate shrublands

  savannas   temperate forests

  tropical woodlands   taiga

  deserts   tundra.

- **Aquatic biomes** include **freshwater biomes** (**lentic** or lake and pond biomes, and **lotic** or stream biomes) and **marine biomes** (shorelines, marshes, coral reefs, open ocean and benthic or bottom-dwelling biomes).

- Ecosystems evolve over long periods of time by a process called **succession**. New communities are started by **pioneers** or **opportunists** which change the habitat so that other organisms can occupy it. After a series of stages or **seres** a stable **climax community** is reached.

- **Primary succession** is the evolution of an ecosystem starting from bare rock, while **secondary succession** starts from bare existing soil.

- A **food chain** represents the feeding relationships between organisms. **Producers** convert solar energy into chemical energy. Producers are eaten by **primary consumers** (herbivores), which in turn are eaten by **secondary consumers** (carnivores). **Decomposers** break down the remains of dead animals and plants. Each link in the chain represents a **trophic level**. Food chains are interconnected in complex **food webs**.

- Feeding relationships can be represented by **pyramids of numbers**, **pyramids of biomass** and most accurately by **pyramids of energy**.

- **Primary production** is the synthesis and storage of organic molecules by producers, and **secondary production** the processing and storage of organic molecules by consumers. Organic material and therefore energy is dispersed at each trophic level, in the processes of respiration and metabolism, such that only about 10% of the energy gained by the organisms at one trophic level is passed on to the organisms of the next trophic level. This limits the number of trophic levels in a food chain.

- **Net primary production** (NPP) of a biome = gross primary production − (respiration + metabolism).

QUESTIONS

1 a What is meant by the term:
  i ecosystem
  ii community
  iii niche
  iv biome?
  b Select three of the major biomes of the Earth, and for each describe its main features.
  c Discuss the interactions between the living and non-living components of any ecosystem.

2 a What is meant by the term 'ecological succession'?
  b What is the difference between primary succession and secondary succession?
  c Describe an example of:
  i primary succession
  ii secondary succession.

3 Explain how energy flows through an ecosystem, using food webs and ecological pyramids in your answer.

6.5 Cycles within ecosystems

An ecosystem is defined as a life-supporting environment which has both living and non-living elements. In section 6.4 we considered the main ecosystems found on Earth, how those ecosystems have evolved by succession, and the transfer of energy through an ecosystem from plants into animals and decomposers. However, for organisms to survive they need other materials as well as the organic compounds we refer to as 'energy'. We shall now consider how inorganic elements cycle through ecosystems and so remain constantly available to all the organisms that require them.

CYCLES OF INORGANIC MATERIALS

Energy is constantly entering ecosystems from the Sun and then being very inefficiently transferred through food chains and webs. The inefficiency does not matter, as there is a constant supply of energy from the Sun. However, the same is not true for other ingredients of life such as water, and also carbon and nitrogen, vital components of the chemicals which make up the structure of living cells. Complex cycles have evolved which result in the chemical constituents of life being used, and then returned to the various ecosystems of the world to be used again. These cycles involve a **biotic phase** where the inorganic ions are incorporated in the tissues of living things, and an **abiotic phase** where the inorganic ions are returned to the non-living part of the ecosystem. Four of the most important cycles are considered here.

THE CARBON CYCLE

As we saw in sections 1.2 and 1.3, carbon is the element which is crucial to the formation of the complex organic molecules – carbohydrates, proteins, lipids

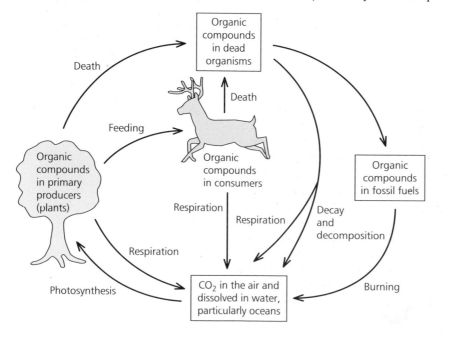

Figure 6.5.1 The carbon cycle. Left to itself the cycle is self regulating – the oceans act as massive reservoirs of carbon dioxide, absorbing excess when it is produced and releasing it when it is in short supply. However, the enormous increase in the production of carbon dioxide by people is now threatening the balance of the carbon cycle, and the results could be disastrous for many forms of life.

and nucleic acids – which are the building blocks of life. There is a massive pool of carbon in the carbon dioxide present in the atmosphere and dissolved in the water that makes up rivers, lakes and oceans. This carbon dioxide is taken up and the carbon incorporated into plants in the process of photosynthesis. The carbon is then passed into animals through food chains. At all stages, carbon dioxide is returned to the atmosphere or water during the process of respiration and by the actions of decomposers. The interactions of the **carbon cycle** can best be summarised in a diagram such as that shown in figure 6.5.1.

Human intervention

The carbon cycle affects not only living things but also the geology and weather of the planet. Over recent centuries the intervention of the human race in the carbon cycle has been increasingly evident. People increase the amounts of carbon dioxide being returned to the atmospheric pool in two main ways. One is by the burning of fossil fuels such as coal, oil and gas in power stations and of petrol in cars. The other is by the cutting down and burning of tropical rainforests to make way for agriculture. This latter activity causes two-fold damage, releasing carbon dioxide gas by the combustion process and also reducing the number of photosynthetic fixers of carbon dioxide. The rise in atmospheric carbon dioxide levels resulting from these activities over the last 20 years can be seen clearly in figure 6.5.2.

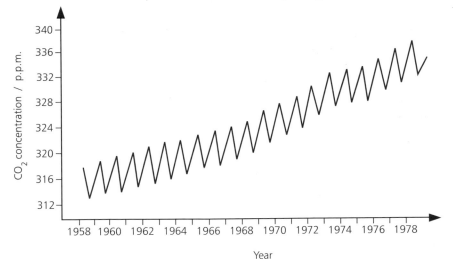

Figure 6.5.2 The burning of fossil fuels and extensive deforestation are leading to a long-term increase in the level of carbon dioxide in the atmosphere. The readings for this graph were taken on a mountaintop in Hawaii. The regular annual fluctuations in the levels of carbon dioxide seem to be the result of seasonal differences in the fixation of carbon dioxide by plants.

The greenhouse effect

The long-term effect of this increase in carbon dioxide levels is not yet known, but it is speculated that temperatures at the surface of the Earth will be raised as a result of the **greenhouse effect**. Carbon dioxide in the atmosphere normally acts as a **greenhouse gas** and prevents heat loss from the surface of the Earth, just as the glass of a greenhouse prevents heat loss from the interior. Increased levels of carbon dioxide will cause this effect to escalate, resulting in increased temperatures which could radically change the climate of the planet. Partial melting of the polar icecaps would result in many countries being at least partly flooded. Climatic changes could also cause many vital crop-producing regions to become virtual deserts. We have yet to see what will happen – whether the Earth will restore its own balance, or whether the worst of the predicted calamities will come about. In the meantime, scientists are almost unanimous in their advice to reduce the burning of fossil fuels in an attempt to prevent a continuing escalation in the carbon dioxide levels of the atmosphere.

Greenhouse gases

Carbon dioxide is not the only greenhouse gas – methane, oxides of nitrogen and chlorofluorocarbons (CFCs) also increase the greenhouse effect. Human actions are adding an excess of 3000 million tonnes of carbon dioxide to the atmosphere each year. Added to this are steadily rising methane emissions resulting from changing agricultural practices, waste disposal and mining. Oxides of nitrogen result from combustion in cars and in power stations, and CFCs are synthetic compounds which have many uses due to their inert nature. The effect of these gases on the Earth is shown in figure 6.5.3.

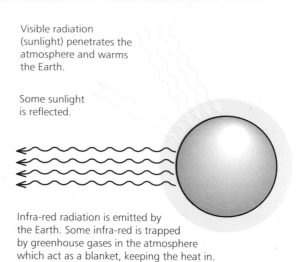

Visible radiation (sunlight) penetrates the atmosphere and warms the Earth.

Some sunlight is reflected.

Infra-red radiation is emitted by the Earth. Some infra-red is trapped by greenhouse gases in the atmosphere which act as a blanket, keeping the heat in.

Figure 6.5.3 Sunlight passes through the atmosphere and warms the Earth. The warm Earth emits infra-red radiation, which is trapped by the greenhouse gases and keeps the Earth warmer.

THE WATER CYCLE

Around two-thirds of the surface of the Earth is covered by water, and this water cycles through the biosphere continuously. Rain or snow (precipitation) fall on the Earth. A small proportion of this water evaporates immediately back into the atmosphere, while the rest reaches either the land or already existing bodies of water.

Some of the water that falls is taken up by living organisms, either through the roots of plants, or by drinking or osmosis in animals. This water will be returned to the atmosphere by evaporation from stomata and respiratory surfaces, and lost into the soil in urine and faeces. The water that is not taken up by living organisms enters lakes, rivers, streams or oceans, or seeps through the soil to reach the water table.

Ultimately water is returned to the atmosphere by evaporation from the surface of the water masses, and it will eventually return to the Earth again as precipitation. This process of the cycling of water through the biosphere can be summed up in the **water cycle**, illustrated in figure 6.5.4.

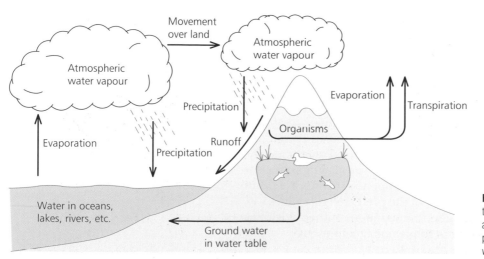

Figure 6.5.4 The water cycle – water evaporates from the oceans and other areas of water into the atmosphere and then falls back to Earth as precipitation, often passing through the bodies of living organisms on the way.

Acid rain

When fuels are burned, carbon dioxide is not the only product. Sulphur dioxide and various nitrogen oxides are also produced. These gases then react with the water in the atmosphere to produce **acid rain**. Sulphur dioxide produces sulphuric acid and nitrogen oxides result in nitric acid. The end result is rain with a greatly lowered pH. In heavily industrialised areas of the north-eastern United States, women have reported holes appearing in their tights when acid raindrops have fallen on them.

In areas with plenty of limestone rocks, the effect of acid rain is small as the acid reacts with the rocks and the pH is maintained at neutral levels. However, in other areas without the limestone buffer lakes become acidic, leading to the death of wildlife. Trees are killed by the direct action of acid rain. In Germany in the early 1980s, up to 40% of fir trees were shown to be sick or dying. The long-term effects of this acid rain pollution cannot yet be calculated. Even more disturbing is the fact that many governments and industries are ignoring the warnings of the scientific community, turning down the opportunity to control the emission of the damaging gases before it is too late. The immediate cost of preventative action is financial, and this is why it is avoided, but the eventual cost for the planet could be incalculably higher.

Country	Mass of sulphur dioxide emitted/ kg head of population $^{-1}$ year $^{-1}$
East Germany	240
Czechoslovakia	201
Hungary	153
Yugoslavia	133
Finland	119
Bulgaria	112
Spain	99
Soviet Union	91
Italy	90
Denmark	89
Britain	83
Belgium	82
Poland	76
France	60
Sweden	60
West Germany	58
Netherlands	31
Switzerland	18

Table 6.5.1 These sulphur dioxide emissions were recorded in 1986. Since then several of the countries listed have either united or fragmented, so that the current picture will be, at first glance at least, very different.

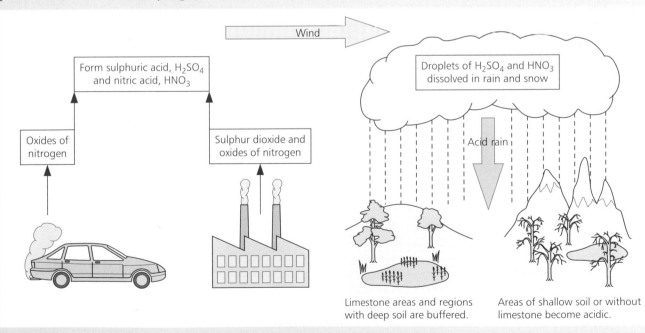

Limestone areas and regions with deep soil are buffered.

Areas of shallow soil or without limestone become acidic.

Figure 6.5.5 The amount of acid rain produced by a country is not necessarily the same as the amount of acid rain which falls upon it. The acidic gases may be carried by prevailing winds and the bulk of the acidic precipitation may occur many miles from the site of the original pollution.

6

THE NITROGEN CYCLE

Nitrogen is probably the limiting factor on the production of new biomass in most ecosystems. It is a vital constituent of amino acids and therefore of proteins. Although almost 80% of the Earth's atmosphere is made up of nitrogen, relatively little is available in a form which can be used by living organisms because nitrogen is relatively inert. It does not naturally combine with other components of the biosphere but is actively 'fixed' by microorganisms. The **nitrogen cycle** involves this 'fixing' of nitrogen from the atmosphere into a form which can be utilised by plants and animals to produce proteins. The nitrogen is eventually released from the tissues of living things and returned to the atmosphere by the action of more microorganisms, the decomposers. The main stages in the nitrogen cycle are summarised in figure 6.5.6.

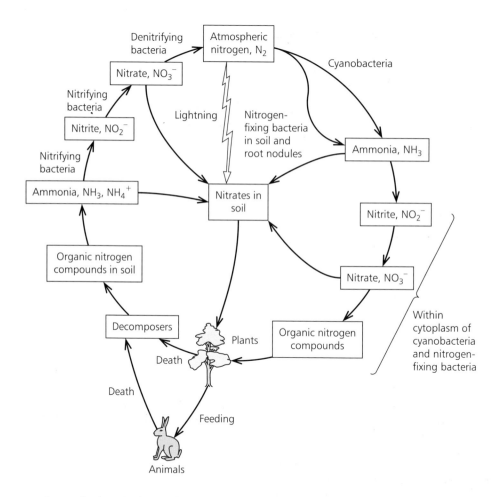

Figure 6.5.6 The nitrogen cycle. Nitrogen from the air is fixed by bacteria, then processed through plants, animals and more bacteria before being returned to the atmosphere.

The majority of plants obtain their nitrates from the soil. The nitrogen in the air is converted or **fixed** into useful nitrates in several ways. This nitrogen fixation is an energy-requiring process – it uses ATP. Nitrogen fixation is carried out by **nitrifying bacteria** and, to a much smaller extent, by the action of lightning on the atmosphere. The bacteria involved include cyanobacteria and organisms such as the *Rhizobium* bacteria in the root nodules of leguminous plants. The product of the initial stages of nitrogen fixation is ammonia. In the soil ammonia is then rapidly converted to nitrites by bacteria such as *Nitrosomas* and then to nitrates by other bacteria such as *Nitrobacter*. In leguminous plants, amino acids are synthesised directly using the ammonia resulting from the initial nitrogen fixation. The nitrates produced by nitrogen fixation are used to form the protein of plant and subsequently animal tissues. The nitrogen is eventually released from the dead and decaying tissues of both plants and animals by the action of yet more bacteria such as *Thiobacillus denitrificans*.

THE PHOSPHORUS CYCLE

Phosphorus in the form of phosphate ions is of enormous importance in living organisms as it occurs in ATP, nucleic acids and enzymes. Phosphate ions are involved in a cycle involving the water, soil, living systems, bones and some exoskeletons and the excreta of birds and reptiles. The movement of phosphate ions through the **phosphorus cycle** is illustrated in figure 6.5.7.

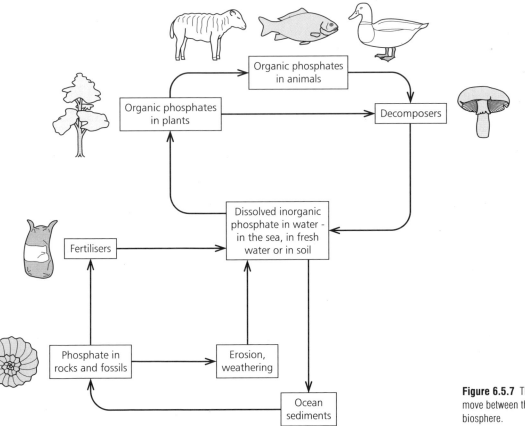

Figure 6.5.7 The phosphorus cycle – phosphates move between the abiotic and biotic elements of the biosphere.

Too many nutrients – pollution of the waterways

Pollution is the contamination of the ecosystem by substances which interfere with the cycles just described in this section. Pollution may be caused by natural substances such as faecal material, or by chemicals synthesised by people. A major problem at present in the developed world is the pollution of rivers and streams by too much nitrate-rich material. Minerals such as nitrates and phosphates may be leached from the soil into the waterways. As a result, populations of algae and photosynthetic bacteria are able to grow very rapidly, making the water appear cloudy and scummy.

The death of these photosynthetic organisms then provides a rich food supply for decomposers. The decomposers use up large quantities of oxygen from the water in the processes of decay, which in turn may deprive other organisms such as fish and crustaceans of oxygen, causing them to die. Their bodies add to the organic material available to the decomposers and so the situation gets worse. This process is known as **eutrophication**, and it may happen quite naturally. However, two major sources of human pollution have made eutrophication increasingly common, and are leading to the death of many rivers, streams and ponds.

Crop farmers are using their land more and more intensively. To replace the nutrients used by the plants, and to compensate for the fact that the crops are not left to rot and return minerals to the soil, farmers apply artificial fertilisers. These fertilisers are particularly rich in the very soluble nitrates, along with phosphates and potassium. When fertilisers are leached from the soil into waterways by rain, eutrophication is the likely result. The other source of pollution which may lead to eutrophication is sewage. If relatively untreated human sewage or animal wastes are pumped into a river or lake, they form a massive food supply for decomposing organisms. These decomposers then use up much of the available oxygen to decompose the sewage and eutrophication – death of the waterway – is again the result.

These problems can be avoided with the use of more organic fertilisers such as manure, which decompose slowly so the nutrients are not readily leached away, and with proper treatment and breakdown of sewage before it is released into our rivers and lakes. But again, old habits die hard and in spite of moves in the direction of conservation, the problem of eutrophication of our waterways remains.

By considering the energy flows within ecosystems and the cycling of essential nutrients, we can build up a picture of some of the factors which influence life within those ecosystems. In section 6.6 we shall consider in more detail the survival of populations within their specific habitats, and the factors influencing their growth or decline.

SUMMARY

- Supplies of inorganic materials on the Earth are finite, and such materials are recycled through ecosystems. These cycles involve a **biotic phase** where the materials are incorporated into the bodies of living organisms, and an **abiotic phase** where they are returned to the non-living part of the ecosystem.

- In the **carbon cycle** the abiotic phase is carbon dioxide in the air and dissolved in the oceans. Photosynthesis fixes carbon in organic compounds, which are then eaten by consumers. Respiration, death of plants and animals and decay by decomposers, and combustion of fossil fuels all recycle carbon dioxide.

- In the **water cycle** the abiotic phase is water in the oceans, lakes and rivers and atmospheric water vapour. Precipitation (condensation of water vapour) provides water on land which may be taken into living organisms. Transpiration and evaporation return water vapour to the air, while runoff returns liquid water to the oceans.

- In the **nitrogen cycle** the main abiotic phase is nitrogen gas in the atmosphere. This is turned via a series of stages involving **nitrogen-fixing microorganisms** to nitrates in the soil. Nitrates are taken in by plants, which form organic nitrogen compounds. Animals eat these, and death and decay of plants and animals results in organic nitrogen compounds in the soil. The action of **denitrifying microorganisms** returns nitrogen to the air, again via a series of stages.

- In the **phosphorus cycle**, the main abiotic phase is dissolved inorganic phosphates in water and in the soil. Phosphates enter sediments and thus rocks. The rocks may be extracted and used as fertilisers which returns the phosphate to the water. Plants take in inorganic phosphate ions and turn them to organic phosphate compounds. These are eaten by animals. Decomposers return phosphates from the bodies of plants and animals to the soil.

QUESTIONS

1 a Describe the cycling of carbon in a named ecosystem.
  b What is meant by the greenhouse effect? How has it come about and why is it a cause for concern?

2 a Describe the nitrogen cycle in nature.
  b Suggest reasons why the felling and removal of forest trees results in changes in the levels of nutrients in the soil.

3 a Describe the water cycle.
  b Explain the various ways in which people pollute water and describe the possible short- and long-term effects of this.

6.6 The ecology of populations

A population is a group of individuals of the same species living in the same habitat. For ecologists to develop effective models of how ecosystems work, studying the growth of populations and the factors affecting that growth is extremely important.

POPULATION GROWTH

As far as geneticists are concerned, a population is a group of interbreeding or potentially interbreeding individuals. For the ecologist, the word 'population' has a rather wider meaning – a group of interbreeding organisms that take up or provide a habitat and produce or consume resources. Members of a population occupy a particular ecological niche.

Populations grow, and also become smaller, at various rates – they may be relatively stable for a long period of time, they may grow rapidly or slowly, or they may become extinct. Such growth can be expressed mathematically. Population growth is affected by a wide variety of factors, and as a result the maximum possible rate of growth is rarely achieved.

Exponential growth

In theory, given ideal conditions, a population will double in size during a constant period of time. This is known as **exponential growth**. The rate of increase in numbers of a population undergoing exponential growth is proportional to the size of the population, and when plotted on a graph of numbers against time such a population gives an **exponential growth curve**. Such optimum growth rarely occurs, although it can be seen occasionally in bacterial populations with carefully managed environments, as shown in figure 6.6.1. Darwin, when discussing the concept of exponential growth, used the example of a pair of elephants. Considering the minimum reproductive rate for a pair of elephants and their offspring (6 infants per pair in a 60-year breeding period), Darwin calculated that over 750 years the descendants of an original pair would number around 19 million elephants. Thus it can be seen that exponential growth is not necessarily a good thing!

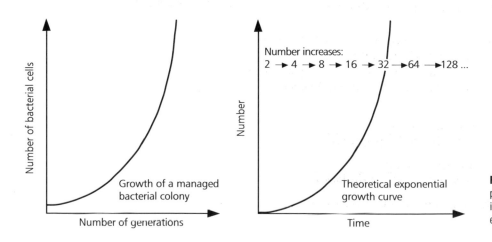

Figure 6.6.1 An exponential growth curve for a population of bacteria. The rate of population growth increases as the number of individual bacterial cells, each capable of dividing, increases.

One of the few situations in which natural populations may undergo exponential growth is when colonising a new habitat. Sooner or later, however, the period of exponential growth ends and the population numbers increase more slowly, or even decrease. A look at figure 6.6.2 showing population growth in a variety of organisms demonstrates clearly how far the situation is from exponential growth. Why is this?

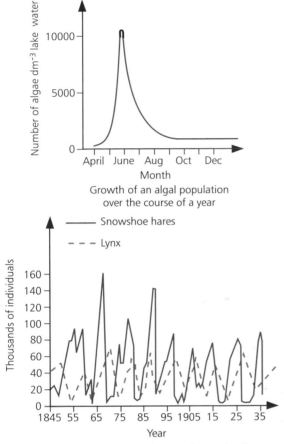

Growth of an algal population over the course of a year

The interrelated populations of snowshoe hares and lynxes in Canada between 1845 and 1935

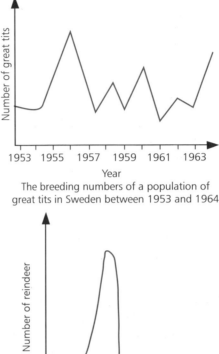

The breeding numbers of a population of great tits in Sweden between 1953 and 1964

Growth and decline of the reindeer population on a small island in the Bering Sea

Changes in population size

Organisms enter a population either when they are born or by immigration into one group from another. They leave by dying or by emigrating to a different group. These changes affect the **population density**, that is the number of individuals per unit area.

The rate at which organisms are born or die changes according to various factors, largely concerned with the food supply and the build-up of wastes. These two factors have a direct effect on both the reproductive activity and the success of a population (its **birth rate** or **natality**) and also on the **death rate** (**mortality**) of the group:

$$\text{Birth rate} = \frac{\text{number of births}}{\text{number of adults in the population}}$$

$$\text{Death rate} = \frac{\text{number of deaths}}{\text{number of adults in the population}}$$

The effects of immigration and emigration on a population are more complex – see the box that follows.

Figure 6.6.2 The population numbers of these organisms fluctuate in patterns that are very different from the exponential pattern of growth. A variety of factors, both biotic and abiotic, are behind these peaks and troughs of population.

Immigration and emigration

It is easy to consider populations as stable entities, affected only by the birth and death rates of the organisms within them. However, the real-life situation is rather different. Organisms leave and enter a population by means other than birth and death. In plants this happens as a result of successful seed dispersal, and in animals the movement of individuals away from their original population or the movement of new individuals into a population can take place for a variety of reasons. For example, in mammalian populations groups of young adult males tend to leave their original population (emigrate) and subsequently enter another population to find mates (immigrate). Immigration and emigration can be affected by many factors, and because of this unpredictability generalisations cannot readily be made. Each population must be considered separately.

The factors influencing the growth of a population can be divided into two categories – abiotic and biotic. The interaction of these elements of an environment determine how many individuals of a population can live in a given area, as illustrated in figure 6.6.3. The upper limit of the population size, which can never be exceeded on a permanent basis, is known as the **carrying capacity** of the environment.

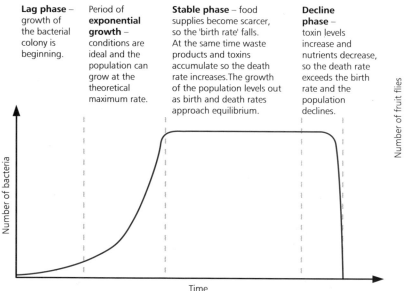

Lag phase – growth of the bacterial colony is beginning.

Period of **exponential growth** – conditions are ideal and the population can grow at the theoretical maximum rate.

Stable phase – food supplies become scarcer, so the 'birth rate' falls. At the same time waste products and toxins accumulate so the death rate increases. The growth of the population levels out as birth and death rates approach equilibrium.

Decline phase – toxin levels increase and nutrients decrease, so the death rate exceeds the birth rate and the population declines.

A typical growth curve for a population of bacteria in nutrient medium. Part of the curve follows the theoretical exponential path, but other factors intervene and the population stabilises and dies.

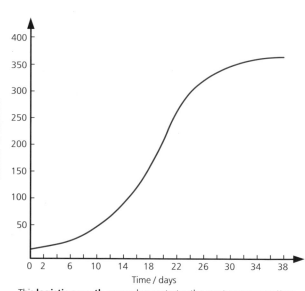

This **logistic growth curve** demonstrates the most common pattern for population growth in a wild population. The early rapid growth phase is replaced by the stable situation, with births and deaths approximately equal. If no other factors intervene the stable phase will occur at the carrying capacity of the environment.

Figure 6.6.3 Common patterns of population growth

A dramatic example of how the growth of a population can be affected by environmental factors occurred in 1944, when United States coastguards introduced 29 reindeer to a small island in the Bering Sea off the coast of Alaska. The reindeer had plenty of food and no predators whatsoever, and as a result the population numbers climbed steadily until in 1963 there were 6000 animals on the island. However, this number of reindeer strained the food resources of the island – particularly the winter fodder plants – to their limit. The death rate rocketed and the birth rate plummeted. As a result the reindeer population was decimated and in 1966 there were only 42 reindeer left (see figure 6.6.2). In this extreme example, as in all populations, it is the complex interaction of a variety of factors that determines the number of individuals which thrive in a particular environment.

Way Kambas and the Javan rhino

The most endangered species of mammal in the world is the Javan rhino. One group of around 50 individuals inhabits a small peninsula in Java. Although the population is slowly increasing, it is vulnerable to rapid extinction from the outbreak of disease or an eruption of nearby Krakatau.

In a combined operation the Indonesian government, the WWF and the IUCN are hoping to relocate a small breeding nucleus of the rhino to another reserve in Indonesia. Way Kambas, a national park of great importance because of the rare habitats and endangered species it already supports, seems likely to be chosen. Javan rhino used to live there but became extinct in the region in 1961 as a result of poaching. Stricter controls mean that poaching is no longer a major threat. However, before an introduction is attempted, a detailed piece of field research was undertaken in 1993. The aim of this was two-fold – to see if Way Kambas does offer a viable second home to these rarest of mammals, and to attempt to estimate the impact of a rhino introduction on the other endangered species there. These include Asian elephants, Malay tapirs, Sumatran tigers, clouded leopards and white-winged wood duck. What type of questions would need to be asked before the project went ahead, to avoid the possibility of an ecological disaster?

Population distributions

Some populations naturally have high densities, such as nematode worms in rich soil, ants in an ants' nest and people in a city. Other species, for example eagles, cheetahs and polar bears, have very low population densities and individuals are found widely scattered through the habitat. The members of populations are generally found distributed in one of three ways in the environment. These are **clumped**, **uniform** and **scattered** distributions, illustrated and explained in figure 6.6.4.

Clumped species are most common – herds of animals or groups of plants and animals that have specific resource requirements and therefore clump in areas where those resources are found, for example a herd of elephants, a school of dolphins or a stand of pines.

Uniform distribution usually occurs when resources are thinly but evenly spread, or when individuals of a species are antagonistic to each other as in the case of many large carnivores. A territory is often defended for example by bears and hawks.

Random dispersal patterns are the result of plentiful resources and no antagonism, for example dandelions on a lawn.

Figure 6.6.4 Dispersal patterns of different populations of organisms

We have already seen in figure 6.6.3 that populations tend to grow most rapidly in the earlier stages of their colonisation of an environment. Dense populations have lower birth rates, higher death rates and slower growth than less dense populations, as illustrated in figure 6.6.5. This suggests that there is an optimum population density. Factors affecting the size and therefore the density of the population are in some cases **density dependent**. Predation, disease, parasitism and most importantly competition increase with the density of the population and so control the size of that population. **Density-independent** factors such as earthquakes, unusual extremes of weather and fires also act to control population sizes, but in a much more random and unpredictable way.

ABIOTIC FACTORS AFFECTING POPULATION SIZE

What are abiotic factors?

As we have seen in section 6.4, the abiotic factors affecting a population are the non-living elements of its environment. These include physical factors such as the amount of sunlight, the wind speed and direction, water currents and seasonal variations in rainfall and temperatures, along with chemical factors such as soil pH and the chemical input into the environment by human beings. The distribution of a species will be determined to a large extent by the abiotic factors in the environment. For example, for both plants and animals temperature is frequently the factor which determines whether or not they can grow in a particular environment. There is a range of temperatures at which a particular organism can grow and successfully reproduce, but above or below that range reproduction does not occur even if the organism survives.

Light

Light has a more direct effect on the growth of plant populations than on that of animals. Plants are dependent on light for photosynthesis, and so populations living in habitats with low light levels have evolved strategies for coping with this situation. Some plants simply reproduce early, which avoids the shade caused by larger plants, as illustrated in figure 6.6.6. Others have evolved to be able to photosynthesise and reproduce effectively in low light levels. Animals are affected by light levels indirectly as a result of the distribution of food plants, and seasonal light changes also affect reproductive patterns, as we have seen in section 5.2 (page 353).

Some plants that live in shaded situations have larger leaves. Nettles growing in a shady woodland have a much larger average leaf surface area than nettles in open sunlight, and this allows the population to thrive in both habitats as shown in figure 6.6.7.

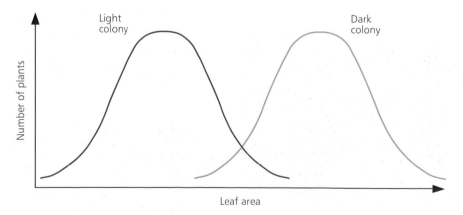

Temperature

The temperature of the environment affects the rate of enzyme-controlled reactions in plants and ectothermic animals. Thus in general populations grow more rapidly in warmer regions. This is demonstrated by the simple

Figure 6.6.5 Population density affects population growth. Many sea birds nest in immense colonies, each with its own very small territorial area. However, if the number of birds in a given area gets too high then reproductive success is affected and egg production drops, as can be seen from the data on great tits in figure 6.6.2.

Figure 6.6.6 Although light levels are generally far higher in summer than they are in early spring, plants growing on the woodland floor are in deep shade in the summer once the tree canopy is fully developed. Thus plants such as bluebells produce leaves very early in the year in spite of low temperatures. They photosynthesise and prepare to reproduce, flowering as insect numbers are increasing with the raised temperatures of late spring. After flowering the plant dies back and becomes dormant through the shaded summer months. This strategy is used by a number of woodland plants to overcome the problem of shading under trees.

Figure 6.6.7 The bell-shaped normal distribution curves show the ranges of leaf areas which might be expected in two different nettle populations. The leaves of nettles grown in the shade are on average considerably larger in area than those of a nettle grown under normal conditions.

experiment of keeping frogspawn in two different situations, some outside and some inside. The spawn inside hatches sooner and the development of the tadpoles is more rapid due to the higher temperature. Similarly, by travelling the length of Britain in early spring the stages of development of plant leaves and flowers can clearly be seen, and the retarding effects of the cooler northern temperatures observed.

The reproductive behaviour of populations is also affected by temperature. Many migratory birds undertake their epic journeys across the world to reach an area where the temperature is suitable for their reproductive patterns.

Wind and water currents

Wind adds to environmental stress, as it increases water and heat loss from the bodies of organisms. In areas with strong prevailing winds the numbers of species which can survive are reduced, and those that do survive have reduced populations. Occasional gales and hurricanes can devastate populations. Whole woodlands may be destroyed along with the communities of plant and animal life within them, as shown in figure 6.6.9.

In water, currents affect the type of organisms that can populate a region, and the numbers within those populations. A strong current will dislodge organisms not well adapted by having a strong attachment or by being extremely good swimmers. Currents are most damaging to populations when the strength increases suddenly, as is the case when flooding occurs.

Water availability

In most aquatic environments the availability of water is not a problem (see section 4.6). In a terrestrial environment the availability of water is affected by a variety of factors including the amount of rainfall, the rate of evaporation and the rate of loss by drainage through the soil. Water is needed for the functioning of the cells of an organism, for excretion and for the transfer of gametes. Any situation of limited water supply will put stress on populations. If the water stress becomes too severe the organisms will die unless, like camels and cacti, they have special adaptations to enable them to cope. Equally, an increase in the availability of water can lead to a massive increase in populations such as that seen in deserts after a long period of drought. After a little rain has fallen, the seeds of many desert plants germinate, grow and flower in a very short space of time, in the phenomenon known as the 'flowering of the desert', as shown in figure 6.6.10.

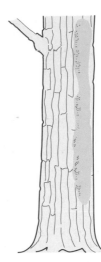

Figure 6.6.8 The higher temperatures on the south side of the tree trunk affect the growth of the mosses and algae colonising the bark – the populations are more successful on the cooler, damper north side. Careful observations in a woodland will show that all tree trunks show a similar pattern of colonisation, the result of the temperature and light differences.

Figure 6.6.9 The wind can cause great damage to populations of both plants and animals, as this picture shows. The damage seen here was caused by very high winds which swept Britain in the autumn of 1987. However, although many populations were greatly depleted or destroyed in the aftermath of the wind, other opportunistic species could take advantage of the new, empty habitats created and their populations increased rapidly.

Figure 6.6.10 The effect of water availability on the ability of populations to survive is graphically illustrated by some of the plants living in the extremely hostile environment of the desert. The seeds survive the hot, dry conditions by remaining dormant for years at a time. Then when water availability suddenly increases after a fall of rain these seeds can progress through germination and growth very rapidly – in a matter of days – to give a large population of reproductive individuals which in turn produce the hardy seeds needed to survive dormant until water becomes available again.

Oxygen availability

There is never a shortage of oxygen in the air, but it can be in short supply in both water and the soil. When water is cold, or fast flowing, sufficient oxygen dissolves in it to support life. If the temperature of the water rises or it becomes still and stagnant, then the oxygen content drops and this affects the survival of populations within it.

Similarly, the environment within the soil is usually well aerated. The spaces between the soil particles contain air and so there is plenty of oxygen for the respiration of plant roots and thus the maintenance of plant populations. If the soil becomes waterlogged, the air spaces are filled with water and the plant roots may be deprived of oxygen.

Soil structure and mineral content

The structure of the soil on which organisms live and grow affects the success of the various populations associated with it. Sand has a loose, shifting structure which allows very little to grow on it. Plant populations which are linked together by massive root and rhizome networks can and do survive, reproducing successfully and also binding the sand together, making it more suitable for colonisation by other species. One of the best known organisms for colonising shifting sand in Britain is marram grass. Marram has an extensive interlinked rhizome network and is also, as we have seen in section 4.6 (page 305), well adapted to survive the physiological drought conditions present on the seashore.

Sandy soils are light, easily worked and easily warmed. However, they are also very easily drained. Water passes through them rapidly, carrying with it minerals useful to plants. This leaching of minerals reduces the population density of plants growing in the soil. Conversely, soils made up predominantly of **clay** particles are heavy, cold, hard to work and readily waterlogged as the drainage is very poor. Leaching of minerals is not a problem in soil of this type, but the populations it will support are still limited. The ideal soil, **loam**, has particles of a wide range of sizes. It is heavier and less prone to leaching than sandy soils, yet easier to warm and work than clay. As a result both the plant populations growing in a loam soil and the populations of animals associated with them will be larger and more successful.

The mineral content of the soil is vital to the successful growth of plants. A variety of minerals are needed for healthy root and shoot development. In a natural ecosystem, many of these mineral requirements are taken up from the soil by plants and incorporated into the material of the plant body. The minerals are then returned to the soil either directly when the plants die and are decomposed, or indirectly as animals eat the plants and they and their waste products are decomposed in turn. This is the basis of the cycles described in section 6.5.

This natural replenishment of resources is undermined when humans grow crops. Most crop plants are grown as **monocultures** – that is, a field full of a single species of plant. Thus all the plants are removing the same minerals from the soil and yet, because the crop is removed long before the plants die and decompose, virtually nothing is returned to the soil. Farmers can apply artificial fertilisers to help supply the mineral needs of their crops. Even so, maintaining this pattern of agriculture over a period of years results in soil becoming relatively barren, with increasing amounts of expensive chemicals being applied to maintain the fertility of the soil.

These problems may be minimised by **crop rotation**. This involves growing a series of different crops in a particular field each year in rotation, thus altering the mineral demands made each year. Also, natural fertilisers such as farmyard manure supply the required minerals in a slow release form as they decay, and they also improve the structure of the soil. By careful management, soil fertility can be maintained and crops taken off at regular intervals.

BIOTIC FACTORS AFFECTING POPULATION SIZE

What are biotic factors?

Biotic factors are all the living elements of the environment of a plant or animal. These include predation, parasitism and disease, and most importantly competition for resources. In many cases it is difficult to distinguish clearly between biotic and abiotic factors affecting a population as they are frequently interlinked – for example, all the cycles covered in section 6.5 involve both abiotic and biotic elements.

Competition

Competition seems to be the major biotic factor determining the density and growth rate of populations. The competition may be directly for resources such as sunlight, minerals and food, or it may be for position, nest sites or mates. **Intraspecific competition** is competition for resources between members of the same population or species (see the example in the box below). When there is competition like this between members of the same population for the same resources, some of the population may not survive, or may not reproduce, and so the growth of the population slows as a consequence. Equally, if resources are plentiful and there is no competition for them, then the numbers of individuals will increase more quickly.

Intraspecific competition in Puerto Rican frogs

In the tropical rainforests of Puerto Rico there is a species of frog known as *Eleutherodactylus coqui*. These frogs feed on insects and are active at night, hiding during the day to avoid predators. It appeared that the amount of food available should support a larger population of frogs than in fact exists, so scientists set about discovering the limiting resource. Competition for space as a factor limiting the population size was investigated, dividing the study area into 100 m^2 plots. In some areas the frogs were provided with many small bamboo shelters to hide in, whilst in others the habitat was left unchanged. All the shelters in the test plots were rapidly occupied and the population density increased accordingly, whilst the population density of the frogs in the control plots remained the same. The population size of the frogs was therefore controlled by competition between them for the one relatively scarce resource – sites in which to hide from predators and breed.

In some cases competition occurs not between members of the same population, but between different species within the community for the same scarce resources. This is known as **interspecific competition**. All the competing populations grow more slowly under these conditions than they would otherwise, and if one species has a high starting density or a superior reproduction rate, then one or more of the competing species may be removed. If this happens on a large scale, extinctions can occur. For example, years ago sailors released goats on Abingdon Island which is part of the Galapagos Archipelego. Goats are relatively large fast-breeding mammals with appetites to match. Over the years the growing goat population consumed so many of the plants previously eaten by the giant tortoises of the island that the reptiles could not survive the competition and in the 1960s became extinct on the island.

Competition within a population may be very obvious, such as blackbirds fighting over territories in a garden or tomcats fighting over a female on heat. The winning competitor gains the resource. This is referred to as **contest competition**. In other cases, many individuals all compete for, and gain some of, a relatively scarce resource – for example, lions squabbling over a carcase, as shown in figure 6.6.11. This is known as **scramble competition**.

Many modern ecologists hypothesise that competition is the major factor in the maintenance of stable communities in natural equilibrium, particularly for those organisms higher up the food web or chain. The role of competition in both population ecology and the structure of complex communities is currently the subject of much research and discussion, but there can be no doubting that it has a major role in the balance of nature.

Predation

Interactions of predators and prey are a factor in the regulation of population size. For the purposes of population studies a predator is any organism which feeds on another living organism, so for ease of discussion herbivores and parasites are regarded as predators too.

The obvious conclusion is that predators must be involved in regulating the population numbers of their prey species. Horses grazing a field must reduce the reproduction of the grass by eating the potentially flower-forming parts, and a fox family must reduce the numbers of the local rabbit population. However, things are rarely as simple as this. A mathematical model has been developed which attempts to describe the relationships between predator and prey populations. The prediction was that the populations would oscillate in a repeating cycle. The reasoning underlying this model is straightforward. As a prey population grows it provides more food for the predators and so, after an interval, the predator population grows too. The predators increase to the point where they are eating the prey species faster than it can reproduce, so the numbers of the prey population fall. In turn, this limits the food supply of the predators and so their numbers will fall as well, allowing the prey to increase again and so on, as shown in figure 6.6.12.

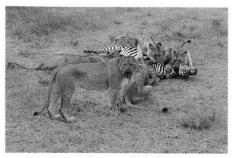

Figure 6.6.11 In scramble competition, all the individuals compete for part of a resource. Although some individuals will fare better than others, in times of relative plenty the whole population will survive. When prey is harder to come by, all the members of the group get less to eat and the population grows more slowly. Only in times of severe shortage will some weaker members of the population get so little to eat that they cannot survive.

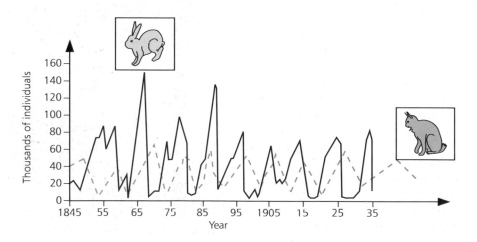

Figure 6.6.12 The data in this graph come from the records of pelts kept by the Hudson's Bay Company in Canada. They appear to confirm the theoretical model of predator and prey populations oscillating in a regular pattern. The peaks and crashes in the lynx population are certainly dependent on the snowshoe hare population. However, the hare population has been shown to follow a similar pattern even in areas where there are no lynxes. The hares are responding to cycles in their own plant 'prey', which themselves seem to reflect climatic variations and changes in insect pest populations. The interrelationships of predator–prey populations are clearly not as simple as it might at first seem.

The difficulty with using predator–prey relationships to understand the population balance of communities is that very few animals rely on only one food source. As we have seen in section 6.4, food webs are complex and many organisms are interlinked and interdependent. Thus one species may prey largely on another in times of plenty, but if the normal prey is in short supply it will switch readily to alternative food sources, thus affecting population numbers in another group but remaining stable itself, as shown in figure 6.6.13. As a result of a variety of studies it appears that the main effect of predators on a prey population is seen when that prey population is in some way weakened, perhaps by lack of food or disease. When this situation arises the inroads made by predators cannot readily be made up by increased reproductive activity and the prey population 'crashes'. In a healthy community predator–prey relationships do have a role to play in maintaining the overall balance of species within the system, but their importance is probably not as great as might at first seem to be the case.

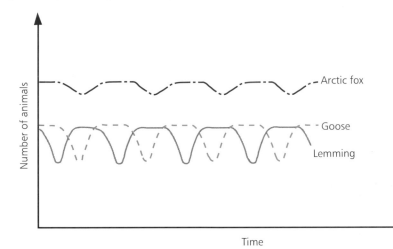

Figure 6.6.13 In a natural environment a crash in the population numbers of one organism may not lead to a delayed crash in the predator population, as the predator may simply switch to another source of food. In this example the major component of the diet of Arctic foxes changes from lemmings to geese.

Parasitism and disease

The size of populations and their growth are both affected by parasitism and disease. Infectious diseases are far more likely to be spread when there is a high population density, as individuals are in much closer proximity to each other. This is well documented in the social history of human diseases. Plague, cholera, smallpox, 'flu and chickenpox are passed on far more readily when people live and work closely with others.

The effect of a disease on populations depends to some extent on the community structure and on the populations within it. A virus which attacks an agricultural monoculture such as fields of corn or a bark beetle invading a woodland with little diversity in the number of species present will have a devastating effect. This is because any other populations within the community will depend heavily on the one affected by the disease. In a richer and more diverse community, although the effect on the particular population attacked by the disease will be as great, the effect on other populations within the habitat will be much less, as they will have many other potential sources of food.

Parasites also play a role in the success or otherwise of populations. Parasites feed off the living body of their host. Some (**endoparasites**) live within the body of their host and others (**ectoparasites**) are external, sometimes living on the outside of the host and sometimes simply visiting when they need a feed.

The major effect of a parasitic infestation on an individual is to weaken it. If all the individuals of a population are heavily infested by parasites then the rate of growth will slow. Many more individuals will be susceptible to predation and disease and reproduction will be less successful – the population as a whole will be checked. As in the case of infectious diseases, parasitic infections are more likely to be passed from one individual to another in situations of high population density than when the individuals are more spaced out, as illustrated in figure 6.6.14.

Figure 6.6.14 Food availability obviously has an effect on the individuals shown here, but more than that the feral dogs will be riddled with parasitic worms which use up many of their resources. Most wild dogs can find sufficient food to survive – it is the parasites which are largely responsible for their emaciated appearance. The same species, kept in a human home and devoid of parasites, utilises the food it is given to produce a well-covered frame.

Parasitism, disease and the human population

Parasitism and disease play an important role in destroying the well-being of the human population of the biosphere. In the developed world the treatment of diseases such as cancer, heart failure and AIDS involves great financial expenditure by both governments and individuals. In the developing world the tropical diseases such as malaria (which affects 267 million people in 103 countries!), yellow fever, river blindness and Chagas' disease cause untold suffering. These diseases, many of which are the result of parasitic infestations, weaken the population of many developing countries thus making it harder for them to utilise their natural resources. They cannot afford the treatments which would keep the population healthy, yet the parasitic infestations make them poorer. It is a vicious circle from which there is at present no escape.

Disease	Cause	Vector	Numbers affected	Symptoms
Malaria	*Plasmodium* – a single-celled protozoan parasite	The female *Anopheles* mosquito	267 million people in 103 countries	'Flu-like symptoms to begin with, then cycle of severe fevers, shaking chills and sweats which cause serious weakness and sometimes death.
Schistosomiasis	Trematode flatworms (flukes)	Infected aquatic snails	200 million people in 76 countries	The body tissues react to eggs which become lodged in them. Causes debilitating diseases of the urinary tract and gut.
Filariasis	Parasitic nematodes	Mosquitoes	90 million people in 76 countries	Blocked lymph vessels – in later stages painful and disfiguring swelling of the limbs known as elephantiasis
Onchocerciasis	Parasitic nematodes	*Simulium* blackflies	17.6 million people in 34 countries	Immature worms lodge in the skin and eyes, eventually causing blindness (river blindness)
Chagas' disease (South American trypanosomiasis)	Protozoan parasites – trypanosomes	The faeces of assassin bugs – large blood-sucking bugs which live in cracks in house walls	16–18 million people in South and Central America	Shortly after infection – by faeces through a bite – fever develops which can be fatal. Followed by symptomless period of months or years when parasite invades most body organs – causing chronic weakness and death.

Table 6.6.1 Although the diseases of the developed world cause suffering to the individuals concerned, it is the parasitic tropical diseases which are the major scourge of the human population.

THE HUMAN POPULATION

Human beings have been present as a species on the Earth for a very small part of the history of the planet. For thousands of years the human population has grown slowly, with increases due to improved weapons for hunting, tool making or crop growing being cancelled out by catastrophes, diseases and war. At the time of the agricultural revolution there were about 133 million people in the world. Since then the human population has ballooned, with the time taken for it to double getting less and less, as shown in figure 6.6.15. The point is now in the foreseeable future when the ecosystems of the Earth will no longer be capable of supporting the numbers of human beings – a situation which in any other species would result in an enormous population crash.

Whether this will be the case for human beings is as yet a case for conjecture. Perhaps we shall control our population growth, or develop an as yet untapped food source. Certainly something will have to be done. Already people are placing enormous strains upon the planet in a variety of ways and to maintain population growth on this scale will require the careful use of all the Earth's resources.

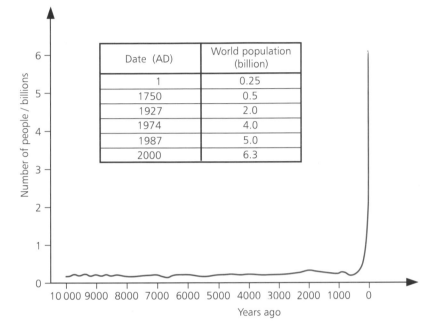

Date (AD)	World population (billion)
1	0.25
1750	0.5
1927	2.0
1974	4.0
1987	5.0
2000	6.3

Figure 6.6.15 The world human population has increased enormously in the last centuries, falling only briefly in the fourteenth century due to the many deaths from the Plague. The question must remain – where do we go from here?

Food production will become a major problem – many people already do not have sufficient to eat. Urbanisation will continue. It is estimated that by the turn of the century, over 50% of the population of the world will live in cities, and this causes enormous social problems of hygiene, shelter and disease. As people take up more and more space, other species will be marginalised and habitats lost.

More people will mean increased use of energy. Reserves of fossil fuels cannot last forever, and if they are burned increasingly rapidly then the gases they produce will add further to the greenhouse effect. Growing food will become more difficult if the climate of the Earth becomes hotter and drier as a result of such human pollution.

If any other species showed a boom in population of the type seen in human beings, natural checks and balances such as we have considered in this section would ensure that the explosion was slowed and stopped if not reversed. In our position as manipulators of nature and controllers of our own environment, will we be able to cope with the results of our own success?

Figure 6.6.16 Alternative energy sources, improved methods of agriculture and limiting the size of families to two (replacement fertility) are all important aspects in the battle to maintain the ecosystems of the Earth, and enable them to continue to support the human population.

The structure of a community within a particular habitat can be seen as the sum of all the different populations within it. Those populations are affected by many factors – their growth is to some extent self limiting, and both biotic and abiotic factors act to increase or limit the numbers of a particular type of organism. Add to this the interactions between the various populations of an ecosystem and the cycling of minerals within it, and ecology can be seen as an immensely complex area of study. As a result of our own role within the biosphere, it may well be the most important area of biological knowledge for the future of the planet.

SUMMARY

- A population that doubles in size in a constant time period shows **exponential growth**. Most populations do not live in ideal conditions and their growth is **limited** by various factors. The upper limit of a population in a habitat is the **carrying capacity** of the habitat, reached after a slow initial **lag phase**, an **exponential phase**, and then a **stable phase** at the carrying capacity. **Decline** follows when resources run out or toxins build up.

- The **population density** is the number of individuals per unit area. The population density is affected by increases in numbers (caused by **birth** or **immigration**) and decreases in numbers (caused by **death** or **emigration**). The **birth rate** is given by the number of births divided by the number of adults in the population, and the **death rate** similarly by the number of deaths per adult in the population.

- The **distributions** of populations may be **clumped**, with organisms living in groups such as herds, **uniform**, with individuals in their own territories, or **scattered**, with individuals randomly dispersed.

- **Density-dependent** factors affecting population growth include predation, disease, parasitism and competition. Density-independent factors include unusual extremes of weather, earthquakes, etc.

- **Abiotic factors** affecting population size include:

light	water availability
temperature	oxygen availability
wind	soil structure and mineral content.
water currents	

- **Biotic factors** affecting population size include **competition** for sunlight, food or other resources. Competition may be **intraspecific**, between members of the same species, or **interspecific**, between different populations. In **contest competition** individuals fight over a resource such as territory, and the winner gains the resource. In **scramble competition** many individuals compete for and gain part of a resource, such as food.

- **Predator–prey relationships** are biotic factors often changing in a cyclical pattern, the prey fluctuating between higher and lower numbers and the predators following a similar pattern but lagging behind slightly. The availability of more than one prey species in a complex food web and seasonal changes in resources complicate this pattern.

- Parasites and disease are biotic factors particularly affecting dense populations, generally weakening them.
- After an initial steady increase, the **human population** has increased extremely rapidly over the last 200 years or so. Measures will have to be taken to conserve the Earth's ecosystems in order to prevent a human population crash.

QUESTIONS

1 a In theory, populations of organisms can increase exponentially. What does this mean and why is it rarely seen in nature?

b What is meant by the 'carrying capacity' of a habitat?

c How can abiotic factors affect the size of populations?

2 Discuss the ways in which competition, predation, parasitism and disease limit the size of natural populations.

3 Of all the different species present on the Earth, human beings have had perhaps the greatest impact. The human population, after a long period of slow growth, is now increasing very rapidly. Why is this happening and what are the implications for the ecology of the planet?

1 **a** Distinguish between a population and a community. (**3 marks**)
   b Choose *three* factors which affect the population density of organisms, and explain how each factor operates. (**12 marks**)
   c Describe how you would measure changes in the population density of a named organism. (**3 marks**)
   (ULEAC January 1992)

2 'Changes in the atmosphere due to human activities will have major biological effects unless reversed.' Discuss this statement with reference to named examples. (**20 marks**)
   (ULEAC January 1992)

3 Compare the life cycles of a moss, a fern and a flowering plant. (**20 marks**)
   (ULEAC June 1991)

4 Discuss the effects of human activities on the environment with particular reference to pollution.

5 **a** Describe the life cycle of a *named* pteridophyte (e.g. a *named* fern). (**10 marks**)
   b Discuss to what extent pteridophytes are:
   i better adapted than bryophytes and
   ii less well adapted than flowering plants to life on land. (**8 marks**)
   (Cambridge June 1990)

6 **a** **i** What is meant by 'net primary production' and what is its ecological significance?
   ii For a *named* ecosystem or habitat, outline how net primary production could be estimated. (**8 marks**)
   b **i** What factors can limit the size of populations? Explain how they can do so.
   ii For a given population, explain briefly how its size and geographical distribution may have evolutionary significance. (**12 marks**)
   (NEAB June 1989)

7 'Crop yields have been improved by the application of biological principles.' Comment on this statement. (**20 marks**)
   (ULEAC June 1990)

8 **a** Write an account of the nitrogen cycle, including reference to relevant organisms. (**8 marks**)
   b This question refers to a flowering plant.
   i In what forms, and how, is nitrogen absorbed by the roots?
   ii Describe how the absorbed nitrogen is converted into amino acids by the leaves.
   iii In what forms, and how, are nitrogen compounds translocated from their site of synthesis to the roots? (**12 marks**)
   (NEAB June 1991)

9 **a** Explain what is meant by the term 'ecosystem'. (**4 marks**)
   b With reference to examples from *named* ecosystems explain what you mean by each of the following:
   i pyramid of biomass
   ii edaphic factors
   iii decomposers
   iv succession. (**16 marks**)
   (ULEAC January 1991)

10 The table below shows the energy content of the faeces and urine of mice, expressed as a percentage of the energy content of a range of ingested foods.

Ingested food	Percentage of energy content	
	Faeces	Urine
Hazelnuts	6.96	1.69
Acorns	18.65	2.75
Beechmast	7.04	3.98
Oatmeal	11.39	3.30

Adapted from Petrusewicz, Secondary Productivity of Terrestrial Ecosystems (Warsaw 1967)

   a What percentage of the energy in hazelnuts was **i** absorbed through the gut wall of the mice and **ii** available for release by respiration in the mice? (**2 marks**)
   b **i** Which of the foods apparently contained the lowest proportion of digestible material? (**1 mark**)
   ii State *two* chemical components of plant material which contribute to its low digestibility. (**2 marks**)
   c Explain why the energy incorporated into the tissues of the mice is not all available to carnivores feeding on the mice. (**3 marks**)
   (ULEAC January 1992)

11 The following table shows the *survivorship patterns* of mountain sheep and two bird species, robins and herring gulls. Survivorship of a species is expressed as the percentage of a generation which survive beyond a given percentage of the maximum recorded life span.
   a **i** Plot these figures using graph paper. (**6 marks**)
   ii Compare and comment on the survivorship patterns of these three species. (**5 marks**)
   iii The maximum recorded life span of a mountain sheep is 15 years. Out of 200 sheep born in the same year, predict how many would be alive after 12 years. Show your working. (**2 marks**)
   b It has been claimed that natural selection has favoured early reproductive maturity in robins, but has favoured delayed reproductive maturity in mountain sheep. Suggest an explanation for this claim. (**4 marks**)

Percentage of maximum recorded life span	Percentage of generation still alive		
	Mountain sheep	Robin	Herring gull
0	100	100	100
20	95	12	55
40	85	3	25
60	60	2	10
80	30	2	5
100	0	0	0

Adapted from Kormondy, Concepts of Ecology (Prentice-Hall 1969)

c Suggest a procedure by which a biologist could investigate the survivorship pattern of a population of wild animals. **(3 marks)**
(Total 20 marks)
(ULEAC January 1992)

12 The diagram below shows a food web for an aquatic ecosystem.
a From the information in the diagram, name **i** a primary consumer and **ii** a tertiary consumer. **(2 marks)**
b Suggest how this community might be altered if the population of water beetles died out. **(3 marks)**
c Only a small percentage of the energy absorbed by the green algae is incorporated into the tissues of pike. Give *three* reasons why this should be so. **(3 marks)**
(Total 8 marks)
(ULEAC January 1991)

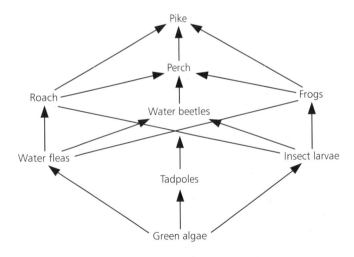

13 The following diagram represents the life cycle of a fungus.
a Which *one* of the lettered alternatives, **A** to **F**, indicates:
i a diploid cell. Explain your answer.
ii a meiotic cell division taking place. Explain your answer.
(4 marks)

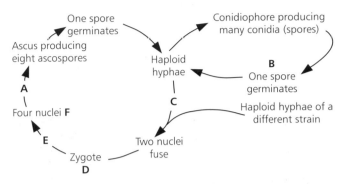

In addition to the 'normal' strain, four different mutant forms of the fungus exist (identified as Mutants 1 to 4). Each mutant is deficient in a **single** but different enzyme involved in the synthetic pathway from **V** to **Y** or **Z**.

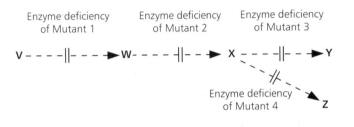

Mutant fungi need the medium on which they are cultivated to be supplemented by the addition of an appropriate substance.
The table shows the growth requirements of different strains of the fungus, with + indicating growth in the presence of the specified supplement, and − indicating no growth.

	Growth with supplement added						Strain
None	Y and Z	V	W	X	Y	Z	
+	+	+	+	+	+	+	
−	+	−	−	−	−	+	
−	+	−	−	+	−	+	
−	+	−	+	+	−	+	

b i Identify the mutant or 'normal' strain involved in each case, writing the appropriate names in the spaces provided in a copy of the table.
ii Explain why the reaction **X**→**Y** in the pathway appears to be irreversible. **(5 marks)**

c i Give *one* experimental method that might be used to increase the number of mutants obtained from a fungus.

  ii Suggest why the mutant forms 1 to 4 rarely, if ever, survive in the wild. **(2 marks)**

d When two strains such as Mutants 1 and 2 are grown together for a while, a new strain may arise which can grow without any supplements being added. Suggest *one* possible reason for this.

(1 mark)

(Total 12 marks)

(NEAB June 1990)

14 An investigation was carried out into nitrogen fixation in soil. Samples of two different soils (A and B) were incubated at 30 °C for 29 days in either aerobic or anaerobic atmospheres containing the same percentage of nitrogen.

Half of the samples, in both the aerobic and the anaerobic atmospheres, were watered with 1% glucose solution. The other half were watered with equal amounts of distilled water.

At the end of the incubation period the amounts of fixed nitrogen were measured, and rates of nitrogen fixation were calculated. The results are shown in the table below.

Soil sample	Watering solution	Rate of nitrogen fixation in mg × 10⁻⁴ nitrogen per g dry soil per day	
		Aerobic	**Anaerobic**
A	Water	0.00	1.63
	Glucose	1.78	3.76
B	Water	1.48	4.14
	Glucose	1.96	8.17

Adapted from Chang and Knowles, Can. J. Microbiol. 11:29–38

a i Explain why some samples in the experiment were watered with water only. **(2 marks)**

  ii Explain why the soil samples were incubated. **(2 marks)**

b Suggest a method by which the rate of nitrogen fixation might be measured. **(3 marks)**

c i Describe the effect of the added glucose on nitrogen fixation in the two soil samples. **(2 marks)**

  ii Suggest an explanation for this effect. **(3 marks)**

d i Compare the effects of aerobic and anaerobic atmospheres on nitrogen fixation in these two soils. **(3 marks)**

  ii The activity of certain enzymes is known to be affected by the presence of oxygen. Suggest an explanation for the different rates of nitrogen fixation in aerobic and anaerobic conditions. **(3 marks)**

e From the results of this investigation, suggest *two* practical ways in which a farmer could increase the rate of nitrogen fixation in the soil. **(2 marks)**

(Total 20 marks)

(ULEAC June 1991)

15 Any habitat may be said to have a *carrying capacity* and this is defined as the maximum population of a given species that can be sustained there.

The graph below shows changes in the numbers of wild sheep on a large island during the 100 years following their introduction to the island.

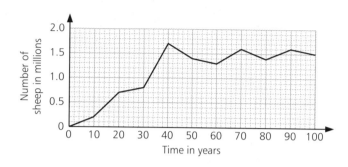

a From the graph, estimate the carrying capacity of this island for wild sheep. **(1 mark)**

b Suggest *two* factors that may determine the carrying capacity of this island for wild sheep. **(2 marks)**

c Comment on the pattern of population change after carrying capacity had been reached. **(3 marks)**

d Suggest *three* factors that may have influenced the length of time taken for the sheep population to reach carrying capacity.

(3 marks)

(Total 9 marks)

(ULEAC June 1990)

16 a Explain the meaning of the following ecological terms:

  i pyramid of biomass **(2 marks)**

  ii community **(2 marks)**

  iii succession. **(2 marks)**

b Woodland forms the natural climax for many areas of the world.

  i What is meant by the term *climax*? **(2 marks)**

  ii Suggest *two* environmental factors that might prevent this climax being reached. **(2 marks)**

(Total 10 marks)

(ULEAC June 1990)

17 a What is meant by the term *parasite*? **(2 marks)**

b The following diagram shows an adult liver fluke.

On a copy, label and annotate *three* features which are related to its mode of life and feeding habits. **(6 marks)**

c Comment on the economic importance of the liver fluke.

(2 marks)

(Total 10 marks)

(ULEAC June 1990)

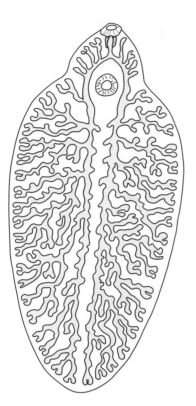

18 Read through the following passage and then answer the questions printed after it.

A polymorphic land snail known as *Cepaea nemoralis* exists in three distinct shades (brown, pink and yellow) on which may be superimposed three different degrees of banding, as shown below.

Unbanded   Mid banded   Five banded

Banding is controlled by two gene loci;
colour is controlled by three multiple alleles.

Studies by Cain and Sheppard in a variety of habitats, such as grassland and woods, showed a high correlation between shell colour and background vegetation type. Yellow shells were common on grassland; here in mid-April the vegetation was predominantly brown, and yellow shells were at a selective disadvantage. By late April, yellow was of neutral survival value relative to the other colours, but by mid-May it was at an advantage. The main daytime predator of *Cepaea nemoralis* is the song thrush.

Banding in the shells also showed a clear correlation with the environment: in woodlands, for instance, the proportion of banded shells in the population was much higher than on grassland.

a What is meant by the following terms and phrases used in the text?

i	*Polymorphic*	**(2 marks)**
ii	*Banding is controlled by 2 gene loci*	**(2 marks)**
iii	*Colour is controlled by multiple alleles*	**(2 marks)**
iv	*Habitat*	**(2 marks)**
v	*Population*	**(2 marks)**
vi	*A high correlation between shell colour and background vegetation type*	**(2 marks)**
vii	*Selective disadvantage*	**(2 marks)**
viii	*Predator*	**(2 marks)**

b Suggest why yellow shells are at a selective disadvantage in mid-April. **(2 marks)**

c Suggest why banded shells are at an advantage in woodlands.
(2 marks)
(Total 20 marks)
(O&C June 1992)

19 An ecological pyramid showing four trophic levels for a pond, expressed in terms of biomass, is given below. Numbers of individuals at each trophic level are also indicated.

a Name the four trophic levels. **(2 marks)**

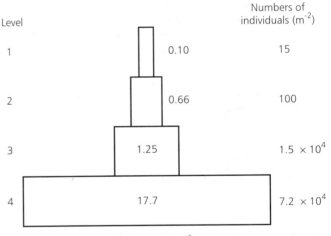

Biomass (dry) g m^{-2}

b Explain why there is a marked decline between trophic levels in:
  i biomass
  ii numbers of individuals. **(4 marks)**

c It has been calculated that of the 1.98×10^9 J m^{-2} per year of solar energy incident on vegetation, 1.95×10^9 J m^{-2} per year are not used in primary production.
  i What is the percentage of incident light which is used in primary production? Show your working. **(2 marks)**
  ii State *two* fates of the incident light which is *not* used in primary production. **(2 marks)**

d In the English Channel, it has been found that the base of the pyramid of biomass is

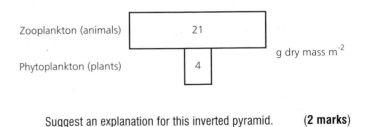

g dry mass m^{-2}

Suggest an explanation for this inverted pyramid. **(2 marks)**
(Total 12 marks)
(O&C June 1992)

Index